Snow Hydrology: Composition and Movement of Snow

Snow Hydrology: Composition and Movement of Snow

Edited by Alfred Mills

□ SYRAWOOD
PUBLISHING HOUSE

New York

Published by Syrawood Publishing House,
750 Third Avenue, 9th Floor,
New York, NY 10017, USA
www.syrawoodpublishinghouse.com

Snow Hydrology: Composition and Movement of Snow
Edited by Alfred Mills

International Standard Book Number: 978-1-64740-118-4 (Hardback)

Cataloging-in-Publication Data

Snow hydrology : composition and movement of snow / edited by Alfred Mills.
   p. cm.
Includes bibliographical references and index.
ISBN 978-1-64740-118-4
1. Snow. 2. Hydrology. 3. Snow mechanics. I. Mills, Alfred.
GB2605 .S66 2022
551.578 4--dc23

TABLE OF CONTENTS

PREFACE

Over the recent decade, advancements and applications have progressed exponentially. This has led to the increased interest in this field and projects are being conducted to enhance knowledge. The main objective of this book is to present some of the critical challenges and provide insights into possible solutions. This book will answer the varied questions that arise in the field and also provide an increased scope for furthering studies.

The scientific field of hydrology that is concerned with the composition, dispersion and movement of snow and ice is referred to as snow hydrology. Important hydrological processes include snowfall, accumulation and melt in watershed at high altitudes and latitudes. Snow melt is useful in many areas as it supplies water to reservoirs and populations, and is also used for agricultural activities. Snow hydrology provides knowledge which is used in weather forecasting. Information of snow composition and movement is gathered through density, depth and composition readings and by using various remote sensing techniques. This book outlines the processes and applications of snow hydrology in detail. It strives to provide a fair idea about this discipline and to help develop a better understanding of the latest advances within this field. It will serve as a valuable source of reference for graduate and post graduate students.

I hope that this book, with its visionary approach, will be a valuable addition and will promote interest among readers. Each of the authors has provided their extraordinary competence in their specific fields by providing different perspectives as they come from diverse nations and regions. I thank them for their contributions.

Editor

Model simulations of the modulating effect of the snow cover in a rain-on-snow event

N. Wever[1,2], T. Jonas[1], C. Fierz[1], and M. Lehning[1,2]

[1]WSL Institute for Snow and Avalanche Research SLF, Flüelastrasse 11, 7260 Davos Dorf, Switzerland
[2]CRYOS, School of Architecture, Civil and Environmental Engineering, EPFL, Lausanne, Switzerland

Correspondence to: N. Wever (wever@slf.ch)

Abstract. In October 2011, the Swiss Alps underwent a marked rain-on-snow (ROS) event when a large snowfall on 8 and 9 October was followed by intense rain on 10 October. This resulted in severe flooding in some parts of Switzerland. Model simulations were carried out for 14 meteorological stations in two affected regions of the Swiss Alps using the detailed physics-based snowpack model SNOWPACK. We also conducted an ensemble sensitivity study, in which repeated simulations for a specific station were done with meteorological forcing and rainfall from other stations. This allowed the quantification of the contribution of rainfall, snow melt and liquid water storage on generating snowpack runoff. In the simulations, the snowpack produced runoff about 4–6 h after rainfall started, and total snowpack runoff became higher than total rainfall after about 11–13 h. These values appeared to be strongly dependent on snow depth, rainfall and melt rates. Deeper snow covers had more storage potential and could absorb all rain and meltwater in the first hours, whereas the snowpack runoff from shallow snow covers reacts much more quickly. However, the simulated snowpack runoff rates exceeded the rainfall intensities in both snow depth classes. In addition to snow melt, the water released due to the reduction of liquid water storage contributed to excess snowpack runoff. This effect appears to be stronger for deeper snow covers and likely results from structural changes to the snowpack due to settling and wet snow metamorphism. These results are specifically valid for the point scale simulations performed in this study and for ROS events on relatively fresh snow.

1 Introduction

For mountain regions, the presence of a snow cover is an important factor in hydrological processes. One type of event that is still poorly understood is the behaviour of a snow cover during rainfall. These rain-on-snow (ROS) events are often accompanied by strong snow melt, due to high latent heat exchange and incoming long-wave radiation (ILWR) that reduces the radiative cooling of the snowpack (Marks et al., 2001; Mazurkiewicz et al., 2008). These effects increase the water available for outflow from the snowpack, which hereafter we will refer to as snowpack runoff. In this way, heavy precipitation can more easily lead to flooding events in mountainous terrain due to the additional snow melt (Pradhanang et al., 2013; Sui and Koehler, 2001). Furthermore, rainfall reduces snowpack stability, resulting in stronger wet snow avalanche activity (Conway and Raymond, 1993).

This study focuses on the dynamical snowpack behaviour during a ROS event in October 2011 in the Swiss Alps. The event is described in detail in Badoux et al. (2013). During 8 October 2011, the passage of a cold front brought significant snowfall amounts on the north side of the Swiss Alps at altitudes above 800 m, accompanied by a strong drop in air temperature. During 9 October, the precipitation faded and cold weather remained. In the night of 9 to 10 October, the passage of a warm front brought new precipitation, mainly rain, this time accompanied by a strong increase in air temperature. The snowfall limit finally increased up to 3000 m on 10 October. This ROS event is very suitable for a case study, because it occurred on a large scale and is well captured at many measurement sites. Furthermore, the fact that it caused widespread flooding shows that the event was extreme. One region, where the snow cover was relatively shallow, was more strongly affected by flooding than a region with a deep snow cover at the onset of rain (Badoux et al., 2013). An important question that arose from the event is whether there is a difference in snowpack behaviour for a shallow and a deep snow cover that may explain the difference in hydrological response in those two areas. The differences in streamflow response (Badoux et al., 2013), in terms of return period, would suggest that deep snowpacks can better dampen peak outflows than shallow ones.

Studies modelling ROS events mostly analyse the daily to weekly timescales and successfully reproduce the temporal evolution of snow water equivalent over several days (Marks et al., 1998, 2001). This suggests that snowpack-related processes during ROS events are sufficiently understood. Consequently, one can estimate rather precisely how much water will be available for snowpack runoff (Marks et al., 1998; Mazurkiewicz et al., 2008), but the temporal dynamics of the release of meltwater on the sub-daily timescales has seldom been investigated in detail. This knowledge is essential, however, to estimate the response in streamflow discharge in catchments and to assess flood risks from ROS events. In Rössler et al. (2014), the meteorological circumstances leading to this event have been studied in combination with a hydrological catchment scale model to simulate streamflow discharge in one of the affected areas. To reproduce the rapid peak discharge in the event, considerable recalibration of the hydrological model setup was required. For example, relatively simple single layer snow models, which are often used in hydrological model frameworks, were unable to follow the snow cover dynamics without significant calibration of snow-related parameters for this particular situation (Rössler et al., 2014).

The exact behaviour of the snow cover during ROS events is governed by complex interactions between several processes. A cold snow cover can store rain water by capillary suction and, to a lesser extent, freezing the liquid water. These processes depend on the state of the snow cover before the onset of rain. As soon as the snow cover becomes wet, strong settling has been observed (Marshall et al., 1999). This settling, combined with a destruction of the snow matrix by melt, reduces the storage capacity, which may increase snowpack runoff. These counteracting processes are difficult to assess without the use of a physics-based snow cover model that includes a representation of the above processes. Here, the detailed multi-layer snow cover model SNOWPACK (Lehning et al., 2002a, b) is used. The SNOWPACK model has been extended recently with a solver for the Richards equation, which provides a demonstrable improvement in modelling liquid water flow through the snow cover, especially on the sub-daily timescale (Wever et al., 2014). This study aims to simulate the snow cover dynamics at individual snow stations during this event, to better understand the snowpack behaviour with respect to the production of snowpack runoff.

2 Methods and data

The results in this study are achieved by simulations with the SNOWPACK model, using measurements from automated meteorological stations in the affected areas. First, the SNOWPACK model will be discussed, focusing in particular on the treatment of snow melt and liquid water flow in the model. Then the available data sets are discussed, followed by how the SNOWPACK model was set up to simulate the event using the measured meteorological data. Finally, we discuss the methods used for carrying out an ensemble sensitivity analysis and subsequent regression analysis to better understand the dynamics of snowpacks in this ROS event as simulated by the model.

2.1 SNOWPACK model

The physics-based snowpack model SNOWPACK was used to simulate the development of the snow cover as a 1D-column, forced with the meteorological conditions as measured by meteorological stations. The model simulates snow cover development, e.g. temperature and density profiles, phase changes, microstructural parameters, liquid water infiltration and snowpack runoff (Lehning et al., 2002a, b). The simulations were done using SNOWPACK version 3.2.0 in which the solver for the Richards equation was introduced (Wever et al., 2014). Furthermore, improvements were made in the treatment of the boundary conditions for the energy balance and accompanying phase changes, which may explain some discrepancies with model results presented in Badoux et al. (2013).

Snow melt is an important source of liquid water in the snowpack. In the SNOWPACK model, snow melt occurs at a specific depth when the local temperature is $0\,°C$ and excess energy is added at this depth either by heat conduction in the snow matrix or by a divergent short-wave radiation flux penetrating the snow. At the top of the snowpack, the model prescribes the energy flux as a Neumann boundary condition in the case of melting conditions in the top snow element or else as a Dirichlet boundary condition, prescribing the measured snow surface temperature. The latter ensures a better assessment of the cold content of the snowpack, although it may result in small discrepancies between changes in internal energy and the diagnosed energy balance. Prescribing the upper or lower boundary temperature may result in changes in internal energy between time steps that are not accounted for by the diagnosed top and bottom energy flux from the preceding time step.

The heat flux that is used to force the model at the top of the snowpack can be expressed as (Lehning et al., 2002a):

$$Q_{sum} = Q_{LWnet} + Q_S + Q_L + Q_P, \tag{1}$$

where Q_{sum} is the prescribed flux ($W\,m^{-2}$) for the Neumann boundary condition at the upper boundary, Q_{LWnet} is the net long-wave radiation ($W\,m^{-2}$), Q_S is the sensible heat

$(\mathrm{W\,m^{-2}})$, Q_L is the latent heat $(\mathrm{W\,m^{-2}})$ and Q_P is the heat advection by liquid precipitation $(\mathrm{W\,m^{-2}})$. The net short-wave radiation absorbed by the snowpack is not incorporated in the Neumann boundary condition for the temperature equation, as it is modelled as a source term in the top layers of the snowpack to reflect the penetration of short-wave radiation in the snowpack.

Water transport in snow is governed by capillary suction and gravitational drainage (Marsh, 2006). Two common model approaches for liquid water flow in snow are the so-called bucket scheme and Richards equation (Wever et al., 2014). In the bucket scheme, downward water transport is determined by the presence of an excess liquid water content above a defined threshold water content in a specific layer. This excess water is transported downwards regardless of the storage capacity of lower layers. In the Richards equation, the balance between gravity and capillary suction is explicitly calculated. It yields performance improvement over the bucket approach on both daily and hourly timescales (Wever et al., 2014). However, solving the Richards equation may be expected to especially improve the simulation of water flow in seasonal snow covers, where snow stratigraphy can have a marked influence on the water flow. Differences in grain sizes can lead to capillary barriers and ice lenses may block the water flow, resulting in ponding (Marsh, 1999; Hirashima et al., 2010). For this ROS event, the snow cover built up in two days, leading to a very homogeneous stratification.

The hydraulic properties of the snowpack, as used for solving the Richards equation, are expected to have changed as follows due to the wet snow metamorphism of the initially fresh, dry snow:

- Fresh, dry snow has generally a dendritic structure and thereby a high capillary suction. Old and wet snow on the other hand has coarse, rounded grains accompanied by lower capillary suction. In the water retention curve proposed by Yamaguchi et al. (2010), dendricity is not explicitly taken into account. However, SNOWPACK initialises new snow layers with small grains and these layers thereby exhibit also higher suction than melt forms or old snow in the simulations.

- The saturated hydraulic conductivity increases with grain size, but decreases with density. In wet snow metamorphism, both grain growth and densification are occurring. However, in the simulations in this study, saturated hydraulic conductivity following Calonne et al. (2012) was increasing during the event (not shown).

- Snow melt destroys the ice matrix locally and in wet snow, also settling and densification occur. When the matrix to store water is decreasing in volume due to snow melt and/or settling, this leads to a decrease in storage capacity and is expected to cause additional snowpack runoff.

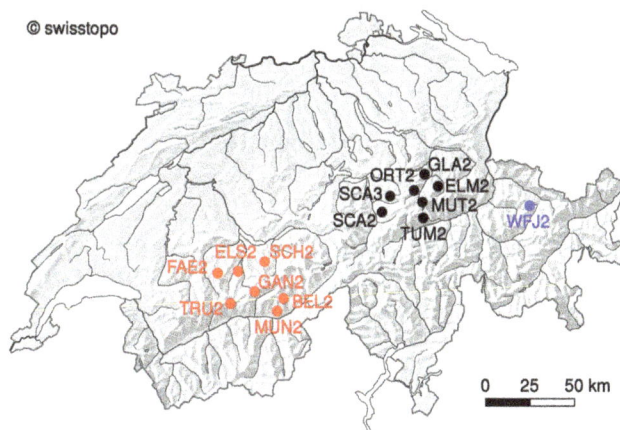

Figure 1. Map of Switzerland showing the locations of the stations used in this study in Bernese Oberland (red), Glarner Alpen (black) and the verification station Weissfluhjoch (blue). Reproduced by permission of swisstopo (JA100118).

2.2 Data

The behaviour of the snow cover during this ROS event is studied for two parts of the Swiss Alps: the Bernese Oberland and Glarner Alpen (Fig. 1). These areas were chosen because in particular the Bernese Oberland and to a lesser extent the Glarner Alpen experienced serious flooding (Badoux et al., 2013). The studied areas are both about $1000\,\mathrm{km^2}$. Both areas are located on the north side of the Alps and extend more or less over a similar altitude range, with glaciated areas in the highest parts. They are about $100\,\mathrm{km}$ apart, and, as will be shown, have experienced different meteorological forcing conditions.

In both areas, several automated weather stations are operated in the IMIS network. The stations measure meteorological and snow cover conditions at $0.5\,\mathrm{h}$ resolution. They are equipped with wind speed and direction, temperature, relative humidity, surface temperature, soil temperature, reflected short-wave radiation and snow height sensors. The stations are also equipped with an unheated rain gauge, which makes the precipitation measurements at the stations unreliable in case of snowfall and mixed precipitation. In both the Bernese Oberland and Glarner Alpen, seven stations were selected for this study (14 in total), as shown in Fig. 1 and listed in Table 1. These particular stations were selected because they represent the altitudinal gradient in the two regions and had limited missing values during the event. The sites are located in relatively flat terrain. The data have been quality checked manually and missing values were interpolated from neighbouring stations (Badoux et al., 2013). Most corrections were needed for wind speed, as the relatively wet snow caused the wind speed sensor to freeze at some stations. For interpreting the results, it is important to note that the average altitude of the analysed stations in the Glarner Alpen is about $270\,\mathrm{m}$ lower than in the Bernese Oberland.

Table 1. List of station abbreviations, station names, station altitudes and statistics for the two study areas and the verification station. The statistics denoted with Event are determined over the period 9 October, 18:00 UTC–11 October, 00:00 UTC. The bracketed sign before the root mean square error (RMSE) of snow height denotes whether modelled snow height is on average higher (+) or lower (−) than measured snow height. Time lag is the lag between the start of rain and the start of snowpack runoff and time runoff > rain denotes the time it took before cumulative snowpack runoff exceeded cumulative rainfall; w.e., water equivalent.

Stn	Name	Altitude (m)	Max snow height 6–14 Oct (cm)	RMSE snow height Event (cm)	Rainfall Event (mm)	Deposition Event (mm w.e.)	Snow melt Event (mm w.e.)	Snowpack runoff Event (mm w.e.)	Cold content (kJ m^{-3})	Time lag (h)	Time runoff > rain (h)
						Bernese Oberland					
FAE2	Faermel	1970	39	(+) 4.6	63	1.0	29	97	6	1.8	4.8
ELS2	Elsige	2140	44	(+) 2.4	59	1.3	34	94	33	3.5	6.2
MUN2	Mund	2210	53	(+) 3.4	34	2.3	45	78	94	3.0	7.0
SCH2	Schilthorn	2360	70	(−) 4.0	88	0.5	14	95	387	4.8	19.5
TRU2	Trubelboden	2480	37	(+) 3.7	84	0.8	22	106	68	3.8	9.8
BEL2	Belalp	2556	51	(+) 6.5	43	0.2	5	44	166	2.8	18.8
GAN2	Gandegg	2717	103	(+) 18.4	75	0.4	2	63	1265	10.0	–
Average all		2348	57	(+) 6.1	64	0.9	21	82	288	4.2	11.0
						Glarner Alpen					
GLA2	Glaernisch	1630	99	(+) 4.4	72	0.7	36	97	294	7.5	14.5
ORT2	Ortstock	1830	108	(−) 4.2	76	3.8	70	142	798	7.5	12.5
SCA2	Schächental	2030	73	(−) 6.4	75	2.5	71	146	330	5.5	9.8
ELM2	Elm	2050	90	(−) 3.1	53	1.0	24	67	559	5.2	13.0
TUM2	Tumpiv	2195	93	(−) 2.0	41	0.6	33	67	662	5.0	10.8
SCA3	Schächental	2330	90	(−) 7.1	81	1.6	23	95	972	5.2	14.8
MUT2	Muttsee	2474	92	(−) 4.8	63	0.7	10	61	932	7.0	–
Average all		2077	92	(−) 4.6	66	1.5	38	96	649	6.1	12.5
						Verification station					
WFJ	Weissfluhjoch	2540	48	(+) 2.4	33	1.1	18	47	799	4.8	11.2

For this study, the model was forced to interpret increases in measured snow height at the IMIS stations as snowfall (following Lehning et al., 1999), deriving the new snow density from a parameterised relationship with wind speed, temperature and relative humidity (Schmucki et al., 2014). The unheated rain gauges at the IMIS stations are not useful during these types of events, so to estimate rainfall, a different approach was followed. The Swiss Federal Office of Meteorology and Climatology (MeteoSwiss) is operating weather stations with a heated rain gauge (SwissMet-Net stations). Combined with several totalisers for precipitation, which are read off once per day, these precipitation measurements are compiled in a gridded data set (RhiresD, MeteoSwiss, 2013) at 2 km resolution with daily precipitation sums (06:00–06:00 UTC). To estimate the liquid precipitation input at an IMIS station, the daily sum derived from the nine grid points closest to the IMIS station in the RhiresD gridded data was distributed over the day by using the relative amounts of precipitation registered by the closest SwissMet-Net station. The rainfall started after 18:00 UTC on 9 October and consequently, all precipitation values before this time are set to zero, as snowfall is determined separately from the snow height measurements.

To validate the model performance for the chosen methods and data preparation procedures, data from the experimental site Weissfluhjoch (WFJ), located at 2540 m altitude in east

Switzerland near Davos (Fig. 1), were also used in this study. The course of the ROS event at this measurement site was quite similar to the 14 chosen IMIS stations, although both snowfall and rainfall amounts were smaller. At WFJ, both incoming and outgoing long- and short-wave radiation are available in addition to the default IMIS-type station setup, enabling a full assessment of the surface energy balance (abbreviated below as full EB). Furthermore, the site is equipped with a heated rain gauge that is part of the SwissMetNet network and a snow lysimeter that measures snowpack runoff (Wever et al., 2014), enabling the validation of simulated snowpack runoff.

2.3 Model setup

The model was initialised with 10 soil layers of 1 cm each. This allows the measured soil temperature at the lower boundary to be prescribed and, thereby, an estimate of the soil heat flux to be achieved. To allow for a spin-up period, the model simulations were started at 2 September, 6 weeks before the event. We consider this to be sufficient time for a soil of 10 cm depth. For soil parameters, typical values for very coarse material were chosen (similar to Wever et al., 2014). Furthermore, a free drainage lower boundary condition was used. This combination prevents liquid water ponding in the soil or snow. We hypothesise that this is generally not happening in the sloped terrain in the Swiss Alps, where liquid water that cannot directly infiltrate the soil is expected to leave the snowpack downslope, instead of ponding inside the snowpack.

A temperature threshold of $0.0\,°C$ is used to determine whether precipitation should be considered rain (from RhiresD) or snow (from the snow height sensors). This threshold is determined by comparing the ventilated and unventilated temperature sensor at WFJ during this particular event. It was found that during the onset of rain, the ventilated sensor was close to $1.2\,°C$ when the unventilated IMIS type sensor was measuring around $0.0\,°C$. This discrepancy may have arisen from bad ventilation due to wet snow collected onto the sensor hut, or condensation from the moist air on the temperature sensor itself. Because the onset of rain was accompanied by a strong and quick increase in air temperature, the influence of the choice of threshold on the results is small. In the case of snowfall, a snow element is added to the model domain for each 2 cm of new snow. In the case of rainfall, the water flux is either added to the top element (bucket scheme) or applied as a Neumann boundary condition (Richards equation).

For calculating the heat fluxes in Eq. (1), a neutral atmospheric stratification was assumed, which is likely an appropriate assumption because of the windy conditions during the event. The turbulent heat fluxes were calculated following a standard Monin–Obukhov parameterisation (Lehning et al., 2002b), using a roughness length z_0 of 0.002 m. The net radiation is approximated by using the measured reflected shortwave radiation and a parameterised albedo (Schmucki et al., 2014). Because the IMIS stations do not measure ILWR, this was approximated by the Omstedt (1990) parameterisation, using an estimated cloud cover. Based on observations of cloudiness at the MeteoSwiss station at Jungfraujoch, cloudiness was set to 1.0 in the period from 7 October to 9 October, 10:00 UTC and from 9 October, 17:00 to 11 October, when there was either solid or liquid precipitation, and 0.5 (half cloudy) for all other times.

2.4 Methods

To investigate the factors influencing the response of the snow cover during the ROS event, we added a sensitivity study by also forcing the SNOWPACK model for each station with the meteorological conditions from all the other stations. Snow melt is mainly governed by atmospheric conditions (temperature, relative humidity and wind speed), whereas liquid precipitation causes relatively little melt. Therefore, we consider atmospheric conditions and liquid precipitation as independent forcings during the event and treat them separately. The meteorological forcing, excluding the liquid precipitation, at the 14 stations represents 14 different melt scenarios. For liquid precipitation, we also have 14 more or less unique scenarios, although the temporal distribution over the day is based on eight SwissMetNet stations

only. However, the scenarios provide differences in rainfall amounts due to the spatial distribution as captured in the RhiresD data set by spatial interpolations and climatological lapse rates. So for each of the 14 stations, with its own unique maximum snow height, we performed an ensemble of 2744 simulations ($14 \times 14 \times 14$) with the SNOWPACK model, with every combination of melt and liquid precipitation scenario for statistical analysis. By replacing time series at a measurement site with a time series from another site, self-consistent series with real meteorological conditions that occurred during the event were created to act on the snow cover existing at the site. The original meteorological measurements at each station were used to force the model up to the moment on which the rainfall started on 9 October. From this specific time onwards, the meteorological and precipitation forcing was replaced by forcings from other stations. The starting time for these replacement series was taken as the moment on which rainfall started at these other stations.

To analyse possible different effects on snowpack runoff for shallow and deep snow covers, the 14 IMIS stations were divided into a shallow and deep snow cover class, depending on being above or below the median of maximum snow height during the event. The stations in Bernese Oberland are all present in the shallow snow cover class, except for GAN2, and all stations in Glarner Alpen are in the deep snow cover class, except SCA2. Per class, we determined a best fit by the linear regression for a given cumulative period using all ensemble simulations in the respective class:

$$Q_{cum} = \alpha P_{cum} + \beta M_{cum} + b, \tag{2}$$

where Q_{cum} is the cumulative snowpack runoff sum (mm w.e.), P_{cum} is the cumulative precipitation sum (mm), α is the linear regression coefficient for precipitation, M_{cum} is the cumulative snow melt sum (mm w.e.), β is the linear regression coefficient for snow melt and b is the intercept. In this context, b can be interpreted as the change in liquid water storage in the snow cover. As a positive value of b describes the snowpack runoff in the absence of any rain or snow melt, it can be assumed to reflect the recession curve and the effect of a decreasing water holding capacity, for example due to snow settling, wet snow metamorphism or changing hydraulic conductivity.

The dependence of the fit coefficients α, β and b over varying cumulative periods can reveal how the snow cover is modulating precipitation input and snow melt when generating snowpack runoff. These coefficients will be used on the original simulations to attribute the individual contributions of snow melt, precipitation and the intercept (change in storage) to the modelled snowpack runoff. The linear regression was done for cumulative periods of 0–1 to 0–24 h with two approaches: (i) taking the onset of rain at the stations as the start of the cumulative period and (ii) taking the onset of snowpack runoff as the start of the cumulative period. The latter was determined by both an increase of snowpack runoff by a factor of 2 compared to the snowpack runoff at

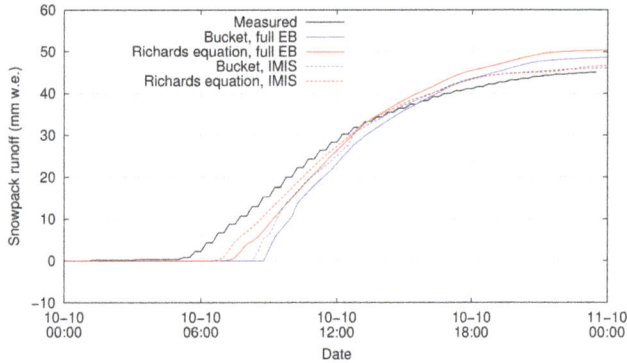

Figure 2. Comparison of modelled snowpack runoff using the bucket scheme or Richards equation for liquid water transport in the snowpack with measured snowpack runoff by a snow lysimeter for the station Weissfluhjoch, for 10 October. Simulations are done with either the full energy balance meteorological forcing (solid lines) or the forcing available for IMIS-type stations (dashed lines).

the onset of rain and a modelled snowpack runoff larger than 0.5 mm in 15 min.

3 Results

3.1 Verification

Before discussing the simulations for the two study regions, the results for the verification station WFJ will be presented. Figure 2 shows the modelled and measured snowpack runoff at the WFJ for the ROS event, starting shortly before the onset of rain. The measured snowpack runoff started shortly after midnight on 10 October, although this involved only marginal amounts, most likely related to snow melt at the snowpack base due to the ground heat flux. The measured snowpack runoff strongly increased just before 06:00 UTC, which we associate with the arrival of the liquid water added to the snowpack by rainfall and snow melt near the snow surface. This particular moment is found 1.5 (for RE with IMIS type setup) to 3.5 h (for bucket with full EB type setup) later in the simulations. Here, the neglect of preferential flow in the model probably plays a role. There is strong observational evidence for preferential flow paths in snow that transport liquid water down efficiently (Kattelmann, 1985; Marsh, 2006; Katsushima et al., 2013), but currently, a modelling concept for this process is not available.

Nevertheless, solving liquid water flow in the snow cover with the Richards equation is providing a closer agreement with observed snowpack runoff than with the bucket scheme concerning the timing of the start of snowpack runoff. Both

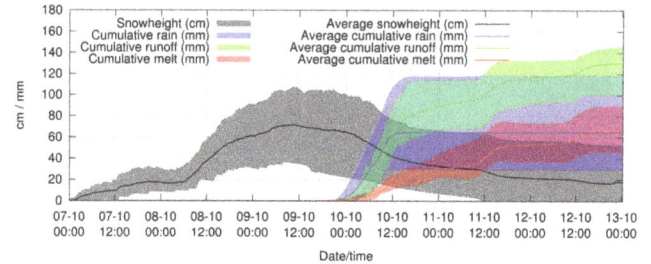

Figure 3. Overview of the simulation results of the temporal evolution of the ROS event between 7 and 13 October. Shown is the range of absolute minimum and maximum modelled snow height, cumulative precipitation, cumulative snowpack runoff and cumulative melt over the 14 stations. The solid lines denote the average values. The accumulation for precipitation and melt was calculated from 9 October, 18:00 UTC onwards.

models are overestimating the snowpack runoff rate, as shown by the steeper cumulative curve, although this overestimation is larger with the bucket scheme. In contrast, the total runoff sum at the end of the day is overestimated more in simulations with the Richards equation than with the bucket scheme. Because of the focus on snowpack runoff dynamics during the event in this study, we chose to do all further calculations with the Richards equation only. It should be noted, however, that several parameterisations are not yet verified with the Richards equation (metamorphism, snow settling, etc.).

In spite of some differences between the full energy balance station from WFJ and the IMIS-type setup from WFJ, it can be seen that the approach of parameterising ILWR and deriving precipitation from the RhiresD data and the Swiss-MetNet stations is providing reasonable results. It shows that the methods used in this study are suitable for analysing the dynamical snowpack behaviour at the IMIS stations in the two study areas. It should be noted, however, that the timing of precipitation for WFJ is very accurate, because a heated SwissMetNet rain gauge is located at this site, whereas for other stations, the closest SwissMetNet station is generally several kilometres away.

3.2 Event description

The event started with snowfall above 800 m altitude on 7 October. Figure 3 shows the temporal development of snow cover height in the two study regions. The snowfall was quite continuous and the maximum snow height averaged over all stations was reached around 9 October, 12:00 UTC. Table 1 shows that the average maximum snow height at the seven IMIS stations in Bernese Oberland was 57 cm, less than the

92 cm in the Glarner Alpen. In Bernese Oberland, snowfall amounts tended to increase with altitude, whereas interestingly, this trend was absent in the Glarner Alpen.

After the maximum snow height was reached, the snow height started decreasing, although rain and surface snow melt had not started yet. This decrease can be attributed mainly to settling of the snowpack and melt at the snowpack base by the ground heat flux. The following precipitation event started after 18:00 UTC on 9 October, and consisted purely of rainfall (except for very high altitudes). It was accompanied by a rapid increase of the 0 °C-isotherm to 3000 m altitude. The rainfall lasted until 15:00 UTC on 10 October, with an average rainfall sum of about 65 mm for both areas (Table 1). This gives a higher precipitation rate during the rainfall period than during the snowfall period.

The rainfall in the first hours was not accompanied by significant snowpack runoff (Fig. 3). This means that liquid water was stored in the snow cover by capillary suction and refreezing inside the snowpack. Refreezing was especially occurring at high altitude and at stations in Glarner Alpen, given the high cold content of those snow covers at the onset of rain (Table 1). Note that for typical snowpack conditions present at the onset of the rainfall, the amount of rain water needed to warm the snowpack to 0 °C is in the order of 5 mm for the typical cold contents reported in Table 1. For this event, this is less than 10 % of the total rainfall amounts. In the model, snowpack runoff started approximately 4–6 h after the onset of rain, depending on snow depth (Table 1). The amount of snow melt during the rainfall period was rather small compared to the rainfall amounts. Table 1 shows that the total amount of snow melt during the event was almost twice as large in Glarner Alpen as in Bernese Oberland, although this is partly caused by the lower average altitude of the stations in the Glarner Alpen and the exact values are strongly dependent on the choice of period. We can see that the average snowpack runoff curve is getting steeper during the rain episode and eventually becomes steeper than the rain curve (Fig. 3). After the rain stopped, snow melt continued due to the sensible heat flux provided by the increased air temperature. The snow melt exhibited a clear daily cycle on 11 and 12 October (Fig. 3), with a peak in the afternoon hours, associated with a high short-wave radiation input and a high air temperature.

To verify the snowpack simulations, the RMSE of measured and modelled snow height was calculated for the period 9 October, 18:00 UTC–11 October, 00:00 UTC (Table 1). Although snow water equivalent would be the preferred way to validate the simulations, as it better reflects the processes of snow melt, rainfall and liquid water flow than snow height, this is not possible due to the lack of validation data. However, measured snow height is generally considered an adequate estimate of snow water equivalent in physics-based models (Sturm et al., 2010). Because the simulations were forced by measured snow height, a high agreement between measured and modelled snow height is present for the ac-

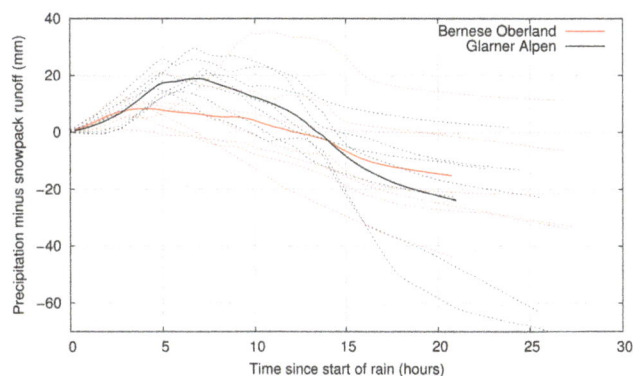

Figure 4. Cumulative rainfall minus cumulative modelled snowpack runoff during the event, starting at the onset of rain for each individual station (dashed lines) and for area averages (solid lines).

cumulation phase. By focusing solely on the ROS period itself, the RMSE values are indicative for the melt phase only. Most stations have an RMSE value below 5 cm, indicating a satisfying agreement between measured and modelled snow height in the melt phase. The largest discrepancy is found for the highest station in the study, where the snow height is overestimated with an RMSE value of 18 cm. Interestingly, the snow height is generally overestimated in the Bernese Oberland, whereas the opposite is found for the Glarner Alpen. Main reasons for this discrepancy may be an underestimation or overestimation, respectively, of snow melt, or an overestimation or underestimation, respectively, of new snow density and thus snow water equivalent.

In Fig. 4, the cumulative difference between rainfall and snowpack runoff is shown, starting from the onset of rain at the individual stations. When the curve is increasing, precipitation amounts exceed snowpack runoff, denoting storage of liquid water in the snowpack. A decreasing curve shows that the snowpack runoff is exceeding precipitation. The model results suggest that the snow cover was storing liquid water after the onset of the rainfall at all stations, dampening the effect of rain in the first few hours of the event. The initially dry and cold snow cover used the latent heat from refreezing rain water to get isothermal and also rain water was stored additionally in the snow cover by capillary suction. The shallow snow cover at the stations in the Bernese Oberland could retain less water than the deeper snow cover at the stations in Glarner Alpen. Furthermore, the tipping point where a net storage of liquid water in the snow cover changed into a net release of liquid water from the snow cover was reached earlier in Bernese Oberland (4 h) than in Glarner Alpen (7 h). Table 1 shows the time needed before cumulative snowpack runoff exceeded cumulative rainfall, which is generally shorter in Bernese Oberland than in Glarner Alpen. However, it still took on average 11–13 h from the start of rainfall before the total snowpack runoff exceeded total rainfall. This shows that the dampening effect of the rainfall by

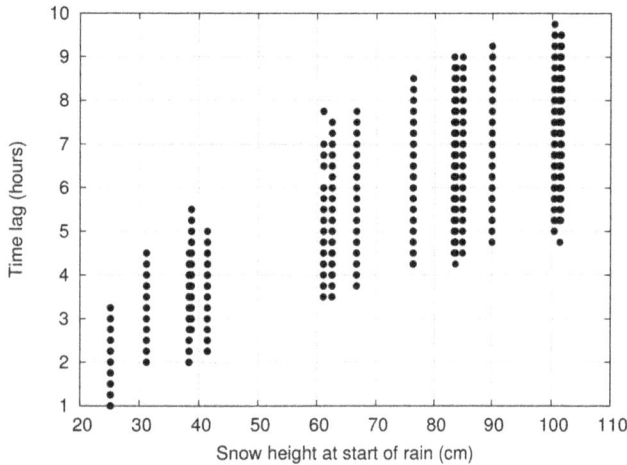

Figure 5. Time lag between the start of rain and the modelled start of snowpack runoff as a function of the snow height at the onset of rain for the ensemble simulations.

the snow cover was quite strong and persisted for several hours. Figure 4 also shows a wide spread between individual stations, related to variations in rainfall and snow melt rates. This motivated us to carry out the ensemble simulations that will be discussed later.

3.3   Energy balance

A net positive energy balance for the surface will first result in a heating of the snowpack (reducing the cold content), followed by melt. Table 2 shows the individual terms of the energy balance at the stations, expressed as mm w.e. melt potential as if the energy would be solely used for snow melt or freezing. The time period denoted as "Event" in Tables 1 and 2 is arbitrarily chosen to contain at least the complete rainfall, but longer or shorter time periods may have a significant effect on the relative contribution of the terms. A comparison with the amount of snow melt provided in Table 1 reveals that at all stations, most energy was used for snow melt. For stations with a high cold content, the net energy is partly used for heating of the snowpack. The contribution of net radiative energy, if not negative, and rain energy is fairly small. Most energy was delivered by heat release due to condensation during the ROS event and sensible heat. Mazurkiewicz et al. (2008) also found a strong contribution of latent heat to snow melt during ROS events. This is in contrast with typical clear sky spring snow melt situations, where the two terms often have opposite sign (Mott et al., 2013). Note that small discrepancies between total heat and snow melt (with more snow melt occurring than total heat provided) are due to small errors in the diagnosed energy balance as a result of the Dirichlet boundary condition at the upper and lower boundaries, as described before.

4   Ensemble simulations

Figure 5 shows the time lag between the onset of rain and the arrival of meltwater at the bottom of the snowpack as a function of snow height for the ensemble simulations. A general tendency of an increasing time lag with deeper snow covers is found, consistent with a longer travel time. However, the spread, caused by variations in rainfall and snow melt amounts, is very large. In Fig. 6, the snow height is divided by the time lag to get an approximation of the simulated velocity of the water movement in the snow cover. There is a clear dependency of the sum of rainfall and snow melt rate on flow velocity. Simulated water flow velocities range from $0.07\,\mathrm{m\,h^{-1}}$ for low rainfall and snow melt rates up to 0.20–$0.25\,\mathrm{m\,h^{-1}}$ for the highest rates. These modelled values and the correlation with rainfall and snow melt rates match well with earlier published results: Jordan (1983) reports experimental values of $0.22\,\mathrm{m\,h^{-1}}$, and also shows that earlier studies found values ranging from 0.04 to $0.6\,\mathrm{m\,h^{-1}}$. The value of $0.22\,\mathrm{m\,h^{-1}}$ was determined for spring snow melt conditions, and is at the upper limit of what was simulated in this model study. The lower values in the simulations are likely associated with the state of the snow cover during this event. The relatively freshly fallen snow is generally fine grained, associated with a lower hydraulic conductivity than for spring snow. The upward trend with increasing rainfall and snow melt rates is associated with higher hydraulic conductivities as a result of a higher saturation inside the snow cover. Furthermore, in the presence of liquid water, wet snow metamorphism is rapid, resulting in grain growth, rounding, and consequently, an increase in hydraulic conductivity. In Singh et al. (1997), a very high velocity of $6\,\mathrm{m\,h^{-1}}$ was found for very high precipitation rates in a study with artificially created rainfall. In that study, it was concluded that the formation of efficient preferential flow paths (not considered in the SNOWPACK model) is likely contributing to this high average velocity.

4.1   Regression analysis

The regression analysis, as described by Eq. (2), was carried out to investigate the temporal evolution of the contribution of the different mechanisms in producing snowpack runoff. Figure 7a shows the regression coefficients of Eq. (2) for both the shallow and deep snow cover class as a function of cumulative period since the start of rain. In the shallow snow cover class, rain is correlated to snowpack runoff after 2 h already, whereas in the deep snow cover class, the first non-zero regression coefficient is found after 5 h. This illustrates that the retardation between rainfall and snowpack runoff is dependent on snow depth. Furthermore, the coefficient for snow melt is higher in the shallow snow cover class than in the deep one in the early hours since the onset of snowpack runoff. The coefficient in the deep snow cover class is below 1.0 for several hours, denoting that 1.0 mm of additional

Table 2. Energy balance at the stations for the period 9 October, 18:00 UTC–11 October, 00:00 UTC. The energy fluxes are expressed as an equivalent snow melt energy in mm w.e. for understanding the magnitude of the energy fluxes, although snow melt should not necessarily have occurred.

Stn	Altitude (m)	Rnet (mm w.e.)	Rain energy (mm w.e.)	Latent heat flux (mm w.e.)	Sensible heat flux (mm w.e.)	Soil heat flux (mm w.e.)	Total energy (mm w.e.)
Bernese Oberland							
FAE2	1970	1	5	8	9	6	28
ELS2	2140	−1	4	10	12	5	30
MUN2	2210	−3	1	17	24	4	44
SCH2	2360	−3	4	4	8	1	14
TRU2	2480	−2	6	6	15	−1	24
BEL2	2556	−5	1	1	3	1	1
GAN2	2717	−6	1	3	6	0	4
average all	2348	−3	3	7	11	2	21
Glarner Alpen							
GLA2	1630	2	6	5	8	11	31
ORT2	1830	−0	6	30	37	12	85
SCA2	2030	−1	6	18	38	6	69
ELM2	2050	−1	3	7	7	6	22
TUM2	2195	−0	3	4	13	11	32
SCA3	2330	−4	3	12	13	1	25
MUT2	2474	−6	1	5	8	1	10
average all	2077	−1	4	12	18	7	39
Verification station							
WFJ	2540	−6	0	8	14	0	17

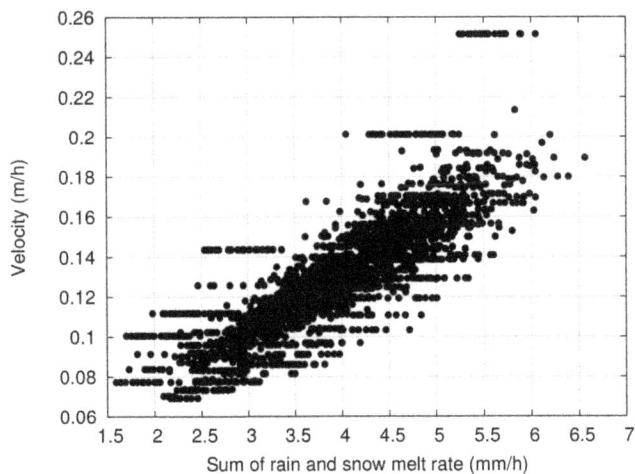

Figure 6. Simulated water velocity in the snow cover as a function of the sum of rain and snow melt in the first 5 h since the start of rain for the ensemble simulations.

snow melt would result in less than 1.0 mm extra snowpack runoff. This is caused by the long travel time needed by the liquid water arising from snow melt that occurred mostly near the surface. Only the part of the total snow melt near the base of the snowpack could have contributed to snowpack runoff in the first hours after the onset of rainfall.

After approximately 15 h from the start of rain, there is almost no difference in regression coefficient for snow melt and rain between the shallow and deep snow cover class. Then, the coefficient for rain is almost equal to 1.0, indicating that 1.0 mm of additional precipitation in this period would result in 1.0 mm extra snowpack runoff and the dampening effect of the snow cover has disappeared. Interestingly, the coefficient for snow melt is about 1.1, suggesting that there was approximately 10 % more snowpack runoff from the snow cover than the amount of snow melt alone. We attribute this to the destruction of the snow matrix by snow melt, which reduced the storage capacity of the snowpack for liquid water. The intercept term clearly demonstrates that the deep snow covers had more storage capacity for meltwater, as the minimum is smaller than in the shallow snow cover class. This results in a longer delay between the onset of rain and the actual snowpack runoff. The intercept term is still negative after 24 h, denoting that the effect of the storage capacity is noticeable over long periods.

In Fig. 7b, the regression coefficients are shown for cumulative periods starting at the onset of snowpack runoff. Expectedly, the intercept term changes sign: once snowpack runoff started, there was a contribution from the intercept.

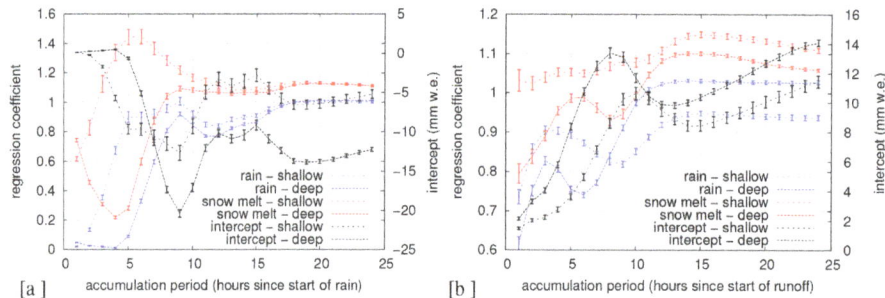

Figure 7. Regression coefficients as a function of cumulative period, since the start of rain (**a**) or start of snowpack runoff (**b**), for rain, snow melt and the intercept (mm w.e.) for both the shallow and deep snow cover class.

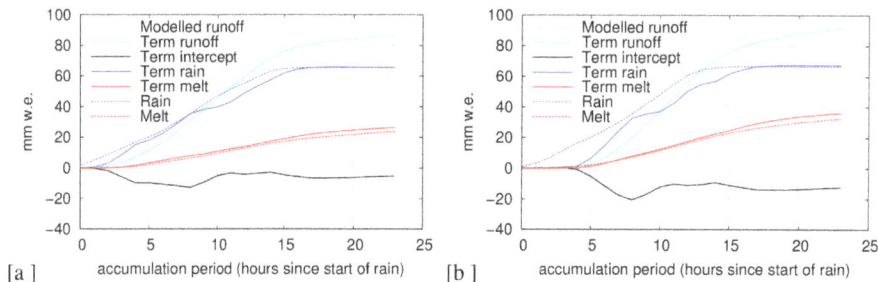

Figure 8. Modelled snowpack runoff, measured rainfall and modelled snow melt together with the terms of the linear regression for both the shallow (**a**) and the deep (**b**) snow cover class. Note that the blue, red and black solid lines sum up to the cyan line.

The intercept term is larger in the deep snow cover class than in the shallow one, indicating that in the simulations, deep snow covers produced more snowpack runoff, independent of snow melt or rainfall. This contribution consists of the snow melt and precipitation prior to the onset of snowpack runoff. Furthermore, settling may cause a reduction in storage capacity of the snow cover.

The stations in the shallow snow cover class have a higher coefficient for precipitation and snow melt in the short cumulative periods, denoting a stronger correlation of both variables with snowpack runoff shortly after the onset of runoff. We suggest that this is caused by short travel times through the snowpack in shallow snow covers. The difference with the deep snow cover class is decreasing with increasing time. After about 13 h, the regression coefficients appear to remain fairly constant. Interestingly, for a deep snow cover, snow melt has a lower regression coefficient than for a shallow one while this is opposite for rain, for which we cannot offer an explanation. Another contrasting effect is that the regression coefficients for precipitation and snow melt show a larger increase with increasing cumulative period in the deep snow cover class than in the shallow one. This points towards a dynamic effect in the snowpack, likely associated with changing snowpack microstructure and associated hydraulic properties.

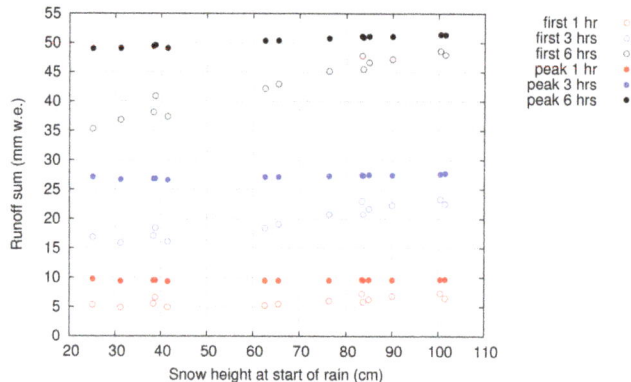

Figure 9. Average maximum cumulative snowpack runoff (denoted peak) and average cumulative snowpack runoff in the first hours after the start of snowpack runoff (denoted first), averaged over all ensemble simulations.

4.2 Attribution

In Fig. 8, the individual terms (precipitation, snow melt and intercept) and the sum (snowpack runoff) of the linear regression (Eq. 2) are shown, using the coefficients for both classes and the average rain and snow melt for each of the respective classes. Also drawn is the average modelled snowpack runoff. The almost perfect match between the modelled snowpack runoff and the sum of the linear regression terms shows that the regression analysis performs well. In both

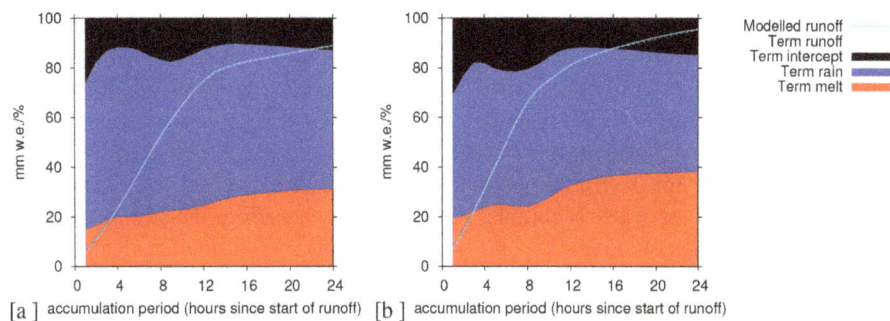

Figure 10. Cumulative snowpack runoff (from model and from regression analysis) and the terms of the linear regression for both the shallow (**a**) and the deep (**b**) snow cover class.

classes, the modelled cumulative snowpack runoff curve is steeper than the cumulative rainfall curve and it is steeper in the deep snow cover class than in the shallow one. From the individual terms of the regression analysis, it can be derived that this is caused not only by the contribution of snow melt, but also by the decrease in the (negative) contribution of the snow storage, denoted by the intercept term. The results also suggest that the rainfall provided a stronger contribution in the deep snow cover class than in the shallow one, as expressed by the steeper curve of the rain term. The intercept term shows that there was a stronger dampening in the deep snow cover class than in the shallow one, although apparently it did not compensate the rainfall and snow melt contributions.

Figure 8 also shows that the time since the start of the rainfall after which the contribution of rainfall flattens out and the increase in snowpack runoff decreases (around 80 mm w.e.) lies for both snow depth classes around 16 h. However, the modelled onset of snowpack runoff, and thus the onset of the individual terms in the regression analysis, was about 3 h later in the deep snow cover class than in the shallow one. This clearly illustrates that the model simulates a higher snowpack runoff rate in the deep snow cover class once snowpack runoff starts. To assess the relationship with snow cover depth, Fig. 9 shows the snowpack runoff sums in the first hours after the start of snowpack runoff and the maximum peak snowpack runoff sum over the first 24 h after the onset of rain, averaged over all 196 simulations per individual station (associated with a specific snow depth). In the first hours of modelled snowpack runoff, snowpack runoff rates show a clear increase with snow depth, whereas the trend is almost absent for the maximum peak snowpack runoff rates. The reason for the latter is that peak snowpack runoff is likely achieved in a kind of steady state situation when incoming rainfall and snow melt are in balance with snowpack runoff. The fact that this value is almost constant with snow depth is a consequence of the ensemble simulation setup, where all precipitation and melt scenarios are present for each station.

The simulations suggest that deep snow covers initially produced more snowpack runoff than shallow snow covers

and that this effect is partly caused by hydraulic effects inside the snowpack and partly by higher snow melt amounts. In Fig. 10, the percentages of respectively intercept, snow melt and rainfall contributions to snowpack runoff are shown for increasing cumulative periods, as determined by the regression analysis. This confirms the earlier conclusions. The contribution of the storage is varying between 15 and 20 % and is higher in the deep snow cover class. The contribution of snow melt is almost doubling from 15 and 20 % to 30 and 38 % between 1 and 24 h cumulative periods for the shallow and deep snow covers, respectively. The higher amount of snow melt experienced at the stations in the deeper snow cover class is likely unrelated to the deeper snow cover, whereas the higher contribution of the intercept term for the deeper snow cover class should be truly connected to the deeper snow cover.

5 Discussion

The response of the snowpack during a ROS event has been studied here using a physics-based snow cover model. The results depict how the SNOWPACK model simulates the influence of rainfall and snow melt on producing snowpack runoff and consequently, the conclusions drawn here are strongly dependent on a sufficient process representation in the SNOWPACK model. The comparison with snow lysimeter measurements at WFJ indicated that average velocity with which liquid water is routed through the snowpack in SNOWPACK was slightly underestimated, most likely due to neglecting preferential flow paths. This would imply that the time lag between the onset of rainfall and the onset of snowpack runoff is overestimated in the model. A preferential flow path formulation for physics-based snowpack models is not yet available and, to our knowledge, preferential flow paths are neglected in most physics-based models. It is difficult to speculate on the influence of preferential flow on the results presented here, in particular for contrasts between shallow and deep snow covers. However, the role of preferential flow in homogeneous layered snowpacks, as was the case in this particular event, may be limited. The snow at the onset of this ROS event had fallen during cold conditions in the 3 days prior to the event. We can thus assume that the initial snow-pack was rather homogeneous with relatively small grains. In laboratory experiments, preferential flow was not observed for small grain sizes (Katsushima et al., 2013). From those experiments, the role of water ponding at strong transitions in snow properties between snow layers for the formation of preferential flow was also identified. These water accumulations may also trigger significant amounts of lateral flow, for example over ice layers inside the snowpack. However, it is unlikely that these inhomogeneities were present in the snow cover in this particular event.

Other discrepancies were found between measured and modelled snow height, which may be caused by underestimations or overestimations of snow settling and/or snow melt. However, typical RMSE for snow height was less than 5 cm, which is around 2 mm w.e., dependent on snow density. Using the ensemble and regression analysis, the individual contributions of snow melt, rainfall and snow storage have been quantified by using melt scenarios from all stations. The higher snowpack runoff rates found in the simulations for deep snow covers were found to be not only dependent on snow melt, but also due to the effects of rainfall on deeper snow covers and the reaction of the snow storage. We therefore conclude that these discrepancies have only a small influence on the general validity of the results, even though the comparison of measured and modelled snow height suggested a consistent underestimation of snow melt in the Bernese Oberland and a consistent overestimation in the Glarner Alpen.

6   Conclusions

Model simulations of a ROS event in October 2011 for 14 meteorological stations in two regions of the Swiss Alps have shown that the snowpack runoff dynamics from the snow cover is strongly dependent on the snow depth at the onset of the rain. Deeper snow covers had more storage and absorbed all rain and meltwater in the first hours, whereas the modelled snowpack runoff from shallow snow covers reacted much more quickly to the onset of rainfall. The modelled time lag between the onset of rain and the onset of snowpack runoff ranged from 2.2 to 10 h, depending on snow depth and cold content of the snowpack at the onset of rainfall, with an average around 4–5 h. In this event, cumulative modelled snowpack runoff became higher than cumulative rainfall as a result of additional snow melt after on average 11–13 h.

An ensemble of simulations was carried out where meteorological and precipitation forcing conditions were interchanged between stations. It was found that the time lag between the onset of rainfall and snowpack runoff in the model study depends not only on snow height but also on the sum of rainfall and melt rates. Simulated flow rates of liquid water in the snowpack were smaller than observations in spring snow, which can be attributed to the structure of the snow-pack that consisted of small grains with a high suction and low hydraulic conductivity.

A regression analysis on the ensemble simulations has shown that deep snow covers generated more snowpack runoff, in the first hours after snowpack runoff started. The analyses suggested that this was caused by a higher release of liquid water from the storage in deep snow covers than in shallow ones. The quicker depletion of the storage in deep snow covers is partly driven by snowpack settling and partly by recession processes.

Note that these conclusions were derived for a ROS event during which the amount of rainfall largely exceeded the storage capacity of the snow, generating large amounts of snowpack runoff, even at locations with a deep snow cover. The effect of initial snow depth may be fundamentally different for ROS events in which rain falls on spring snow, where most settling has already occurred and liquid water is already present in the snowpack.

Given that the snow cover was deeper in Glarner Alpen than in Bernese Oberland, these differences in snowpack behaviour in terms of time lag between the onset of rain and the onset of snowpack runoff may have contributed to the differences found in streamflow discharge. In Bernese Oberland, streams reacted quickly on the onset of rain, whereas in Glarner Alpen, where the snow cover was thicker, flooding occurred mainly in the late afternoon of 10 October after most rainfall occurred (Badoux et al., 2013). On the other hand, the model results in this study have shown that once the snowpack produces runoff, the modelled snowpack runoff is higher in the deep snow cover class than in the shallow one. The validity of this conclusion depends on the adequacy of the representation of liquid water flow in the SNOWPACK model, as snow lysimeter measurements to support this result are lacking.

Acknowledgements. Part of this research was financed by the Swiss Federal Office for the Environment FOEN. Funding was also provided from the IRKIS project, supported by the Office for Forests and Natural Hazards of the Swiss Canton of Grisons (Chr. Wilhelm), the region of South Tyrol (Italy) and the community of Davos. We also would like to thank Jan Magnusson for his comments and suggestions for the analysis.

Edited by: H.-J. Hendricks Franssen

References

Badoux, A., Hofer, M., and Jonas, T. (Eds.): Hydrometeorologische Analyse des Hochwasser-ereignisses vom 10. Oktober 2011, Birmensdorf, Swiss Federal Institute for Forest, Snow and Landscape Research WSL; Davos, WSL-Institute for Snow and Avalanche Research SLF; Zürich, Federal Office of Meteorology and Climatology MeteoSwiss; Bern, geo7 geowissenschaftliches Büro; Bern, Federal Office for the Environment FOEN, in German, 2013.

Calonne, N., Geindreau, C., Flin, F., Morin, S., Lesaffre, B., Rolland du Roscoat, S., and Charrier, P.: 3-D image-based numerical computations of snow permeability: links to specific surface area, density, and microstructural anisotropy, The Cryosphere, 6, 939–951, doi:10.5194/tc-6-939-2012, 2012.

Conway, H. and Raymond, C. F.: Snow stability during rain, J. Glaciol., 39, 635–642, 1993.

Hirashima, H., Yamaguchi, S., Sato, A., and Lehning, M.: Numerical modeling of liquid water movement through layered snow based on new measurements of the water retention curve, Cold Reg. Sci. Technol., 64, 94–103, doi:10.1016/j.coldregions.2010.09.003, 2010.

Jordan, P.: Meltwater movement in a deep snowpack: 1. Field observations, Water Resour. Res., 19, 971–978, doi:10.1029/WR019i004p00971, 1983.

Katsushima, T., Yamaguchi, S., Kumakura, T., and Sato, A.: Experimental analysis of preferential flow in dry snowpack, Cold Reg. Sci. Technol., 85, 206–216, doi:10.1016/j.coldregions.2012.09.012, 2013.

Kattelmann, R.: Macropores in snowpacks of Sierra Nevada, Ann. Glaciol., 6, 272–273, 1985.

Lehning, M., Bartelt, P., Brown, B., Russi, T., Stöckli, U., and Zimmerli, M.: SNOWPACK calculations for avalanche warning based upon a new network of weather and snow stations, Cold Reg. Sci. and Technol., 30, 145–157, doi:10.1016/S0165-232X(99)00022-1, 1999.

Lehning, M., Bartelt, P., Brown, B., Fierz, C., and Satyawali, P.: A physical SNOWPACK model for the Swiss avalanche warning Part II. Snow microstructure, Cold Reg. Sci. Technol., 35, 147–167, doi:10.1016/S0165-232X(02)00073-3, 2002a.

Lehning, M., Bartelt, P., Brown, B., and Fierz, C.: A physical SNOWPACK model for the Swiss avalanche warning Part III: Meteorological forcing, thin layer formation and evaluation, Cold Reg. Sci. Technol., 35, 169–184, doi:10.1016/S0165-232X(02)00072-1, 2002b.

Marks, D., Kimball, J., Tingey, D., and Link, T.: The sensitivity of snowmelt processes to climate conditions and forest cover during rain-on-snow: a case study of the 1996 Pacific Northwest flood, Hydrol. Process., 12, 1569–1587, doi:10.1002/(SICI)1099-1085(199808/09)12:10/11<1569::AID-HYP682>3.0.CO;2-L, 1998.

Marks, D., Link, T., Winstral, A., and Garen, D.: Simulating snowmelt processes during rain-on-snow over a semi-arid mountain basin, Ann. Glaciol., 32, 195–202, doi:10.3189/172756401781819751, 2001.

Marsh, P.: Snowcover formation and melt: recent advances and future prospects, Hydrol. Process., 13, 2117–2134, doi:10.1002/(SICI)1099-1085(199910)13:14/15<2117::AID-HYP869>3.0.CO;2-9, 1999.

Marsh, P.: Encyclopedia of Hydrological Sciences., chap. 161: Water Flow Through Snow and Firn, 1–14, John Wiley & Sons, Ltd, Chichester, England, doi:10.1002/0470848944.hsa167, 2006.

Marshall, H., Conway, H., and Rasmussen, L.: Snow densification during rain, Cold Reg. Sci. Technol., 30, 35–41, doi:10.1016/S0165-232X(99)00011-7, 1999.

Mazurkiewicz, A. B., Callery, D. G., and McDonnell, J. J.: Assessing the controls of the snow energy balance and water available for runoff in a rain-on-snow environment, J. Hydrol., 354, 1–14, doi:10.1016/j.jhydrol.2007.12.027, 2008.

MeteoSwiss: Documentation of MeteoSwiss Grid-Data Products, Daily Precipitation (final analysis): RhiresD, Tech. rep., Federal Office of Meteorology and Climatology MeteoSwiss, Zürich, Switzerland, 2013.

Mott, R., Gromke, C., Grünewald, T., and Lehning, M.: Relative importance of advective heat transport and boundary layer decoupling in the melt dynamics of a patchy snow cover, Adv. Water Resour., 55, 88–97, doi:10.1016/j.advwatres.2012.03.001, 2013.

Omstedt, A.: A coupled one-dimensional sea ice-ocean model applied to a semi-enclosed basin, Tellus A, 42, 568–582, doi:10.1034/j.1600-0870.1990.t01-3-00007.x, 1990.

Pradhanang, S. M., Frei, A., Zion, M., Schneiderman, E. M., Steenhuis, T. S., and Pierson, D.: Rain-on-snow runoff events in New York, Hydrol. Process., 27, 3035–3049, doi:10.1002/hyp.9864, 2013.

Rössler, O., Froidevaux, P., Börst, U., Rickli, R., Martius, O., and Weingartner, R.: Retrospective analysis of a nonforecasted rain-on-snow flood in the Alps –a matter of model limitations or unpredictable nature?, Hydrol. Earth Syst. Sci., 18, 2265–2285, doi:10.5194/hess-18-2265-2014, 2014.

Schmucki, E., Marty, C., Fierz, C., and Lehning, M.: Evaluation of modelled snow depth and snow water equivalent at three contrasting sites in Switzerland using SNOWPACK simulations driven by different meteorological data input, Cold Reg. Sci. Technol., 99, 27–37, doi:10.1016/j.coldregions.2013.12.004, 2014.

Singh, P., Spitzbart, G., Hübl, H., and Weinmeister, H.: Hydrological response of snowpack under rain-on-snow events: a field study, J. Hydrol., 202, 1–20, doi:10.1016/S0022-1694(97)00004-8, 1997.

Sturm, M., Taras, B., Liston, G. E., Derksen, C., Jonas, T., and Lea, J.: Estimating Snow Water Equivalent Using Snow Depth Data and Climate Classes, J. Hydrometeor., 11, 1380–1394, doi:10.1175/2010JHM1202.1, 2010.

Sui, J. and Koehler, G.: Rain-on-snow induced flood events in Southern Germany, J. Hydrol., 252, 205–220, doi:10.1016/S0022-1694(01)00460-7, 2001.

Wever, N., Fierz, C., Mitterer, C., Hirashima, H., and Lehning, M.: Solving Richards Equation for snow improves snowpack meltwater runoff estimations in detailed multi-layer snowpack model, The Cryosphere, 8, 257–274, doi:10.5194/tc-8-257-2014, 2014.

Yamaguchi, S., Katsushima, T., Sato, A., and Kumakura, T.: Water retention curve of snow with different grain sizes, Cold Reg. Sci. Technol., 64, 87–93, doi:10.1016/j.coldregions.2010.05.008, 2010.

Representing glacier geometry changes in a semi-distributed hydrological model

Jan Seibert[1,2]**, Marc J. P. Vis**[1]**, Irene Kohn**[3]**, Markus Weiler**[3]**, and Kerstin Stahl**[3]

[1]Department of Geography, University of Zurich, Zurich, 8057, Switzerland
[2]Department of Aquatic Sciences and Assessment, Swedish University of Agricultural Sciences, Uppsala, Sweden
[3]Faculty of Environment and Natural Resources, University of Freiburg, 79098 Freiburg, Germany

Correspondence: Jan Seibert (jan.seibert@geo.uzh.ch)

Abstract. Glaciers play an important role in high-mountain hydrology. While changing glacier areas are considered of highest importance for the understanding of future changes in runoff, glaciers are often only poorly represented in hydrological models. Most importantly, the direct coupling between the simulated glacier mass balances and changing glacier areas needs feasible solutions. The use of a complex glacier model is often not possible due to data and computational limitations. The Δh parameterization is a simple approach to consider the spatial variation of glacier thickness and area changes. Here, we describe a conceptual implementation of the Δh parameterization in the semi-distributed hydrological model HBV-light, which also allows for the representation of glacier advance phases and for comparison between the different versions of the implementation. The coupled glacio-hydrological simulation approach, which could also be implemented in many other semi-distributed hydrological models, is illustrated based on an example application.

1 Introduction

Glacier meltwater makes an important contribution to discharge in high-mountain catchments (Köplin et al., 2013; Miller et al., 2012) and can sustain summer streamflow in many large river basins (Hagg et al., 2007; Stahl et al., 2017). When modelling the hydrology of such catchments for longer periods (> 10 years), the changing glacier area has to be considered, especially when climate change is causing glacier retreat. The simplest approach is to update the hydrological model with an externally simulated glacier extent, but this is unsatisfactory, as the mass balance simulated by the hydrological model might not agree with the updated glacier extent. The use of coupled glacio-hydrological models allows the glacier extent to be linked directly to the simulated glacier mass balance and is, thus, better suited for modelling catchments with changing glacier areas (Huss et al., 2008; Stahl et al., 2008). However, modellers are faced with the question of which degree of complexity is needed to represent glaciers and glacier evolution in hydrological models. Several fully distributed, physically based glacier models which consider mass balance, subglacial drainage, ice flow dynamics etc. have been developed over the past decades (Frans et al., 2016; Naz et al., 2014; Pattyn, 2002; Stroeven et al., 1989). While there are studies in which such complex glacier models have been coupled with hydrological models (Frans et al., 2016; Naz et al., 2014), a simpler approach might be useful in many cases as the limited data availability would not allow the application of complex models, in particular their parameterization and validation. The use of such a complex model is also often too computationally expensive for use in a combined glacio-hydrological model for which an entire catchment has to be considered. Many semi-distributed hydrological models use simplified representations of catchment hydrology using a limited number of conceptual buckets (reservoirs), and coupling such a model with a more complex glacier model would lead to a mismatch in terms of physical and spatial representation. Hence, for hydrological modelling studies there is a need for glacier models that use a similar degree of complexity and data demand as other components of the hydrological model but which are still able to represent the important glacier processes.

Recently an increasing number of hydrological models have incorporated glacier evolution models, using for example an equilibrium line altitude (ELA) shift (e.g. Linsbauer et al., 2012), volume–area scaling (e.g. Luo et al., 2013; Radić et al., 2008), volume–area scaling and morphological image analysis (e.g. Stahl et al., 2008), other simple schemes without ice flow (e.g. Bongio et al., 2016), or more complex approaches focusing on glacier modelling (e.g. Immerzeel et al., 2012). One approach with limited glacier input data requirements, which is mass-conserving and well suited for hydrological modelling studies, is the Δh parameterization, which describes the glacier thickness change at a certain elevation in response to an overall change in ice mass (Huss et al., 2010). Initially, Huss et al. (2008) introduced the Δh parameterization as part of their Glacier Evolution Runoff Model (GERM), while a more detailed presentation of the approach, including the derivation of generalized empirical functions applicable to unmeasured glaciers, is given in Huss et al. (2010). Since then, the Δh parameterization has been applied in global-scale modelling by Huss and Hock (2015) as well as in numerous studies applying GERM to simulate individual glaciers or glacierized regions in the Swiss Alps (Farinotti et al., 2012; Finger et al., 2013; Gabbi et al., 2012; Huss et al., 2014; Huss and Fischer, 2016) and in central Asia (Sorg et al., 2014). Several other glacio-hydrological models were coupled with glacier retreat simulations following the Δh approach (Addor et al., 2014; Gabbi et al., 2014; Linsbauer et al., 2013; Ragettli et al., 2013; Salzmann et al., 2012; Vincent et al., 2014). However, details on its practical implementation in the respective conceptual hydrological models have been provided by only a few studies, for instance those by Li et al. (2015) and Duethmann et al. (2015).

As the Δh parameterization is an empirical approximation to describe glacier retreat, it is subject to uncertainty and several limitations in terms of accurate glaciological modelling at the scale of individual glaciers (discussed in Huss et al., 2010; Linsbauer et al., 2013; Vincent et al., 2014). Nevertheless, for the purpose of transient hydrological modelling, particularly for regional studies covering large samples of glacierized catchments, the Δh approach represents an efficient state-of-the-art alternative to more complex glacier evolution models (Huss et al., 2010; Li et al., 2015). Originally, Huss et al. (2010) derived the Δh parameterization for periods dominated by negative mass balances and glacier retreat. The missing representation of glacier advance is related to uncertainties in regions with indications of the presence of recent glacier advance (Ragettli et al., 2013). Moreover, it represents a major drawback for long-term hydrological modelling covering past periods, for example the period with positive mass balance in the European Alps during the 1970s. A simplified scheme to incorporate short-term glacier change in case of advance as an extension of the original Δh approach is presented by Huss and Hock (2015).

Here, we describe a conceptual implementation of the Δh parameterization in the semi-distributed hydrological model HBV-light (Seibert and Vis, 2012), which also allows the representation of glacier advance phases, and we compare different versions of the implementation. This approach has recently been used to model a century of glacier runoff for 49 alpine catchments of the Rhine basin (Stahl et al., 2017). We present results from one of these catchments for illustration. This technical note aims at describing our implementation of the Δh parameterization in such a way that researchers using other hydrological models also could follow the same approach. This follows the quest for reproducible science as recently emphasized for hydrological modelling (Hutton et al., 2016).

2 Materials and methods

2.1 New glacier routine

2.1.1 HBV model and data requirements

The HBV model is a semi-distributed conceptual precipitation–runoff model. It has continued to be developed in Scandinavia since the 1970s (Bergström, 1976; Lindström et al., 1997) and has become a standard tool which is widely used in different model variants, particularly for modelling snow-dominated catchments. Required input data are daily temperature, precipitation and potential evaporation time series. Additionally, for the new glacier routine, information on the initial glacier areas and ice thickness values, both as a function of elevation, is required. For the estimation of these initial conditions, glaciologists have developed a number of approaches as recently reviewed by Farinotti et al. (2017). One possible method is described in Appendix A1.

In the HBV model the hydrological processes within a catchment are modelled by four different routines, a snow–glacier routine, a soil moisture routine, a response routine, and, finally, a streamflow routing routine. Here, we describe the recent integration of a glacier evolution approach into the HBV-light software, a user-friendly and freely available version of HBV (Seibert and Vis, 2012).

2.1.2 Snow and ice accumulation, melt and runoff

The glacier area within a catchment is conceptually simulated by two reservoirs representing glacier ice and the liquid water contained within the glacier. There can be a snowpack on top of the glacier, which also consists of a solid (snow) and a liquid (water content) reservoir. The snow and glacier routine calculations are performed at each simulation time step for each elevation zone, for which elevation intervals of 100 to 200 m are typically used. The elevation zones can be further subdivided according to three aspect classes (N, S, and W/E). Depending on the temperature in relation to the threshold temperature, precipitation falls either as snow or rain. In the case of rain, the precipitation is added to the water content of the snow if a snow layer is present or otherwise to the water content of the glacier. If the temperature is above the threshold temperature, melt takes place in the snowpack based on a degree-day factor, and the melted snow is added to the water content of the snowpack. In the case that the water content exceeds the snow water holding capacity, the amount exceeding the snow water holding capacity flows out and is added to the liquid water reservoir of the glacier. If the temperature is below the threshold temperature, part of the water content in the snow layer refreezes. The use of aspect classes allows both the faster and slower snowmelt in certain parts of the catchment to be considered by applying an additional aspect factor to the degree-day equation (Hagg et al., 2007; Hottelet et al., 1993), which taken together leads to a prolonged but less intense melt period at the catchment scale compared to the situation when not using different aspect classes.

For ice melt of the glacier a degree-day method is used as well, but ice melt is only simulated at times when there is no snow layer on the glacier. For temperatures above the threshold temperature, glacier melt is calculated using the degree-day factor multiplied by a glacier correction factor, which represents the different albedo of ice compared to snow and typically has values of about 1 to 2 (Hock, 2003). The ice melt is added to the liquid component of the glacier, from which the outflow is computed individually for each elevation zone as suggested by Stahl et. al. (2008), extending earlier concepts by Moore (1993), to account for the enlargement of glacial conduits over the melt season.

$$Q(t) = S(t)(K_{min} + K_{range} \cdot e^{-A_G \cdot S_{WE}(t)}) \tag{1}$$

Q is the outflow, S the liquid water content of the glacier, the parameters K_{min} and K_{range} the minimum outflow coefficient and maximum range of outflow coefficient values, and A_G a calibration parameter controlling the outflow response dependent on S_{WE}, which is the water equivalent of the snowpack on the glacier. To represent the transition from snow to firn in a simple way, at the end of each time step a certain fraction of the snow on top of the glacier is converted into firn and equally distributed over the whole glacier area. Typical values for this model parameter are 0.001–0.003, which implies that the conversion of snow to firn on average takes about 1 to 3 years (Luo et al., 2013). The further transition from firn to ice takes place over much longer time periods from 10 to over 100 years. For the glacier modelling presented here, however, firn is considered as a part of the accumulated glacier mass.

Snow redistribution by wind and avalanches can be important to consider in modelling alpine catchments as recently reviewed by Freudiger et al. (2017). Therefore, in our modelling approach snow redistribution can optionally be applied at the end of each time step to avoid unrealistic multi-year snow accumulation, the so-called "snow tow-

ers". As snow redistribution was not the focus of this study, we used a simple approach. During the snow redistribution, the snow (i.e. snowpack and snow water content) of all non-glacier areas above a certain user-specified elevation, H_{redist}, and after reaching a certain user-specified S_{WE} threshold, is redistributed evenly over the non-glacier and glacier areas within a user-specified elevation range below H_{redist} as well as the glacier areas above H_{redist}. Here we used an elevation of 2500 m a.s.l. for H_{redist}, 500 mm for the S_{WE} threshold, and 1900 m a.s.l. as the lower boundary for receiving redistributed snow. These values were motivated by the assumption that non-glacierized areas at high elevations correspond to the main snow erosion areas, that snow in these areas should melt away each summer, and that redistribution gains occur mainly in the snow zones below the high elevations.

2.1.3 Glacier mass and area changes

The technical details of the implementation of the new Glacier Area Change Routine (GACR) in HBV-light are outlined in a flowchart (Fig. 1). To translate glacier mass changes into area changes, a single-valued relation between glacier mass and glacier area needs to be established. This relationship is technically represented in the model by a lookup table, which provides the glacier areas for the different elevation zones for certain glacier mass values. Here we suggest that the relationship (and lookup table) is computed based on an initial variation of glacier thickness values with elevation (termed "initial glacier profile" in the following) and the Δh parameterization method described in Huss et al. (2010), scaling the relative elevations to those of the study catchment (Fig. 2). For these calculations each elevation zone (of typically 100–200 m) in the model application is further subdivided into elevation bands (of typically 10 m) to ensure smooth changes. The use of a lookup table enables the representation of periods of glacier advance (though not further than the initial glacier extent).

The basic idea is that the total glacier volume, M, is defined by integration of the initial glacier profile (Eq. 2):

$$M = \sum_{i=1}^{N} a_i \cdot h_i. \tag{2}$$

M is the total glacier mass in mm water equivalent relative to the entire catchment area, and for each elevation band i, the area a_i (expressed as a proportion of the catchment area) and water equivalent h_i in mm. To generate the lookup table the glacier is then melted in steps of ΔM. For each of these steps the Δh parameterization method of Huss et al. (2010) is applied. For each elevation band the normalized elevation, $E_{i,norm}$, is computed from the absolute elevation E_i of the corresponding elevation band i, as well as the maximum and

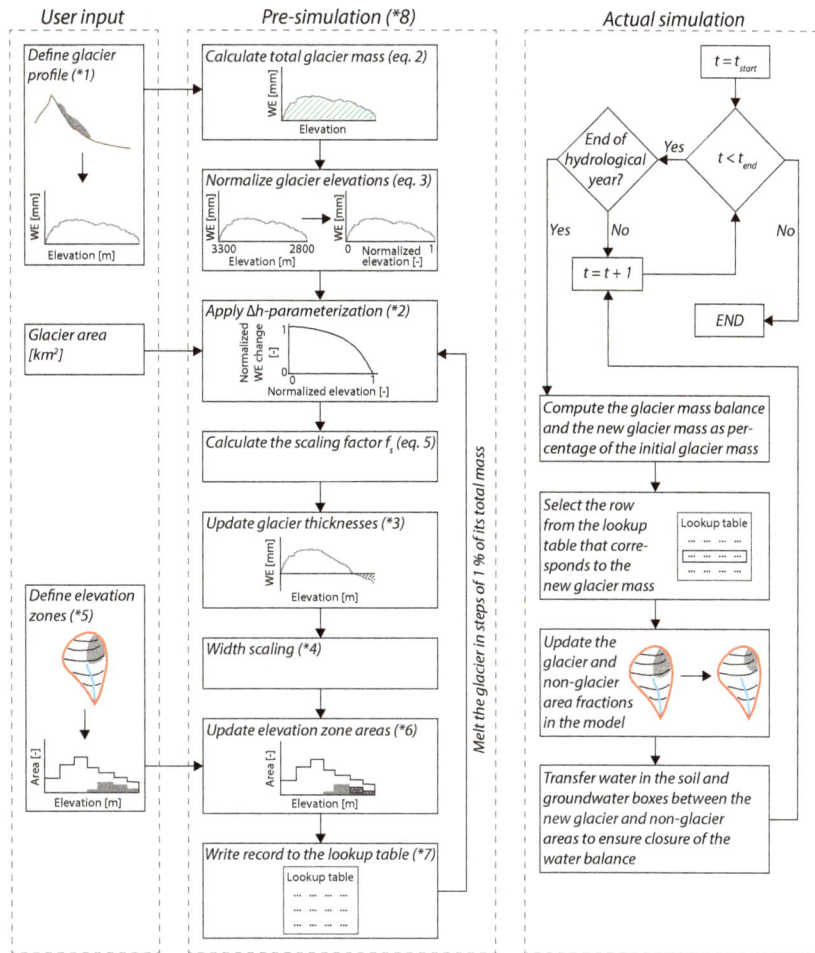

Figure 1. Flowchart describing the update of the glacier geometry depending on glacier mass balance changes in HBV-light. Additional information is given in the following notes (numbers refer to corresponding numbers in the flowchart).

(*1) Elevation bands and corresponding water equivalent are given, with elevation bands at a finer resolution than the elevation zones. While the areal distribution of a static glacier is specified in HBV-light by means of elevation and aspect zones, for establishing the relationship between glacier mass and glacier area, a glacier profile, which defines the initial thickness (in mm water equivalent) and areal distribution of the glacier at a finer resolution, is needed as model input data. Note that the resolution of the glacier routine simulations largely depends on the number of elevation bands per elevation zone; i.e. all glacier area within each band is either covered by a glacier or not, and the percentage of glacierized area within a certain elevation zone is based on the state of the individual elevation bands within that elevation zone. Elevation zones typically have resolutions of 100 to 200 m, whereas for the elevation bands a resolution of 10 m is commonly used.

(*2) Depending on the glacier area, select one of the three parameterizations suggested by Huss et al. (2010) (see Eq. 4).

(*3) For each elevation band reduce the glacier water equivalent according to the empirical functions from Huss et al. (2010) (Eq. 4) to compute the glacier geometry for the reduced mass (see Eq. 6). If the computed thickness change is larger than the remaining glacier thickness (most likely to occur at the glacier tongue; see the area that is marked in red in the figure), the glacier thickness is reduced to zero, resulting in a glacier-free elevation band, and the portion of the glacier thickness change that would have resulted in a negative glacier thickness is included in the next iteration step (i.e. the next 1 % melt).

(*4) The Δh approach distributes the change in glacier mass over the different elevation zones though it results in glacier-free areas mainly at the lowest elevations. The width scaling within each elevation band relates a decrease in glacier thickness to a reduction of the glacier area within the respective elevation band. In other words, this approach also allows for glacier area shrinkage at higher elevations, which mimics the typical spatial effect of the downwasting of glaciers.

(*5) Define elevation zones and compute the fractions of glacier and non-glacier area (relative to the catchment area) for each elevation zone.

(*6) Sum the total (width-scaled) areas for all respective elevation bands which are covered by glaciers (i.e. glacier water equivalent ≥ 0) for each elevation zone.

(*7) M (in % of initial M) is in the first column, followed by one column for each elevation zone with the areal glacier cover area (in % of catchment area).

(*8) Run once before the actual simulation of the time series starts (automatically within the HBV-light software).

minimum elevations of the glacier, $E_{\max}$ and $E_{\min}$ (Eq. 3).

$$E_{i,\text{norm}} = \frac{E_{\max} - E_i}{E_{\max} - E_{\min}} \qquad (3)$$

The normalized water equivalent change is then computed for each of the normalized elevations using the following function (Huss et al., 2010):

$$\Delta h_i = \left(E_{i,\text{norm}} + a\right)^{\gamma} + b\left(E_{i,\text{norm}} + a\right) + c, \qquad (4)$$

where Δh_i is the normalized (dimensionless) ice thickness change of elevation band i and a, b and c and γ are empirical coefficients. Based on the initial total glacier area (in km^2) that needs to be specified in addition to the initial glacier thickness profile, one of the three empirical parameterizations applicable for unmeasured glaciers from Huss et al. (2010) is used (Figs. 1 and 2a).

In the next step a scaling factor f_S (mm), which scales the dimensionless Δh, is computed based on the glacier volume change ΔM and on the area and normalized water equivalent change for each of the elevation bands:

$$f_S = \frac{\Delta M}{\sum\limits_{i=1}^{N} a_i \cdot \Delta h_i}. \qquad (5)$$

The new water equivalent $h_{i,k+1}$ is then computed for each elevation band, starting from the user-specified initial glacier thickness profile for $k = 0$ as

$$h_{i,k+1} = h_{i,k} + f_S \, \Delta h_i, \qquad (6)$$

where $h_{i,k}$ is the water equivalent of elevation band i after reducing the glacier mass k times by ΔM. Exemplary results of this step-wise melt process based on the Δh parameterization are visualized in Fig. 2b.

Once the new water equivalent values have been computed for each elevation band, the glacier area is updated for each elevation zone. The relative glacier area for a certain elevation zone is defined as the cumulative area of the glacier covered elevation bands within that elevation zone, divided by the total area of the elevation zone. Thus the model described so far is essentially a 2-D representation of glacier retreat. However, glaciers have an uneven distribution of ice at a particular elevation, with a thinner ice layer along the edges. In order to take the area reduction that results from this uneven distribution into account, a simplified representation of the 3-D glacier geometry is used to scale the area within a certain elevation band (Eq. 7) following the relation between glacier width and glacier thickness given in Bahr et al. (1997), as also applied by Huss and Hock (2015):

$$a_{i,\text{scaled}} = a_i \cdot \sqrt{h_i / h_{i,\text{initial}}}. \qquad (7)$$

The reduction in glacier area over elevation resulting from the application of the Δh parameterization following Eqs. (2)–(6) in combination with the glacier width scaling (Eq. 7) is illustrated in Fig. 2c. The resulting relationship between glacier area and glacier mass is stored in the lookup table at steps of 1 % of the initial glacier mass. This means that the lookup table consists of glacier areas per elevation zone for 101 different glacier mass situations, ranging from the initial glacier mass to zero (Fig. 1). It should be noted that this approach, similar to the original Δh parameterization method of Huss et al. (2010), neglects any delays in the response of glacier areas to mass balance changes.

During the actual simulation in HBV-light, the glacier extent is updated at the beginning of each hydrological year (1 October). The total water equivalent of the glacier is computed. Based on the percentage of glacier water equivalent in comparison to the total glacier water equivalent in the initial glacier profile definition, the corresponding record is extracted from the glacier lookup table and the corresponding glacier areas are applied to the different elevation zones. In the case that the glacier water equivalent exceeds its maximum, the areas corresponding to 100 % are applied (i.e. the glacier can never grow larger than defined by the user in the glacier profile definition). Optionally, simulations can start, however, with a reduced glacier size, by specifying the initial glacier fraction in the glacier profile file (as fraction of water equivalent). The initial glacier profile definition should thus contain the maximum extent of the glacier during the full simulation period. For each glacierized part of an elevation zone in HBV-light, the corresponding non-glacierized part is used to exchange the area for which the state changed from glacier to non-glacier and vice versa. In order to ensure the water balance is correct, "bookkeeping" is done between the corresponding glacierized and non-glacierized zones. Soil moisture and snow, for example, are moved between the corresponding zones as far as these water storages correspond to the area exchanged.

2.2 Sensitivity test of different model variants

To illustrate the new Glacier Area Change Routine (GACR) and its different components on the simulation results, we applied the HBV-light model for one example catchment, the Alpbach catchment in the Swiss Alps. This catchment is one of the glacierized headwater catchments in the Rhine River basin, located in central Switzerland. The catchment area is about 21 km^2 and elevations range from 1022 m up to 3192 m a.s.l., with a mean elevation of 2194 m. The catchment consists of two main valleys with the glacier Glatt Firn extending into both of them. According to the glacier inventory for the year 2010 the glacierized area was estimated to be 4.03 km^2 (Fischer et al., 2014), whereas the estimate was 4.54 km^2 for 2003 (Paul et al., 2011), corresponding to a catchment glacier coverage of about 20 %. The initial glacier profile for 1900 was estimated as described in Appendix A1.

To demonstrate the effect of the different parts of the GACR, four different versions of the GACR were used,

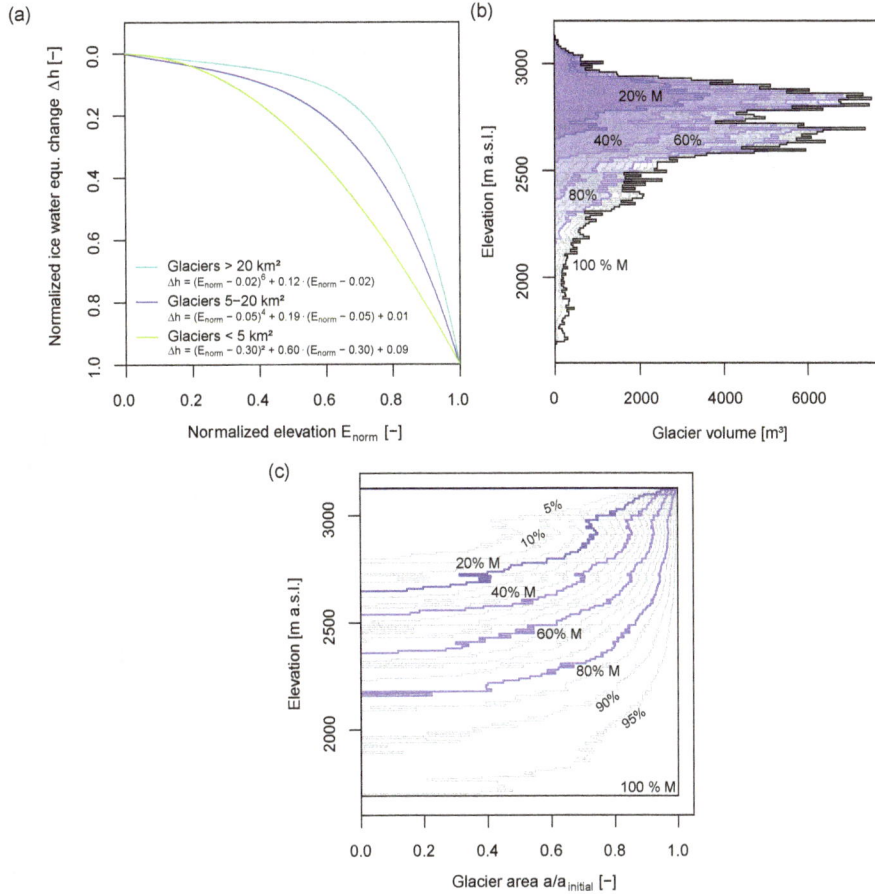

Figure 2. The Δh parameterization and its implementation in HBV-light: **(a)** empirical Δh parameterization functions for three glacier size classes from Huss et al. (2010), **(b, c)** pre-simulation application of the Δh parameterization for a medium glacier size to the example glacier profile data of the Alpbach catchment by melt in steps of $\Delta M = 1\% M$ to generate the lookup table. Panel **(b)** shows the absolute glacier volume per elevation band; panel **(c)** shows the relative glacier area per elevation band as relative fraction of the initial glacier area a_initial of the elevation interval.
For each elevation interval, the resulting glacier water equivalent/glacier volume **(b)** and glacier area **(c)** are shown as grey lines; for visibility, only results of steps of $\Delta M = 5\%$ are shown here. The initial profile ($100\% M$) and profiles for a glacier volume reduction by 20, 40, 60, and 80% are highlighted by coloured labelled lines.

where for three versions certain components of the new glacier routine were disabled:

1. *Stationary glacier area (no GACR).* Only the static part of the glacier routine is used; i.e. the complete dynamic part of the glacier routine is disabled; the glacier area is not adjusted but stays exactly as defined by the user in the model set-up during the whole simulation.

2. *Full new GACR (GACR).* The full version of the model as described in Sect. 2.1, with the static and dynamic part of the glacier routine included, is employed.

3. *GACR without glacier width scaling (GACR-w).* The application of glacier width scaling (Eq. 7) by elevation band is disabled. In practice, this corresponds to a 2-D representation of glacier area change. A change in glacier area is only realized when the mean glacier water equivalent of an elevation band (Eq. 6) reaches zero, which will in most cases only occur at the glacier terminus.

4. *GACR without glacier advance (GACR-a).* This only considers glacier retreat. The original method described by Huss et al. (2010) only considers the parameterization of glacier retreat and not glacier advance. In the new GACR, glacier advance up to the initial state is enabled by means of the lookup table generation. To demonstrate the effect of neglecting temporary glacier advance, we used a version that only applies glacier retreat. In periods with a positive annual glacier mass balance the glacier area is kept constant.

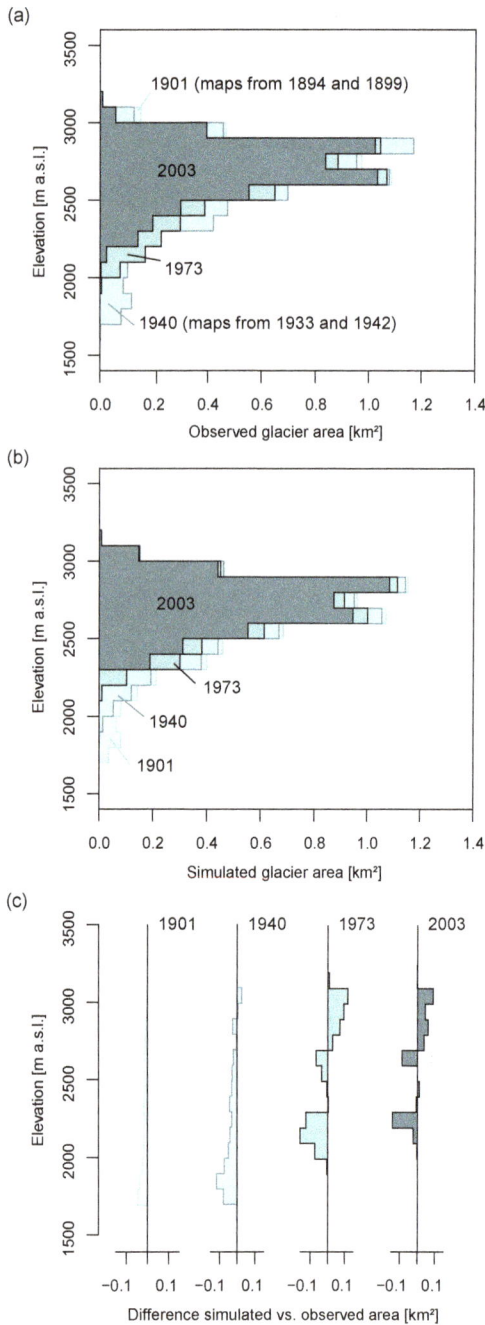

Figure 3. Observed and simulated glacier areas per elevation zone for years for which historical maps or glacier inventories are available: (**a**) glacier areas for the different elevations derived from maps or remote sensing, (**b**) glacier areas for the different elevations as simulated with the full new GACR model version, and (**c**) differences between glacier areas from simulation and reference datasets.

The original method described by Huss et al. (2010) only considers the parameterization of glacier retreat and not glacier advance. In the new GACR, glacier advance up to the initial state is enabled by means of the lookup table gen-

eration. To demonstrate the effect of neglecting temporary glacier advance, we used a version that only applies glacier retreat. In periods with a positive annual glacier mass balance the glacier area is kept constant.

For each of these four versions, we calibrated the model 10 times, using a genetic algorithm (Seibert, 2000) with 3500 model runs per calibration trial. The 10 independent calibration trials allowed parameter uncertainty effects to be considered by taking several optimized parameter sets into account. The simulation period was 1 January 1901 to 31 December 2006 and was preceded by a 3-year warm-up period. As an objective function, the average of the Nash–Sutcliffe efficiency for daily discharge, the relative volume error of the total discharge, the root mean squared error of the snow cover simulations, and the absolute mean relative error of the glacier volume estimates were used. The estimates of glacier volume were based on different glacier cover datasets for three particular years during the simulation period as described below.

The simulation period (1901–2006) is a period in which glaciers of the European Alps retreated considerably; yet this period also covers diverse climate conditions including a period between the 1960s and the 1980s that was characterized by rather balanced conditions or temporarily by glacier advance. For the set-up and the calibration of the model in terms of glacier conditions, several observation-based datasets from diverse sources were used: the glacier area for the state around the years of 1901 (start of simulation period) and 1940 was based on digitized historical topographic maps, known in Switzerland as "Siegfriedkarte" (Freudiger et al., 2018). For both years, 1901 and 1940, the glacier area of the Alpbach catchment is reconstructed from two adjacent map sheets. To describe the glacier area around the start of the simulation in 1901, maps from the years 1894 and 1899 were used, and to describe the glacier area around 1940, maps from the years 1933 and 1942 were used. Glacier areas for the years 1973, 2003, and 2010 were extracted from the glacier inventories by Müller et al. (1976), Paul et al. (2011), and Fischer et al. (2014), respectively. For the years 1973 and 2010 gridded datasets of estimated glacier thickness based on the method presented in Huss and Farinotti (2012) were also used (unpublished data provided by Matthias Huss). In addition, discharge observations (Erstfeld station, Bodenberg, period 1960–2006) from the Swiss Federal Office for the Environment (FOEN) and a gridded snow water equivalent (SWE) climatology product from the Institute of Snow and Avalanche Research (WSL-SLF, covering November–May for the period 1972–2006) were used to calibrate the model. More details on the underlying data sources and the applied multi-criteria calibration can be found in Stahl et al. (2017).

To set up the HBV-light model for the Alpbach catchment, the spatial modelling units were discretized as follows: firstly, the glacierized and non-glacierized catchment area fractions for the state at the start of the simulation in

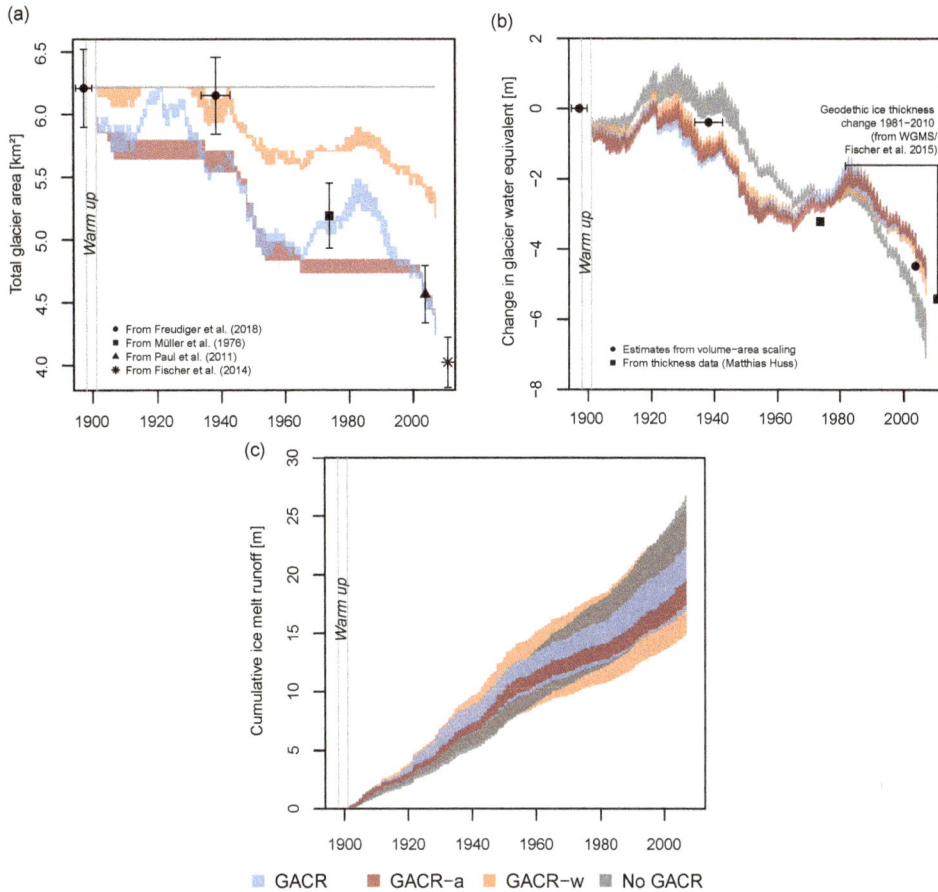

Figure 4. Comparison of the simulations by the different versions of the glacier routine: **(a)** total glacier area, **(b)** change in glacier storage, and **(c)** cumulative glacier ice melt runoff (this is the runoff originating from ice melt, which is tracked through the model; snowmelt on the glacier is not included). The range of simulation results represents the results from 10 model (equally suitable) parameterizations for each of the different versions of the Glacier Area Change Routine (GACR) and for the version without Glacier Area Change Routine (see Sect. 2.2 for the definition of model-variant abbreviations). All simulations started with the same glacier volume and area and differences in 1901 were caused by differences in the simulations during the 3-year warm-up period. The uncertainties for the observed glacier volumes and areas were best estimates based on the available information in the respective publications. The geodetic ice thickness change from 1981 to 2010 (Fischer et al., 2015) was not used in model calibration but added here for comparison.

1901 were distinguished. Therefore all areas within the Alpbach catchment that were glacier-covered according to the underlying map or glacier inventory for a specific year were summed up as one model glacier. Both the non-glacierized and the glacierized model areas were further divided into area fractions per elevation zones and then further differentiated within each elevation zone into area fractions for three aspect classes.

For the application of the Δh parameterization, in addition to the main model set-up the initial glacier profile needs to be defined by the user (Fig. 1). As no data on glacier thickness for the state at 1901 (start of simulation) were available, an initial glacier profile had to be reconstructed; details for the method, which was chosen in this application, are described in the Appendix. The reconstructed glacier profile used for

model initialization is shown in Fig. 2b (black line for $M = 100\%$) and Fig. A1.

3 Results

Figure 3a shows the reference glacier profile for the initial state at the start of the simulation in the year 1901 as well as for the three different years for which data were available from which the glacier profile could be derived. The observed decrease of glacier area occurred at all elevations. Figure 3b shows the glacier profile for the simulation with the full new GACR model version. With this version, glacier retreat also occurred at all elevations. This is due to the combination of the Δh approach and the implemented width scaling. In order to compare the simulated and observed glacier profiles, Fig. 3c shows the differences between simulated and observed glacier area for the different elevation zones. The Δh approach by definition results in zero change in glacier thickness at the very top of the glacier. The lower the position on the glacier is, the larger the change in thickness can be. This pattern can be seen in Fig. 3b, where there is, contrary to the observed data of Fig. 3a, hardly any change in glacier area in the higher elevation zones. As a result, the difference between simulated and observed areas in Fig. 3c is positive for the higher elevation zones (for the years 1973 and 2003). This is compensated for by a negative difference between simulated and observed glacier areas for the lower elevation zones. Overall, the new GACR is able to depict the major pattern of long-term glacier area change over the elevation zones in the example catchment.

The simulations using the full GACR also correspond in general with the reference datasets in terms of total catchment glacier area (Fig. 4a), but one has to recognize the considerable uncertainties in the glacier volume estimates used for comparison. In addition, Fig. 4 illustrates the differences in the four different model versions to simulate the changes in total catchment glacier area (Fig. 4a) and the resulting effects on the change in glacier water equivalent and cumulative ice melt runoff (Fig. 4b and c), which are relevant within the scope of hydrological modelling. Among all model versions the new full GACR is best in representing the pattern of change in total glacier area based on the comparison with available reference data (Fig. 4a). Whereas there is a considerable mismatch of the simulated and observed glacier area around the year 1940, for the later years the simulated and observed glacier areas are in good agreement. The model version that does not incorporate glacier advance is just as effective in reaching the final state of the glacier area in the year 2003 as the full version. In terms of glacier area the results of both versions, GACR-a and GACR, are only different during phases dominated by positive glacier mass balance. As soon as the annual glacier water equivalent (glacier volume) decreases to its previous minimum again, the reduction in glacier area continues. For glaciers with a net negative mass balance over time, differences can therefore be rather small. If there are more and longer periods of glacier advance, differences might become more apparent. However, in the case of overall net positive glacier mass balance, the fact that the maximum glacier extent cannot exceed what is specified in the glacier profile becomes an obvious limitation of the new GACR routine. For the version GACR-w the glacier stays at its maximum size a bit longer than for the full new GACR and the version GACR-a, since elevation bands need to be melted completely before the glacier area starts to reduce. In contrast, in the full new GACR and GACR-a simulations width scaling is applied as soon as the glacier mass balance becomes slightly negative, and therefore a reduction in glacier area can be observed immediately. It should be noted that in all variants it is assumed that mass balance changes directly cause area changes while there might be some delay in the area response in reality. For simulations with only

the static glacier routine (no GACR) the glacier area stays constant (horizontal grey line in Fig. 4a).

The constant area with the no GACR version allows for (much) higher melt rates in comparison to the other model versions once the glacier has partly melted, since a larger area, which is also located at lower elevations and thus becomes snow-free earlier in the season, is contributing to the overall melt than in the version including the GACR. This can also be clearly observed in Fig. 4b, where the model version with a stationary glacier area shows much stronger glacier water equivalent changes. As a result the cumulative ice melt runoff (Fig. 4c) is highest for simulations with no GACR, especially during the second half of the simulation period when the difference in glacier area in comparison to the other versions is more notable. Generally, the larger the glacier area is, the more runoff is generated by the glacier. The stationary glacier area model (no GACR) results in the potentially largest amount of glacier runoff, followed by the simulations without width scaling (GACR-w), for which the 10 different model calibrations resulted in the largest spread. The difference between the versions GACR and the GACR-a is minor, with the latter likely resulting in an underestimation of generated glacier runoff, due to the smaller area during phases of glacier advance.

4 Discussion and conclusion

The glaciological part of the coupled model as described above is a simple representation of glacier processes, but it allows glacier geometry changes to be considered at a level of complexity which is similar to the hydrological model. In most current hydrological models no representation of changing glacier areas is realized, which basically implies an infinitely thick glacier. The approach described here, which allows for area changes as a result of simulated mass changes, is certainly a more realistic representation and the changing area clearly affects variables such as the simulated runoff. Some previous studies used the simple volume–area scaling approach (e.g. Luo et al., 2013). This method does not consider any catchment-specific information, whereas the Δh parameterization allows elevation distributions and the ice thickness profile to be considered. In volume–area scaling any volume change directly translates to an area change, although this may not always be the case. The Δh parameterization also allows the glacier area changes to be attributed to the different elevation zones, which would not be directly possible with simple volume–area scaling, which does not allow the region of glacier shrinkage to be assigned (see also the discussion by Stahl et al., 2008). As discussed by Huss et al. (2010) the Δh approach is a simple but still physically based approach to consider changing glaciers as a result of the simulated mass balance change.

A major simplification of the approach presented here is that only one glacier is considered in each subcatchment, which means that if there are several glaciers these are simulated as one virtually aggregated glacier. Principally this could be solved by using as many subcatchments as there are glaciers. However, this would not solve the issue of a glacier which splits up into several glaciers at some point during the simulation. The representation of all glaciers in a catchment as one virtually aggregated glacier might, thus, be a suitable representation. The Δh parameterization approach of Huss et al. (2010) and the use of their empirical functions were found to be suitable. This reduces the need for new calibration parameters. The Δh parameterization could also be based on data for specific glacier(s) (as done, for instance, by Duethmann et al., 2015), which would better represent local conditions. However, for practical reasons for the incorporation into a hydrological model as HBV-light, the use of established empirical parameterizations will usually be preferred because this facilitates straightforward applications and is assumed to represent the glacier area changes sufficiently well for the majority of typical hydrological modelling applications. A re-evaluation of the empirical Δh parameterizations, which included glaciers from different parts of the world, rendered mainly satisfying results (Huss and Hock, 2015).

Several adoptions were needed for the implementation in a semi-distributed model. Most importantly the use of a lookup table to represent the mass–area relationship allows for the inclusion of advancing glaciers. It should be noted that the lookup table alternatively could also be derived from any other glaciological model. This means that this approach presents a technical solution that potentially allows flexible implementations of appropriate glacier geometry change models in hydrological catchment models. Furthermore, the geometric width scaling for individual elevation bands allows for the representation of a decreasing glacier area with decreasing thickness in an elevation band. The example simulations shown in this technical note illustrate the effect of these modifications, which maintain the conceptual model approach. In all variants it is assumed that glacier mass changes immediately translate into area changes and that glacier retreat and glacier advance follow the same (but reverse) pattern. While this is not the case in reality, it is assumed to be an acceptable simplification for use in hydrological catchment models, for which the focus is a realistic simulation of glacier ice melt. Allowing for advancing glaciers and changing areas due to glacier thinning makes a difference in the simulations (Fig. 4). Both these aspects are also important as they enable a comparison between simulated and observed glacier area (see Figs. 3a and 2). This is crucial for model calibration and validation as glacier areas and glacier lengths are much more frequently available than other glacier observations. The simulations demonstrate that the new glacier evolution routine is, in general, capable of simulating reasonable area changes. However, given the limited data this should not be taken as proof that the model is correct, even if the simulations appear glacio-hydrologically

reasonable. The validation of any glacier model or routine against observations is challenging due to limited suitable datasets and is beyond the scope of this technical note.

Besides its simplicity, the presented GACR implementation also has other limitations. One challenge is to obtain initial thickness distributions along the glacier. While this estimation of initial glacier conditions certainly adds uncertainties, information on initial ice thicknesses is needed for any approach that aims at simulating changing glacier areas. In the approach presented here, glacier advance is only possible up to the initial state. In most cases this is not a major limitation as long as suitable information on early glacier extents is available as most climate data and scenarios lead to retreating glaciers. If needed, a larger initial glacier extent (with some thickness profile) can be provided to establish the mass–area relation to create the lookup table. In this case the actual simulations would start at a certain fraction of this hypothetical maximum situation.

The Δh parameterization represents an approach, which allows changing glacier areas to be considered in an approximate but realistic way. The conceptually stringent implementation presented in this technical note could in principle also be used by other semi-distributed hydrological models. In many hydrological model applications of partially glacierized catchments that do not specifically target the contributions of glaciers to runoff, glacier areas are not directly updated. Studies with a coupled glacio-hydrological approach often describe little detail of the glacier routine, especially when it comes to the question of whether simulated mass balance changes are translated into glacier area changes and, if so, how this is done. In a recent review on hydrological modelling of glacierized catchments in central Asia (Chen et al., 2016), for instance, this issue is not discussed at all. The main advantage of the coupled glacio-hydrological approach as described in this technical note is that glacier mass and area changes are consistent with the hydrological model. This also allows the model to be used to simulate future scenarios. While the GACR described in this technical note is a rather simple representation of glacier processes, it enables this important representation of changing glacier areas in high-mountain catchments.

Data availability. Meteorological data input used was the HYRAS interpolation product made available by the German Weather Service DWD and the Bundesanstalt für Gewässerkunde BfG and a HYRAS-REC reconstruction by Stahl et al. (2017). Climate station data were provided by MeteoSwiss. Model calibration used hydrometric data from the Swiss Federal Office for the Environment FOEN, snow data of the "SLF Schneekartenserie Winter 1972–2012" from the SLF (WSL Institute for Snow and Avalanche Research), glacier data provided by Matthias Huss, and data on glacier areas from the Siegfriedkarte by Swisstopo. The data can be accessed from the respective agencies.

Appendix A: Reconstruction of initial glacier geometry

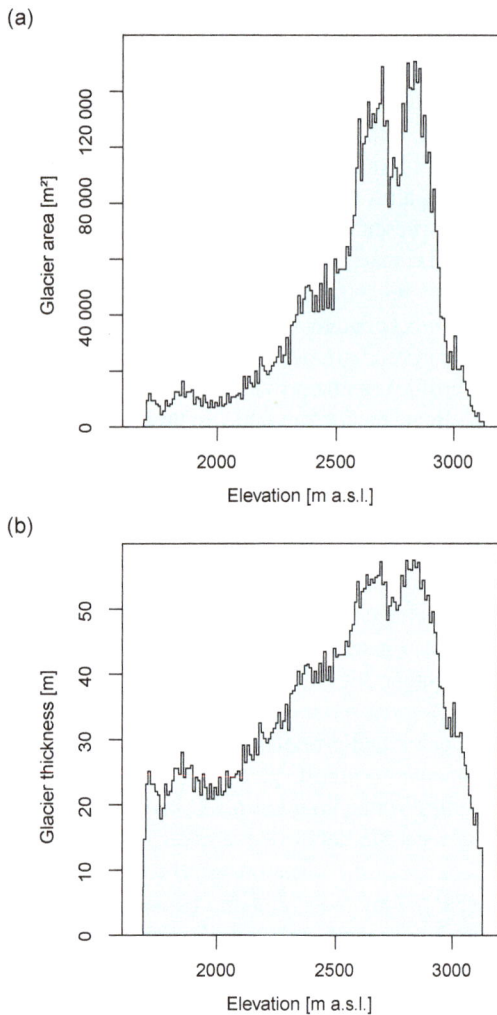

(a)

(b)

Figure A1. Estimated initial glacier geometry as a function of elevation: (a) areal extent and (b) glacier thickness.

A challenging requirement for the application of the new HBV-light GACR, as for any modelling of temporally changing glacier geometry, is the definition of the initial state of the glacier in terms of total volume and ice thickness distribution, briefly termed "initial glacier profile" in the following. Approaches to tackle this, as recently reviewed by Farinotti et al. (2017), strongly depend on the available glacier survey data. For the case of the Alpbach catchment a reconstruction of the initial glacier profile for the state around the start of the model simulation in the year 1901 was required. Table A1 summarizes all available primary glacier datasets with reference to their origin as well as derived data used for the reconstruction of the initial glacier profile.

The glacier profile finally needed in the HBV-light set-up consists of glacier area and thickness per elevation band. Whereas such data are available for the more recent years

1973 and 2010, for 1901 glacier area was the only information available. Generally, the approach to estimate the initial ice thickness distribution was based on two physically based glacier scaling relationships taken from Bahr et al. (1997): (i) the widely applied general volume–area scaling relation (Eq. A1) and (ii) a proportionality of glacier width and the square root of glacier thickness. The latter relationship assumes a parabolic cross section as being characteristic of valley glaciers and was also used for the implementation of the new GACR (Eq. 7 in the main text). In detail, for the reconstruction of the initial ice thickness distribution, the total glacier volume around 1901 was estimated based on

$$V = c \cdot A^{\gamma}, \tag{A1}$$

where V is the total glacier volume (m^3), A is the total glacier area (m^2), c is a glacier-specific scaling parameter (m), and γ is the scaling exponent (–), which was fixed to its theoretically defined value (Bahr et al., 2015) of $\gamma = 1.375$. The multiplicative scaling parameter c for both glacier volume–area pairs (Table A1), for the years 1973 and 2010, was obtained. The average of both values of the multiplicative scaling parameter c was then used to estimate the total glacier volume for the start of the simulation in 1901 using the known glacier area (Table A1). To reconstruct the glacier thickness distributions over the elevation bands (10 m resolution in the example of the Alpbach), the proportionality of glacier width and the square root of glacier thickness were then applied to the elevation bands. The glacier width of an elevation interval can be used to approximate the glacier area of the elevation interval i with

$$A_i = p_i \cdot \sqrt{H_i}, \tag{A2}$$

where A_i is the glacier area (m^2), H_i is the glacier thickness (m), and p_i is a scaling parameter (m$^{1.5}$). Based on Eq. (A2) the glacier-specific and elevation-band-specific glacier width scaling parameters p_i were determined for the "glacier profiles" (A_i and H_i for all elevation bands i) for the years 1973 and 2010, for which ice thickness data are available. A power-law function was fitted with the values for the year 1973 to estimate the glacier width scaling parameter p_i as a function of A_i. The obtained function was then used to estimate the initial glacier thickness $H_{i,1901}$ for all elevation bands based on $A_{i,1901}$. Finally the resulting estimated glacier thickness values were corrected by a factor to enforce that the resulting total glacier volume $\left(\sum A_{i,1901} \cdot H_{i,1901}\right)$ equals the total glacier volume estimate derived for the year 1901 from Eq. (A1) above (Fig. A1). With that, the glacier area $A_{i,1901}$ taken from the historical map (Table A1), and the estimated glacier thickness $H_{i,1901}$, the tabular glacier profile to initialize the HBV-light model simulations was generated. For use in HBV-light, the glacier area A_i needs to be expressed as a fraction of total catchment area (a_i, (–), glacier thickness H_i is converted to water equivalents (h_i,

Table A1. Glacier datasets with reference and derived data for the reference years 1900, 1973, and 2010 used for the reconstruction of initial glacier geometry for the Alpbach catchment.

Reference year (ca.)	Reference	Original data	Derived data
1901	Freudiger et al. (2018)	Glacier outlines[a]	Total glacier area Glacier area per elevation band[c]
1973 and 2010	Matthias Huss (unpublished data)	Gridded ice thickness data[b]	Total glacier area Glacier area per elevation band[c] Total glacier volume Mean thickness per elevation band[c]

[a] Digitization from historical topographic maps ("Siegfriedkarte") provided by Swisstopo. [b] Computed ice thickness based on the approach by Huss and Farinotti (2012) using glacier outline inventories from Maisch et al. (2000), originally Müller et al. (1976), and from Fischer et al. (2014). [c] All GIS analyses based on the same digital elevation model (25 m × 25 m) for recent conditions.

mm) by applying an ice density of $900\,\mathrm{kg\,m^{-3}}$, and the elevation bands i (10 m intervals) are assigned to the corresponding elevation zones (100 m intervals) of the HBV-light catchment discretization.

One should note that the presented procedure to estimate the initial glacier geometry is subject to several uncertainties and limitations. These are, for instance, related to the uncertainties of the underlying data sources, the combination of glacier volume datasets derived from differing methodologies, the treatment of several glacier parts or branches as one aggregated glacier, the application of the average of the glacier scaling parameter c for the years 1973 and 2010 to estimate the glacier volume in 1901, the negligence of changes in surface elevation, or the fact that results obtained from glacier scaling applications on individual glaciers should always be regarded as an order of magnitude estimate only. However, though this is a way to get a rough estimate for glacier initialization, it may still be considered a feasible and reasonable approach for many hydrological model applications in glacierized catchments and in particular large catchment sample modelling studies facing a lack of detailed glacier survey data. In particular, the combination of the volume estimates from volume–area scaling and the ice thickness data through inverting glacier surface topography (approach presented by Huss and Farinotti, 2012) has to be regarded critically and cannot be recommended as a standard procedure. The reason this approach had to be used here was that we were given the challenge to estimate the initial glacier geometry for the early state at the beginning of the 20th century within the scope of a long-term modelling study (Stahl et al., 2017). For the objectives of this study and also for the demonstration of the new GACR for one example catchment, the presented method was considered as an acceptable solution in the technical note here. If the required data for other approaches (e.g. Farinotti et al., 2017) were available, the combination of data derived from differing approaches would be avoided.

Competing interests. The authors declare that they have no conflict of interest.

Acknowledgements. We thank Daphné Freudiger and Damaris De for their contributions including the digitization of glacier outlines from historical maps. We are also thankful to Matthias Huss, who provided details on the original Δh method and kindly shared his knowledge and unpublished information regarding the estimation of the glacier profiles as well as ice thickness data. The model code extension was made possible with funding from the University of Zürich. The method developments were made within the ASG-Rhein project (The snow and glacier melt components of the streamflow of the River Rhine and its tributaries considering the influence of climate change) funded by the International Commission for the Hydrology of the Rhine Basin (CHR). Valuable comments of the editor and two anonymous reviewers helped to clarify the text.

Edited by: Jim Freer

References

Addor, N., Rössler, O., Köplin, N., Huss, M., Weingartner, R., and Seibert, J.: Robust changes and sources of uncertainty in the projected hydrological regimes of Swiss catchments, Water Resour. Res., 50, 1–22, https://doi.org/10.1002/2014WR015549, 2014.

Bahr, D., Meier, M., and Peckham, S.: The physical basis of glacier volume-area scaling, J. Geophys. Res., 102, 20355, https://doi.org/10.1029/97JB01696, 1997.

Bahr, D. B., Pfeffer, W. T., and Kaser, G.: A review of volume-area scaling of glaciers, Rev. Geophys., 53, 95–140, https://doi.org/10.1002/2014RG000470, 2015.

Bergström, S.: Development and application of a conceptual runoff model for Scandinavian catchments, SMHI, Norrköping, Sweden, No. RHO 7, 134 pp., 1976.

Bongio, M., Avanzi, F., and De Michele, C.: Hydroelectric power generation in an Alpine basin: Future water-energy scenarios in a run-of-the-river plant, Adv. Water Res., 94, 318–331, https://doi.org/10.1016/j.advwatres.2016.05.017, 2016.

Chen, Y., Li, W., Fang, G., and Li, Z.: Review article: Hydrological modeling in glacierized catchments of central Asia – status and challenges, Hydrol. Earth Syst. Sci., 21, 669–684, https://doi.org/10.5194/hess-21-669-2017, 2017.

Duethmann, D., Bolch, T., Farinotti, D., Kriegel, D., Vorogushyn, S., Merz, B., Pieczonka, T., Jiang, T., Su, B., and Güntner, A.: Attribution of streamflow trends in snow and glacier melt-dominated catchments of the Tarim River, Central Asia, Water Resour. Res., 51, 4727–4750, https://doi.org/10.1002/2014WR016716, 2015.

Farinotti, D., Usselmann, S., Huss, M., Bauder, A., and Funk, M.: Runoff evolution in the Swiss Alps: projections for selected high-alpine catchments based on ENSEMBLES scenarios, Hydrol. Proc., 26, 1909–1924, https://doi.org/10.1002/hyp.8276, 2012.

Farinotti, D., Brinkerhoff, D. J., Clarke, G. K. C., Fürst, J. J., Frey, H., Gantayat, P., Gillet-Chaulet, F., Girard, C., Huss, M., Leclercq, P. W., Linsbauer, A., Machguth, H., Martin, C., Maussion, F., Morlighem, M., Mosbeux, C., Pandit, A., Portmann, A., Rabatel, A., Ramsankaran, R., Reerink, T. J., Sanchez, O., Stentoft, P. A., Singh Kumari, S., van Pelt, W. J. J., Anderson, B., Benham, T., Binder, D., Dowdeswell, J. A., Fischer, A., Helfricht, K., Kutuzov, S., Lavrentiev, I., McNabb, R., Gudmundsson, G. H., Li, H., and Andreassen, L. M.: How accurate are estimates of glacier ice thickness? Results from ITMIX, the Ice Thickness Models Intercomparison eXperiment, The Cryosphere, 11, 949–970, https://doi.org/10.5194/tc-11-949-2017, 2017.

Finger, D., Hugentobler, A., Huss, M., Voinesco, A., Wernli, H., Fischer, D., Weber, E., Jeannin, P.-Y., Kauzlaric, M., Wirz, A., Vennemann, T., Hüsler, F., Schädler, B., and Weingartner, R.: Identification of glacial meltwater runoff in a karstic environment and its implication for present and future water availability, Hydrol. Earth Syst. Sci., 17, 3261–3277, https://doi.org/10.5194/hess-17-3261-2013, 2013.

Fischer, M., Huss, M., Barboux, C., and Hoelzle, M.: The new Swiss Glacier Inventory SGI2010: relevance of using high-resolution source data in areas dominated by very small glaciers, Arctic, Antarct. Alp. Res., 46, 933–945, https://doi.org/10.1657/1938-4246-46.4.933, 2014.

Fischer, M., Huss, M., and Hoelzle, M.: Surface elevation and mass changes of all Swiss glaciers 1980–2010, The Cryosphere, 9, 525–540, https://doi.org/10.5194/tc-9-525-2015, 2015.

Frans, C., Istanbulluoglu, E., Lettenmaier, D. P., Clarke, G., Bohn, T. J., and Stumbaugh, M.: Implications of decadal to century scale glacio-hydrological change for water resources of the Hood River basin, OR, USA, Hydrol. Proc., 30, 4314–4329, https://doi.org/10.1002/hyp.10872, 2016.

Freudiger, D., Kohn, I., Seibert, J., Stahl, K., and Weiler, M.: Snow redistribution for the hydrological modeling of alpine catchments, Wiley Interdiscip. Rev. Water, 4, 1–16, https://doi.org/10.1002/wat2.1232, 2017.

Freudiger, D., Mennekes, D., Seibert, J., and Weiler, M.: Historical glacier outlines from digitized topographic maps of the Swiss Alps, Earth Syst. Sci. Data, in press, 2018.

Gabbi, J., Farinotti, D., Bauder, A., and Maurer, H.: Ice volume distribution and implications on runoff projections in a glacierized catchment, Hydrol. Earth Syst. Sci., 16, 4543–4556, https://doi.org/10.5194/hess-16-4543-2012, 2012.

Gabbi, J., Carenzo, M., Pellicciotti, F., Bauder, A., and Funk, M.: A comparison of empirical and physically based glacier surface melt models for long-term simulations of glacier response, J. Glaciol., 60, 1199–1207, https://doi.org/10.3189/2014JoG14J011, 2014.

Hagg, W., Braun, L. N., Kuhn, M., and Nesgaard, T. I.: Modelling of hydrological response to climate change in glacierized Central Asian catchments, J. Hydrol., 332, 40–53, https://doi.org/10.1016/j.jhydrol.2006.06.021, 2007.

Hock, R.: Temperature index melt modelling in mountain areas, J. Hydrol., 282, 104–115, https://doi.org/10.1016/S0022-1694(03)00257-9, 2003.

Hottelet, C., Braun, L. N., Leibundgut, C., and Rieg, A.: Simulation of Snowpack and Discharge in an Alpine Karst Basin, in Snow and Glacier Hydrology, Proceedings of the Kathmandu Symposium, November 1992, IAHS Publ., 218, 249–260, 1993.

Huss, M. and Farinotti, D.: Distributed ice thickness and volume of all glaciers around the globe, J. Geophys. Res., 117, 1–10, https://doi.org/10.1029/2012JF002523, 2012.

Huss, M., Farinotti, D., Bauder, A., and Funk, M.: Modelling runoff from highly glacierized alpine drainage basins in a changing climate, Hydrol. Process., 22, 3888–3902, https://doi.org/10.1002/hyp, 2008.

Huss, M. and Fischer, M.: Sensitivity of Very Small Glaciers in the Swiss Alps to Future Climate Change, Front. Earth Sci., 4, 1–17, https://doi.org/10.3389/feart.2016.00034, 2016.

Huss, M. and Hock, R.: A new model for global glacier change and sea-level rise, Front. Earth Sci., 3, 1–22, https://doi.org/10.3389/feart.2015.00054, 2015.

Huss, M., Jouvet, G., Farinotti, D., and Bauder, A.: Future high-mountain hydrology: a new parameterization of glacier retreat, Hydrol. Earth Syst. Sci., 14, 815–829, https://doi.org/10.5194/hess-14-815-2010, 2010.

Huss, M., Zemp, M., Joerg, P. C., and Salzmann, N.: High uncertainty in 21st century runoff projections from glacierized basins, J. Hydrol., 510, 35–48, https://doi.org/10.1016/j.jhydrol.2013.12.017, 2014.

Hutton, C., Wagener, T., Freer, J., Han, D., Duffy, C. J., and Arheimer, B.: Most computational hydrology is not reproducible, so is it really science?, Water Resour. Res., 52, 7548–7555, https://doi.org/10.1002/2016WR019285, 2016.

Immerzeel, W. W., Shrestha, A. B., Bierkens, M. F. P., Beek, L. P. H., and Konz, M.: Hydrological response to climate change in a glacierized catchment in the Himalayas, Clim. Change, 110, 721–736, https://doi.org/10.1007/s10584-011-0143-4, 2012.

Köplin, N., Schädler, B., Viviroli, D., and Weingartner, R.: The importance of glacier and forest change in hydrological climate-impact studies, Hydrol. Earth Syst. Sci., 17, 619–635, https://doi.org/10.5194/hess-17-619-2013, 2013.

Li, H., Beldring, S., Xu, C.-Y., Huss, M., Melvold, K., and Jain, S. K.: Integrating a glacier retreat model into a hydrological model – Case studies of three glacierised catchments in Norway and Himalayan region, J. Hydrol., 527, 656–667, https://doi.org/10.1016/j.jhydrol.2015.05.017, 2015.

Lindström, G., Johansson, B., Persson, M., Gardelin, M., and Bergström, S.: Development and test of the distributed HBV-96 hydrological model, J. Hydrol., 201, 272–288, 1997.

Linsbauer, A., Paul, F., and Haeberli, W.: Modeling glacier thickness distribution and bed topography over entire mountain ranges with GlabTop: Application of a fast and robust approach, J. Geophys. Res., 117, F03007, https://doi.org/10.1029/2011JF002313, 2012.

Linsbauer, A., Paul, F., Machguth, H., and Haeberli, W.: Comparing three different methods to model scenarios of future glacier change in the Swiss Alps, Ann. Glaciol., 54, 241–253, https://doi.org/10.3189/2013AoG63A400, 2013.

Luo, Y., Arnold, J., Liu, S., Wang, X., and Chen, X.: Inclusion of glacier processes for distributed hydrological modeling at basin scale with application to a watershed in Tianshan Mountains, northwest China, J. Hydrol., 477, 72–85, https://doi.org/10.1016/j.jhydrol.2012.11.005, 2013.

Maisch, M., Wipf, A., Denneler, B., Battaglia, J., and Benz, C.: Die Gletscher der Schweizer Alpen: Gletscherhochstand 1850, Aktuelle Vergletscherung, Gletscherschwundszenarien, vdf Hochschulverlag, Zürich, 2000.

Miller, J. D., Immerzeel, W. W., and Rees, G.: Climate Change Impacts on Glacier Hydrology and River Discharge in the Hindu Kush – Himalayas A Synthesis of the Scientific Basis, Mt. Res. Dev., 32, 461–467, https://doi.org/10.1659/MRD-JOURNAL-D-12-00027.1, 2012.

Moore, R. D.: Application of a conceptual streamflow model in a glacierized drainage basin, J. Hydrol., 150, 151–168, https://doi.org/10.1016/0022-1694(93)90159-7, 1993.

Müller, F., Caflish, T., and Müller, G.: Firn und Eis der Schweizer Alpen, Gletscherinventar, Geographisches Institut, vdf-Verlag, ETH Zürich, 174 pp., 1976.

Naz, B. S., Frans, C. D., Clarke, G. K. C., Burns, P., and Lettenmaier, D. P.: Modeling the effect of glacier recession on streamflow response using a coupled glacio-hydrological model, Hydrol. Earth Syst. Sci., 18, 787–802, https://doi.org/10.5194/hess-18-787-2014, 2014.

Pattyn, F.: Transient glacier response with a higher-order numerical ice-flow model, J. Glaciol., 48, 467–476, https://doi.org/10.3189/172756502781831278, 2002.

Paul, F., Frey, H., and Le Bris, R.: A new glacier inventory for the European Alps from Landsat TM scenes of 2003: challenges and results, Ann. Glaciol., 52, 144–152, https://doi.org/10.3189/172756411799096295, 2011.

Radić, V., Hock, R., and Oerlemans, J.: Analysis of scaling methods in deriving future volume evolutions of valley glaciers, J. Glaciol., 54, 601–612, https://doi.org/10.3189/002214308786570809, 2008.

Ragettli, S., Pellicciotti, F., Bordoy, R., and Immerzeel, W. W.: Sources of uncertainty in modeling the glaciohydrological response of a Karakoram watershed to climate change, Water Resour. Res., 49, 6048–6066, https://doi.org/10.1002/wrcr.20450, 2013.

Salzmann, N., Machguth, H., and Linsbauer, A.: The Swiss Alpine glaciers' response to the global "2 °C air temperature target", Environ. Res. Lett., 7, 44001, https://doi.org/10.1088/1748-9326/7/4/044001, 2012.

Seibert, J.: Multi-criteria calibration of a conceptual runoff model using a genetic algorithm, Hydrol. Earth Syst. Sci., 4, 215–224, https://doi.org/10.5194/hess-4-215-2000, 2000.

Seibert, J. and Vis, M. J. P.: Teaching hydrological modeling with a user-friendly catchment-runoff-model software package, Hydrol. Earth Syst. Sci., 16, 3315–3325, https://doi.org/10.5194/hess-16-3315-2012, 2012.

Stahl, K., Moore, R. D., Shea, J. M., Hutchinson, D., and Cannon, A. J.: Coupled modelling of glacier and streamflow response to future climate scenarios, Water Resour. Res., 44, W02422, 1–13, https://doi.org/10.1029/2007WR005956, 2008.

Stahl, K., Weiler, M., Kohn, I., Seibert, J., Vis, M., and Gerlinger, K.: The snow and glacier melt components of streamflow of the river Rhine and its tributaries considering the influence of climate change, Final report to the International Commission for the Hydrology of the Rhine Basin (CHR), available at: http://www.chr-khr.org/en/publications, (last access: 21 March 2018), 2017.

Stroeven, A., Van de Wal, R., and Oerlemans, J.: Historic front variations of the Rhone glacier: simulation with an ice flow model, Glacier Fluctuations, Clim. Chang., 391–405, 1989.

Vincent, C., Harter, M., Gilbert, A., Berthier, E., and Six, D.: Future fluctuations of Mer de Glace, French Alps, assessed using a parameterized model calibrated with past thickness changes, Ann. Glaciol., 55, 15–24, https://doi.org/10.3189/2014AoG66A050, 2014.

Assessing the benefit of snow data assimilation for runoff modeling in Alpine catchments

Nena Griessinger[1,2]**, Jan Seibert**[2]**, Jan Magnusson**[3]**, and Tobias Jonas**[1]

[1]WSL Institute for Snow and Avalanche Research SLF, Davos, Switzerland
[2]Department of Geography, University of Zurich, Zurich, Switzerland
[3]Norwegian Water Resources and Energy Directorate (NVE), Oslo, Norway

Correspondence to: Nena Griessinger (nena.griessinger@slf.ch)

Abstract. In Alpine catchments, snowmelt is often a major contribution to runoff. Therefore, modeling snow processes is important when concerned with flood or drought forecasting, reservoir operation and inland waterway management. In this study, we address the question of how sensitive hydrological models are to the representation of snow cover dynamics and whether the performance of a hydrological model can be enhanced by integrating data from a dedicated external snow monitoring system. As a framework for our tests we have used the hydrological model HBV (Hydrologiska Byråns Vattenbalansavdelning) in the version HBV-light, which has been applied in many hydrological studies and is also in use for operational purposes. While HBV originally follows a temperature-index approach with time-invariant calibrated degree-day factors to represent snowmelt, in this study the HBV model was modified to use snowmelt time series from an external and spatially distributed snow model as model input. The external snow model integrates three-dimensional sequential assimilation of snow monitoring data with a snowmelt model, which is also based on the temperature-index approach but uses a time-variant degree-day factor. The following three variations of this external snow model were applied: (a) the full model with assimilation of observational snow data from a dense monitoring network, (b) the same snow model but with data assimilation switched off and (c) a downgraded version of the same snow model representing snowmelt with a time-invariant degree-day factor. Model runs were conducted for 20 catchments at different elevations within Switzerland for 15 years. Our results show that at low and mid-elevations the performance of the runoff simulations did not vary considerably with the snow model version chosen. At higher elevations, however, best performance in terms of simulated runoff was obtained when using the snowmelt time series from the snow model, which utilized data assimilation. This was especially true for snow-rich years. These findings suggest that with increasing elevation and the correspondingly increased contribution of snowmelt to runoff, the accurate estimation of snow water equivalent (SWE) and snowmelt rates has gained importance.

1 Introduction

Snowmelt provides a dominant contribution to runoff and groundwater storages in mountainous regions. In such areas, modeling snow processes is crucial for resource management as well as for flood and drought forecasting. Snow accumulates and acts as a temporary storage of water that is released as soon as snowmelt occurs. Since erroneous simulations of snow accumulation can bias the amount and timing of simulated snowmelt, accurately modeling both processes is important for runoff predictions. Problems for modelers may occur not only due to the great heterogeneity and variability of these processes, but also due to the limited availability of necessary observational data (Adam et al., 2009; Viviroli and Weingartner, 2004; Viviroli et al., 2011), including erroneous precipitation input data at higher altitudes (Wiesinger, 1993). Additionally, computational resources often constrain operational applications as timely model outputs are required. To cope with these challenges, many hydrological models make use of the temperature-index (TI) melt method instead of the energy-balance approach, which has higher input data requirements and is also computationally more demanding (Vehviläinen, 1992; Kumar et al., 2013). TI models can result in sufficient model performance if evaluated at a daily resolution and at the catchment scale (Lang and Braun, 1990; Hock, 2003), provided they use a reasonable parameterization (such as degree-day factor (DDF) and threshold temperature). The basic concept of TI models is to use air temperature as a proxy for the three energy sources that contribute to snowmelt: incoming longwave radiation, absorbed global radiation and sensible heat flux (Ohmura, 2001). The methods differ in their number of parameters, such as threshold values, to parameterize snowfall and melt, ranging from implementations using 2–5, as in the HBV (Hydrologiska Byråns Vattenbalansavdelning) model (Bergström, 1976), to 11 (Irannezhad et al., 2015) parameters. Inappropriate calibration of parameters will fail to accurately describe accumulation and melt rates and lead to a biased duration of the snow season and incorrect melt-out dates (Seibert, 2003). Identifying catchment characteristics that impact hydrological responses (i.e., geology, soil types or land use types) is also critical (Fontaine et al., 2002). Snow models of high complexity have been developed for a great variety of applications and their development is still ongoing. For avalanche research or snow studies on a small scale, simulating detailed processes within the snowpack is of great interest and importance. Otherwise, for operational purposes, which require short computation time and therefore cannot represent snowpack processes in great detail, different approaches are used to simulate snow accumulation and melt. Recently, various methods to assimilate observational snow data for snow cover models have been developed. At the point scale, model improvements due to assimilation of snow water equivalent data from observations were already shown (Magnusson et al., 2014). At the catchment scale and for operational purposes, several studies evaluated the effect of additional information from snow observations with different approaches. Franz et al. (2014) evaluated data assimilation based on a small number of ground-based observation sites within a hindcasting framework. In contrast to predictions of runoff under low-flow conditions, the overall skill of the forecasts could not be significantly improved. Jörg-Hess et al. (2015) improved snow water and runoff volume predictions by replacing simulated snow water equivalent at model initialization with data from measurements. Integrating snow data sets within the calibration procedures is an additional method to improve hydrological models as shown by Finger et al. (2015). A multiple objective calibration based on daily runoff data and snow depth data converted to spatially snow cover data, as introduced by Parajka et al. (2007), could improve snow cover simulations, but not runoff simulations compared to a single objective calibration based on daily runoff data only. Andreadis and Lettenmaier (2006) showed that the assimilation of remotely sensed snow cover area data did not significantly improve the model performance during

accumulation, whereas for the snowmelt season small improvements were found. The authors concluded that assimilating snow water equivalent data from observations might be a more successful approach. Therefore, as the main objective of this study, we evaluated the sensitivity of a conceptual runoff model (conceptual in terms of the linear reservoir concept) to the external input of snowmelt data from three different snow models of different complexities. Particularly, we examined the benefit of snow water equivalent data assimilation for hydrological applications in mountainous regions.

2 Data

To cover a wide range of elevations and different climatic regions, for this study we chose 20 catchments spread over Switzerland. All of them were at most minimally affected by human activities, such as water regulation or abstraction. A further crucial selection criterion was the availability of the required data. Since, especially at high elevations, the runoff regime of many catchments in Switzerland is affected by man-made installations, the number of possible catchments was highly limited. Catchments analyzed in this study varied in size from 17 to $473 \, km^2$ and the mean elevations of these catchments ranged between 560 and 2656 m a.s.l. (Table 1 and Fig. 1). We grouped the catchments for our analysis based on their mean elevation into three elevation classes: below 1000 m a.s.l., 1000 to 2000 m a.s.l. and above 2000 m a.s.l. Runoff data measured at the catchment outlets of these 20 catchments was provided and checked for plausibility by FOEN (Federal Office of the Environment). According to the temporal resolution of the model output, we aggregated the hourly runoff records into daily sums. For the data assimilation for the full snow model used in this study we considered daily snow depth measurements from both manual and automatic monitoring stations (see red stars in Fig. 1 for locations). All 320 stations used were part of either the MeteoSwiss (Federal Office of Meteorology and Climatology) or the SLF (WSL Institute for Snow and Avalanche Research) snow station networks in Switzerland, covering elevations between 210 and 2950 m a.s.l. and located on open, flat terrain. Out of approximately 600 available stations, only 320 were used after a careful selection process to avoid sites that were influenced by wind or frequent sensor failures, or known to systematically deviate from representative measurements. Daily data from the morning measurements between 1 September 1998 and 31 August 2013 were carefully checked for missing values or erroneous readings and corrected where necessary. These values were replaced using a stochastic gap filling model that accounts for data from the same station before and after the gap, as well as for data from neighboring stations at similar elevations. Temperature data were obtained from 220 stations and interpolated using an inverse distance weighting approach as described in Magnusson et al. (2014), which considers both horizontal and

Table 1. Characteristics of 20 Swiss catchments in this study.

Number	Station name	Area [km^2]	Min elevation [m a.s.l.]	Max elevation [m a.s.l.]	Mean elevation [m a.s.l.]	Elevation class	Begin snowmelt [month-day]	End snowmelt [month-day]
2202	Ergolz – Liestal	276	305	1087	577	1	01-01	03-01
2126	Murg – Wängi	77	501	911	640	1	01-14	03-14
2034	Broye – Payerne, Caserne d'aviation	416	450	1402	721	1	01-14	03-14
2343	Langeten – Huttwil, Häberenbad	61	592	1032	757	1	01-14	03-14
2374	Necker – Mogelsberg, Aachsäge	89	649	1359	948	1	02-14	04-14
2321	Cassarate – Pregassona	74	286	1809	954	1	02-14	04-14
2603	Ilfis – Langnau	188	699	1695	1040	2	02-21	04-21
2634	Kleine Emme – Emmen	473	440	2261	1044	2	02-21	04-21
2179	Sense – Thörishaus, Sensematt	355	609	2028	1072	2	03-01	05-01
2609	Alp – Einsiedeln	82	845	1577	1096	2	02-21	04-21
2409	Emme – Eggiwil, Heidbüel	127	770	2007	1296	2	02-21	04-21
2300	Minster – Euthal, Rüti	59	918	1994	1345	2	03-07	05-07
2203	Grande Eau – Aigle	130	579	2830	1546	2	03-14	05-14
2605	Verzasca – Lavertezzo, Campioi	188	546	2590	1656	2	03-14	05-14
2276	Grosstalbach – Isenthal	43	931	2682	1794	2	03-14	05-14
2232	Allenbach – Adelboden	31	1360	2587	1907	2	03-14	05-14
2366	Poschiavino – La Rösa	17	1920	3005	2316	3	04-14	06-14
2304	Ova dal Fuorn – Zernez, Punt la Drossa	56	1797	2903	2337	3	04-14	06-14
2327	Dischmabach – Davos, Kriegsmatte	42	1772	2869	2349	3	04-14	06-14
2256	Rosegbach – Pontresina	67	1833	3721	2686	3	05-01	07-01

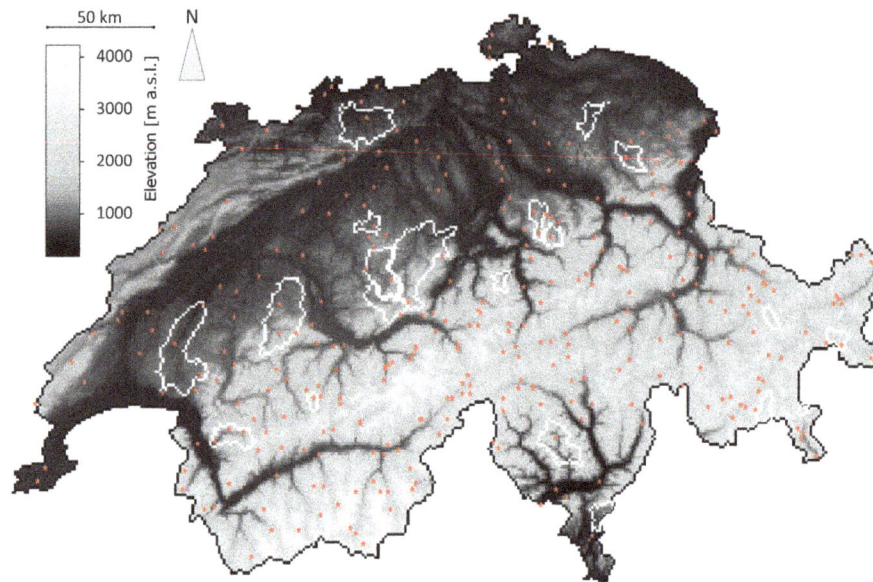

Figure 1. Locations of snow observation stations (red stars) and 20 studied catchments (white border lines) in Switzerland.

vertical distances between measurement stations and interpolated grid cells. A variable weighting factor was used to determine the influence of horizontally near but vertically distant stations. The resolution of the resulting temperature grid data set was 1 km × 1 km. Precipitation data were also required as a gridded input data set. We used a daily product (RhiresD) with a spatial resolution of 2 km × 2 km available from MeteoSwiss. The product is based on a dense precipitation gauge network with approximately 500 stations within Switzerland. Methodological details are described in Frei and Schär (1998), Frei et al. (2006) and Isotta et al. (2014).

3   Methods

3.1   Hydrological model

The hydrological model HBV (Bergström, 1976, 1992, 1995; Lindström et al., 1997) in the version HBV-light (Seibert and Vis, 2012) was used to simulate runoff at the 20 selected catchments. HBV requires a time series of precipitation, air temperature and potential evaporation to simulate runoff for a specific catchment. Potential evaporation was calculated following the methods of Priestley and Taylor (1972). In the HBV snow routine, precipitation is expected to be solid below a certain temperature threshold and multiplied by a correction factor to account for possible undercatch and to compensate for the missing snow interception. Snowmelt is usually calculated using the same threshold temperature and a DDF. Up to a certain fraction, liquid water can be stored in the snowpack and refreezes if temperatures are below the threshold temperature. In our study, however, we disabled this snow routine of the HBV model and replaced snowmelt as well as rain input with data coming both from the external snow model. Groundwater recharge and actual evaporation were simulated in a soil routine depending on the actual water storage. A response routine consisting of three linear reservoirs and a routing routine using a triangular weighting function follow. Runoff data observed at the outlet of all catchments considered in this study were used for calibration and validation of the model. More details are available in Seibert and Vis (2012). To evaluate the performance of the hydrological model in response to the input from different variants of the external snowmelt model, we focused our analysis on the main melt period, denoted below as snowmelt season. Although the onset and duration of the snowmelt season vary from year to year, we have determined a fixed snowmelt season for each individual catchment (Table 1), based on the average timing of the first snowmelt runoff in spring and the average duration until 75 % of the snow has melted. Two approaches were chosen to split the available runoff data into separate data sets for calibration and validation. The first approach was to use all years for calibration except one, which was used for validation. This so-called leave-one-out procedure was repeated so that each year was used for validation once. The second approach was differential split sampling (Klemeš, 1986), where the snow-poor and normal years were used for calibration and the snow-rich years were used for validation. This separation into different snow year groups was done individually for each catchment. To optimize the parameter set of the hydrological model for each catchment and each of the input data sets within the calibration period, we ran a genetic calibration algorithm as described in Seibert (2000) with 5000 model runs and 1000 runs for local optimization. This was done individually for each of the above model configurations, as well as for the benchmark model. As the objective function, we used the

Nash–Sutcliffe model efficiency (Nash and Sutcliffe, 1970) computed for the catchment-specific snowmelt season.

3.2   Snow model

The external snow model framework, which we used in this study instead of the snow routine built in the HBV model, also simulates snowmelt by a TI approach but in addition allows for integration of observational snow data using a data assimilation scheme. While some details on the external snow model framework are given below, a full description of model and data assimilation methods is available in Magnusson et al. (2014). We applied three versions of this model, denoted M1 to M3. Version M1 includes the full model and data assimilation scheme (an approach unavailable in the internal snow routine of HBV), whereas M2 an M3 are downgraded versions of M1 as described below. Several characteristics are common to all model versions described below. First, a threshold temperature differentiates whether precipitation falls as snowfall or rain. However, the models allow for mixed precipitation in a range close to the threshold temperature (see Eq. (10) and the corresponding description in Magnusson et al., 2014). Second, fractional snow-covered area (SCF) is parameterized using modeled snow depth and terrain parameters that were derived from a 25 m digital elevation model according to Helbig et al. (2015). Third, all three model versions allow for the snow cover to hold a fraction of liquid water. Fourth, all model versions consider the influence of topography on snow distribution and redistribution in mountainous terrain. Slope- and aspect-dependent correction functions were trained using a set of high-resolution snow depth maps from airborne lidar acquisitions in the European Alps as presented in Grünewald and Lehning (2015), and applied at a subgrid 25 m spatial resolution. This procedure ensured accurate inference of areal mean snow depths from snow and precipitation measurements on flat field sites. In the following section, we describe the three versions of the snow model used in this study:

– TI snowmelt model with data assimilation and time-varying DDF (M1): this model is the same as that described in detail in Magnusson et al. (2014). Using an elaborated TI approach, daily snowmelt at each grid cell was calculated if a certain threshold temperature is exceeded. The DDF defines the possible melt rate per day and per degree temperature above the threshold. For M1, the DDF varied as a function of season between a minimal [$1.0 \, \text{mm} \, {}^{\circ}\text{C}^{-1} \, \text{day}^{-1}$] and maximal [$4.5 \, \text{mm} \, {}^{\circ}\text{C}^{-1} \, \text{day}^{-1}$] value using a sinusoidal function (see Eq. (12) in Magnusson et al., 2014). The DDF is independent of elevation. For the data assimilation, the daily measured snow depth data at all stations were first converted to snow water equivalents (SWE) using a snow density model, which is based on the methods of Jonas et al. (2009) and Martinec and Rango (1991). Second, by applying an optimal interpolation

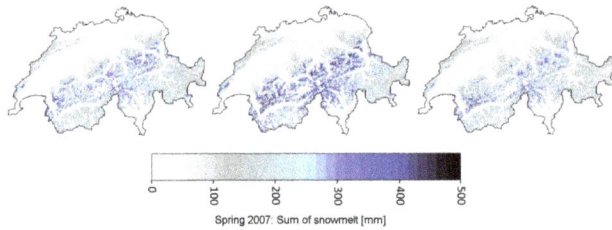

Figure 2. Cumulative snowmelt during the snowmelt season 2007 as calculated by the snow model method M1 (full model with data assimilation, left), M2 (full model without data assimilation, middle) and M3 (simplified model, right). The sums between the three model methods differ depending on the use of observational snow data assimilation and the use of different DDFs.

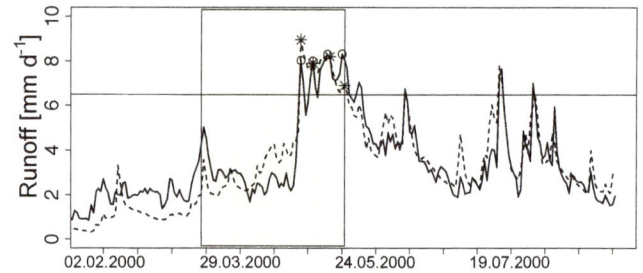

Figure 3. Graphical explanation of how to calculate E_{PF}. The yellow background shows a catchment-specific snowmelt season window within which the efficiency criteria were computed. The horizontal line indicates the threshold of 1.5 times the mean observed runoff (blue line) above which measured peak flow events (blue circles) are detected. Red stars present corresponding events of the simulated runoff (dashed red line). See Sect. 3.3 for details.

approach, the SWE data were used to correct the computed snowfall amounts. Finally, the simulated melt rates and model state variables (SWE and liquid water content) were updated using the ensemble Kalman filter with the same SWE data. Both the optimal interpolation scheme and the ensemble Kalman filter were set up using spatially correlated error statistics. With such an approach, often called three-dimensional data assimilation, the point snow observations influence the gridded simulation results even at locations lacking observations. For more details about the model, and the data assimilation method in particular, see Magnusson et al. (2014).

– TI snowmelt model with time-varying DDF without data assimilation (M2): in this version, the same elaborated TI approach as in M1 was applied to simulate snow accumulation and melt at each grid cell based on the same input data grids as in M1. The DDF seasonal variations are equal to those in M1. The only difference concerns the data assimilation procedures, which were switched off in M2, such that observed SWE data were not used to update the initial estimates on snow accumulation and melt rates.

– TI snowmelt model using a constant DDF without data assimilation (M3): this version differs from M2 with respect to the DDF. Here the DDF does not show seasonal variations but is assumed to be constant over the season. The average DDF of $2.5\,\mathrm{mm}\,{}^{\circ}\mathrm{C}^{-1}\,\mathrm{day}^{-1}$ was chosen, which is a good compromise if used for the full winter season. For comparison only, complementary analyses were performed with the constant DDF of $4.0\,\mathrm{mm}\,{}^{\circ}\mathrm{C}^{-1}\,\mathrm{day}^{-1}$, which is more appropriate if used for a late snowmelt season only. Note that M3 represents the type of snow routine used in HBV except for that DDF is a model parameter determined by calibration in HBV, whereas it is a pre-defined value in M3.

Replacing a TI model with another TI model, and not with an energy-balance or snowpack-physics model, may appear

unusual at first glance. However, if concerned with conceptual hydrological modeling at a daily timescale, the TI model framework used here constituted an ideal testing environment. To provide daily snowmelt rates, the dynamic data assimilation framework within M1 represents current state-of-the art methodology in operational snow hydrological monitoring. Since it accounts for measured snow depletion rates at hundreds of monitoring sites, it provides the best possible input to the hydrological model. Even with data assimilation switched off (M2), if validated against snow lysimeter data at daily time steps, the performance is almost on par with the output of top-notch energy-balance models (Magnusson et al., 2015). Only the concept of using a constant DDF (M3) could result in a severely downgraded performance, as already seen by Lang and Braun (1990). Hence, the triplet, M1, M2, M3, provides a ranked set of input options, which allows for an evaluation of the sensitivity of conceptual hydrological modeling on the input from different types of snow models. This ultimately was the purpose of the study, rather than testing the performance of a specific runoff model (i.e., HBV). As mentioned above, HBV originally uses a TI snowmelt routine, which is similar to our external model version M3. However, as part of HBV, the constant DDF is a free parameter to be optimized during calibration of the snowmelt season. Hence, to provide a benchmark for our performance tests, we also ran the HBV model with the original snow routine switched on. We used these runs as an upper benchmark, since the HBV snow routine was tuned by calibration to allow for the maximum possible performance of the runoff model for each individual catchment. In contrast, we created a lower benchmark by assuming all precipitation to be rain, i.e., a no-snow-model scenario. These two benchmarks allowed for scaling of the performances, which were achieved when using M1 to M3 to provide input to HBV. All model variants were run for the whole study period on a daily time step at 1 km spatial resolution. During the snowmelt season, the three snow model methods created individual spatial pattern of simulated snowmelt. As an illustrative example, the cumulative sums of snowmelt between 1 February 2007 and 30 April 2007 are shown in Fig. 2. As expected for the snowmelt season, M2 yielded higher amounts of snowmelt compared to M3 due to differences in the DDF. In this particular year, the observations used for the assimilation did not support the high melt rates as predicted by M2, resulting in M1 to calculate lesser amounts of snowmelt.

3.3 Validation methods

Timing of snowmelt onset and of runoff events due to snowmelt affects the availability of water resources and influences flooding and droughts (Semmens and Ramage, 2013). Therefore, it is crucial to simulate and to evaluate the timing of streamflow accurately when comparing snowmelt models. Several efficiency criteria are used in the literature for evaluating hydrological models and should be selected carefully depending on the aim of the validation (Krause et al., 2005). To assess the performance of the hydrological model in combination with the input options from our set of snow models, we chose the following two criteria. First, since we were interested in how precise single peak flow events due to snowmelt could be simulated when integrating data from the different snow model approaches, we used the "peak flow efficiency for snowmelt season" E_{PF}. Figure 3 illustrates the procedure to calculate this measure. Observed peak flow events during the snowmelt season (yellow period in Fig. 3) that exceed a certain threshold (defined as 1.5 times of the mean runoff during snowmelt season; horizontal line in Fig. 3) were picked and denoted as $Q_{peak\,obs\,i}$ (blue circles in Fig. 3). The maximum simulated runoff in a time window of 1 day before and after each of the n observed peak flow events was taken as simulated reference value $Q_{peak\,sim\,i}$ (red stars in Fig. 3). The reference values did not necessarily have to be local peaks or greater than a certain threshold (Eq. (1); Seibert, 2003).

$$E_{PF} = 1 - \frac{\sum_{i=1}^{n} \left| Q_{peak\,obs\,i} - Q_{peak\,sim\,i} \right|}{\sum_{i=1}^{n} Q_{peak\,obs\,i}} \qquad (1)$$

Additionally, the frequently used Nash–Sutcliffe efficiency of runoff E_Q (Eq. 2) according to Nash and Sutcliffe (1970), which is also supposed to be sensitive to peak flow events (Krause et al., 2005), was chosen and applied to the defined snowmelt season.

$$E_Q = 1 - \frac{\sum_{i=1}^{m} (Q_{obs\,i} - Q_{sim\,i})^2}{\sum_{i=1}^{m} (Q_{obs\,i} - mean(Q_{obs}))^2}, \qquad (2)$$

where i represents all (1 to m) days within the snowmelt season, and $Q_{obs\,i}$ and $Q_{sim\,i}$ are observed and simulated runoff at day i, respectively. This was also used as the objective function for the genetic calibration algorithm (GAP-optimization) within the hydrological model framework.

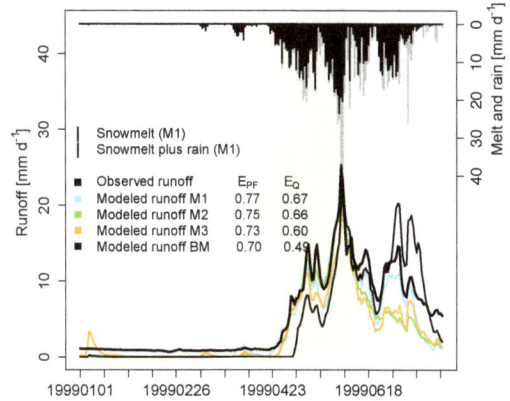

Figure 4. Observed and modeled runoff for the Dischma catchment for 1999, as well as water input from snowmelt and rain modeled with method M1. The upper benchmark model BM in red.

4 Results and discussion

Both efficiency metrics were calculated for (a) each catchment and (b) each of the two calibration experiments. The performance statistics are discussed separately for each of the three groups of catchments depending on mean elevation.

4.1 Example of runoff simulation for a representative catchment

To look for differences between the three snow model methods, individual catchments and years were selected. Representing a catchment at high elevations, results for the Dischma catchment (EZG 2327, gauge Davos Kriegsmatte) with a mean elevation of 2349 m a.s.l. are shown in Fig. 4. The yellow background displays the catchment-specific snowmelt season during which the bulk of the snowmelt typically occurs. The blue and gray lines at top of the graph indicate the snowmelt input to the hydrological model from M1 excluding and including rain, respectively, in this example for the record-high snow year 1999. The observed runoff is shown by the black curve, while the different colored curves indicate the simulations with M1, M2 and M3. The curves as well as the performance metrics achieved by the differential split-sample experiment demonstrate that for this catchment, the M1 model as input to the hydrological framework provided the best runoff simulations, even though the differences are small. Note however, that in this example M1 particularly outperforms the other models in the month of July, which is outside the standard evaluation period.

4.2 Model performance across elevation classes: leave-one-out sample

First, we used the leave-one-out approach to calibrate the hydrological model. The leave-one-out approach represents a typical scenario in operational conceptual runoff modeling,

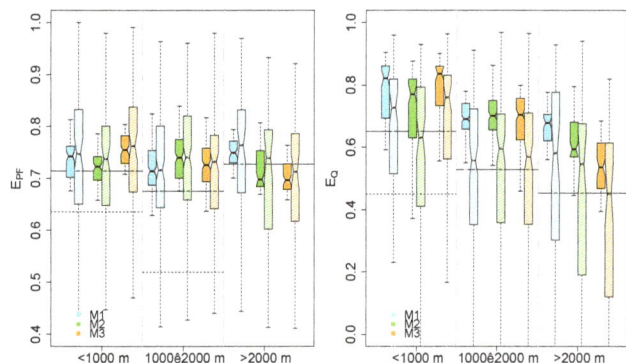

Figure 5. Results of the leave-one-out approach. E_{PF} (left panel) and E_Q (right panel) for each elevation class and snowmelt model. For the individual elevation classes and melt models, the left box plots (darker colors) show the results for the calibration period, and the right box plots (lighter colors) show the results for the validation period. The whisker boxes represent the median (center line), the interquartile range (25th–75th percentile; box outline) and highest/lowest performance within the interquartile range (± 1.5 times of the interquartile range; whiskers). The benchmark performance is denoted by a solid red line (upper benchmark) and a dashed red line (lower benchmark), and the latter only displayed if within the range of the axis limits.

i.e., to use as much data as possible for calibration and to apply the resulting parameter values to the current season. Results grouped according to mean catchment height are presented below (Fig. 5). Using this calibration procedure for catchments with mean elevation below 1000 m a.s.l., the hydrological model showed good results independent of which snow model was used as input to the hydrological model framework. Even without using a snow model at all (i.e., the lower benchmark), the runoff model resulted in lower but still positive performance values, indicating that the choice of snow model within a conceptual runoff modeling framework is of less importance when dealing with catchments at lower elevations. Similarly for catchments with mean elevation between 1000 and 2000 m a.s.l. the differences between the three model runs were small. While E_{PF} levels were maintained relative to our assessment for catchments below 1000 m a.s.l., they were separated more clearly from the benchmark model runs, which dropped in performance. E_Q values, on the other hand, decreased for all the M1, M2, M3 and the benchmark model runs. Only for the highest elevation class did the results based on M1 significantly outperform the other model runs, and even reached better E_{PF} values than most simulations at lower elevation classes. Even the model runs based on M2 performed better than those based on M3. This shows that the benefit of better snowmelt input data for conceptual runoff modeling only seems to pay off if considering catchments above a certain elevation. At lower elevation, differences between the model input options could be mitigated by way of the calibration procedure. Fur-

ther, while results based on M1 showed a relatively constant performance across all elevation classes in both E_{PF} and E_Q, this was not the case for results based on M2 and M3, which deteriorated with increasing elevation. Looking at all elevation classes, the median performance of the M1 runs was always higher than the upper benchmark. This was also mostly the case for M2 and M3. This result shows that all versions of the external snow model performed unexpectedly well in combination with the hydrological framework even though they were not included in the calibration procedure. Finding instances where even M3 (which uses a prescribed DDF) outperforms the upper benchmark model (which relies on a calibrated DDF) may appear counter-intuitive. However, note that M1, M2, M3 have been particularly trained for an optimal performance in the Swiss Alps, e.g., regarding the representation of processes like liquid water content, refreezing, cold content dynamics, the partitioning of rain and snow, and redistribution of snow in steep terrain. Further, calibrating HBV for the melt season only could result in a DDF that is too high during the snow accumulation period, which would inhibit an accurate timing of the meltwater release (c.f. Fig. 4). On the contrary, M3 features a more moderate DDF of 2.5 mm $°C^{-1}$ day^{-1}, allowing for a more balanced performance over the entire snow season. The above results demonstrate a benefit of using an advanced snowmelt modeling system in the context of conceptual hydrological modeling, even if the benefit seems comparably small and restricted to catchments above a certain elevation. Other studies that evaluated the influence of integrating snow water equivalent data into hydrological models showed similar results (Finger et al., 2015; Jörg-Hess et al., 2015). Only a few studies have used direct assimilation of ground-based snow data. Due to limited availability of ground observations, assimilating remotely sensed snow data is a more common practice but requires further inversion methods, which is quite challenging to implement and induces additional uncertainties (Andreadis and Lettenmaier, 2006). Several studies used satellite observations of snow cover extent in different assimilation schemes to update snow models. Clark et al. (2006) as well as Thirel et al. (2013) could slightly improve runoff predictions by assimilation of snow-covered area using the ensemble Kalman filter and the particle assimilation filter, respectively. As in the above studies, we focused on a catchment-specific snowmelt season and used two performance measures that evaluated the ability of the models to capture peak flow events, among other characteristics of the hydrograph. Simulating such events is of great importance, especially for operational flood forecasting purposes. While the performance of well-calibrated models may be adequate independent of model complexity (Hock, 2003; Magnusson et al., 2015), we are particularly interested in the model performance in extreme years, when the snowmelt contribution greatly increases flood risks. This is why in the second set of modeling experiments we singled out snow-rich years as a validation data set to generate both a more challenging and

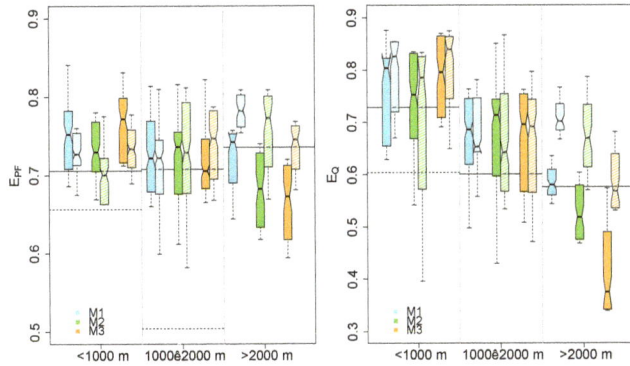

Figure 6. Results of the differential split-sample approach. E_{PF} (left panel) and E_Q (right panel) for each elevation class and snowmelt model. For the individual elevation classes and melt models, the left box plots (darker colors) show the results for the calibration period, and the right box plots (lighter colors) show the results for the validation period. The whisker boxes represent the median (center line), the interquartile range (25th–75th percentile; box outline) and highest/lowest performance within the interquartile range ($\pm$1.5 times of the interquartile range; whiskers). The benchmark performance is denoted by a solid red line (upper benchmark) and a dashed red line (lower benchmark), and the latter only displayed if within the range of the axis limits.

more relevant test scenario. For the snow-rich years, we selected the 6 years with the highest cumulative snowmelt individually for each catchment.

4.3 Model performance across elevation classes: differential split sample

For the differential split-sample approach, snow-rich years were used to validate the runoff models. As expected, the analysis using the differential split-sample approach revealed similar performance patterns compared to the leave-one-out approach, but with increased differences between model runs (Fig. 6). As seen before, at low and mid-elevation classes the differences between the three model versions as well as between calibration and validation were comparably small. The median values of efficiencies for each model version ranged between 0.7 and 0.8 (E_{PF}) respectively 0.75 and 0.85 (E_Q). As seen before, at high elevations, model results based on M1 were superior (significantly for E_Q) to those based on M2, which in turn outperformed the model runs based on M3. However, the differences between the three runs were considerably larger than those seen with the leave-one-out approach. Another notable difference between both calibration methods was that the differential split-sample approach led to significantly higher E_Q for validation years compared to calibration years, while the opposite was the case when using the leave-one-out approach. Both findings strongly suggest that the benefit of advanced snowmelt input data for conceptual runoff modeling is particularly valuable in situations that feature a strong snowmelt component (high elevation, snow-rich years). Both E_{PF} and E_Q for M1-based model runs show an exceptional performance at high elevation for validation years, which highlights the value of snow data assimilation when concerned with forecasting snowmelt related floods. An additional analysis was performed with M3 using a DDF of 4.0 mm °C^{-1} day^{-1} (results not included in figures). This is a typical value found in the literature for high elevations with melting conditions later in the season (Martinec et al., 1983). As expected, compared to the standard DDF of 2.5 mm °C^{-1} day^{-1} in M3, the additional model runs resulted in slightly better performance metrics at high elevations with a later onset of snowmelt (catchments above 2000 m a.s.l.), but considerably worse performance in all other model runs.

4.4 Model performance for high elevation catchments: leave-one-out sample

The validation of the differential split-sample experiment showed that the three external snow models provided the best runoff simulations for snow-rich years, specifically for catchments with a mean elevation of above 2000 m a.s.l. In a further analysis, we ordered the single validation years individually by catchment for the leave-one-out approach from snow poor to snow rich based on peak SWE. This procedure allowed testing of whether there was a trend in the runoff performance metrics associated with the snow amount of single years. Such a trend was indeed evident, as seen in Fig. 7. Independent of the snow model used, the best results were achieved when validating the model performance during snow-rich years regarding both E_{PF} and E_Q. The performance measures discussed above were computed for a catchment-specific pre-defined fixed snowmelt season, which was based on the typical timing of observed snowmelt runoff. Extending the snowmelt season to 120 days gave similar results (data not shown) with the same relative differences between M1, M2, M3, but with a lower overall performance due to the decreasing relevance of snowmelt as the snow-covered area declines. While our approach allowed us to focus on the sensitivity of runoff modeling to different approaches for estimating snowmelt, it has four main implications to the interpretation of the results. First, E_Q values tend to be lower if calculated over a short period, and values may not be comparable to E_Q data from assessment of multi-year or multi-season data sets, in particular if analyzing daily runoff data that do not encompass diurnal variations. Second, within a pre-defined season, the variation of a time-varying DDF as used in M2 is small. Especially at low elevations and early in the year, the DDF of M2 and M3 do not differ much and therefore produces similar runoff simulations with comparable performance. According to Lang and Braun (1990) and Magnusson et al. (2015), a clearer benefit of using a flexible instead of a fixed DDF would have been expected if used within a longer time window. Third, at low elevations snowmelt may occur sporadically and not necessarily

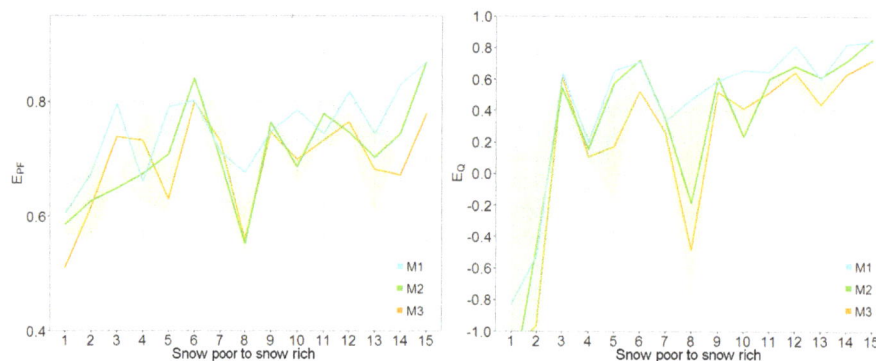

Figure 7. Results of the leave-one-out approach for catchments with mean elevation above 2000 m a.s.l. Median (solid lines) and interquartile (25th–75th percentile; shading) range of E_{PF} (left panel) and E_Q (right panel) for validation years ordered from snow-poor (index $= 1$) to snow-rich (index $= 15$) years.

within a pre-defined season. At high elevations, it is also possible that the main melt does not occur within the catchment-specific snowmelt season due to longer melt-out duration of extremely snow-rich years. Consequently, if snowmelt occurred outside of the validation period, it would not affect the performance statistics. This may have partly suppressed differences between the three different snow models. Finally, note that seasonal E_{PF} and E_Q statistics are two metrics out of several possible evaluation criteria. While we also tested other metrics, these were not further integrated to the discussion, given that the results were similar compared to the performance data presented above.

5   Conclusions

Based on daily runoff data measured over a period of 15 years at 20 catchments in Switzerland, we evaluated the sensitivity of a conceptual hydrological modeling framework to snowmelt input from snow models of different complexity. The most complex snow model integrated three-dimensional sequential assimilation of snow monitoring data with a snowmelt model based on the temperature-index approach. In contrast, the simplest snow model represented snowmelt with a constant degree-day factor, and did not include any data assimilation. The snow models were combined with the HBV model in the version HBV-light (Seibert and Vis, 2012) to produce a runoff record. The performance of the HBV runs based on snowmelt data from the snow models was assessed by way of performance metrics evaluated during the snowmelt season only. Our results showed that advanced methods to calculate snowmelt as input to conceptual runoff models only improved model performance if considering snow-dominated catchments. At low elevations, differences between the model input options were found to be minor. For higher elevation catchments, however, snowmelt input from the data assimilation framework consistently provided the best results. Further analysis demonstrated con-

siderably higher performance metrics for snow-rich years as compared to years with little snow. In contrast to earlier studies, which have shown that assimilation of snow-covered area only has limited impact on runoff simulations, our results indicate that the assimilation of snow water equivalent data can have a larger benefit for accurate streamflow predictions. This finding highlights the value of choosing the appropriate snow data assimilation methods, and perhaps even more important, selecting the correct variable for assimilation when concerned with operational forecasting of snowmelt related floods.

Acknowledgements. This study was partly funded by the Federal Office of the Environment (FOEN). We thank MeteoSwiss for access to the meteorological data and FOEN for providing river-runoff observations used in this study. Thanks to Manfred Stähli and Massimiliano Zappa for helpful discussions and to Nathalie Chardon for reviewing the English of this article.

Edited by: R. Woods

References

Adam, J. C., Hamlet, A. F., and Lettenmaier, D. P.: Implications of global climate change for snowmelt hydrology in the twenty-first century, Hydrol. Process., 23, 962–972, 2009.

Andreadis, K. M. and Lettenmaier, D. P.: Assimilating remotely sensed snow observations into a macroscale hydrology model, Adv. Water Resour., 29, 872–886, 2006.

Bergström, S.: Development and application of a conceptual runoff model for Scandinavian catchments, Lund Institute of Technology, University of Lund, Sweden, Bulletin Series A, 52, 134 pp., 1976.

Bergström, S.: The HBV model: Its structure and applications, Swedish Meteorological and Hydrological Institute, 35 pp., 1992.

Bergström, S.: The HBV model, edited by: Singh, V., Computer Models of Watershed Hydrology, Water Resources Publications, Highlands Ranch, Colorado, USA, 443–476, 1995.

Clark, M. P., Slater, A. G., Barrett, A. P., Hay, L. E., McCabe, G. J., Rajagopalan, B., and Leavesley, G. H.: Assimilation of snow covered area information into hydrologic and land-surface models, Adv. Water Resour., 29, 1209–1221, 2006.

Finger, D., Vis, M., Huss, M., and Seibert, J.: The value of multiple data set calibration versus model complexity for improving the performance of hydrological models in mountain catchments, Water Resour. Res., 51, 1939–1958, 2015.

Fontaine, T., Cruickshank, T., Arnold, J., and Hotchkiss, R.: Development of a snowfall–snowmelt routine for mountainous terrain for the soil water assessment tool (SWAT), J. Hydrol., 262, 209–223, 2002.

Franz, K. J., Hogue, T. S., Barik, M., and He, M.: Assessment of SWE data assimilation for ensemble streamflow predictions, J. Hydrol., 519, 2737–2746, 2014.

Frei, C. and Schär, C.: A precipitation climatology of the Alps from high-resolution rain-gauge observations, Int. J. Climatol., 18, 873–900, 1998.

Frei, C., Schöll, R., Fukutome, S., Schmidli, J., and Vidale, P. L.: Future change of precipitation extremes in Europe: Intercomparison of scenarios from regional climate models, J. Geophys. Res.-Atmos., 111, D06105, doi:10.1029/2005JD005965, 2006.

Grünewald, T. and Lehning, M.: Are flat-field snow depth measurements representative? A comparison of selected index sites with areal snow depth measurements at the small catchment scale, Hydrol. Process., 29, 1717–1728, 2015.

Helbig, N., van Herwijnen, A., Magnusson, J., and Jonas, T.: Fractional snow-covered area parameterization over complex topography, Hydrol. Earth Syst. Sci., 19, 1339–1351, doi:10.5194/hess-19-1339-2015, 2015.

Hock, R.: Temperature index melt modelling in mountain areas, J. Hydrol., 282, 104–115, 2003.

Irannezhad, M., Ronkanen, A.-K., and Kløve, B.: Effects of climate variability and change on snowpack hydrological processes in Finland, Cold Reg. Sci. Technol., 118, 14–29, 2015.

Isotta, F. A., Frei, C., Weilguni, V., Perčec Tadić, M., Lassègues, P., Rudolf, B., Pavan, V., Cacciamani, C., Antolini, G., Ratto, S. M., et al.: The climate of daily precipitation in the Alps: development and analysis of a high-resolution grid dataset from pan-Alpine rain-gauge data, Int. J. Climatol., 34, 1657–1675, 2014.

Jonas, T., Marty, C., and Magnusson, J.: Estimating the snow water equivalent from snow depth measurements in the Swiss Alps, J. Hydrol., 378, 161–167, 2009.

Jörg-Hess, S., Griessinger, N., and Zappa, M.: Probabilistic Forecasts of Snow Water Equivalent and Runoff in Mountainous Areas*, J. Hydrometeorol., 16, 2169–2186, 2015.

Klemeš, V.: Operational testing of hydrological simulation models, Hydrol. Sci. J., 31, 13–24, 1986.

Krause, P., Boyle, D., and Bäse, F.: Comparison of different efficiency criteria for hydrological model assessment, Adv. Geosci., 5, 89–97, 2005.

Kumar, M., Marks, D., Dozier, J., Reba, M., and Winstral, A.: Evaluation of distributed hydrologic impacts of temperature-index and energy-based snow models, Adv. Water Resour., 56, 77–89, 2013.

Lang, H. and Braun, L.: On the information content of air temperature in the context of snow melt estimation, IAHS Publ, 190, 347–354, 1990.

Lindström, G., Johansson, B., Persson, M., Gardelin, M., and Bergström, S.: Development and test of the distributed HBV-96 hydrological model, J. Hydrol., 201, 272–288, 1997.

Magnusson, J., Gustafsson, D., Hüsler, F., and Jonas, T.: Assimilation of point SWE data into a distributed snow cover model comparing two contrasting methods, Water Resour. Res., 50, 7816–7835, 2014.

Magnusson, J., Wever, N., Essery, R., Helbig, N., Winstral, A., and Jonas, T.: Evaluating snow models with varying process representations for hydrological applications, Water Resour. Res., 51, 2707–2723, 2015.

Martinec, J. and Rango, A.: Indirect evaluation of snow reserves in mountain basins, International Association of Hydrological Sciences. IAHS/AISH Publ., 602, 111–119, 1991.

Martinec, J., Rango, A., and Major, E.: The snowmelt-runoff model (SRM) user's manual, NASA Ref. Publ. 1100, 1983.

Nash, J. E. and Sutcliffe, J. V.: River flow forecasting through conceptual models part I – A discussion of principles, J. Hydrol., 10, 282–290, 1970.

Ohmura, A.: Physical basis for the temperature-based melt-index method, J. Appl. Meteorol., 40, 753–761, 2001.

Parajka, J., Merz, R., and Blöschl, G.: Uncertainty and multiple objective calibration in regional water balance modelling: case study in 320 Austrian catchments, Hydrol. Process., 21, 435–446, 2007.

Priestley, C. and Taylor, R.: On the assessment of surface heat flux and evaporation using large-scale parameters, Mon. Weather Rev., 100, 81–92, 1972.

Seibert, J.: Multi-criteria calibration of a conceptual runoff model using a genetic algorithm, Hydrol. Earth Syst. Sci., 4, 215–224, doi:10.5194/hess-4-215-2000, 2000.

Seibert, J.: Reliability of model predictions outside calibration conditions, Hydrol. Res., 34, 477–492, 2003.

Seibert, J. and Vis, M. J. P.: Teaching hydrological modeling with a user-friendly catchment-runoff-model software package, Hydrol. Earth Syst. Sci., 16, 3315–3325, doi:10.5194/hess-16-3315-2012, 2012.

Semmens, K. A. and Ramage, J. M.: Recent changes in spring snowmelt timing in the Yukon River basin detected by passive microwave satellite data, The Cryosphere, 7, 905–916, doi:10.5194/tc-7-905-2013, 2013.

Thirel, G., Salamon, P., Burek, P., and Kalas, M.: Assimilation of MODIS snow cover area data in a distributed hydrological model using the particle filter, Remote Sens., 5, 5825–5850, 2013.

Vehviläinen, B.: Snow cover models in operational watershed forecasting, National Board of Waters and the Environment Helsinki, Finland, 112 pp., 1992.

Viviroli, D. and Weingartner, R.: The hydrological significance of mountains: from regional to global scale, Hydrol. Earth Syst. Sci., 8, 1017–1030, doi:10.5194/hess-8-1017-2004, 2004.

Viviroli, D., Archer, D. R., Buytaert, W., Fowler, H. J., Greenwood, G. B., Hamlet, A. F., Huang, Y., Koboltschnig, G., Litaor, M. I., López-Moreno, J. I., Lorentz, S., Schädler, B., Schreier, H., Schwaiger, K., Vuille, M., and Woods, R.: Climate change and mountain water resources: overview and recommendations for

4

Evaluating uncertainties in modelling the snow hydrology of the Fraser River Basin, British Columbia, Canada

Siraj Ul Islam and Stephen J. Déry

Environmental Science and Engineering Program, University of Northern British Columbia, 3333 University Way, Prince George, BC, V2N 4Z9, Canada

Correspondence to: Stephen J. Déry (sdery@unbc.ca)

Abstract. This study evaluates predictive uncertainties in the snow hydrology of the Fraser River Basin (FRB) of British Columbia (BC), Canada, using the Variable Infiltration Capacity (VIC) model forced with several high-resolution gridded climate datasets. These datasets include the Canadian Precipitation Analysis and the thin-plate smoothing splines (ANUSPLIN), North American Regional Reanalysis (NARR), University of Washington (UW) and Pacific Climate Impacts Consortium (PCIC) gridded products. Uncertainties are evaluated at different stages of the VIC implementation, starting with the driving datasets, optimization of model parameters, and model calibration during cool and warm phases of the Pacific Decadal Oscillation (PDO).

The inter-comparison of the forcing datasets (precipitation and air temperature) and their VIC simulations (snow water equivalent – SWE – and runoff) reveals widespread differences over the FRB, especially in mountainous regions. The ANUSPLIN precipitation shows a considerable dry bias in the Rocky Mountains, whereas the NARR winter air temperature is 2 °C warmer than the other datasets over most of the FRB. In the VIC simulations, the elevation-dependent changes in the maximum SWE (maxSWE) are more prominent at higher elevations of the Rocky Mountains, where the PCIC-VIC simulation accumulates too much SWE and ANUSPLIN-VIC yields an underestimation. Additionally, at each elevation range, the day of maxSWE varies from 10 to 20 days between the VIC simulations. The snow melting season begins early in the NARR-VIC simulation, whereas the PCIC-VIC simulation delays the melting, indicating seasonal uncertainty in SWE simulations. When compared with the observed runoff for the Fraser River main stem at Hope, BC, the ANUSPLIN-VIC simulation shows considerable under-

estimation of runoff throughout the water year owing to reduced precipitation in the ANUSPLIN forcing dataset. The NARR-VIC simulation yields more winter and spring runoff and earlier decline of flows in summer due to a nearly 15-day earlier onset of the FRB springtime snowmelt.

Analysis of the parametric uncertainty in the VIC calibration process shows that the choice of the initial parameter range plays a crucial role in defining the model hydrological response for the FRB. Furthermore, the VIC calibration process is biased toward cool and warm phases of the PDO and the choice of proper calibration and validation time periods is important for the experimental setup. Overall the VIC hydrological response is prominently influenced by the uncertainties involved in the forcing datasets rather than those in its parameter optimization and experimental setups.

1 Introduction

While advances in computational power and ongoing developments in hydrological modelling have increased the reliability of hydrologic simulations, the issue of adequately addressing the associated uncertainty remains challenging (Liu and Gupta, 2007). There is a growing need for proper estimation of uncertainties associated with hydrological models and the observations required to drive and evaluate their outputs. Hydrological simulations of snow processes and related hydrology depend critically on the input climate forcing datasets, particularly the precipitation and air temperature (Reed et al., 2004; Mote et al., 2005; Tobin et al., 2011). Both of these input forcings regulate the quantity and phase of modelled precipitation and affect the response of simulated snow accumulation and runoff. The model results therefore rely heavily on the quality of these forcings as the uncertainty (measurement errors, etc.) in such data will propagate through all hydrological processes during simulations (Wagener and Gupta, 2005; Anderson et al., 2007; Tapiador et al., 2012). Studies such as Essou et al. (2016a) compared hydrological simulations of different observed datasets over the continental United States (US). They reported that there are significant differences between the datasets, although all the datasets were essentially interpolated from almost the same climate databases. Furthermore, Essou et al. (2016b) compared the hydrological response of three reanalysis datasets over the US and found precipitation biases in all reanalyses, especially in summer and winter in the southeastern US. The uncertainties in hydrological simulations also arise from the model parameters, model structure and in the objective function and the calibration variable that are used for model calibration. Hence the reliability of input forcings along with the capability of the hydrological model and the experimental setup ultimately determine the fate of hydrological variables essential for water resource management.

Several observed gridded climate datasets of precipitation and air temperature (Mesinger et al., 2006; Hopkinson et al., 2011), based on available observational data, post-processing techniques and, in some cases, climate modelling, are currently available over the Canadian landmass to facilitate climate and hydrological simulations. These datasets provide long-term gridded precipitation and air temperature records on hourly and daily bases, making them especially useful for hydrological simulations, particularly over areas where in situ station densities are low. However, these datasets, being spatially interpolated or assimilated to grid cells, rely mainly on the spatial density of the observational network, which is often quite low in mountainous regions (Rinke et al., 2004). Observational data incorporated into gridded datasets may also contain measurement errors and missing records that translate into the data interpolation and contribute to the overall uncertainty in gridded data products. Such uncertainties are assessed in many studies focusing on the forcing data (Horton et al., 2006; Graham et al., 2007; Kay et al., 2009; Eum et al., 2014).

The quality of hydrological modelling depends on how well a model simulates the regional detail and topographic characteristics of the region, especially in mountainous regions. However, most mountainous regions exhibit higher errors in gridded datasets because they are usually based on an uneven number of stations that are mostly located at lower elevations (Eum et al., 2012). This is true for most large basins in western Canada that exhibit highly variable elevation ranges and strong climatological heterogeneity. One such large basin is British Columbia's (BC's) Fraser River Basin (FRB), which is vital for Canada's environment, economy and cultural identity. Its mountainous snowpack serves as a natural reservoir for cold-season precipitation, providing snowmelt driven flows in summer. Evaluating uncertainties

in modelling the FRB's hydrology is crucial for informed decision-making and water resource management. This includes the communication of the uncertainties, propagated into the model predictions, in an appropriate manner to decision makers or stakeholders, thereby allowing confidence in the model results.

Although the currently available gridded datasets (reanalysis and interpolated) over the FRB are derived from observational stations using various interpolation and assimilation techniques, they may still have systematic biases because of their grid resolution, the density of the surface station network used for data assimilation, and the topographic characteristics of the FRB. In the FRB, 23 % of the basin exceeds 1500 m in elevation, whereas roughly 5 % of the in situ meteorological stations surpasses this elevation (Shrestha et al., 2012). Such mismatch between station densities at different elevations makes the precipitation interpolation at higher elevations excessively influenced by the lower elevation stations (Stahl et al., 2006; Rodenhuis et al., 2009; Neilsen et al., 2010). Therefore, despite extensive implementation of hydrologic modelling with single observed forcings (e.g. Shrestha et al., 2012; Kang et al., 2014, 2016), evaluation of the uncertainties in forcing datasets remains a critical and challenging issue for the FRB. As such, the first step is to evaluate available observation-based forcing datasets for their suitability to be used in hydrological modelling over the FRB.

In Canada, numerous studies have assessed the performance of hydrologic simulations driven by only one particular driving dataset (Pietroniro et al., 2006; Choi et al., 2009; Bennett et al., 2012; Kang et al., 2014). Sabarly et al. (2016) used four reanalysis datasets to assess the terrestrial branch of the water cycle over Quebec with satisfactory results over 1979–2008. Eum et al. (2014) recently compared hydrological simulations driven by several high-resolution gridded climate datasets over western Canada's Athabasca watershed and found significant differences across the simulations. While BC's snowpacks and hydrology are well studied in the literature (Danard and Murty, 1994; Choi et al., 2010; Thorne and Woo, 2011; Déry et al., 2012; Shrestha et al., 2012; Kang et al., 2014, 2016; Islam et al., 2017; Trubilowicz et al., 2016), detailed inter-comparisons of available observational forcing in terms of their hydrological response are not thoroughly analysed, particularly over the FRB's complex topography. In this study, we therefore investigate the simulated hydrological response of uncertainties associated with air temperature and precipitation forcing on the FRB's mountainous snowpack and runoff. To achieve this, four forcing datasets, namely the Canadian Precipitation Analysis and the thin-plate smoothing splines (ANUSPLIN hereafter; Hopkinson et al., 2011), North American Regional Reanalysis (NARR hereafter; Mesinger et al., 2006), University of Washington (UW hereafter; Shi et al., 2013) and Pacific Climate Impacts Consortium (PCIC hereafter; Shrestha et al., 2012) gridded products are applied to the FRB. These datasets are explored across three different regions and multiple elevation ranges. The PCIC and UW datasets are used by Shrestha et al. (2012) and Kang et al. (2014, 2016), respectively, to drive the VIC hydrological model over the FRB, whereas the NARR and ANUSPLIN datasets are not yet evaluated over this region. However, the NARR dataset is used in studies focusing on other regions of Canada (Woo and Thorne, 2006; Choi et al, 2009; Ainslie and Jackson, 2010; Eum et al., 2014; Trubilowicz et al., 2016). To our knowledge, this is the first comprehensive study that collectively examines the spatial and elevation-dependent hydrological response of these datasets for the FRB.

Along with forcing datasets, many studies have focused their attention either on model structure (Wilby and Harris, 2006; Jiang et al., 2007; Poulin et al., 2011; Velazquez et al., 2013) or on calibration parameters (Teutschbein et al., 2011; Bennett et al., 2012). Arsenault and Brissette (2014) estimated the uncertainty due to parameter set selection using a hydrological model over several basins in Quebec. They showed that parameter set selection can play an important role in model implementation and predicted flows. For parameter uncertainty, a hydrological model can have many equivalent local optima within a realistic parameter space (Poulin et al., 2011). Therefore, several different parameter sets may be available for the same "optimal" measure of efficiency during the optimization process (i.e. parameter non-uniqueness; Beven, 2006). Here we evaluate the parameter uncertainties involved in the model calibration process, i.e. calibration optimizer sensitivity to parameter initial limits. Moreover we focus on another unique aspect of modelling uncertainty related to the selection of time periods for model calibration and validation under changing climatic conditions on decadal timescales. Studies such as Klemeš (1986) and Seiller et al. (2012) highlighted the issue of calibration and validation of hydrological modelling under different climatological conditions. In this study, we estimate the hydrological model sensitivity to different climatological conditions by focusing on the FRB's air temperature and precipitation teleconnections with cool and warm phases of the Pacific Decadal Oscillation (PDO).

Overall, the main goals of this study are (i) to compare and identify the most reliable available gridded forcing datasets for hydrological simulations over the FRB; (ii) to evaluate hydrological modelling responses of different driving datasets over a range of FRB elevations; (iii) to assess the uncertainty involved in the model calibration process by focusing on the optimizer used for parameter optimization; and (iv) to evaluate the calibration process under changing climatic conditions. To achieve these four objectives, the macroscale Variable Infiltration Capacity (VIC) hydrological model (Liang et al., 1994, 1996) is used as the simulation tool. The VIC model conserves surface water and energy balances for large-scale watersheds such as the FRB (Cherkauer et al., 2003). It has been successfully implemented, calibrated

and evaluated over the FRB (Shrestha et al., 2012; Kang et al., 2014; Islam et al., 2017).

The remainder of this paper is structured as follows. Section 2 discusses the FRB, the driving datasets, the VIC model and the experimental setup. Section 3 describes the forcings inter-comparison, hydrological simulations, parameter sensitivity and uncertainty related to the PDO. Section 4 summarizes and concludes this study.

2 Study area, model and methodology

2.1 Fraser River Basin (FRB)

The FRB is one of the largest basins of western North America, spanning $240\,000\,\text{km}^2$ of diverse landscapes with elevations varying from sea level to 3954 m a.s.l. (above sea level) at Mt. Robson, its tallest peak (Benke and Cushing, 2005). It covers the mountainous terrain of the Coast and Rocky Mountains along with dry central plateaus (Fig. 1). The FRB's headwaters are in the Rocky Mountains, with its major tributaries being the Stuart, Nechako, Quesnel, Chilcotin, Thompson, and Harrison rivers. The Fraser River runs 1400 km through the whole basin before reaching Hope, BC, where it veers westward to drain into the Salish Sea and the Strait of Georgia at Vancouver, BC (Benke and Cushing, 2005; Schnorbus et al., 2010).

In winter, considerable amounts of snow usually accumulate at higher elevations, except in coastal areas. In late spring and early summer, snowmelt from higher elevations induces peak flows in the main stem of the Fraser River and its many tributaries (Moore and Wondzell, 2005), which rapidly decline in late summer following the depletion of snowmelt. Owing to its complex mountainous ranges, the FRB's hydrologic response varies considerably across the basin, differentiating it into snow-dominant, hybrid (rain and snow), or rain-dominant regimes (Wade et al., 2001). Glaciers cover only 1.5 % of the FRB (Shrestha et al., 2012) and provide only a modest contribution to streamflow, primarily in late summer (August/early September).

2.2 Datasets

Along with recent developments in hydrological models, several observation-based gridded datasets are now available to drive the models such as ANUSPLIN, NARR, UW and PCIC. These meteorological forcing datasets are developed using high-resolution, state-of-the-art data interpolation and (for NARR only) assimilation techniques. This is to improve the quality of forcing data to analyse a model's hydrological response over any particular basin.

The ANUSPLIN dataset, developed by Natural Resources Canada (NRCan), contains gridded data of daily maximum and minimum air temperature (°C), and total daily precipitation (mm) for the Canadian landmass south of 60° N at $\sim 10\,\text{km}$ resolution (NRCan, 2014). This Canadian dataset

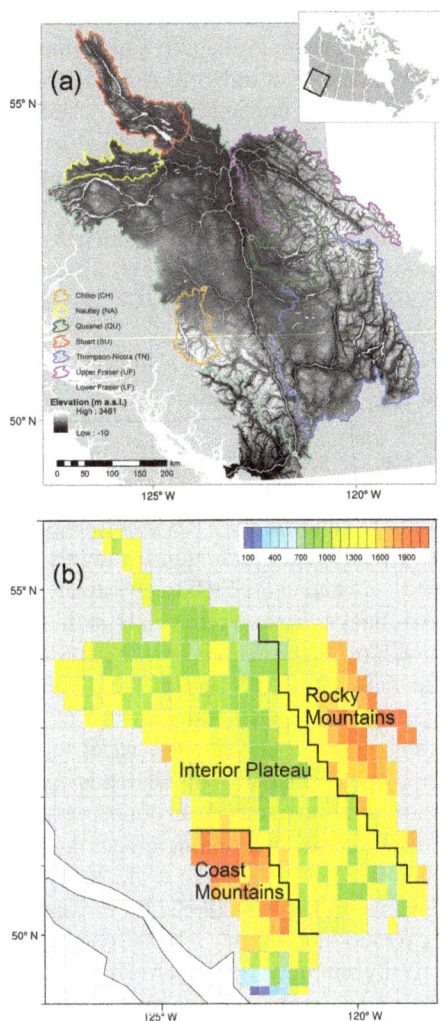

Figure 1. (a) High-resolution digital elevation map of the FRB with identification of major sub-basins, including the Fraser River main stem at Hope, BC. **(b)** FRB mean elevation (m) per VIC model grid cell. The location of the hydrometric gauge on the Fraser River's main stem at Hope is highlighted with a light green circle in panel (a).

uses a trivariate thin-plate smoothing spline technique referred to as ANUSPLIN (Hutchinson et al., 2009) with recent modifications (Hopkinson et al., 2011). Eum et al. (2014) used the ANUSPLIN dataset for hydrological modelling over Alberta's Athabasca watershed and reported underestimations in simulated runoff, owing to a dry bias in ANUSPLIN precipitation.

NARR was developed at 32 km spatial and 3-hourly temporal resolution to improve the National Centers for Environmental Prediction (NCEP)/National Center for Atmospheric Research (NCAR) global reanalysis data by employing the Eta Data Assimilation system for the North American domain for the period from 1979 to the current year. The interannual variability of the NARR seasonal precipitation

and accuracy of its temperature and winds are found to be superior to earlier versions of the NCEP/NCAR reanalysis datasets (Mesinger et al., 2006; Nigam and Ruiz-Barradas, 2006). Choi et al. (2009) investigated the applicability of air temperature and precipitation data from NARR for hydrological modelling of selected watersheds in northern Manitoba. They found that NARR air temperature and precipitation data are in much better agreement with observations than the NCEP–NCAR Global Reanalysis-1 dataset (Kalnay et al., 1996; Kistler et al., 2001). Woo and Thorne (2006) used air temperature and precipitation data from two global reanalysis datasets and from NARR as input to a hydrological model for the Liard River Basin in western subarctic Canada and reported significant improvement in its hydrological simulations. NARR output has also been used in regional water budget calculations (Luo et al., 2007; Ruane, 2010; Sheffield et al., 2012). Choi et al. (2009) and Keshta and Elshorbagy (2011) reported that NARR output is suitable for hydrologic modelling, especially when other observations are unavailable. However, they focused on the Canadian Prairies, where the topography is not complex.

The UW dataset of daily precipitation, maximum and minimum air temperature, and average wind speed is based on the extended gridded UW dataset (Shi et al., 2013; Adam et al., 2006; Adam and Lettenmaier, 2008). Monthly precipitation originates from the University of Delaware observed land surface precipitation product (Matsuura and Willmott, 2009), which was converted to daily data using the high temporal precipitation dataset from Sheffield et al. (2006). To improve the precipitation estimates, the monthly data were adjusted to account for gauge undercatch by using the methods outlined by Adam and Lettenmaier (2008). Such adjustment is important since gauge-based precipitation measurements may underestimate solid precipitation in winter by 10–50 % (Adam and Lettenmaier, 2003). Daily wind speeds are extracted from the NCEP/NCAR reanalysis datasets (Kalnay et al., 1996).

The PCIC dataset of precipitation, maximum and minimum temperature, and wind speed was derived primarily from Environment and Climate Change Canada (ECCC) climate station observations, with additional inputs from the United States Co-operative Station Network, the BC Ministry of Forests, Lands and Natural Resource Operations, the BC Ministry of Environment's automated snow pillow network, and BC Hydro's climate network (Schnorbus et al., 2011; Shrestha et al., 2012). These data are available at ∼ 6 km resolution and were corrected for point precipitation biases and elevation effects (Schnorbus et al., 2011).

The ANUSPLIN, NARR, UW and PCIC datasets are available at 10, 32, 25 and 6 km spatial resolution, respectively, and at a daily timescale. To facilitate comparison, the ANUSPLIN, NARR and PCIC datasets were regridded to 25 km resolution using bilinear interpolation to match the scale of the current VIC implementation. The NARR (32 km) dataset was interpolated from coarse-resolution curvilinear grids to slightly higher (25 km) resolution rectilinear grids. On the other hand, both the PCIC (6 km) and ANUS-PLIN (10 km) datasets were interpolated to a coarser resolution (25 km). The elevation correction, which is important when interpolating from coarser to higher spatial resolutions (Dodson and Marks, 1997), was not used to correct the orographic effects for the NARR dataset. Interpolating the NARR dataset from a 32 km to a 25 km spatial resolution induces negligible elevation-dependent uncertainties as elevation changes remain below ±20 % in the FRB, with most of the grid cells having nearly no difference in orography. Thus the relationship of atmospheric variables such as air temperature with elevation remains nearly identical at both resolutions.

Daily wind speeds, a required VIC input variable, are not available for the ANUSPLIN dataset. We therefore used the PCIC-based wind speeds in the ANUSPLIN driven VIC simulations. The PCIC wind speeds are sourced from the Environment and Climate Change Canada station product (Schnorbus et al., 2011).

To calibrate and validate the VIC model simulated flows, we used daily streamflow data from ECCC's Hydrometric Database (HYDAT; Water Survey of Canada, 2014). These data were extracted and compiled into a comprehensive streamflow dataset for the FRB spanning 1911–2010 (Déry et al., 2012).

In addition, we compared the simulated SWE with observations from the BC River Forecast Centre's network of snow pillow sites (BC Ministry of Forests, Lands and Natural Resource Operations, 2014). The snow pillow stations record the mass of the accumulated snowpack (SWE) on a daily basis. Based on the availability of data, we used SWE observations from four sites located at Yellowhead (ID: 1A01P) and McBride (ID: 1A02P) in the upper Fraser and at Mission Ridge (ID: 1C18P) and Boss Mountain Mine (ID: 1C20P) in the middle Fraser. Due to data availability, we used the 1996–2006 time period for the Yellowhead, Mission Ridge and Boss Mountain Mine snow pillows and 1980–1986 for the McBride location. Detailed information about these sites is available in Kang et al. (2014) and Déry et al. (2014).

2.3 Variable Infiltration Capacity (VIC) model

The VIC model resolves energy and water balances and therefore requires a large number of parameters, including soil, vegetation, elevation, and daily meteorological forcings, at each grid cell. To evaluate hydrological responses over complex terrain, the model simulates the subgrid variability in topography and precipitation by dividing each grid cell into a number of snow elevation bands (Nijssen et al., 2001a). The model utilizes a mosaic-type representation by partitioning elevation bands into a number of topographic tiles that are based on high-resolution spatial elevations and fractional area. The snow model embedded in the VIC model is then applied to each elevation tile separately (Gao et al., 2009).

The VIC model is widely used in many hydrological applications including water availability estimation and climate change impact assessment in North America (Maurer et al., 2002; Christensen and Lettenmaier, 2007; Adam et al., 2009; Cuo et al., 2009; Elsner et al., 2010; Gao et al., 2010; Wen et al., 2011; Oubeidillah et al., 2014) and around the world (Nijssen et al., 2001a, b; Haddeland et al., 2007; Zhou et al., 2016). It is also commonly used to simulate hydrologic responses in snowmelt-dominated basins (Christensen and Lettenmaier, 2007; Hidalgo et al., 2009; Cherkauer and Sinha, 2010; Schnorbus et al., 2011).

2.3.1 The VIC implementation

The VIC model, as set up by Kang et al. (2014) and Islam et al. (2017) for the FRB, is employed for evaluating the model's ability to simulate the FRB's hydrological response when driven by different observational forcings. The model was previously applied to the FRB to investigate its observed and projected changes in snowpacks and runoff. In this study, we performed model integrations over the entire FRB using grid cells spanning 48–55° N and 119–131° W. The model is configured at 0.25° spatial resolution using a daily time step, three soil layer depths and 10 vertical snow elevation bands. Once an individual VIC simulation is completed, the runoff for the basin is extracted at an outlet point of the given sub-basin, using an external routing model that simulated a channel network (adapted from Wu et al., 2011) with several nodes (Lohmann et al., 1996, 1998a, b). Streamflow is converted to areal runoff by dividing it by the corresponding sub-basin area. Daily runoff at the outlet cell is integrated over time to obtain total water year runoff for a selected basin. Other than the calibration parameters, the soil and vegetation parameters, leaf area index (LAI) and albedo data are kept identical as per the Kang et al. (2014) VIC model implementation to the FRB.

2.3.2 Calibration

To explore the feasible parameter space, we used the University of Arizona multi-objective complex evolution (MOCOM-UA) optimizer for the VIC calibration process (Yapo et al., 1998; Shi et al., 2008). MOCOM-UA searches a set of VIC input parameters using the population method to maximize the Nash–Sutcliffe efficiency (NSE) coefficient (Nash and Sutcliffe, 1970) between observed and simulated runoff. Six soil parameters are used in the optimization process, i.e. b_infilt (a parameter of the variable infiltration curve), Dsmax (the maximum velocity of base flow for each grid cell), Ws (the fraction of maximum soil moisture where nonlinear base flow occurs), D2 and D3 (the depths of the second and third soil layers), and Ds (the fraction of the Dsmax parameter at which nonlinear baseflow occurs). These calibration parameters were selected based on the manual calibration experience from previous studies by

Table 1. Description of VIC inter-comparison experiments performed using observational forcings.

VIC model driving data	Data description	VIC configuration
ANUSPLIN	The Canadian Precipitation Analysis and the thin-plate smoothing splines (Hopkinson et al., 2011)	Domain = 48–55° N and 119–131° W Resolution = 25 km × 25 km Time step: daily
NARR	North American Regional Reanalysis (Mesinger et al., 2006)	Soil layers: 3 Vertical elevation bands: 10
PCIC	Pacific Climate Impacts Consortium (Shrestha et al., 2012)	Time period: 1979–1990 (calibration), 1991–2006 (validation)
UW	University of Washington (Shi et al., 2013)	Ensemble runs: 5*

* Ensemble validation runs are initiated five times with different initial conditions.

Nijssen et al. (1997), Su et al. (2005), Shi et al. (2008), Kang et al. (2014, 2016) and Islam et al. (2017). VIC is a physically based hydrologic model that has many (about 20, depending on how the term "parameter" is defined) parameters that must be specified. However, the usual implementation approach involves the calibration of only these six soil parameters. Such parameters have the largest effects on the hydrograph shape and are the most sensitive parameters in the water balance components (Nijssen et al., 1997; Su et al., 2005). These parameters must be estimated from observations, via a trial and error procedure that leads to an acceptable match of simulated discharge with observations.

For the snow calibration, the values of thresholds for maximum (at which snow can fall) and minimum (at which rain can fall) air temperature were fixed as 0.5 and −0.5 °C, respectively. These values were adjusted based on the region's climatology and were kept constant for all simulations in the global control file. Parameters related to the snow albedo were adjusted using the traditional VIC algorithm based on the US Army Corps of Engineers empirical snow albedo decay curves for transitions from snow accumulation to ablation.

Final values of these six calibrated parameters were estimated for each forcing dataset by a number of simulation iterations minimizing the difference between the simulated and observed monthly flow.

While the MOCOM-UA automated optimization process utilizes monthly streamflow during calibration, we evaluated the overall model performance on daily timescales using NSE and correlation performance metrics.

The VIC model calibration is applied to the Fraser River's main stem at Hope, BC, and the FRB's major sub-basins, namely the upper Fraser at Shelley (UF), Stuart (SU), Nautley (NA), Quesnel (QU), Chilko (CH) and Thompson-Nicola (TN) basins (Fig. 1a and Table S1 in the Supplement). These sub-basins contribute 75 % of the annual observed Fraser River discharge at Hope, BC, with the largest contributions from the TN, UF and QU sub-basins (Déry et al., 2012).

2.3.3 Experiments

A series of different VIC experiments was performed to (i) compare the VIC model's response when driven by different forcings, (ii) evaluate the uncertainties related to the VIC optimizer, and (iii) investigate the effect of PDO teleconnections on the VIC calibration and validation time periods. For objective (i), we used all four datasets to run VIC simulations to facilitate detailed comparison of different datasets and their hydrological response. In objectives (ii) and (iii), rather than the inter-comparison of datasets, our goal is to evaluate the uncertainties in the model implementation, particularly in its calibration process. We therefore only used the UW dataset to force the VIC model as this dataset along with our VIC model implementation is examined extensively over the FRB in Kang et al. (2014, 2016). The experiments are categorized as follows.

Inter-comparison runs: the VIC model was driven by each forcing dataset for 28 years (1979 to 2006) with 1979–1990 as the calibration period and 1991–2006 as the validation period using the MOCOM-UA optimizer (Table 1). The VIC simulations driven by ANUSPLIN, UW and PCIC forcings are initiated 5 years prior to the year 1979 to allow model spin-up time. Since NARR is not available until 1979, its VIC simulations were recursively looped for 5 years using the year 1979 as the forcing data. After calibration, the model validation runs were initialized with five different state files to produce five ensemble members. The ANUSPLIN, NARR, UW and PCIC driven ensemble mean VIC simulations are referred to as ANUSPLIN-VIC, NARR-VIC, UW-VIC and PCIC-VIC, respectively. These ensemble simulations were run for the whole FRB and its UF, SU, NA, QU, CH and TN sub-basins.

Optimizer uncertainty runs: here we only used the UW forcing data for VIC model simulations to investigate the uncertainties in the model calibration process for the 1979–1990 time period. Our primary goal is to evaluate optimizer sensitivity to a unique set of parameter limits. We want to see how the MOCOM optimizer results in different optimized parameters and change the overall simulated hydrograph in the calibration process. We performed the optimization of six soil parameters, i.e. b_infilt, Dsmax, Ws, D2, D3 and Ds, in five experimental setups using different initial ranges of parameter limits. The VIC calibration experiments (OPT1, OPT3, OPT4 and OPT5) were run using four narrow ranges selected from the maximum limits of calibration parameters. The same experiment is then run with maximum limits of the calibration parameters (OPT2). Calibration parameters, their initial ranges and final optimized values for all the experiments are given in Table 3. The OPT1, OPT2, OPT3, OPT4 and OPT5 simulations were run over the whole FRB only.

PDO uncertainty runs: we used the UW dataset to drive long-term (1950–2006) VIC simulations. This is to capture the decadal variability of cool and warm phases of the PDO. Five different experiments, namely PDO1, PDO2, PDO3, PDO4 and PDO5, were performed with calibration periods of 1981–1990, 1956–1965, 1967–1976, 1977–1987 and 1991–2001 and with corresponding validation periods of 1991–2001, 1966–1976, 1977–1987, 1967–1976 and 1981–1990, respectively (Table 4). Each time period was selected to capture cool or warm PDO phases, i.e. its cool (1956–1965 and 1967–1976) and warm (1981–1990, 1991–2001 and 1977–1987) phases. For each calibration experiment in one particular phase of the PDO, the MOCOM-UA was used to optimize calibration parameters. The NSE was calculated for the calibration and validation periods using the daily observed streamflow data for the Fraser River at Hope. All PDO simulations were run over the whole FRB only.

2.4 Analysis strategy

The analyses were performed for three FRB hydro-climatic regimes: the Interior Plateau, the Rocky Mountains and the Coast Mountains (Moore, 1991). These three regions were chosen given their distinct physiography and hydro-climatic conditions. The grid-cell partitioning of these three regions and their elevations are shown in Fig. 1b. Results in this study mainly focused on the Fraser River main stem at Hope, BC, since it covers 94 % of the basin's drainage area and has a continuous streamflow record over the study periods. However, the inter-comparison runs were also compared over the FRB's major sub-basins. The total runoff was calculated using the sum of baseflow and runoff. Seasonal variations were assessed by averaging December–January–February (DJF), March–April–May (MAM), June–July–August (JJA) and September–October–November (SON) months for winter, spring, summer and autumn, respectively.

In the SWE analysis, the snowmelt was calculated by taking the difference between maximum and minimum SWE over the water year (1 October to 30 September of the following calendar year). The corresponding day of the water year having maximum SWE (maxSWE) is referred to as maxSWE-day.

Although glacier dynamics are not included in the VIC model physics, the model produces a perennial snowpack in several grid cells in its output. We compared those cells to baseline thematic mapping (BTM) and found that the glaciating cells match the location of observed glaciers. We therefore masked those grid cells in the SWE analysis considering that the effects of glaciers may not change our results significantly due to the ∼ 25 km model grid cell resolution (625 km^2 area per grid cell) used in this study.

The Mann–Kendall test (Mann, 1945; Kendall, 1970) was used to estimate monotonic trends in the input forcing data and the simulated hydrological variables. This non-parametric trend test has been used in several other studies to detect changing hydrological regimes (Lettenmaier et al., 1994; Ziegler et al., 2003; Déry et al., 2005, 2016; Kang et al., 2014). Trends were considered to be statistically significant when $p < 0.05$ with a two-tailed test.

3 Results and discussion

We first examine the ANUSPLIN, NARR, UW and PCIC datasets to investigate how substantial the differences in precipitation and air temperature are at several temporal and spatial scales across the FRB and its sub-regions. The VIC simulations, driven by these forcing datasets, are then discussed to evaluate uncertainties in simulated SWE and runoff. This is followed by the discussion of uncertainty in the VIC calibration process.

3.1 Forcings dataset inter-comparison

The daily mean air temperature of ANUSPLIN, NARR, UW and PCIC datasets remains below 0 °C from November to March and rises above 0 °C in early spring over all three FRB sub-regions (Fig. 2). While the inter-dataset seasonal variability of air temperature is quite similar, the winter in NARR is ∼ 2 °C warmer compared to the remaining datasets. The grid-scale seasonal differences (PCIC minus ANUSPLIN, NARR and UW) of mean air temperature spatially quantify the inter-dataset disagreements (Fig. S1 in the Supplement). While the PCIC–ANUSPLIN and PCIC–UW differences are within ±1 °C, the PCIC–NARR difference exceeds 2 °C over most of the FRB in DJF and SON, revealing NARR air temperatures to be quite warmer than in the PCIC dataset.

The magnitudes of daily mean precipitation vary markedly amongst datasets. Winter precipitation, which begins in November and persists until April, shows greater inter-dataset differences, particularly over the Rocky and Coast

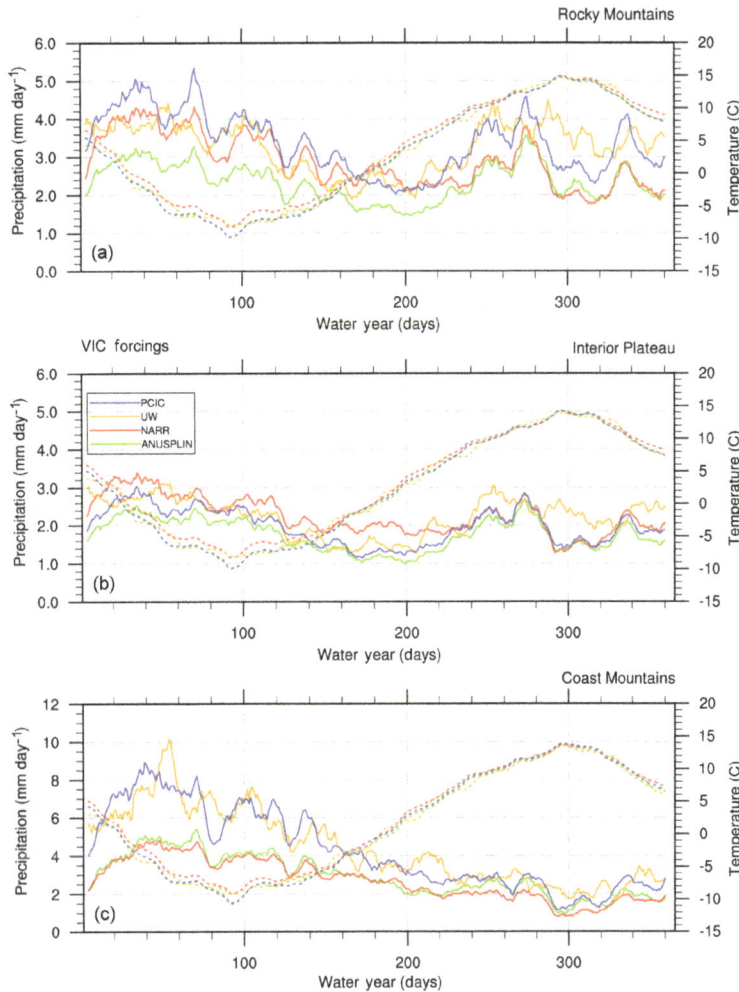

Figure 2. Area-averaged time series of mean daily air temperature (dotted lines) and daily precipitation (solid lines) over the **(a)** Rocky Mountains, **(b)** Interior Plateau, and **(c)** Coast Mountains for the ANUSPLIN, NARR, UW and PCIC forcing datasets, water years 1979–2006. Water year starts on 1 October and ends on 30 September of the following calendar year.

Mountains. Compared to the PCIC and UW datasets, the ANUSPLIN precipitation is underestimated in all three regions, with nearly 2.0 to 5.0 mm day^{-1} differences in the Rocky and Coast Mountains, respectively. This underestimation is more evident in the PCIC-ANUSPLIN spatial difference, revealing up to 5 mm day^{-1} difference over the mountainous regions (Fig. S2). The precipitation differences in the Interior Plateau approach zero for all datasets. The maximum intraseasonal variability arises in the Coast Mountains, ranging from 10.0 mm day^{-1} of precipitation in winter to nearly zero in summer. The range of inter-dataset spread for peak precipitation varies from 5.0 to 10.0 mm day^{-1} during winter for the Coast Mountains. Precipitation in the Coast Mountains is more variable due to its proximity to the Pacific Ocean, where the interaction between steep elevations and storm track positions is quite complex. In the Coast Mountains, the NARR precipitation is underestimated and is comparable to ANUSPLIN.

The underestimation of the ANUSPLIN mountainous precipitation is probably due to the thin plate smoothing spline surface fitting method used in its preparation. For NARR, air temperature and precipitation uncertainties may have been induced by the climate model used to assimilate and produce the reanalysis product.

3.2 Hydrological simulations

The ANUSPLIN-VIC, NARR-VIC, UW-VIC and PCIC-VIC simulation performance was evaluated using the NSE and correlation coefficients by calibrating and validating against observed daily streamflow for the Fraser River at Hope (Table 2). The NSE scores are much higher for the PCIC-VIC and UW-VIC simulations compared to the ANUSPLIN-VIC and NARR-VIC. The lower NSE score in the ANUSPLIN-VIC simulation reflects a dry precipitation bias in the ANUSPLIN dataset. As the model configuration, resolution, and

Table 2. Daily performance metrics for the VIC inter-comparison runs. Calibration (1979–1990) and validation (1991–2006) for the Fraser River main stem at Hope, BC, are evaluated using the Nash–Sutcliffe efficiency (NSE) coefficient and correlation coefficient (r, all statistically significant at $p < 0.05$).

Experiment names	1979–1990 Daily calibration		1991–2006 Daily validation	
	NSE	r	NSE	r
ANUSPLIN-VIC	0.54	0.91	0.55	0.94
NARR-VIC	0.67	0.85	0.81	0.90
PCIC-VIC	0.90	0.96	0.90	0.95
UW-VIC	0.82	0.94	0.80	0.92

soil data were identical for all VIC simulations, different NSE values reveal uncertainty associated only with each observational forcing dataset. Despite the low NSE score of the ANUSPLIN-VIC simulation, the correlation coefficient is significantly high. The bias in the simulated streamflow is contributing to the lower NSE coefficient, whereas the phase of seasonal flow is quite similar to the observed flow in the ANUSPLIN-VIC simulation. There may be additional sources of uncertainty due to the method used to assess simulation accuracy. For example, instead of using NSE, other model evaluation metrics such as the Kling–Gupta efficiency (KGE) coefficient (Gupta et al., 2009) may produce different levels of model accuracy.

The ANUSPLIN-VIC, NARR-VIC, UW-VIC and PCIC-VIC simulated SWE and snowmelt, areally averaged over the FRB's three sub-regions, show similar seasonal variability but considerably different magnitudes, especially over mountainous regions. Figure 3a shows these differences for the Rocky Mountains revealing the range of peak SWE from 400 mm for ANUSPLIN to > 600 mm for PCIC. The dry bias in ANUSPLIN precipitation forcing induces lower SWE magnitudes in the ANUSPLIN-VIC simulation. The lower SWE in the NARR-VIC simulation is probably due to the warmer air temperature during winter and spring (Fig. 2b). Winter temperatures being warmer in the NARR dataset may alter the phase of precipitation partitioning with more rainfall than snowfall and hence less SWE in the NARR-VIC simulation. Such differences in SWE are reflected in the associated snowmelt (Fig. 3b) where the NARR-VIC simulation shows earlier snowmelt. This is further investigated by VIC sensitivity experiments and is discussed later in the text. Grid-scale differences in simulated SWE (Fig. 4) and runoff (Fig. S3) arise most notably over the mountainous regions. In the interior FRB, the simulation differences between PCIC-VIC and ANUSPLIN-VIC mean SWE are within a 10 mm range, whereas such differences exceed 50 to 100 mm for the NARR-VIC and UW-VIC simulations.

In the FRB's mountainous regions, the VIC model can lead to inaccurate snowpack estimates if the elevation dependence

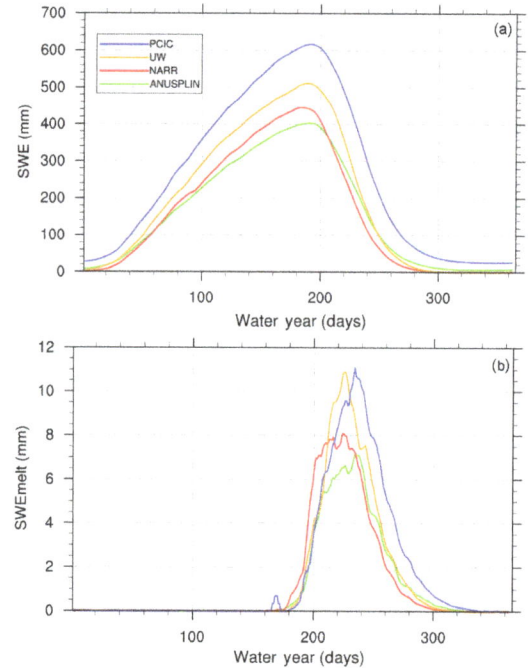

Figure 3. Area-averaged time series of daily mean **(a)** SWE and **(b)** SWE_{melt} for the ANUSPLIN-VIC, NARR-VIC, UW-VIC and PCIC-VIC simulations averaged over the Rocky Mountains, water years 1979–2006. Water year starts on 1 October and ends on 30 September of the following calendar year.

on snow accumulation and ablation is not modelled properly. As mentioned in Sect. 2.3, we used 10 elevation bands in our VIC implementation so that each band's mean elevation was used to lapse the grid-cell average air temperature and precipitation to produce more reliable estimates. We clustered the elevation distribution within 10 bands into different elevation ranges. This allowed in-depth analysis of the elevation-dependent variation of mean SWE that is of particular importance for the Rocky and Coast Mountains regions of the FRB. We examined the magnitude of maxSWE and the corresponding maxSWE-day of the water year between all simulations and elevation ranges (Fig. 5). The difference in maxSWE between all VIC simulations increases with elevation, particularly the Rocky Mountains, where higher elevations (> 1400 m) show large disagreement between simulated maxSWE (Fig. 5a). In the Interior Plateau, the NARR-VIC simulated maxSWE exceeds 300 mm, whereas all other simulations are within 200 mm. The maxSWE elevation-dependent variation is quite complex in the Coast Mountains. However, the simulation differences at elevations > 1400 m are smaller compared to the lower elevations below 1000 m. Apart from maxSWE magnitude, the maxSWE-day variation differs considerably across the VIC simulations. Generally, the maxSWE-day varies by nearly 2 months between lower and higher elevations as snow onset occurs later in autumn. While the maxSWE-day variation is quite com-

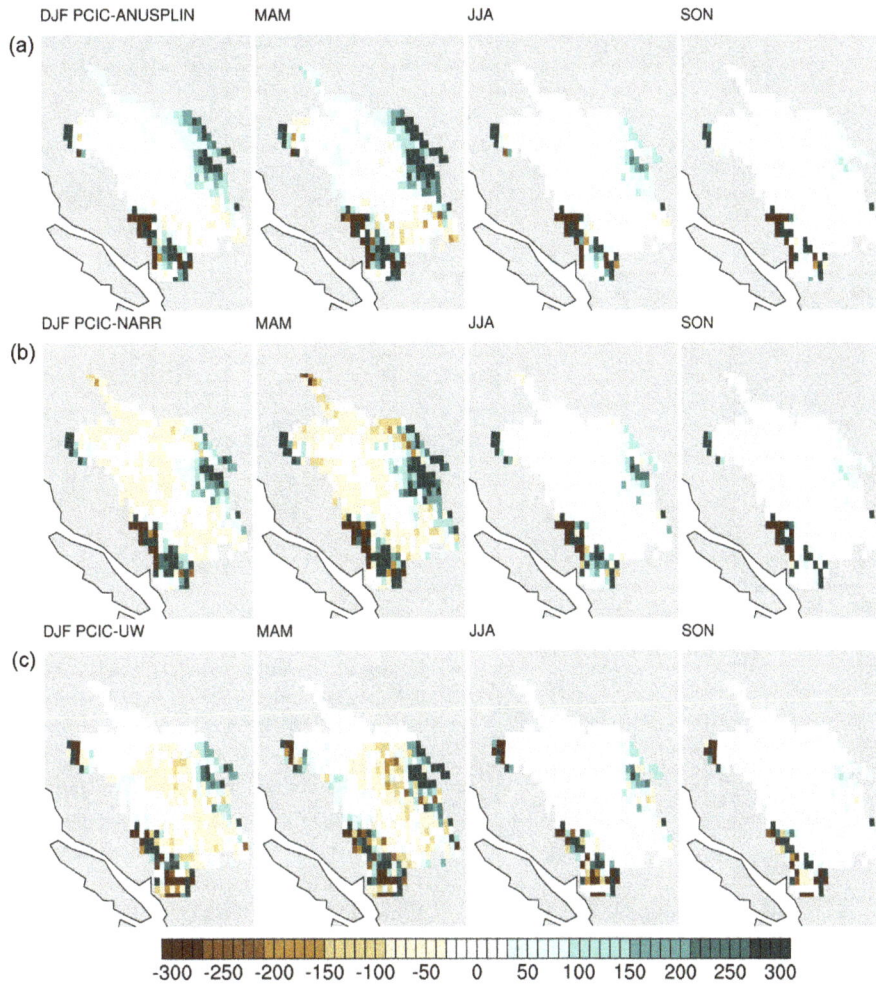

Figure 4. Spatial differences of mean seasonal SWE (mm) based on PCIC-VIC minus (row **a**) ANUSPLIN-VIC, (row **b**) NARR-VIC and (row **c**) UW simulations, water years 1979–2006. DFJ, MAM, JJA and SON correspond to winter, spring, summer and autumn, respectively.

plex within each elevation range, the NARR-VIC maxSWE-day is earliest, whereas PCIC-VIC delays the snow accumulation over the 600–2000 m elevation range in the Rocky Mountains. There are nearly 20 days of simulated variation in maxSWE-day at the Rocky Mountains' highest elevation range. Such variation highlights the uncertainties in seasonality of the VIC simulated snowpacks. For the Interior Plateau and the Coast Mountains, no consistent pattern of maxSWE-day variation exists for any particular simulation.

3.2.1 Comparison of observed vs. simulated SWE

As mentioned earlier, all gridded climate forcing datasets are based on station observations. The density of stations in the FRB's mountainous regions remains quite low and therefore induces higher uncertainties in the observational gridded products. It is important to quantify the spatial discrepancy between the simulated (0.25° grid cell) and observed (snow pillow station dataset) SWE that may lead to an uncertainty in snow estimations by models (Elder et al., 1991; Tong et al.,

2010). We used observed SWE from BC snow pillow sites and the VIC simulated SWE data over the same elevation and overlapping continuous time periods at four different locations in the upper and middle Fraser, where a high volume of SWE accumulates seasonally.

The daily time series of VIC simulated SWE (Fig. S4) follows the observed interannual variability in snow accumulation but with considerable differences across simulations. The PCIC-VIC simulation accumulates too much SWE compared to observations in the grid cell corresponding to the Yellowhead location. This overestimation is further explored for this site by expanding the time series back to 1979 (not shown), which reveals issues with PCIC precipitation data only during 1996–2004 with considerable above normal anomalies at Yellowhead. While ANUSPLIN-VIC shows lower SWE amounts, the NARR-VIC and UW-VIC simulations reproduce the observed variation quite reasonably for Yellowhead. For McBride, all simulations are more or less comparable except ANUSPLIN-VIC, showing a SWE under-

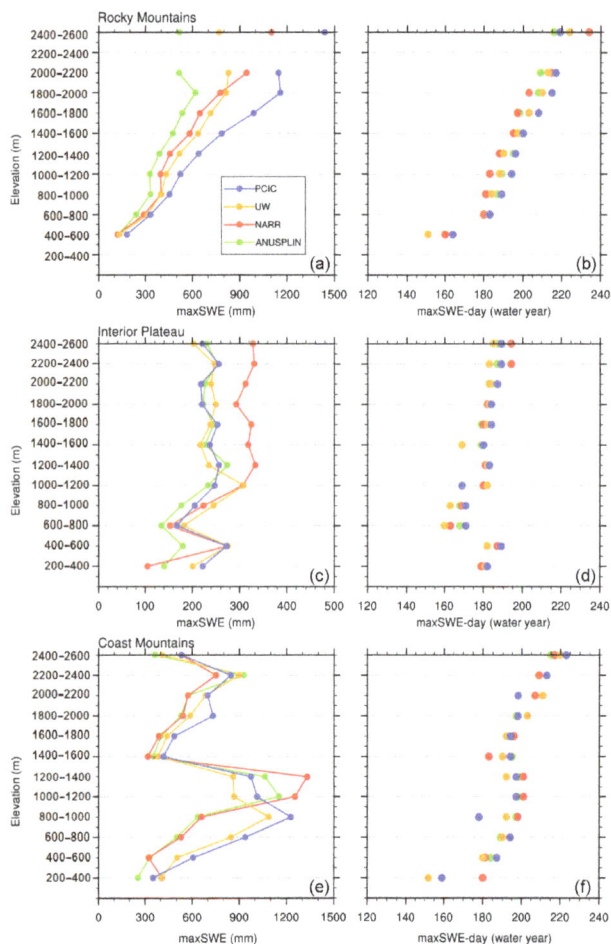

Figure 5. Variation of (**a, c, e**) maxSWE and corresponding (**b, d, e**) maxSWE-day for the ANUSPLIN-VIC, NARR-VIC, UW-VIC and PCIC-VIC simulations averaged over the (**a, b**) Rocky Mountains, (**c, d**) Interior Plateau and (**e, f**) Coast Mountains, water years 1979–2006.

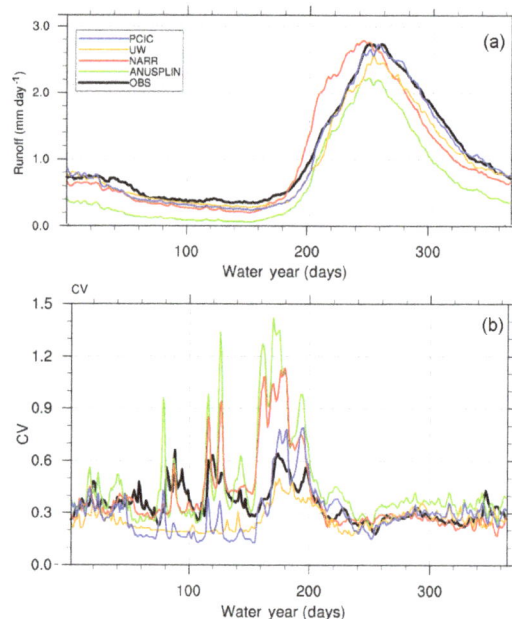

Figure 6. The simulated and observed daily (**a**) runoff and (**b**) coefficient of variation (CV) for the Fraser River at Hope averaged over water years 1979–2006. An external routing model is used to calculate runoff for the ANUSPLIN-VIC, NARR-VIC, UW-VIC and PCIC-VIC simulations. Water year starts on 1 October and ends on 30 September of the following calendar year.

estimation compared to observations. In the middle Fraser, the UW-VIC simulation is quite comparable to observations, whereas the PCIC-VIC simulation underestimates SWE at Mission Ridge. Both ANUSPLIN-VIC and NARR-VIC underestimate SWE in the middle Fraser locations. The observed SWE values in the lower Fraser locations are not well captured by VIC, perhaps owing to the region's coastal influence and strong sensitivity to air temperatures (not shown). These results highlight the importance of accurate precipitation forcings to simulate SWE. Along with this, even small perturbations in air temperature can change the phase of precipitation, which directly contributes to changes in SWE accumulation.

3.2.2 Comparison of observed vs. simulated runoff

The VIC simulated flows are routed to produce hydrographs for the Fraser River at Hope, BC (Fig. 6a). Comparison of

simulated runoff with observations shows the highly consistent model performance for PCIC-VIC and UW-VIC simulations, whereas the runoff is considerably lower for the ANUSPLIN-VIC simulation. The NARR-VIC hydrograph is comparable in magnitude with observations, but the runoff timing is considerably shifted ($\sim$15 days), yielding more winter and spring runoff and earlier decline of flows in summer. The shift in the hydrograph is probably caused by the 2 °C warmer air temperatures causing earlier snowmelt. This finding was confirmed by a VIC sensitivity experiment where the air temperature was perturbed by 2 °C while keeping the precipitation unchanged. Similar to the case of NARR-VIC results, the simulated SWE and runoff decreases with 2 °C rises in air temperatures (Fig. S5). The coefficient of variation in daily runoff for all four datasets reveals that variability in the PCIC-VIC and UW-VIC simulations is similar to observations (Fig. 6b). We further produced the hydrographs for the FRB's six major sub-basins to compare VIC simulation runs of each basin (Fig. 7). Similar to the hydrograph of the Fraser River at Hope, the ANUSPLIN-VIC runoff shows considerable disagreement with the observed hydrograph, especially in the UF, QU and TN basin, owing to the dry bias in its precipitation forcing. Moreover, NARR-VIC runoff is overestimated in the SU, NA and CH sub-basins, whereas for UF, QU and TN, the simulated runoff underestimates observed flows. Consistent with spatial differences of mean air temperature and runoff (Figs. S1 and S3), the warmer NARR

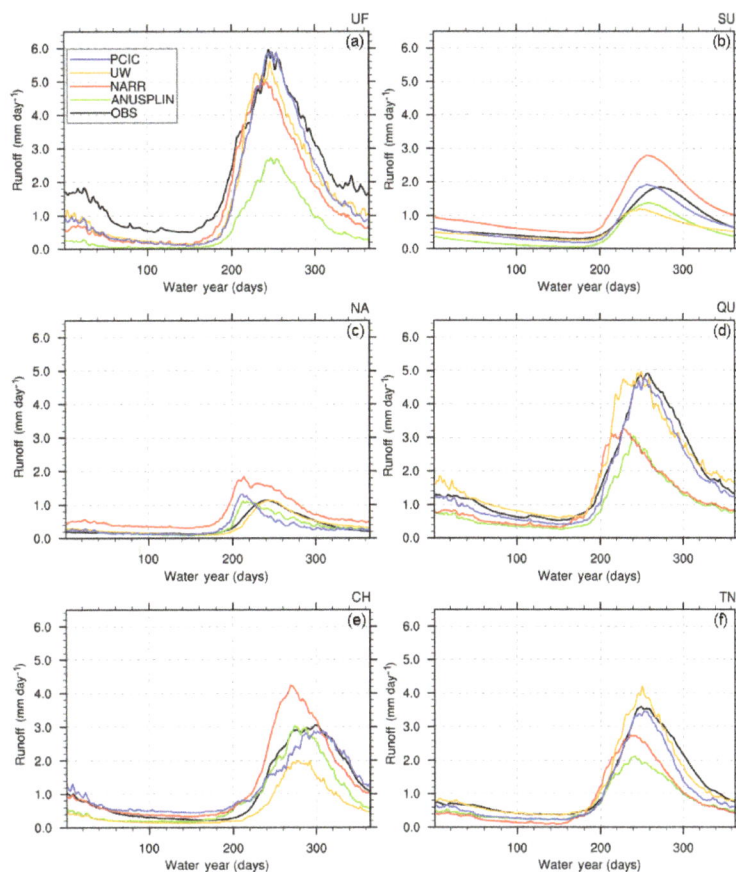

Figure 7. Same as Fig. 6a but for the FRB's six major sub-basins: **(a)** Fraser-Shelley (UF), **(b)** Stuart (SU), **(c)** Nautley (NA), **(d)** Quesnel (QU), **(e)** Chilko (CH) and **(f)** Thompson-Nicola (TN).

air temperatures (compared to PCIC) over the SU, NA and CH sub-basins in winter and spring induce more snowmelt and hence overestimate runoff. In contrast, over the UF, QU and TN, the NARR air temperature is comparatively cooler in winter. This may reduce the snowmelt driven runoff, causing underestimation over these sub-basins. The PCIC-VIC hydrographs are better in most of the basins with high NSE scores (Table S2).

Differences seen in the FRB's flow magnitude and timing clarify the impact of forcing uncertainties on the simulations. Such variation in simulated runoff, especially during the snow-melting period (April–July), is either due to the uncertain amount of precipitation or the magnitude of air temperature in the forcing datasets.

We further investigated differences in forcings and their VIC simulation based on their climatic trends. The monthly climate trends in air temperature, precipitation and simulated runoff (Fig. S6) show relatively similar warm air temperatures (up to 3 °C in December) and the declined precipitation (mainly snowfall) during winter for all four forcing datasets. The magnitude of trends in the NARR dataset is somewhat lower for air temperature and higher for precipitation compared to the other three datasets. In the simulated runoff, the

monthly variation of trends generally agrees among simulations, but the trends are weak in the ANUSPLIN-VIC and UW-VIC simulations, whereas the PCIC-VIC and NARR-VIC simulations exhibit strong trends. In the NARR-VIC simulations, runoff trends are affected by lower air temperature and higher precipitation trends, yielding increasing runoff. Grid-scale trends show widespread differences in the NARR-VIC runoff, particularly in the interior of the FRB when compared to ANUSPLIN-VIC, UW-VIC and PCIC-VIC monthly trends (Fig. S7). All four simulations exhibit strong positive runoff trends in April followed by declining trends in May in the Rocky Mountains (the UF and TN sub-basins).

The inter-comparison analysis shows that the uncertainties in forcing datasets contribute substantially to the performance of the VIC model. This is consistent with studies reporting that the uncertainties in model structure contribute less to snowpack and runoff simulations (Troin et al., 2015, 2016), whereas the uncertainties in forcing datasets are the predominant sources of uncertainties (Kay et al., 2009; Chen et al., 2011). Using the NARR dataset, the systematic biases in simulations and the substantial effect of lateral boundary conditions on the performance of the regional model have also been identified in many other studies (de Elia et al., 2008; Eum et al., 2012).

While the small differences in precipitation are acceptable, the air temperature uncertainties play an especially important role in the hydrological simulations. In the FRB, air temperature controls summer water availability, making regional snowpacks more vulnerable to temperature-induced effects, rather than precipitation. Thus uncertainties in air temperatures are crucial for the runoff timing in hydrological simulations over the FRB rather than those in precipitation.

3.3   Uncertainty in the calibration optimizer

We further investigated the uncertainty in the optimization of parameters during the calibration process. Many studies have evaluated the parameter uncertainties by adding random noise to the calibration parameters. We used a different approach by estimating the uncertainty in the MOCOM-UA optimizer used in the calibration of parameters. This was to estimate the optimizer uncertainty during the VIC calibration process using different values of initial parameter limits. The optimization process for the OPT1, OPT2, OPT3, OPT4 and OPT5 experiments required 39, 89, 61, 52 and 56 iterations, respectively, to optimize the b_inf, Ds, Ws, D2, D3 and Dsmax parameters to their final values (Table 3). The corresponding mean monthly (as the optimizer cannot utilize daily data) runoff for the Fraser River at Hope in the OPT1, OPT2, OPT3, OPT4 and OPT5 experiments is quite different when compared to observations (Fig. 8). The NSE scores reveal different accuracies for the five simulations even when the parameters' initial range in the OPT1, OPT3, OPT4 and OPT5 experiments is a subset of the OPT2 experiment. The optimization process for parameter calibration would require an expert's experience to set the initial parameter ranges to converge them to their optimal values. Note that if the initial parameter uncertainty distribution is set as wide as is physically meaningful, then the optimization will require more computational time to converge toward the Pareto optimum. However, to set the initial parameter limits, subjective judgement and skill based on experience are needed.

While we performed many sets of experiments with different initial parameters, only OPT1's initial limits produced higher NSE and utilized less computational time. The estimation of hydrologic model parameters depends significantly on the availability and quality of the precipitation and observed streamflow data along with the accuracy of the routing model used. It is therefore important to consider bias correction of forcing datasets as part of automatic calibration. The observed streamflow data used to calibrate the model are often based on water levels that are converted to discharge by the use of a rating curve, which can also induce uncertainty in the observed discharge data. The overall conclusion of this analysis is that the automated optimizers used to converge calibration parameters still rely on the hydrologist's experi-

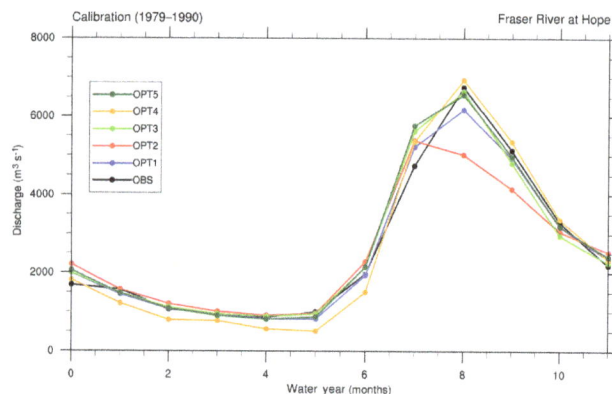

Figure 8. UW-VIC simulations using five different parameter sets (labelled as OPT1, OPT2, OPT3, OPT4 and OPT5; see text and Table 3 for details) are compared for mean monthly discharge for the Fraser River at Hope during the calibration period 1979–1990. The black curve represents observed monthly discharge.

ence and some manual adjustment of initial calibration parameter ranges.

3.4   Uncertainty in calibration due to PDO phases

The FRB streamflow varies from year to year as well on decadal timescales depending on the timing and magnitude of precipitation and air temperatures during the preceding winter and spring. Given that the FRB air temperature and precipitation are influenced by cool and warm phases of the PDO (Mantua et al., 1997; Fleming and Whitfield, 2010; Whitfield et al., 2010; Thorne and Woo, 2011), the choice of VIC calibration and validation periods may induce uncertainty in calibration. The influence of PDO phases in the forcing dataset can produce different snowpack and runoff levels in the hydrological simulation. The long-term UW-VIC simulations (1949–2006) show higher mean SWE and runoff levels in a cool PDO phase (1949–1976) and lower mean values in a warm PDO phase (1979–2006) (Fig. S8). The interannual variations show earlier peak flows characterized by a warm PDO, in response to warmer basin conditions, increased rainfall, and earlier snowmelt. The VIC model calibrations may be biased towards hydrologic conditions of the warm and cold PDO phases and may induce uncertainties in the results. The model performance could be improved by calibrating and validating the model in the same PDO phase (experiments PDO1, PDO2 and PDO5), i.e. the NSE coefficient is similar in the calibration and validation periods (Table 4). If the calibration is performed in the cool PDO phase and validation in the warm PDO phase (experiment PDO3), the NSE score decreases to 0.79 for the validation period since the model calibration is biased towards the cool conditions, simulating higher flows for the Fraser River at Hope owing to more snow and later snowmelt. The same is true if the calibration and validation is performed in the warm

Table 3. Parameters used to optimize during the calibration process for mean daily runoff for the Fraser River at Hope. OPT1, OPT2, OPT3, OPT4 and OPT5 are different experiments using the same forcing data but with a different initial range for each calibration parameter.

Calibration parameters (units)	Description	Initial range (final optimized parameters)				
		Experiment OPT1	Experiment OPT2	Experiment OPT3	Experiment OPT4	Experiment OPT5
b_inf	Controls the partitioning of precipitation (or snowmelt) into surface runoff or infiltration	0.2–0.00001 (0.07)	0.3–0.00001 (0.16)	0.25–0.10 (0.10)	0.1–0.0001 (0.08)	0.16–0.12 (0.12)
Ds	Fraction of maximum baseflow velocity	0.1–0.000001 (0.05)	0.9–0.00001 (0.09)	0.30–0.04 (0.05)	0.6–0.0001 (0.19)	0.09–0.03 (0.05)
Ws	Fraction of maximum soil moisture content of the third soil layer at which nonlinear baseflow occurs	0.6–0.20 (0.33)	1.0–0.1 (0.49)	0.65–0.20 (0.50)	0.5–0.3 (0.42)	0.35–0.20 (0.31)
D2 (m)	The second soil layer thicknesses, which affect the water available for transpiration	1.0–0.7 (0.82)	3.0–0.7 (1.02)	0.80–0.70 (0.76)	2.8–1.0 (1.07)	0.80–0.70 (0.78)
D3 (m)	The third soil layer thicknesses, which affect the water available for baseflow	2.5–0. 7 (1.66)	5.5–0.7 (2.70)	2.00–1.00 (1.82)	3.0–1.0 (1.38)	1.8–1.2 (1.76)
Dsmax (mm day^{-1})	Maximum baseflow velocity	18.0–12.0 (16.0)	30.0–12.0 (22.71)	23.0–12.0 (14.28)	18–12 (16.22)	16–13 (14.11)
Monthly NSE	–	0.93	0.84	0.92	0.89	0.91

Table 4. Daily performance metrics for the UW forcing driven PDO runs. Calibration and validation for the Fraser River main stem at Hope, BC, are evaluated using the NSE coefficient using the dataset. See text for the detail of PDO experiments.

Experiment name	Calibration		Validation	
	NSE (time period)	PDO phase (flows)	NSE (time period)	PDO phase (flows)
PDO1	0.84 (1981–1990)	Warm (low flows)	0.84 (1991–2001)	Warm (low flows)
PDO2	0.84 (1957–1966)	Cool (high flows)	0.85 (1967–1976)	Cool (high flows)
PDO3	0.84 (1967–1976)	Cool (high flows)	0.79 (1977–1987)	Warm (low flows)
PDO4	0.86 (1977–1987)	Warm (low flows)	0.80 (1967–1976)	Cool (high flows)
PDO5	0.89 (1991–2001)	Warm (low flows)	0.87 (1981–1990)	Warm (low flows)

and cool PDO phases, respectively (experiment PDO4). For each set of calibration experiments, the calibration parameters are different, which affects the formation of the snow-pack and the timing of snowmelt. Figure 9 shows observed and simulated runoff for the Fraser River at Hope, revealing lower observed peak flows $\sim 2.7\,\mathrm{mm\,day}^{-1}$ in a warm

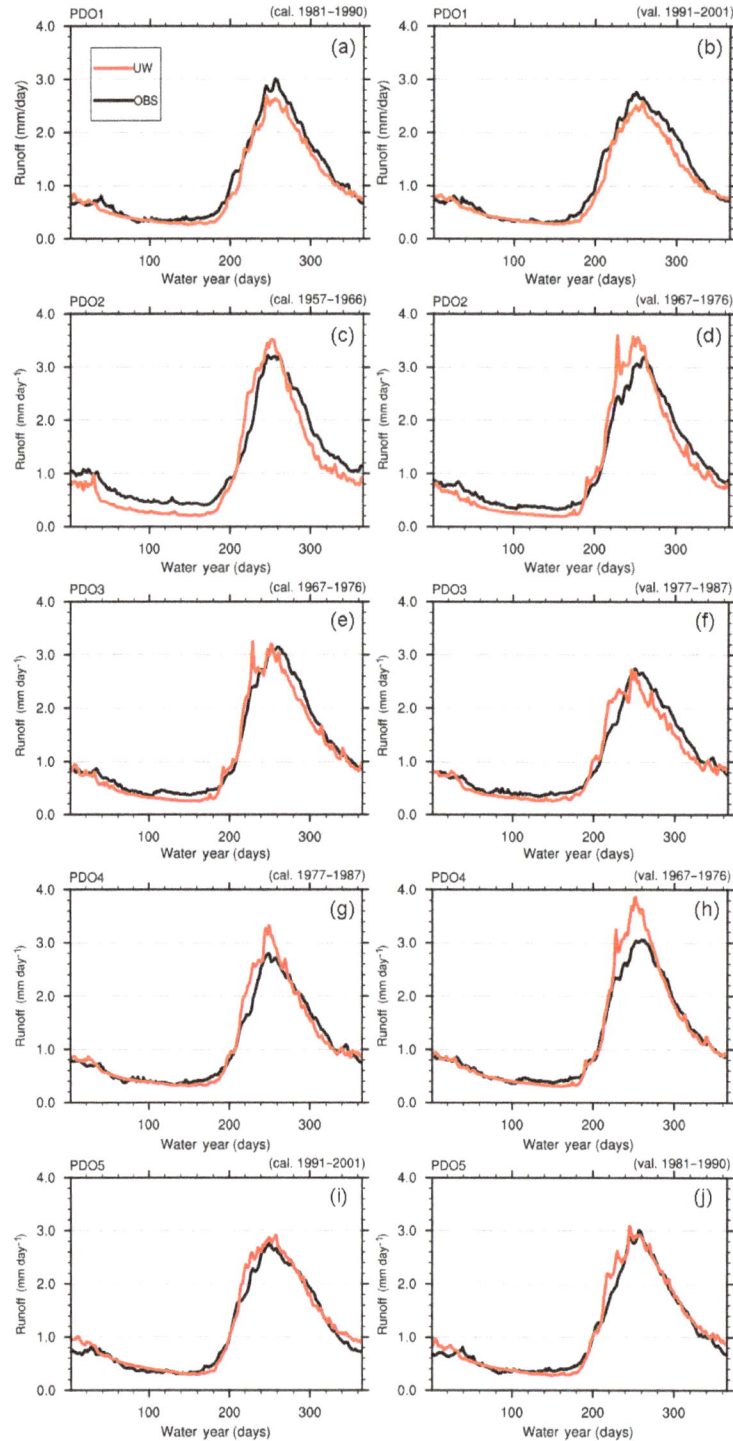

Figure 9. UW-VIC simulated daily runoff during calibration (cal.) and validation (val.) for the Fraser River at Hope, BC. PDO1, PDO2, PDO3, PDO4 and PDO5 refer to the VIC experiments performed during different experimental setups (see text and Table 4 for details). Water year starts on 1 October and ends on 30 September of the following calendar year.

PDO phase (PDO1) and higher peak flows $\sim 3.3\,\mathrm{mm\,day^{-1}}$ in a cool PDO phase (PDO2). Interestingly the UW driven PDO simulations underestimate peak flows in the warm PDO phase and overestimate them in the cool PDO phase, whereas the NSE coefficient for both the cool and warm PDO phases is almost equivalent (Table 4). The PDO4 and PDO5 experiments further support these findings.

This analysis reveals that the hydrological model performance changes considerably with different climatic conditions and the choice of the calibration and validation time periods, an important factor in hydrological simulations. The proper implementation of a hydrological model requires a careful calibration strategy to produce reliable hydrological information important for water resource management.

4 Conclusions

This study utilized ANUSPLIN, NARR, UW and PCIC observation-based gridded datasets to evaluate systematic inter-dataset uncertainties and their VIC simulated hydrological response over the FRB. The uncertainties involved in the optimization of model parameters and model calibration under cool and warm phases of the PDO were also examined.

The air temperatures in the PCIC and UW datasets were comparable, while the PCIC precipitation remains quite high in the Rocky Mountains compared to the UW and NARR datasets. The ANUSPLIN precipitation forcing had a considerable dry bias over mountainous regions of the FRB compared to the NARR, UW and PCIC datasets. The NARR winter air temperature was 2 °C warmer than the other datasets over most of the FRB. The PCIC-VIC and UW-VIC simulations had higher NSE values and more reasonable hydrographs compared with observed flows for the Fraser River at Hope. Their performance for many of the FRB's major sub-basins remained satisfactory. The PCIC-VIC simulation revealed higher SWE compared to other datasets, probably due to its higher precipitation amounts. The ANUSPLIN-VIC simulation had considerably lower runoff and NSE values along with less SWE and snowmelt amounts owing to its reduced precipitation. The NARR dataset showed warm winter air temperatures, which influenced its hydrological response by simulating less SWE and decreased snowmelt, and hence lower runoff. The monthly trend analysis distinguished the NARR dataset by showing decreased trends in air temperature and increased trends in precipitation and its VIC driven runoff. The elevation dependence of maxSWE showed disagreements over the higher elevations of the Rocky Mountains between simulations where the PCIC-VIC simulation overestimated SWE and ANUSPLIN-VIC resulted in underestimation. Furthermore the elevation-dependent variation of the maxSWE-day fluctuated considerably between simulations.

The parametric uncertainty in the VIC calibration process revealed that the choice of the initial parameter range plays a crucial role in defining the model performance. During the PDO phases, choice of the calibration and validation time periods plays a crucial role in defining the model hydrological response for the FRB. Model calibration was biased towards hydrologic conditions of the warm and cold PDO phases. The UW-VIC PDO simulations underestimated and overes-

timated the peak flows in the warm and cool PDO phases, respectively.

This study's inter-comparison revealed spatial and temporal differences amongst the ANUSPLIN, NARR, UW and PCIC datasets over the FRB, which is essential to capture the uncertainties in modelling hydrologic responses. Overall, the PCIC and UW datasets had reliable results for the FRB snow hydrology, whereas the ANUSPLIN and NARR datasets had issues with either precipitation or with air temperature. The FRB snow-dominated hydrology and its complex elevation profile require highly accurate meteorological station densities to increase the reliability of the high-resolution gridded datasets. While the air temperature plays a dominant role in the hydrological simulations, improving the quality of precipitation data can lead to more accurate hydrological responses in the FRB. Considerable precipitation bias can substantially degrade the model performance. There is the need for concrete methods to deal with the increasing uncertainty associated with the models themselves, and with the observations required for driving and evaluating the models.

In this study, the FRB hydrological response varied considerably under different forcing datasets, modelling parameters and remote teleconnections. However, there are other sources of uncertainties not discussed here that may establish a range of possible impacts on hydrological simulations. First, the hydrological model used in this study runs at a daily time step, which can be increased to hourly to refine the model performance. The lack of the representation of glaciers in the current version of the VIC model may induce uncertainties in model results. Along with these, the VIC simulations are also affected by intrinsic uncertainties in its parameterizations such as, for example, the representation of cold processes (e.g. snowpacks and soil freezing). The in situ soil moisture observations that are not necessarily representative of the model grid scale may also contribute to the overall uncertainties in the results. Finally, hydrological simulations are mainly validated using comparisons between simulated and observed flows, which depend on routing models that may contain structural uncertainties. Our future work will investigate such uncertainties using high temporal and spatial resolution hydrological models over the FRB.

Competing interests. The authors declare that they have no conflict of interest.

Acknowledgements. This work was supported by the NSERC-funded Canadian Sea Ice and Snow Evolution (CanSISE) Network. This paper was motivated by the Eric Wood Symposium held at Princeton University on 2–3 June 2016. The authors are grateful to colleagues from the Pacific Climate Impacts Consortium (PCIC) for providing ongoing assistance with this research and to Dennis Lettenmaier at UCLA for providing assistance in the VIC model implementation. The authors are thankful to Michael Allchin

(UNBC) for plotting Fig. 1a, Xiaogang Shi (Xi'an Jiaotong-Liverpool University) for development and improvements of the UW dataset, and Do Hyuk Kang (NASA GSFC) for helping in setting up the VIC model over the FRB. Thanks to the anonymous referees and the handling editor for constructive comments that greatly improved the paper.

Edited by: M. Bierkens

References

Adam, J. C. and Lettenmaier, D. P.: Adjustment of global gridded precipitation for systematic bias, J. Geophys. Res., 108, 4257, doi:10.1029/2002JD002499, 2003.

Adam, J. C. and Lettenmaier, D. P.: Application of new precipitation and reconstructed streamflow products to streamflow trend attribution in Northern Eurasia, J. Climate, 21, 1807–1828, doi:10.1175/2007JCLI1535.1, 2008.

Adam, J. C., Clark, E. A., Lettenmaier, D. P., and Wood, E. F.: Correction of global precipitation products for orographic effects, J. Climate, 19, 15–38, doi:10.1175/JCLI3604.1, 2006.

Adam, J. C., Hamlet, A. F., and Lettenmaier, D. P.: Implications of global climate change for snowmelt hydrology in the twenty-first century, Hydrol. Process., 23, 962–972, doi:10.1002/hyp.7201, 2009.

Ainslie, B. and Jackson, P. L.: Downscaling and bias correcting a cold season precipitation climatology over coastal southern British Columbia using the regional atmospheric modeling system (RAMS), J. Appl. Meteorol. Clim., 49, 937–953, doi:10.1175/2010JAMC2315.1, 2010.

Anderson, J., Chung, F., Anderson, M., Brekke, L., Easton, D., Ejeta, M., Peterson, R., and Snyder, R.: Progress on incorporating climate change into management of California's water resources, Climatic Change, 87, 91–108, doi:10.1007/s10584-007-9353-1, 2007.

Arsenault, R. and Brissette, F. P.: Continuous streamflow prediction in ungauged basins: The effects of equifinality and parameter set selection on uncertainty in regionalization approaches, Water Resour. Res., 50, 6135–6153, doi:10.1002/2013WR014898, 2014.

BC Ministry of Forests, Lands and Natural Resource Operations: Automated Snow Pillow Data, available at: http://bcrfc.env.gov.bc.ca/data/asp/, last access: 1 December 2014.

Benke, A. C. and Cushing, C. E.: Rivers of North America, Elsevier Press, New York, 607–732, 2005.

Bennett, K. E., Werner, A. T., and Schnorbus, M.: Uncertainties in hydrologic and climate change impact analyses in headwater basins of British Columbia, J. Climate, 25, 5711–5730, doi:10.1175/JCLI-D-11-00417.1, 2012.

Beven, K.: A manifesto for the equifinality thesis, J. Hydrol., 320, 18–36, 2006.

Chen, J., Brissette, F. P., Poulin, A., and Leconte, R.: Overall uncertainty study of the hydrological impacts of climate change for a Canadian watershed, Water Resour. Res., 47, W12509, doi:10.1029/2011WR010602, 2011.

Cherkauer, K. A. and Sinha, T.: Hydrologic impacts of projected future climate change in the Lake Michigan region, J. Great Lakes Res., 36, 33–50, doi:10.1016/j.jglr.2009.11.012, 2010.

Cherkauer, K. A., Bowling, L. C., and Lettenmaier, D. P.: Variable infiltration capacity cold land process model updates, Global Planet. Change, 38, 151–159, doi:10.1016/S0921-8181(03)00025-0, 2003.

Choi, G., Robinson, D. A., and Kang, S.: Changing northern hemisphere snow seasons, J. Climate, 23, 5305–5310, doi:10.1175/2010JCLI3644.1, 2010.

Choi, W., Kim, S. J., Rasmussen, P. F., and Moore, A. R.: Use of the North American Regional Reanalysis for hydrological modelling in Manitoba, Can. Water Resour. J., 34, 17–36, doi:10.4296/cwrj3401017, 2009.

Christensen, N. S. and Lettenmaier, D. P.: A multimodel ensemble approach to assessment of climate change impacts on the hydrology and water resources of the Colorado River Basin, Hydrol. Earth Syst. Sci., 11, 1417–1434, doi:10.5194/hess-11-1417-2007, 2007.

Cuo, L., Lettenmaier, D. P., Alberti, M., and Richey, J. E.: Effects of a century of land cover and climate change on the hydrology of the Puget Sound basin, Hydrol. Process., 23, 907–933, doi:10.1002/hyp.7228, 2009.

Danard, M. and Murty, T. S.: On recent climate trends in selected salmon-hatching areas of British Columbia, J. Climate, 7, 1803–1808, doi:10.1175/1520-0442(1994)007<1803:ORCTIS>2.0.CO;2, 1994.

de Elía, R., Caya, D., Côté, H., Frigon, A., Biner, S., Giguère, M., Paquin, D., Harvey, R., and Plummer, D.: Evaluation of uncertainties in the CRCM-simulated North American climate, Clim. Dynam., 30, 113–132, doi:10.1007/s00382-007-0288-z, 2008.

Déry, S. J., Stieglitz, M., McKenna, E. C., and Wood, E. F.: Characteristics and trends of river discharge into Hudson, James, and Ungava Bays, 1964–2000, J. Climate, 18, 2540–2557, doi:10.1175/JCLI3440.1, 2005.

Déry, S. J., Hernández-Henríquez, M. A., Owens, P. N., Parkes, M. W., and Petticrew, E. L.: A century of hydrological variability and trends in the Fraser River Basin, Environ. Res. Lett., 7, 024019, doi:10.1088/1748-9326/7/2/024019, 2012.

Déry, S. J., Knudsvig, H. K., Hernández-Henríquez, M. A., and Coxson, D. S.: Net snowpack accumulation and ablation characteristics in the Inland Temperate Rainforest of the Upper Fraser River Basin, Canada, Hydrology, 1, 1–19, doi:10.3390/hydrology1010001, 2014.

Déry, S. J., Stadnyk, T. A., MacDonald, M. K., and Gauli-Sharma, B.: Recent trends and variability in river discharge across northern Canada, Hydrol. Earth Syst. Sci., 20, 4801–4818, doi:10.5194/hess-20-4801-2016, 2016.

Dodson, R. and Marks, D.: Daily air temperature interpolated at high spatial resolution over a large mountainous region, Clim. Res., 8, 1–20, 1997.

Elder, K., Dozier, J., and Michaelsen, J.: Snow accumulation and distribution in an Alpine Watershed, Water Resour. Res., 27, 1541–1552, doi:10.1029/91WR00506, 1991.

Elsner, M. M., Cuo, L., Voisin, N., Deems, J. S., Hamlet, A. F., Vano, J. A., Mickelson, K. E. B., Lee, S. Y., and Lettenmaier, D. P.: Implications of 21st century climate change for the hydrology of Washington State, Climatic Change, 102, 225–260, doi:10.1007/s10584-010-9855-0, 2010.

Essou, G. R. C., Arsenault, R., and Brissette, F. P.: Comparison of climate datasets for lumped hydrological modeling over the continental United States, J. Hydrol., 537, 334–345, doi:10.1016/j.jhydrol.2016.03.063, 2016a.

Essou, G. R. C., Sabarly, F., Lucas-Picher, P., Brissette, F., and Poulin, A.: Can precipitation and temperature from meteorological reanalyses be used for hydrological modeling?, J. Hydrometeorol., 17, 1929–1950, doi:10.1175/JHM-D-15-0138.1, 2016b.

Eum, H., Gachon, P., Laprise, R., and Ouarda, T.: Evaluation of regional climate model simulations versus gridded observed and regional reanalysis products using a combined weighting scheme, Clim. Dynam., 38, 1433–1457, doi:10.1007/s00382-011-1149-3, 2012.

Eum, H., Dibike, Y., Prowse, T., and Bonsal, B.: Intercomparison of high-resolution gridded climate data sets and their implication on hydrological model simulation over the Athabasca Watershed, Canada, Hydrol. Process., 28, 4250–4271, doi:10.1002/hyp.10236, 2014.

Fleming, S. W. and Whitfield, P. H.: Spatiotemporal mapping of ENSO and PDO surface meteorological signals in British Columbia, Yukon, and southeast Alaska, Atmos.-Ocean, 48, 122–131, doi:10.3137/AO1107.2010, 2010.

Gao, H., Tang, Q., Shi, X., Zhu, C., Bohn, T. J., Su, F., Sheffield, J., Pan, M., Lettenmaier, D. P., and Wood, E. F.: Water Budget Record from Variable Infiltration Capacity (VIC) Model. In Algorithm Theoretical Basis Document for Terrestrial Water Cycle Data Records, unpublished, 2009.

Gao, H., Tang, Q., Ferguson, C. R., Wood, E. F., and Lettenmaier, D. P.: Estimating the water budget of major US river basins via remote sensing, Int. J. Remote Sens., 31, 3955–3978, doi:10.1080/01431161.2010.483488, 2010.

Graham, L. P., Hagemann, S., Jaun, S., and Beniston, M.: On interpreting hydrological change from regional climate models, Climatic Change, 81, 97–122, doi:10.1007/s10584-006-9217-0, 2007.

Gupta, H. V., Kling, H., Yilmaz, K. K., and Martinez, G. F.: Decomposition of the mean squared error and NSE performance criteria: Implications for improving hydrological modelling, J. Hydrol., 377, 80–91, doi:10.1016/j.jhydrol.2009.08.003, 2009.

Haddeland, I., Skaugen, T., and Lettenmaier, D. P.: Hydrologic effects of land and water management in North America and Asia: 1700–1992, Hydrol. Earth Syst. Sci., 11, 1035–1045, doi:10.5194/hess-11-1035-2007, 2007.

Hidalgo, H. G., Das, T., Dettinger, M. D., Cayan, D. R., Pierce, D. W., Barnett, T. P., Bala, G., Mirin, A., Wood, A. W., Bonfils, C., Santer, B. D., and Nozawa, T.: Detection and attribution of streamflow timing changes to climate change in the western United States, J. Climate, 22, 3838–3855, doi:10.1175/2009JCLI2470.1, 2009.

Hopkinson, R. F., Mckenney, D. W., Milewska, E. J., Hutchinson, M. F., Papadopol, P., and Vincent, L. A.: Impact of aligning climatological day on gridding daily maximum-minimum temperature and precipitation over Canada, J. Appl. Meteorol. Clim., 50, 1654–1665, doi:10.1175/2011JAMC2684.1, 2011.

Horton, P., Schaefli, B., Mezghani, A., Hingray, B., and Musy, A.: Assessment of climate-change impacts on alpine discharge regimes with climate model uncertainty, Hydrol. Process., 20, 2091–2109, 2006.

Hutchinson, M. F., McKenney, D. W., Lawrence, K., Pedlar, J. H., Hopkinson, R. F., Milewska, E., and Papadopol, P.: Development and testing of Canada-wide interpolated spatial models of daily minimum–maximum temperature and precipitation for 1961–2003, J. Appl. Meteorol. Clim., 48, 725–741, doi:10.1175/2008JAMC1979.1, 2009.

Islam, S. U., Déry, S. J., and Werner, A. T.: Future climate change impacts on snow and water resources of the Fraser River Basin, British Columbia, J. Hydrometeorol., 18, 473–496, doi:10.1175/JHM-D-16-0012.1, 2017.

Jiang, T., Chen, Y. D., Xu, C., Chen, X., Chen, X., and Singh, V. P.: Comparison of hydrological impacts of climate change simulated by six hydrological models in the Dongjiang Basin, South China, J. Hydrol., 336, 316–333, doi:10.1016/j.jhydrol.2007.01.010, 2007.

Kalnay, E., Kanamitsu, M., Kistler, R., Collins, W., Deaven, D., Gandin, L., Iredell, M., Saha, S., White, G., Woollen, J., Zhu, Y., Chelliah, M., Ebisuzaki, W., Higgins, W., Janowiak, J., Mo, K. C., Ropelewski, C., Wang, J., Leetmaa, A., Reynolds, R., Jenne, R., and Joseph, D.: The NCEP/NCAR 40-year reanalysis project, B. Am. Meteorol. Soc., 77, 437–471, doi:10.1175/1520-0477(1996)077<0437:TNYRP>2.0.CO;2, 1996.

Kang, D. H., Shi, X., Gao, H., and Déry, S. J.: On the changing contribution of snow to the hydrology of the Fraser River Basin, Canada, J. Hydrometeorol., 15, 1344–1365, doi:10.1175/JHM-D-13-0120.1, 2014.

Kang, D. H., Gao, H., Shi, X., Islam, S. U., and Déry, S. J.: Impacts of a rapidly declining mountain snowpack on streamflow timing in Canada's Fraser River Basin, Sci. Rep., 6, 19299, doi:10.1038/srep19299, 2016.

Kay, A. L., Davies, H. N., Bell, V. A., and Jones, R. G.: Comparison of uncertainty sources for climate change impacts: Flood frequency in England, Climatic Change, 92, 41–63, doi:10.1007/s10584-008-9471-4, 2009.

Kendall, M. G.: Rank Correlation Methods, 4th Edn., Griffin, London, 202 pp., 1970.

Keshta, N. and Elshorbagy, A.: Utilizing North American Regional Reanalysis for modeling soil moisture and evapotranspiration in reconstructed watersheds, Phys. Chem. Earth, 36, 31–41, doi:10.1016/j.pce.2010.12.001, 2011.

Kistler, R., Collins, W., Saha, S., White, G., Woollen, J., Kalnay, E., Chelliah, M., Ebisuzaki, W., Kanamitsu, M., Kousky, V., van den Dool, H., Jenne, R., and Fiorino, M.: The NCEP-NCAR 50-year reanalysis: Monthly means CD-ROM and documentation, B. Am. Meteorol. Soc., 82, 247–267, doi:10.1175/1520-0477(2001)082<0247:TNNYRM>2.3.CO;2, 2001.

Klemeš, V.: Operational testing of hydrological simulation models, Hydrolog. Sci. J., 31, 13–24, doi:10.1080/02626668609491024, 1986.

Lettenmaier, D. P., Wood, E. F., and Wallis, J. R.: Hydro-climatological trends in the continental United States, 1948–88, J. Climate, 7, 586–607, doi:10.1175/1520-0442(1994)007<0586:HCTITC>2.0.CO;2, 1994.

Liang, X., Lettenmaier, D. P., Wood, E. F., and Burges, S. J.: A simple hydrologically based model of land surface water and energy fluxes for general circulation models, J. Geophys. Res.-Atmos., 99, 14415–14428, doi:10.1029/94JD00483, 1994.

Liang, X., Wood, E. F., and Lettenmaier, D. P.: Surface soil moisture parameterization of the VIC-2L Model: Evaluation and modifications, Global Planet. Change, 13, 195–206, 1996.

Liu, Y. and Gupta, H. V.: Uncertainty in hydrologic modeling: Toward an integrated data assimilation framework, Water Resour. Res., 43, W07401, doi:10.1029/2006WR005756, 2007.

Lohmann, D., Nolte-Holube, R., and Raschke, E.: A large-scale horizontal routing model to be coupled to land surface parametrization schemes, Tellus A, 48, 708–721, doi:10.1034/j.1600-0870.1996.t01-3-00009.x, 1996.

Lohmann, D., Raschke, E., Nijssen, B., and Lettenmaier, D. P.: Regional scale hydrology: I. Formulation of the VIC-2L model coupled to a routing model, Hydrolog. Sci. J., 43, 131–141, doi:10.1080/02626669809492107, 1998a.

Lohmann, D., Raschke, E., Nijssen, B., and Lettenmaier, D. P.: Regional scale hydrology: II. Application of the VIC-2L model to the Weser River, Germany, Hydrolog. Sci. J., 43, 143–158, doi:10.1080/02626669809492108, 1998b.

Luo, Y., Berbery, E. H., Mitchell, K. E., and Betts, A. K.: Relationships between land surface and near-surface atmospheric variables in the NCEP North American Regional Reanalysis, J. Hydrometeorol., 8, 1184–1203, doi:10.1175/2007JHM844.1, 2007.

Mann, H. B.: Nonparametric tests against trend, Econometrica, 13, 245–259, doi:10.1017/CBO9781107415324.004, 1945.

Mantua, N. J., Hare, S. R., Zhang, Y., Wallace, J. M., and Francis, R. C.: A Pacific interdecadal climate oscillation with impacts on salmon production, B. Am. Meteorol. Soc., 78, 1069–1079, doi:10.1175/1520-0477(1997)078<1069:APICOW>2.0.CO;2, 1997.

Matsuura, K. and Willmott, C. J.: Terrestrial precipitation: 1900–2008 gridded monthly time series (version 2.01). Center for Climatic Research, Department of Geography, University of Delaware, Newark, DE, digital media, available at: http://climate.geog.udel.edu/~climate/ (last access: 1 September 2014), 2009.

Maurer, E. P., Wood, A. W., Adam, J. C., Lettenmaier, D. P., and Nijssen, B.: A long-term hydrologically based dataset of land surface fluxes and states for the conterminous United States, J. Climate, 15, 3237–3251, doi:10.1175/1520-0442(2002)015<3237:ALTHBD>2.0.CO;2, 2002.

Mesinger, F., DiMego, G., Kalnay, E., Mitchell, K., Shafran, P. C., Ebisuzaki, W., Jović, D., Woollen, J., Rogers, E., Berbery, E. H., Ek, M. B., Fan, Y., Grumbine, R., Higgins, W., Li, H., Lin, Y., Manikin, G., Parrish, D., and Shi, W.: North American regional reanalysis, B. Am. Meteorol. Soc., 87, 343–360, doi:10.1175/BAMS-87-3-343, 2006.

Moore, R. and Wondzell, S. M.: Physical hydrology and the effects of forest harvesting in the Pacific Northwest: A review, J. Am. Water Resour. Assoc., 41, 763–784, doi:10.1111/j.1752-1688.2005.tb03770.x, 2005.

Moore, R. D.: Hydrology and water supply in the Fraser River Basin, in: Water in sustainable development: exploring our common future in the Fraser River Basin, edited by: Dorcey, A. H. J. and Griggs, J. R., Wastewater Research Centre, University of British Columbia, Vancouver, British Columbia, Canada, 21–40, 1991.

Mote, P. W., Hamlet, A. F., Clark, M. P., and Lettenmaier, D. P.: Declining mountain snowpack in western North America, B. Am. Meteorol. Soc., 86, 39–49, doi:10.1175/BAMS-86-1-39, 2005.

Nash, J. E. and Sutcliffe, J. V.: River flow forecasting through conceptual models part I - A discussion of principles, J. Hydrol., 10, 282–290, doi:10.1016/0022-1694(70)90255-6, 1970.

Neilsen, D., Duke, G., Taylor, B., Byrne, J., Kienzle, S., and van der Gulik, T.: Development and verification of daily gridded climate surfaces in the Okanagan Basin of British Columbia, Can. Water Resour. J., 35, 131–154, doi:10.4296/cwrj3502131, 2010.

Nigam, S. and Ruiz-Barradas, A.: Seasonal hydroclimate variability over North America in global and regional reanalyses and AMIP simulations: Varied representation, J. Climate, 19, 815–837, doi:10.1175/JCLI3635.1, 2006.

Nijssen, B., Lettenmaier, D. P., Liang, X., Wetzel, S. W., and Wood, E. F.: Streamflow simulation for continental-scale river basins, Water Resour. Res., 33, 711–724, 1997.

Nijssen, B., Schnur, R., and Lettenmaier, D. P.: Global retrospective estimation of soil moisture using the Variable Infiltration Capacity land surface model, 1980–93, J. Climate, 14, 1790–1808, doi:10.1175/1520-0442(2001)014<1790:GREOSM>2.0.CO;2, 2001a.

Nijssen, B., O'Donnell, G. M., Lettenmaier, D. P., Lohmann, D., and Wood, E. F.: Predicting the discharge of global rivers, J. Climate, 14, 3307–3323, doi:10.1175/1520-0442(2001)014<3307:PTDOGR>2.0.CO;2, 2001b.

NRCan: Regional, national and international climate modeling, available at: http://cfs.nrcan.gc.ca/projects/3?lang=en_CA (last access: 1 December 2016), 2014.

Oubeidillah, A. A., Kao, S. C., Ashfaq, M., Naz, B. S., and Tootle, G.: A large-scale, high-resolution hydrological model parameter data set for climate change impact assessment for the conterminous US, Hydrol. Earth Syst. Sci., 18, 67–84, doi:10.5194/hess-18-67-2014, 2014.

Pietroniro, A., Leconte, R., Toth, B., Peters, D. L., Kouwen, N., Conly, F. M., and Prowse, T. D.: Modelling climate change impacts in the Peace and Athabasca catchment and delta: III – Integrated model assessment, Hydrol. Process., 20, 4231–4245, doi:10.1002/hyp.6428, 2006.

Poulin, A., Brissette, F., Leconte, R., Arsenault, R., and Malo, J. S.: Uncertainty of hydrological modelling in climate change impact studies in a Canadian, snow-dominated river basin, J. Hydrol., 409, 626–636, doi:10.1016/j.jhydrol.2011.08.057, 2011.

Reed, S., Koren, V., Smith, M., Zhang, Z., Moreda, F., and Seo, D. J.: Overall distributed model intercomparison project results, J. Hydrol., 298, 27–60, 2004.

Rinke, A., Marbaix, P., and Dethloff, K.: Internal variability in Arctic regional climate simulations: Case study for the SHEBA year, Clim. Res., 27, 197–209, doi:10.3354/cr027197, 2004.

Rodenhuis, D., Bennett, K., Werner, A., Murdock, Q., and Bronaugh, D.: Hydro-climatology and future climate impacts in British Columbia. Pacific Climate Impacts Consortium, University of Victoria, Victoria, BC, 132 pp., 2009.

Ruane, A. C.: NARR's atmospheric water cycle components. Part I: 20-year mean and annual interactions, J. Hydrometeorol., 11, 1205–1219, doi:10.1175/2010JHM1193.1, 2010.

Sabarly, F., Essou, G., Lucas-Picher, P., Poulin, A., and Brissette, F.: Use of four reanalysis datasets to assess the terrestrial branch of the water cycle over Quebec, Canada, J. Hydrometeorol., 17, 1447–1466, doi:10.1175/JHM-D-15-0093.1, 2016.

Schnorbus, M., Bennett, K., and Werner, A.: Quantifying the water resource impacts of mountain pine beetle and associated salvage harvest operations across a range of watershed scales: Hydrologic modelling of the Fraser River Basin, Information Report: BC-X-423, Natural Resources Canada, Canadian Forestry Service, Pacific Forestry Centre, Victoria, BC, 79 pp., 2010.

Schnorbus, M., Bennett, K., Werner, A., and Berland, A. J.: Hydrologic impacts of climate change in the Peace, Campbell and Columbia sub-basins, British Columbia, Canada, Pacific Climate Impacts Consortium, University of Victoria, Victoria, BC, 157 pp., 2011.

Seiller, G., Anctil, F., and Perrin, C.: Multimodel evaluation of twenty lumped hydrological models under contrasted climate conditions, Hydrol. Earth Syst. Sci., 16, 1171–1189, doi:10.5194/hess-16-1171-2012, 2012.

Sheffield, J., Goteti, G., and Wood, E. F.: Development of a 50-yr high-resolution global dataset of meteorological forcings for land surface modeling, J. Climate, 19, 3088–3111, 2006.

Sheffield, J., Livneh, B., and Wood, E. F.: Representation of terrestrial hydrology and large-scale drought of the continental United States from the North American Regional Reanalysis, J. Hydrometeorol., 13, 856–876, doi:10.1175/JHM-D-11-065.1, 2012.

Shi, X., Wood, A. W., and Lettenmaier, D. P.: How essential is hydrologic model calibration to seasonal streamflow forecasting?, J. Hydrometeorol., 9, 1350–1363, doi:10.1175/2008JHM1001.1, 2008.

Shi, X., Déry, S. J., Groisman, P. Y., and Lettenmaier, D. P.: Relationships between recent pan-arctic snow cover and hydroclimate trends, J. Climate, 26, 2048–2064, doi:10.1175/JCLI-D-12-00044.1, 2013.

Shrestha, R. R., Schnorbus, M. A., Werner, A. T., and Berland, A. J.: Modelling spatial and temporal variability of hydrologic impacts of climate change in the Fraser River basin, British Columbia, Canada, Hydrol. Process., 26, 1841–1861, doi:10.1002/hyp.9283, 2012.

Stahl, K., Moore, R. D., Floyer, J. A., Asplin, M. G., and McKendry, I. G.: Comparison of approaches for spatial interpolation of daily air temperature in a large region with complex topography and highly variable station density, Agr. Forest Meteorol., 139, 224–236, doi:10.1016/j.agrformet.2006.07.004, 2006.

Su, F., Adam, J. C., Bowling, L. C., and Lettenmaier, D. P.: Streamflow simulations of the terrestrial Arctic domain, J. Geophys. Res., 110, D08112, doi:10.1029/2004JD005518, 2005.

Tapiador, F. J., Turk, F. J., Petersen, W., Hou, A. Y., García-Ortega, E., Machado, L. A. T., Angelis, C. F., Salio, P., Kidd, C., Huffman, G. J., and de Castro, M.: Global precipitation measurement: Methods, datasets and applications, Atmos. Res., 104–105, 70–97, doi:10.1016/j.atmosres.2011.10.021, 2012.

Teutschbein, C., Wetterhall, F., and Seibert, J.: Evaluation of different downscaling techniques for hydrological climate-change impact studies at the catchment scale, Clim. Dynam., 37, 2087–2105, doi:10.1007/s00382-010-0979-8, 2011.

Thorne, R. and Woo, M. K.: Streamflow response to climatic variability in a complex mountainous environment: Fraser River Basin, British Columbia, Canada, Hydrol. Process., 25, 3076–3085, doi:10.1002/hyp.8225, 2011.

Tobin, C., Nicotina, L., Parlange, M. B., Berne, A., and Rinaldo, A.: Improved interpolation of meteorological forcings for hydrologic applications in a Swiss Alpine region, J. Hydrol., 401, 77–89, doi:10.1016/j.jhydrol.2011.02.010, 2011.

Tong, J., Déry, S. J., Jackson, P. L., and Derksen, C.: Testing snow water equivalent retrieval algorithms for passive microwave remote sensing in an alpine watershed of western Canada, Can. J. Remote Sens., 36, S74–S86, doi:10.5589/m10-009, 2010.

Troin, M., Arsenault, R., and Brissette, F.: Performance and uncertainty evaluation of snow models on snowmelt flow simulations over a Nordic catchment (Mistassibi, Canada), Hydrology, 2, 289–317, doi:10.3390/hydrology2040289, 2015.

Troin, M., Poulin, A., Baraer, M., and Brissette, F.: Comparing snow models under current and future climates: Uncertainties and implications for hydrological impact studies, J. Hydrol., 540, 588–602, doi:10.1016/j.jhydrol.2016.06.055, 2016.

Trubilowicz, J. W., Shea, J. M., Jost, G., and Moore, R. D.: Suitability of North American Regional Reanalysis (NARR) output for hydrologic modelling and analysis in mountainous terrain, Hydrol. Process., 30, 2332–2347, doi:10.1002/hyp.10795, 2016.

Velazquez, J. A., Schmid, J., Ricard, S., Muerth, M. J., Gauvin St-Denis, B., Minville, M., Chaumont, D., Caya, D., Ludwig, R., and Turcotte, R.: An ensemble approach to assess hydrological models' contribution to uncertainties in the analysis of climate change impact on water resources, Hydrol. Earth Syst. Sci., 17, 565–578, doi:10.5194/hess-17-565-2013, 2013.

Wade, N. L., Martin, J., and Whitfield, P. H.: Hydrologic and climatic zonation of Georgia Basin, British Columbia, Can. Water Resour. J., 26, 43–70, doi:10.4296/cwrj2601043, 2001.

Wagener, T. and Gupta, H. V.: Model identification for hydrological forecasting under uncertainty, Stoch. Environ. Res. Risk A., 19, 378–387, doi:10.1007/s00477-005-0006-5, 2005.

Water Survey of Canada: HYDAT database, available at: http://www.ec.gc.ca/rhc-wsc/, last access: 1 December 2014.

Wen, L., Lin, C. A., Wu, Z., Lu, G., Pomeroy, J., and Zhu, Y.: Reconstructing sixty year (1950–2009) daily soil moisture over the Canadian Prairies using the Variable Infiltration Capacity model, Can. Water Resour. J., 36, 83–102, doi:10.4296/cwrj3601083, 2011.

Whitfield, P. H., Moore, R. D., Fleming, S. W., and Zawadzki, A.: Pacific Decadal Oscillation and the hydroclimatology of Western Canada – Review and prospects, Can. Water Resour. J., 35, 1–28, doi:10.4296/cwrj3501001, 2010.

Wilby, R. L. and Harris, I.: A framework for assessing uncertainties in climate change impacts: Low-flow scenarios for the River Thames, UK, Water Resour. Res., 42, W02419, doi:10.1029/2005WR004065, 2006.

Woo, M. K. and Thorne, R.: Snowmelt contribution to discharge from a large mountainous catchment in subarctic Canada, Hydrol. Process., 20, 2129–2139, 2006.

Wu, H., Kimball, J. S., Mantua, N., and Stanford, J.: Automated upscaling of river networks for macroscale hydrological modeling, Water Resour. Res., 47, W03517, doi:10.1029/2009WR008871, 2011.

Yapo, P. O., Gupta, H. V., and Sorooshian, S.: Multi-objective global optimization for hydrologic models, J. Hydrol., 204, 83–97, doi:10.1016/S0022-1694(97)00107-8, 1998.

Zhou, T., Nijssen, B., Gao, H., and Lettenmaier, D. P.: The contribution of reservoirs to global land surface water storage variations, J. Hydrometeorol., 17, 309–325, doi:10.1175/JHM-D-15-
0002.1, 2016.

Ziegler, A. D., Sheffield, J., Maurer, E. P., Nijssen, B., Wood,

E. F. and Lettenmaier, D. P.: Detection of intensification in
global- and continental-scale hydrological cycles: Temporal
scale of evaluation, J. Climate, 16, 535–547, doi:10.1175/1520-
0442(2003)016<0535:DOIIGA>2.0.CO;2, 2003.

5

The role of glacier changes and threshold definition in the characterisation of future streamflow droughts in glacierised catchments

Marit Van Tiel[1,2,a], **Adriaan J. Teuling**[2], **Niko Wanders**[3,4], **Marc J. P. Vis**[5], **Kerstin Stahl**[6], and **Anne F. Van Loon**[1]

[1]School of Geography, Earth and Environmental Sciences, University of Birmingham, Birmingham, UK
[2]Hydrology and Quantitative Water Management Group, Wageningen University and Research, Wageningen, the Netherlands
[3]Department of Civil and Environmental Engineering, Princeton University, Princeton, NJ, USA
[4]Department of Physical Geography, Utrecht University, Utrecht, the Netherlands
[5]Department of Geography, University of Zurich, Zurich, Switzerland
[6]Faculty of Environment and Natural Resources, University of Freiburg, Freiburg, Germany
[a]now at: Faculty of Environment and Natural Resources, University of Freiburg, Freiburg, Germany

Correspondence: Marit Van Tiel (marit.van.tiel@hydrology.uni-freiburg.de)

Abstract. Glaciers are essential hydrological reservoirs, storing and releasing water at various timescales. Short-term variability in glacier melt is one of the causes of streamflow droughts, here defined as deficiencies from the flow regime. Streamflow droughts in glacierised catchments have a wide range of interlinked causing factors related to precipitation and temperature on short and long timescales. Climate change affects glacier storage capacity, with resulting consequences for discharge regimes and streamflow drought. Future projections of streamflow drought in glacierised basins can, however, strongly depend on the modelling strategies and analysis approaches applied. Here, we examine the effect of different approaches, concerning the glacier modelling and the drought threshold, on the characterisation of streamflow droughts in glacierised catchments. Streamflow is simulated with the Hydrologiska Byråns Vattenbalansavdelning (HBV-light) model for two case study catchments, the Nigardsbreen catchment in Norway and the Wolverine catchment in Alaska, and two future climate change scenarios (RCP4.5 and RCP8.5). Two types of glacier modelling are applied, a constant and dynamic glacier area conceptualisation. Streamflow droughts are identified with the variable threshold level method and their characteristics are compared between two periods, a historical (1975–2004) and future (2071–2100) period. Two existing threshold approaches to define future droughts are employed: (1) the threshold from the historical period; (2) a transient threshold approach, whereby the threshold adapts every year in the future to the changing regimes. Results show that drought characteristics differ among the combinations of glacier area modelling and thresholds. The historical threshold combined with a dynamic glacier area projects extreme increases in drought severity in the future, caused by the regime shift due to a reduction in glacier area. The historical threshold combined with a constant glacier area results in a drastic decrease of the number of droughts. The drought characteristics between future and historical periods are more similar when the transient threshold is used, for both glacier area conceptualisations. With the transient threshold, factors causing future droughts can be analysed. This study revealed the different effects of methodological choices on future streamflow drought projections and it highlights how the options can be used to analyse different aspects of future droughts: the transient threshold for analysing future drought processes, the historical threshold to assess changes between periods, the constant glacier area to analyse the effect of short-term climate variability on droughts and the dynamic glacier area to model more realistic future discharges under climate change.

1 Introduction

Glaciers and snow packs are an important freshwater resource, supplying water to more than one-sixth of the Earth's population (Barnett et al., 2005). Glaciers play an essential role in the global water cycle as hydrologic reservoirs on various timescales (Jansson et al., 2003; Vaughan et al., 2013). They, for example, reduce the interannual variability by storing water in cold and wet years and releasing it in warm and dry years (Jansson et al., 2003; Koboltschnig et al., 2007; Zappa and Kan, 2007; Viviroli et al., 2011). Also, on seasonal timescales, glacier storage and release are important: the glacier melt peak in summer sustains discharge during otherwise low flow conditions (due to low precipitation or high evapotranspiration; e.g. Fountain and Tangborn, 1985; Miller et al., 2012; Bliss et al., 2014) and especially during low flow conditions downstream (Huss, 2011). Fluctuations in the summer glacier melt peak may therefore be an important driver of streamflow drought.

Drought is defined as a below-normal water availability (Tallaksen and Van Lanen, 2004; Sheffield and Wood, 2012) and streamflow drought (also called hydrological drought) is a drought in river discharge. According to this definition, we defined streamflow droughts in this study as anomalies (or deficiencies) from the hydrological regime, including the important high flow melt season. Streamflow droughts are a recurring and worldwide phenomenon (Tallaksen and Van Lanen, 2004) which can have severe impacts on river ecology, water supply and energy production (e.g. Jonsdottir et al., 2005; van Vliet et al., 2016). Hydrological drought is often caused by meteorological drought (deficit in precipitation) which propagates through the hydrological cycle (Tallaksen and Van Lanen, 2004; Van Loon, 2015). In cold climates, where snow and ice are an important part of the seasonal water balance, streamflow drought can also be caused by anomalies in temperature (Van Loon et al., 2015). In glacierised catchments, "glacier melt droughts", defined as a deficiency in the glacier melt peak and caused by below-normal temperatures in the summer season (Van Loon et al., 2015), can be important to downstream water users.

Climate change is expected to have large influences on both glaciers and streamflow droughts due to a reduction in the water storage capacity of glaciers and snow packs. This will have major consequences for the water supply downstream (e.g. Kaser et al., 2010; Immerzeel et al., 2010; Huss, 2011; Finger et al., 2012). The Intergovernmental Panel on Climate Change (IPCC) reports with high confidence that glaciers worldwide are shrinking and that current glacier extents are out of balance with the current climate, indicating that glaciers will continue to shrink (Vaughan et al., 2013). Retreating glaciers affect the discharge regimes in glacierised catchments. Déry et al. (2009) and Bard et al. (2015) found a shift in the melt peak towards an earlier moment in the season in trend studies of observed streamflow, in British Columbia, Canada, and in the European Alps, respectively. Also, for the future, changes in the timing of

the melt peak are expected, together with a more dominant role of rainfall and less snow accumulation (Horton et al., 2006; Jeelani et al., 2012, for the Swiss Alps and Western Himalayas, respectively). Two recent studies showed that retreating glaciers can have contrasting effects on the hydrology. Ragettli et al. (2016) project rising flows with limited shifts in the seasonality for the Langtang catchment in Nepal and a reduced and shifted peak in streamflow for the Juncal catchment in Chile. The latter was also found by Lutz et al. (2016) for the Upper Indus Basin. Farinotti et al. (2012) show the combined responses with increasing and then decreasing annual discharges for several glacierised catchments in Switzerland by modelling the period 1900–2100. What these projected changes in glacial hydrology mean for streamflow droughts has, however, not been explicitly modelled. From global- and continental-scale drought studies, we expect streamflow droughts to become more severe in the future (Bates et al., 2008; Van Huijgevoort et al., 2014), with an increase in average streamflow drought duration and deficit volume expected for the globe (Van Huijgevoort et al., 2014; Wanders and Van Lanen, 2015). Also, for Europe, Feyen and Dankers (2009) and Forzieri et al. (2014) found that many river basins are likely to experience more severe streamflow drought.

These projections are, however, strongly dependent on the methodology applied in the analysis and for both future glacier modelling and future drought analysis many options exist. In order to make projections for hydrology in glacierised catchments under climate change, a glaciohydrological model is needed. Especially in highly glacierised catchments and when modelling long time periods, a realistic representation of the glacier in the model is crucial. However, complex ice flow models require a lot of input data (e.g. glacier bathymetry and density estimates; see also Immerzeel et al., 2012; Naz et al., 2014) which are often not available, and they are in general not applicable for hydrological modelling (Huss, 2011). Different types of glacier geometry conceptualisations are therefore used in hydrological studies. For example, past studies by Klok et al. (2001), Verbunt et al. (2003) and Schaefli et al. (2005) used a simple infinite and constant glacier reservoir in their hydrological model. Also, e.g. Akhtar et al. (2008), Tecklenburg et al. (2012) and Sun et al. (2015) used a constant glacier area in their modelling studies as a benchmark to compare with model runs where the glacier area is adjusted. Juen et al. (2007) simulate future glacier extent assuming a new steady state in the future obtained by reducing the glacier area gradually until the future annual mass balance is zero. Stahl et al. (2008) used a volume–area relation to rescale the glacier based on modelled glacier mass balances, however, distributing the area reduction only conceptually in space. Huss et al. (2010) used a more detailed glacier representation in their model by introducing the $\triangle h$ parameterisation to calculate the transient evolution of the glacier surface elevation and area. Huss et al.

(2010) found that the simulation of glacier evolution with this $\triangle h$-parameterisation method was comparable to the results of a 3-D finite-element ice flow model. Li et al. (2015) used the approach of Huss et al. (2010) in combination with the well-known Hydrologiska Byråns Vattenbalansavdelning (HBV) model (Bergström and Singh, 1995; Seibert and Vis, 2012). The effect of these different glacier area conceptualisations on streamflow drought characterisation remains to be investigated.

For the analysis of future streamflow drought, methodological questions have been raised in the literature that relate to the definition of drought as a below-normal water availability. To quantify below-normal discharge, often a threshold method is used that defines the "normal" based on a baseline period. In the large-scale drought studies mentioned above (Feyen and Dankers, 2009; Wanders and Van Lanen, 2015; Van Huijgevoort et al., 2014; Forzieri et al., 2014) and, e.g. also in Wong et al. (2011), Lehner et al. (2006) and Arnell (1999), a threshold based on a historical period was used to define streamflow droughts in the future. It can be questioned if this historical threshold is a good indicator of the "normal water availability" in the future (see Wanders et al., 2015; Wanders and Wada, 2015; Van Loon et al., 2016). Especially in cold climates, expected regime shifts lead to the identification of severe droughts when evaluated against a historical threshold (Van Huijgevoort et al., 2014). This is particularly relevant in studies on future changes in streamflow drought in glacierised catchments where we expect fast-changing regimes due to the retreat of glaciers (e.g. Horton et al., 2006; Lutz et al., 2016). Wanders et al. (2015) therefore developed a transient threshold approach that takes into account changing regimes under climate change. This transient threshold assumes adaptation to long-term changes in the hydrological regime and hence identifies future streamflow droughts with reference to changed normal conditions. Wanders et al. (2015) applied this method to identify future streamflow droughts on a global scale and found that it reduces the area for which increases in drought duration and deficit volume are expected from 62 to 27 %. The transient threshold approach has, however, never been tested at the catchment scale and more specifically not in glacierised catchments.

This study aims to systematically test the effect of different methodological choices in simulating and analysing streamflow drought in glacierised catchments and elucidate which method to use for which purposes. We focus on two options for glacier modelling in a hydrological model (constant and dynamic glacier areas) and two different drought threshold approaches (historical and transient thresholds) resulting in four combinations. We test these combinations in two contrasting case study catchments in Norway and Alaska and discuss the implications for projections of future streamflow drought in glacierised basins in general.

2 Study areas and data

2.1 Study areas

Two catchments, one in Alaska (the Wolverine catchment) and one in Norway (the Nigardsbreen catchment), are used as case study in this research (Fig. 1) because of their good data availability, especially regarding glaciological data. The catchments are highly glacierised, i.e. 67 % (for the Wolverine catchment) and 70 % (for the Nigardsbreen catchment). The Wolverine glacier is a so called "benchmark glacier", where a long-term glacier monitoring program is maintained by the United States Geological Survey (USGS, 2015). Annual mass balances of the Wolverine glacier have been negative since 1990. The glacier has a southerly aspect. The area of the Wolverine catchment is 25 km^2, and the catchment elevation range is 360–1700 m. It is located in the Kenai Mountains in Alaska and close to the ocean at 60° N. It experiences a maritime climate (O'Neel et al., 2014). Long-term average monthly temperatures range from −6.7 to +8.8 °C. The catchment receives most of its annual precipitation (2700 mm) in autumn (410 mm in September) and precipitation is lowest in summer (100 mm in June). The Nigardsbreen glacier in Norway is one of the largest outlet glaciers of the Jostedalsbreen, which is the largest glacier in Europe. The Nigardsbreen glacier shows alternating negative and positive annual mass balances; however, the cumulative mass balance series is positive and has shown an increasing trend since around 1990. The main aspect of the glacier is south-east. The catchment area is 65 km^2 and it has a large elevation range of 260–1950 m. The climate of this catchment is also maritime. Long-term average monthly temperatures range from −6.6 to +6.6 °C. Precipitation amounts are highest in winter (450 mm in December) and lowest in spring (130 mm in May). Annual precipitation is around 3300 mm. The discharge station is located at the outlet of the Nigardsbreen lake.

2.2 Climate and hydrometric data

Observations of temperature (T_{obs}) and precipitation (P_{obs}) were used to force the model in the calibration and validation periods and to validate the climate model data in the historical period. Daily T_{obs} and P_{obs} data of the Nigardsbreen catchment were taken from a gridded dataset based on interpolation of observations from different gauging stations. From this dataset, the catchment average precipitation and temperature were calculated by the Norwegian Water Resources and Energy Directorate (NVE). Data were available for the period 1957–2014. Daily T_{obs} and P_{obs} data of the Wolverine catchment were obtained from USGS and were available for the period 1967–2015. The data come from a weather station close to the margin of the Wolverine glacier. However, the Wolverine catchment is a windy site, where wind speeds up to 100 km h^{-1} can occur during precipitation

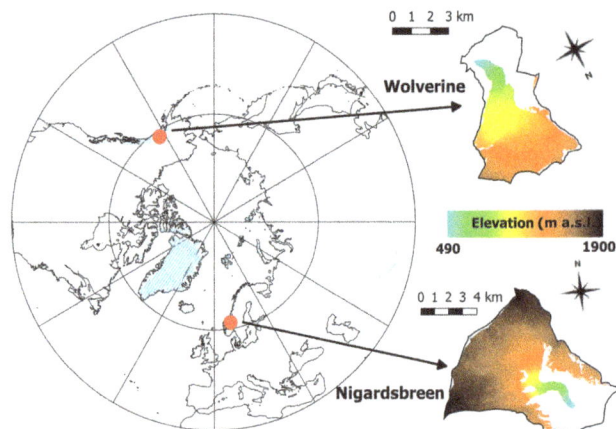

Figure 1. Location of case study catchments in Alaska (Wolverine catchment) and Norway (Nigardsbreen catchment). The coloured parts in the catchments indicate the glacier areas of 2006 (Wolverine) and 2009 (Nigardsbreen) and the elevation of the glaciers. The light blue colour in the overview map shows glaciated areas.

events, which can result in an undercatch problem. Therefore, after comparison with ERA-Interim precipitation data (Dee et al., 2011), observed precipitation amounts were increased by a factor of 2.5, to account for this precipitation undercatch in the Wolverine catchment. This was verified during the calibration process where the model forced with increased precipitation amounts resulted in a better fit with observed discharge than when using the original precipitation values. Gaps in the T_{obs} time series (7 %) of the Wolverine catchment were filled in with linear interpolation (for < 10 days of missing data) or, for longer than 10 days of missing data, with data from surrounding National Oceanic and Atmospheric Administration (NOAA) stations (taking into account altitude differences) or, when no data were available from surrounding stations, with long-term average daily temperatures. Gaps in the P_{obs} time series (7 %) of Wolverine were filled based on surrounding NOAA stations, again accounting for elevation differences.

For the future projections, daily P and T data from a set of climate models were used (P_{cm} and T_{cm}). Additionally, the model in the historical period was forced with climate model data, in order to compare discharge and droughts between the historical and future periods. The climate model data are output from global climate model – regional climate model (GCM-RCM) model combinations from the World Climate Research Program Coordinated Regional Climate Downscaling Experiment (CORDEX; Giorgi et al., 2009). For Norway, data from EURO-CORDEX, and for Alaska, data from the North American CORDEX (Jacob et al., 2014) were available. The resolution of the data over Norway is 0.11° and for Alaska 0.22°. Nearest neighbour interpolation to the centre point of the catchments was used to obtain catchment aver-

age P_{cm} and T_{cm} from the climate models. Climate model data for the period 1975–2004 (historical period) were used as reference data and compared with P_{obs} and T_{obs}. For the period 2006–2100, the climate model outcomes for two climate scenarios were used, i.e. the RCP4.5 and RCP8.5 scenarios. For Norway, bias-corrected (with E-OBS; Haylock et al., 2008) climate model output data from eight GCM-RCM model combinations were available for the RCP4.5 scenario and nine for the historical period and the RCP8.5 scenario. For Alaska, only data from one GCM-RCM model combination were available without bias correction. Therefore, the empirical quantile mapping method was applied to perform bias correction on the Alaskan data by using the observations from the weather station in the Wolverine catchment (Teutschbein and Seibert, 2012). This method was chosen because it is the same method as was used for the Norwegian climate data.

Observed discharge (Q_{obs}) was used for calibration and validation of the model and was provided by NVE and USGS, for Nigardsbreen and Wolverine (USGS Waterdata, 2016), respectively. The discharge was measured at the outlet of the catchments. Daily discharge data were available for 1963–2013 for Nigardsbreen and 1969–2015 for the Wolverine catchment. In the Wolverine discharge time series, gaps were present for several years. These years were excluded from the analysis.

2.3 Glaciological data

Seasonal glacier-wide mass balances of both glaciers were also obtained from USGS (O'Neel et al., 2016) and NVE (Andreassen and Engeset, 2016). The mass balances were used for calibration of the HBV-light model. Geodetically adjusted seasonal mass balances (winter and summer mass balances) were available for the Wolverine glacier and a homogenised seasonal mass balance series was available for this study for the Nigardsbreen glacier (Van Beusekom et al., 2010; O'Neel, 2014; Andreassen and Engeset, 2016).

Glacier outlines were used to define the glacier fraction in the catchments. These glacier outlines were obtained from the Randolph Glacier Inventory (RGI version 5.0; Pfeffer et al., 2014) and from NVE (Winsvold et al., 2014; Andreassen et al., 2012). The glacier outlines were also used in combination with ice thickness data to define the volume of the glaciers. The ice thickness maps were available at a spatial resolution of 100×100 m for Nigardsbreen and 25×25 m for Wolverine. The information on distributed ice thickness of the glaciers from the maps was used for the dynamic glacier area modelling. For the Wolverine glacier, the ice thickness data of Huss and Farinotti (2012), and for the Nigardsbreen glacier the data of Andreassen et al. (2015) were used.

## 3	Methods

### 3.1	General modelling approach

The main variable of interest in this research is the river discharge. Since we are interested in the future, streamflow is modelled using a coupled glaciohydrological model (see Sect. 3.2). Streamflow droughts are studied in two periods, a historical period (1975–2004) and a future period (2071–2100), in order to assess changes in drought characteristics between both periods (see Fig. 2). To systematically test the effect of the glacier dynamics and threshold approach on future streamflow droughts and their characteristics, four scenarios, in which the glacier dynamics and threshold approach options are combined, are used to characterise streamflow droughts in these two periods (Fig. 3). The historical variable threshold (HVT) and the constant glacier area conceptualisation (C) represent the baseline conditions which we compare with the changing conditions: the transient variable threshold (TVT) and dynamic glacier area conceptualisation (D) (see Fig. 3).

The two threshold approaches that are tested and compared in our glacierised case study catchments are the more often used HVT method, based on a fixed reference period in the past, and the recently introduced TVT method, based on a changing reference period, thereby taking into account changes in the hydrological regime. The calculations of the thresholds are explained in Sect. 3.5. The glacier modelling options that are evaluated include a static and infinite glacier reservoir and a glacier geometry change conceptualisation using the Δh parameterisation of Huss et al. (2010). These two glacier modelling options, in the following referred to as "constant" and "dynamic" glacier modelling options, are further explained in Sect. 3.2. Although the constant glacier modelling option will be unrealistic in transient mode, we include this option in our analysis because dynamical glacier modelling is not yet included in all (large)-scale hydrological models (e.g. Zhang et al., 2013) and it is an interesting benchmark, also frequently used in other studies (Akhtar et al., 2008; Stahl et al., 2008; Tecklenburg et al., 2012). The effect on streamflow drought characterisation has not yet been assessed. P_{obs}, T_{obs} and Q_{obs} were used to calibrate and validate the model, and were compared with simulated discharge (Q_{sim}) obtained by forcing with observations (Q_{sim_o}) and climate model data ($Q_{sim_{cm}}$) in the historical period to address the uncertainty in both components. Future runs start in 2006 with a 4-year spin-up period, so that discharge is modelled for the period 2010–2100, to include the transient evolution of the glacier area during the 21st century (Fig. 2). The model is forced with RCP4.5 and RCP8.5 climate change scenario data during the future simulations.

### 3.2	Conceptual model

The model used in this study is the conceptual HBV-light

model with extended glacier routine (Seibert and Vis, 2012; Seibert et al., 2017). It is a version of the original HBV model developed at the Swedish Meteorological and Hydrological institute (Bergström and Singh, 1995). The model is semi-distributed, based on elevation zones, vegetation zones and aspect classes. Daily temperature, precipitation and daily or long-term monthly potential evapotranspiration are needed as input variables. The model simulates discharge and also calculates the contributions of the different components (rain, glacier ice (Q_g) and snow) to the total discharge. A glacier profile, in which the ice volume in the different elevation zones is defined, is needed in order to run the model with a dynamic glacier area.

The model consists of different routines. The glacier, snow and soil moisture routines are semi-distributed, whereas the groundwater and routing routines are lumped. The model simulates discharge at a daily time step. Based on a threshold temperature, precipitation will fall either as snow or rain. A snowfall correction factor is used in the model to compensate for systematic errors in snowfall measurements and for evaporation/sublimation from the snowpack (not explicitly modelled). In the snow and glacier routines, the melt is computed by a degree-day method. A different degree-day factor is used for snow and glacier because of the lower albedo of glacier ice. Snow redistribution is not taken into account. For a detailed model description, we refer to Seibert and Vis (2012). The calibrated parameter values of the snow and glacier routines are presented in Appendix A. The glacier in the model is represented by two components: a glacier ice reservoir and a glacier water content reservoir. A small fraction (0.001) of the snow on the glacier is transformed into ice each time step. When the glacier is not covered by snow, glacier melt is taking place for temperatures above the threshold temperature. Glacier melt is added to the glacier water content reservoir, just like water from snow on the glacier which melts and rain falling on the glacier. From the glacier water content reservoir, water is flowing directly into the routing routine. The amount of discharge from the glacier is based on an outflow coefficient which varies in time because it depends on the snow water equivalent of the snowpack on the glacier. It represents the development of glacial drainage systems (Stahl et al., 2008). In the non-glaciated part of the catchment, snowmelt and rainfall flow into the soil routine. From here, water can evaporate or be added to the groundwater reservoirs. Peak flow, intermediate flow and baseflow discharge components are generated within the groundwater routine, which is followed by the routing routine, in which the total discharge of one time step is distributed over one or multiple time steps according to a weighting function.

The glacier routine in the HBV-light model can be used as a static or dynamic conceptualisation of the glacier in the catchment. In the static conceptualisation, the glacier area

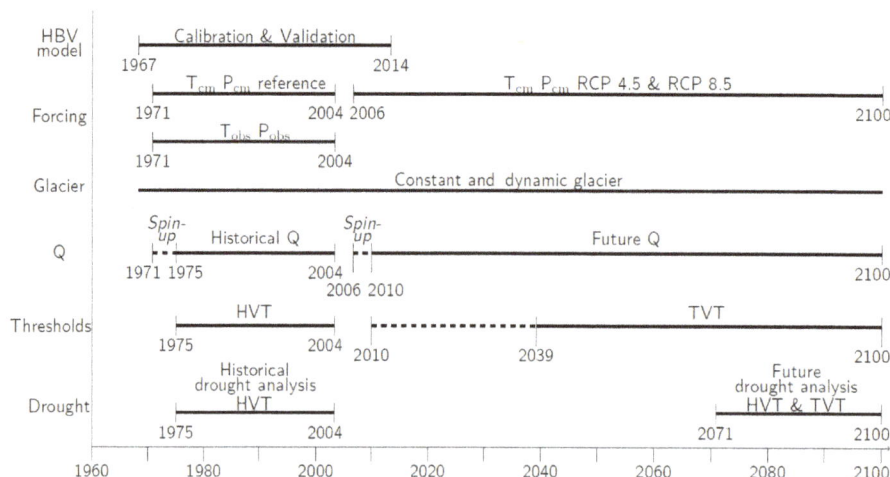

Figure 2. Timeline indicating the simulation periods, forcings and periods for threshold derivation and application. The different glacier area conceptualisations, constant and dynamic, are used in all simulation periods.

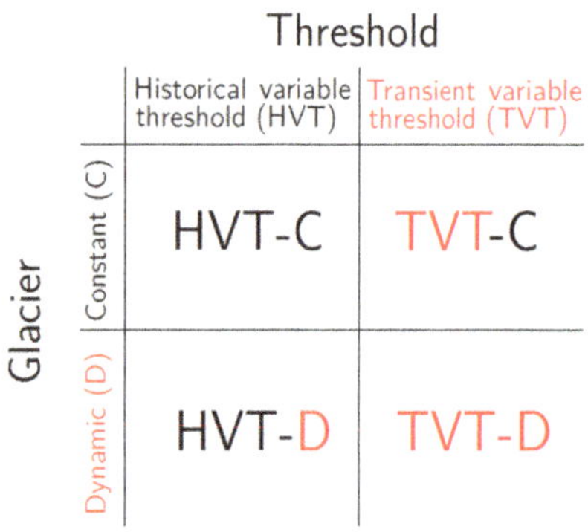

Figure 3. Four analysis scenarios. The matrix shows the combination of the two threshold approaches with the two different glacier area conceptualisations, resulting in four possible combinations. The baseline options are indicated in black and the options where changes are taken into account are shown in red.

is constant over time, while in the dynamic conceptualisation, the area of the glacier is adjusted every year. The dynamic glacier conceptualisation in the HBV-light model is based on Huss et al. (2010), who proposed a simple parameterisation to calculate the change in glacier surface elevation and area ($\triangle h$ parameterisation), so that future glacier geometry change can be approximated without using complex ice flow modelling. The $\triangle h$ parameterisation describes the spatial distribution of the glacier surface elevation change in response to a change in mass balance and has also been used in other studies, e.g. Salzmann et al. (2012); Farinotti et al. (2012); Li et al. (2015); Duethmann et al. (2015). The implementation of various dynamic glacier change options into HBV-light based on the $\triangle h$ parameterisation is described and tested in Seibert et al. (2017). In HBV-light, one out of three possible type curves for different glacier sizes can be chosen (Huss et al., 2010). Furthermore, a glacier profile, in which

the water equivalent and area of the glacier for each elevation band (elevation bands are subdivisions of the elevation zones) are specified, is required by HBV-light as input for the dynamic glacier conceptualisation. Before the actual model simulation starts, the glacier profile is melted in steps of 1 % of the total glacier volume, and for each step the $\triangle h$ parameterisation of Huss et al. (2010) is applied to compute the areal change for each elevation zone. This information is stored by the HBV-light model in a lookup table of percentage of melt and corresponding glacier areas. This table is then used to dynamically change the glacier during the actual model simulation. Each hydrological year, the area of the glacier is updated by calculating the percentage of glacier volume change from the modelled mass balance and selecting the corresponding glacier areas from the lookup table.

3.3 Model set-up

For daily temperature and precipitation input, we used observations or output from climate models. The HBV-light model requires a climate station at a certain elevation for the input of P and T. For the Wolverine catchment, the HBV climate station elevation was set to the elevation of the weather station in the catchment for T_{obs}, P_{obs}, T_{cm} and P_{cm}. For the Nigardsbreen catchment, the average catchment elevation was used for the HBV climate station elevation for T_{obs} and P_{obs} and the average elevation of the RCM model grids for the T_{cm} P_{cm}. P and T values for each elevation zone are calculated based on precipitation and temperature lapse rates, which are calibration parameters. Monthly evapotranspiration (E) was calculated for all simulation periods with the Blaney–Criddle method by using monthly average temperatures in order to get E values for both the historical and future simulations (Xu and Singh, 2001; Brouwer and Heibloem, 1986). The monthly values were linearly interpolated to retrieve daily values which were used as input to HBV-light. Each catchment was divided into several elevation zones, with elevation bins of 100 or 200 m depending on the elevation range in the catchment. Each elevation zone was split up into three aspect classes (north, south, east–west). The mean elevation of each elevation zone and the area of each elevation zone and aspect class were determined from the ASTER digital elevation model (DEM). Missing values present in the ASTER DEM of Nigardsbreen were filled in by interpolation. The lake present in the Nigardsbreen catchment was defined as a separate model unit.

To determine the glacier area in each elevation zone, glacier outlines of 2006 were used. For the static glacier conceptualisation, these areas were used in all model runs, independent of time. However, in order to run the model with the dynamical glacier settings, initial glacier areas and glacier profiles were adapted to the largest glacier extent within the specific simulation period. For the future simulation period, it was assumed that the glacier extent will be largest at the start of the period (2006). Therefore, initial glacier areas and the glacier profile based on ice thickness maps and 2006 glacier outlines needed no adaptation. For the other simulation periods (historical period, calibration period and validation period), the largest glacier extent was determined from area information from USGS for the Wolverine glacier and from homogenised area data from NVE for Nigardsbreen glacier (Andreassen and Engeset, 2016). The 2006 glacier areas and glacier profiles were adapted to these largest glacier extents.

For the construction of the 2006 glacier profile, each glacier elevation zone was subdivided into smaller elevation bands, with elevation bins of 20 or 50 m, depending on the size of the elevation zone. For each elevation band, the average ice thickness was determined from the ice thickness maps and converted into millimetre water equivalent (mm w.e.q.) by multiplying with the ratio of the densities from ice to water (0.917). The adjustment of the glacier profile to another glacier extent was done based on volume–area scaling (Bahr et al., 1997; Andreassen et al., 2015) to calculate the needed increase in ice thickness/water equivalent to match the new volume based on the new largest glacier extent. When the largest glacier extent did not occur at the beginning of the simulation period, an initial glacier fraction was defined in the glacier profile which was also calculated with the volume–area scaling method.

3.4 Calibration

The models were calibrated against (selected periods of) observed discharge and seasonal mass balances using the automatic calibration tool genetic algorithm and Powell (GAP) optimisation in HBV-light (Seibert and Vis, 2012; Seibert, 2000). Including mass balances in the calibration is known to improve the model performance significantly (Konz and Seibert, 2010; Mayr et al., 2013; Engelhardt et al., 2014). For each catchment, the model was calibrated with a constant glacier area conceptualisation and a dynamic glacier

conceptualisation, so that a different parameter set was obtained for both glacier area conceptualisations. To calibrate on mass balances, the dates of maximum and minimum mass balances were used for the winter balance and the summer balance, respectively, for the Wolverine catchment, and the actual measurement dates of the summer and winter balances for the Nigardsbreen catchment (metadata from NVE). A calibration period of at least 10 years was used for both catchments. The objective function that was maximised during the calibration is

$$R = 0.4 \times R_{\text{effG}} + 0.4 \times R_{\text{effS}} + 0.2 \times R_{\text{effP}}, \tag{1}$$

with

$$R_{\text{eff}} = 1 - \left(\frac{\sum (\text{Obs} - \text{Sim})^2}{\sum (\text{Obs} - \overline{\text{Obs}})^2} \right),$$

where R is the model performance, R_{effG} the calibration on glacier mass balances, R_{effS} the calibration on the discharge from April to September and R_{effP} the calibration on the peak discharges. Obs and Sim are observed and simulated (seasonal) discharge or glacier mass balances, respectively. A R_{eff} value of 1 indicates a perfect fit for that variable.

After the calibration was performed, model performance was evaluated with the Kling–Gupta efficiency (KGE) which is defined as

$$\text{KGE} = 1 - \sqrt{(r - 1)^2 + (\alpha - 1)^2 + (\beta - 1)^2}. \tag{2}$$

In Eq. (2), r is the Pearson product–moment correlation coefficient, α the ratio of the SDs of simulated and observed discharge and β the ratio between the means of simulated and observed discharge (Gupta et al., 2009). A KGE value of 1 indicates a perfect fit between modelled and observed discharge.

3.5 Drought thresholds

A variable threshold level method was used to identify droughts and to determine their characteristics (Hisdal et al., 2000; Fleig et al., 2006; Van Loon, 2013). A drought occurs when a variable (in our study discharge) falls below the threshold. We used a daily variable threshold that is derived from a 30-day moving average discharge time series. The moving average time series was used to compute the daily flow duration curves and to determine the 80th percentile for use as a drought threshold (Van Loon et al., 2014). Usually threshold levels between the 70th and 95th percentiles are applied in drought studies (Fleig et al., 2006). Using another threshold or different moving window size will result in slightly different drought characteristics, but the percentile choice has less effect on the results when only looking at changes in drought characteristics and comparing different approaches, as was done in this study.

This variable threshold was calculated for both catchments and glacier conceptualisations separately. The historical variable threshold was calculated from the discharge in the historical period (1975–2004). For the future period, two threshold approaches were used: (1) the variable threshold from the historical period (HVT) following the work of Wanders and Van Lanen (2015) and Van Huijgevoort et al. (2014) and (2) a TVT that assumes adaptation in the future based on reduced or increased water availability of the preceding 30-year period (Wanders et al., 2015). Hence, each year in the future has a different TVT, calculated from the previous 30 years of discharge as described above. The same HVT was used in the historical period and the future period for both climate change scenarios. The TVT was used in the future period, but for both climate change scenarios (RCP4.5 and RCP8.5) a different transient threshold was calculated. For the Nigardsbreen catchment, the multi-model mean $Q_{sim_{cm}}$ was used for calculation of the thresholds and the drought analysis.

We computed the drought duration, deficit and intensity to characterise changes in drought characteristics. The drought duration is defined as the consecutive number of days that the discharge is below the threshold. Droughts with a duration of 3 days or shorter were not taken into account (Fleig et al., 2006). The drought deficit volume is computed by taking the cumulative difference between the drought threshold and the discharge for each drought event. Drought intensity is defined as the deficit divided by the duration. We analysed drought processes by studying temperature, precipitation, snow water equivalent (SWE) and the different discharge components together with the total discharge following the approach of Van Loon and Van Lanen (2012) and Van Loon et al. (2015). The thresholds for these variables were computed in the same way as was done for the discharge, except for temperature for which we used the median as threshold.

4 Results

4.1 Calibration and validation of model and data

The KGE values of the calibration and validation periods are generally high (Table 1). Especially the Nigardsbreen catchment shows a very good agreement between modelled and observed discharge (KGE of 0.94). The KGE is slightly lower in the validation period of the Nigardsbreen catchment and somewhat higher for the Wolverine catchment. The latter might be caused by the very short validation period of Wolverine. The type of glacier area modelling does not influence the model performance with respect to discharge in both the calibration and validation periods. The individual R_{eff} values of Eq. (1) range between 0.51 and 0.90 for the seasonal calibration and between 0.15 and 0.60 for the peak flow calibration for the two catchments. The R_{eff} values for

the mass balance calibration are 0.51 and 0.83 for the dynamic glacier simulations of Nigardsbreen and Wolverine, respectively. The hydrological regimes of observed and modelled discharge also match well for Nigardsbreen for the historical period (Fig. 4a), for both types of forcing: observations and climate model data. For the Wolverine catchment, only 3 years of observed data were available in the historical period, resulting in a more uncertain observed regime compared to the simulated regimes in Fig. 4d. The inset in Fig. 4d shows the matching observed regime and the simulated regime forced by observations for the calibration period. Besides matching regimes, the model is also able to simulate a similar interannual variability in discharge compared to the observations for Nigardsbreen (Fig. 4b, historical period) and Wolverine (Fig. 4e, calibration period).

We also compared modelled and observed glacier mass balances for the dynamic glacier area (see Fig. 4c and 4f). During the calibration period, the negative trend in cumulative mass balance is simulated very well by the model for the Wolverine catchment (Fig. 4f). Winter mass balances and the total volume change are slightly underestimated. In the Nigardsbreen catchment, the model simulates negative cumulative mass balances at the start of the calibration period, while observed cumulative mass balances are positive. In this period, the model did not capture the sign of the almost balanced conditions right. However, during the second half of the calibration period, the positive trend in mass balance is the same in the observations and simulations. The intra-annual differences in summer and winter balances are smaller in the simulations in both catchments.

Finally, we verified the streamflow drought characteristics of observed and simulated discharge in the calibration period (Table 2). The number of droughts for Nigardsbreen is a bit higher in the simulations than in the observations. However, in general, drought characteristics of observed and simulated discharge agree well for both catchments.

4.2 Glacier area conceptualisations and their effect on discharge

During the constant glacier area runs, the model used a glacier area from 2006, both in the historical and future periods (Fig. 5). Assuming that glaciers will shrink in the future, this area is too big during the future period and too small during the historical period because both glaciers had a larger area in the past compared to 2006. With a dynamic glacier area conceptualisation, this mismatch should not occur. In the Wolverine catchment, the glacier area in the historical period for the dynamic settings is indeed higher than the glacier area in the constant settings and the glacier area at the end of the historical period agrees with the constant area (observed glacier area in 2006) used throughout the whole modelled time period (Fig. 5). However, in the Nigardsbreen catchment, the average modelled glacier area at the end of the historical period (2004) is smaller than the observed glacier

Table 1. Model performance for the two catchments. Performance is expressed by KGE between observed and modelled discharge and shown for the calibration and validation periods and the dynamic (D) and constant (C) glacier area conceptualisations.

Catchment	Calibration			Validation		
	C	D	Period	C	D	Period
Nigardsbreen	0.94	0.94	1967–2003	0.90	0.90	2004–2013
Wolverine	0.82	0.83	2005–2014	0.89	0.87	1973–1977

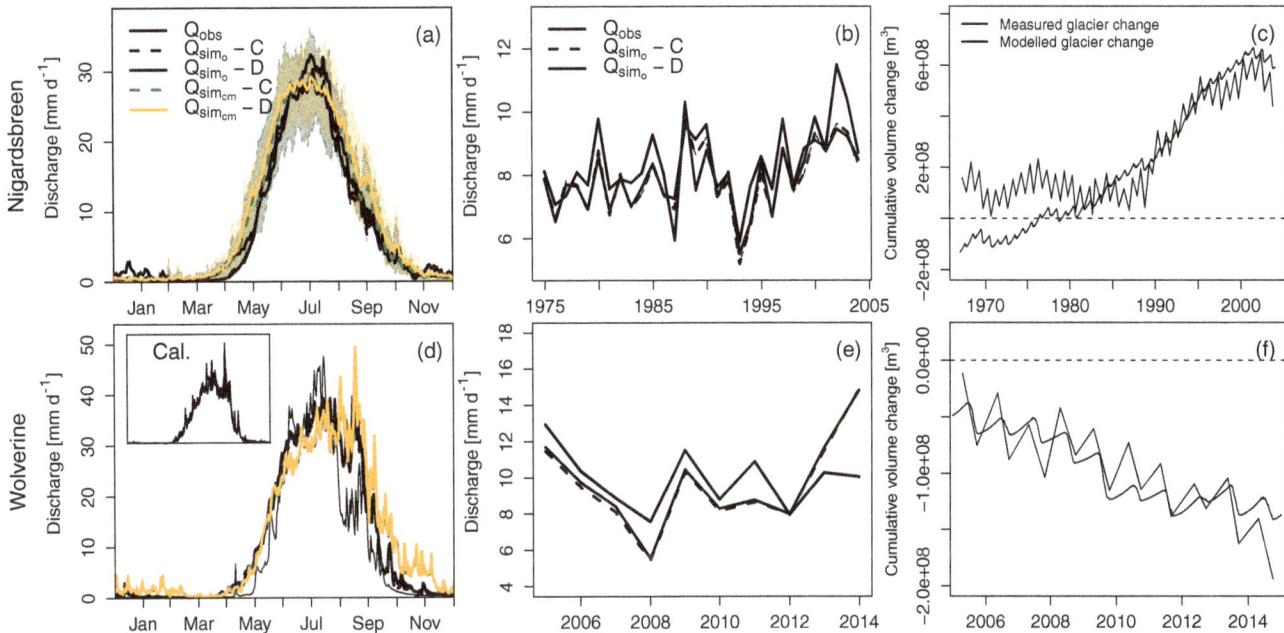

Figure 4. Model validation. The hydrological regime (**a** and **d**), annual discharges (**b** and **e**) and mass balances (**c**) and (**f**) are shown for the Nigardsbreen catchment (**a**, **b** and **c**) and for the Wolverine catchment (**d**, **c** and **f**). Panels (**a**) and (**d**) show results for the historical period (1975–2004) in order to compare observed discharge with both ($Q_{\mathrm{sim_o}}$) and ($Q_{\mathrm{sim_{cm}}}$) for both glacier model conceptualisations. The coloured areas in panel (**a**) indicate the range of discharge outputs as a result of the different climate model forcings. The inset in panel (**d**) shows the agreement between Q_{obs} and $Q_{\mathrm{sim_o}}$ for the Wolverine catchment during the calibration period. In panel (**b**), $Q_{\mathrm{sim_{cm}}}$ is not shown because climate models only statistically represent historic climate. The interannual variability is shown for the historical period for Nigardsbreen and for the calibration period for Wolverine (due to Q_{obs} availability). Panels (**c**) and (**f**) show the observed and measured glacier volume changes (water equivalent) for the calibration period of the Nigardsbreen and Wolverine glaciers, respectively.

area in 2006 (the constant glacier area). The model simulates a glacier area that decreases too much, or a too-small glacier extent was used at the start of the historical period, and therefore there is a small jump between the average glacier area at the end of the historical period and the start of the future period (2006–2100) (the model periods are not coupled) (Fig. 5). The model simulates a glacier disappearance in the Wolverine catchment in the future when dynamic glacier areas are used, first in the RCP8.5 scenario and later also in the RCP4.5 scenario. In the Nigardsbreen catchment, the glacier area develops similarly in both climate scenarios until 2060, after which the glacier is projected to shrink more quickly in the RCP8.5 scenario. The spread in glacier area evolution projections for the Nigardsbreen catchment is however large.

One climate model forcing even gives hardly any decrease in glacier area.

The different options for glacier area modelling have an effect on the future water availability (Fig. 6). The constant glacier area causes an amplification of the hydrological regime and increasing annual discharges in the future in both catchments. On the other hand, the dynamic glacier area causes a drastic change in the regime in the Wolverine catchment in the future period (2071–2100) (Fig. 6c). The regime in the Nigardsbreen catchment changes as well: the magnitude of the high flow period is smaller, the rising limb starts earlier and the recession limb starts later and is less steep than during the historical period. For both catchments, the changes compared to the historical period are larger for the RCP8.5 scenario. Annual discharges are projected to de-

Table 2. Streamflow drought characteristics of observed and simulated discharge. Drought characteristics are shown for observed discharge (obs) and simulated discharge (sim) with constant (C) and dynamic (D) glacier area conceptualisation in the calibration period of Nigardsbreen and Wolverine.

Catchment	Discharge	Number	Avg. duration (d)	Avg. deficit (mm)	Avg. intensity (mm d^{-1})
Nigardsbreen	obs	357	12.21	16.48	1.39
(1967–2003)	sim-C	565	9.66	12.27	1.23
	sim-D	484	10.92	13.40	1.27
Wolverine	obs	99	13.49	25.97	2.80
(2005–2014)	sim-C	114	13.89	19.28	2.02
	sim-D	99	12.95	25.73	2.64

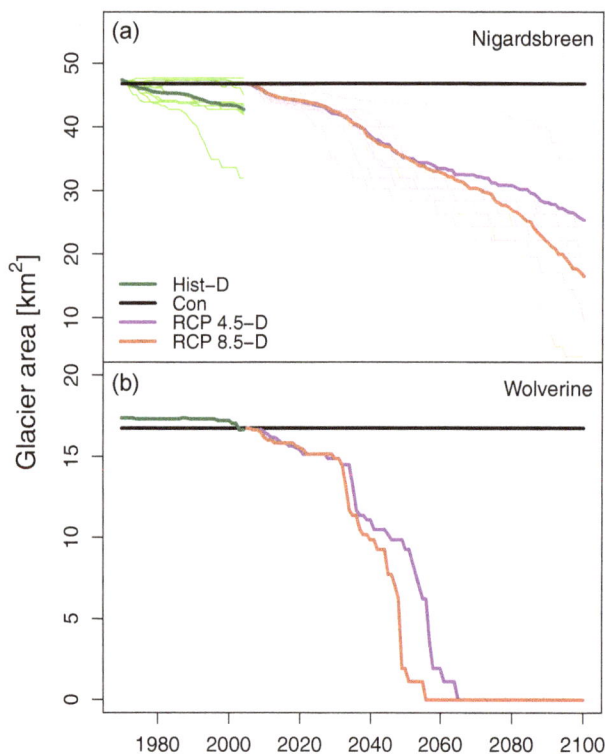

Figure 5. Temporal evolution of glacier areas for the historical and future periods. Panel **(a)** shows the glacier areas for Nigardsbreen and **(b)** for Wolverine. The glacier area in both glacier conceptualisations is shown. The lighter coloured lines in the Nigardsbreen graph for the historical period and the two RCP scenarios show the results of glacier area evolution for the different climate model forcings.

crease in the Wolverine catchment with dynamic glacier area. The changes in multi-model mean annual discharges for Nigardsbreen are not so clear and the spread among the discharges forced by the different climate models increases in the future.

4.3 Drought thresholds: the result of different glacier conceptualisations and threshold methods

Four approaches were used for the determination of the drought thresholds and future drought analysis, based on combinations of the threshold options and the glacier area conceptualisations. For both catchments, the HVT-C and HVT-D thresholds are quite comparable (Fig. 7), except in the rising and recession limb of Nigardsbreen, where the HVT-D is above the HVT-C. The transient thresholds, however, vary in time. The magnitude of the high flow season in the TVT-C increases, while with the TVT-D it decreases each year in the future. In the Nigardsbreen catchment, the TVT-D threshold has a higher peak during the first decade compared to the historical period, after which the peak in the threshold becomes lower. All future TVT-D have, however, a longer high flow season than the historical threshold has. In the Wolverine catchment, the TVT-D only shows a higher peak in August and September in the first years in the future compared to the HVT-D. Moreover, a shift is visible for the rising limb in the TVT-D towards an earlier moment in the spring season for Nigardsbreen. The TVT-C develops in both catchments differently; in Nigardsbreen, the peak shifts to earlier in the season, while for Wolverine, the TVT-C peak shifts to later in the season.

The transient threshold does not adapt at a constant rate, shown by the different spaces between the lines (Fig. 7). The threshold follows the climate. The RCP8.5 scenario gives similar results (not shown), but there is even more difference between consecutive thresholds. This is due to a faster changing climate and discharge. For the Wolverine catchment, the changes in the transient threshold are more extreme than Nigardsbreen, especially in the first half of this future period (2039–2070) of the TVT-D, in which the glacier is rapidly shrinking. Furthermore, due to the drastically changing regime, the transient threshold in the Wolverine catchment changes also rapidly in the historical low flow periods (winter), in contrast with Nigardsbreen where the threshold stays low in the historical low flow periods.

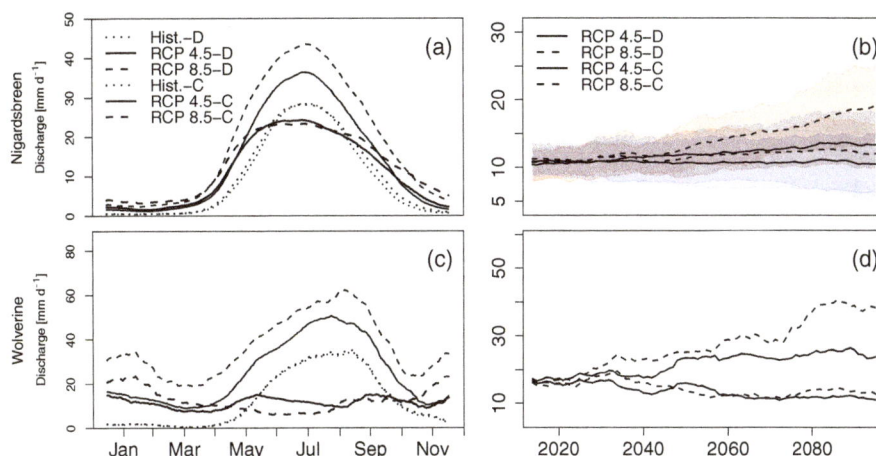

Figure 6. Future water availability. Panels **(a)** and **(c)** show the hydrological regime (30-day moving window of the daily average of 2071–2100) for Nigardsbreen **(a)** and Wolverine **(c)**, and panels **(b)** and **(d)** the annual average discharges for the future period (2010–2100) with a 10-year moving window, for **(b)** Nigardsbreen and **(d)** Wolverine. Discharge is shown for both glacier area conceptualisations (colours) and both climate change scenarios (line type). The shaded areas in panel **(b)** indicate the spread in annual average discharges among the different climate model forcings.

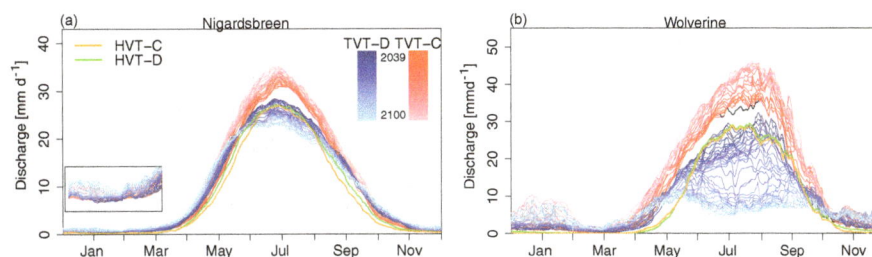

Figure 7. Drought thresholds for the four different scenarios (HVT-C, HVT-D, TVT-C and TVT-D). The colour gradient for both transient thresholds (blue and red) indicates the adaptation of the threshold each year (for 2039–2100). The thresholds are shown for the Nigardsbreen **(a)** and Wolverine **(b)** catchments for climate scenario RCP4.5. The inset in panel **(a)** zooms in to the low flow period of Nigardsbreen.

4.4 Effect of thresholds on the identification and characterisation of future droughts

Applying the different thresholds to the discharge time series shows when droughts (below threshold discharges) occur during the year (Fig. 8 shows an example for the Nigardsbreen catchment). The HVT-C and TVT-C are applied to the discharge output of the model simulated with a constant glacier area conceptualisation and the HVT-D and TVT-D to the output produced with a dynamic glacier area conceptualisation. Applying the threshold of the past to the discharge of the future with a constant glacier area (HVT-C) results in (almost) no droughts (Fig. 8) due to increased glacier melt. If the threshold of the past is applied to discharge with a dynamic glacier area conceptualisation (HVT-D), severe droughts occur at the period of the threshold high flow season and in the recession limb of the discharge curves due to a lower peak flow and a shift in the hydrological regime (Fig. 8).

Using the transient threshold results in future droughts with much smaller deficit volume, compared to droughts determined with HVT-D (Fig. 8). Droughts do not only occur in the peak flow period but are more distributed over the season and occur in the rising limb and low flow period as well, in both the TVT-C and TVT-D cases. In Fig. 8, streamflow droughts look more severe (higher deficits) in the TVT-D settings than in TVT-C settings, while in both cases the threshold has adapted. This is probably caused by the contribution of glacier melt to discharge. In the TVT-D, the threshold is based on 30 previous years when the glacier was larger than the year to which the threshold is applied, resulting in droughts partly caused by glacier retreat. The TVT-C, on the other hand, is based on 30 previous years in which the climate was colder than the year the threshold is applied, resulting in less melt from the glacier compared to the year the threshold is applied (glacier area is constant), and consequently less droughts are observed in the high flow season compared to TVT-D.

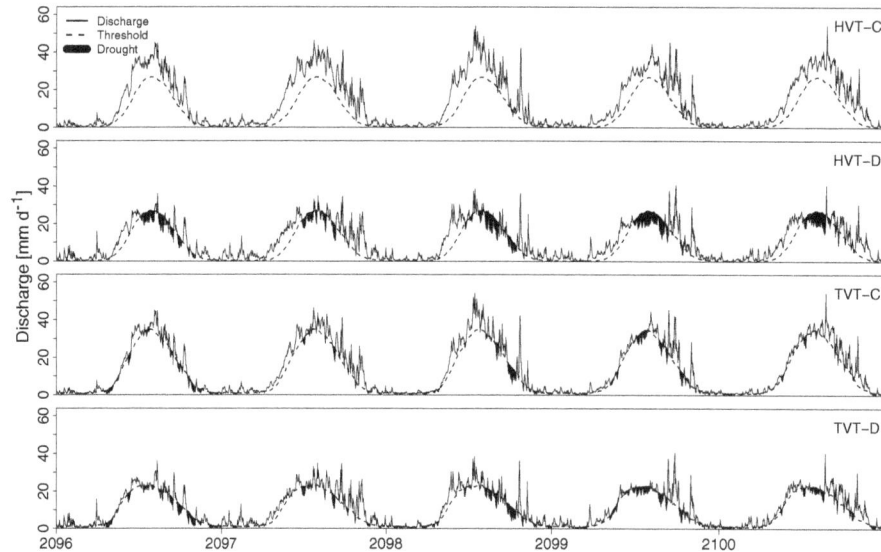

Figure 8. Example time series of possible timing and deficit volume of droughts in the four scenarios (HVT-C, HVT-D, TVT-C and TVT-D). The droughts are shown for the Nigardsbreen catchment and the RCP4.5 climate scenario for the period 2096–2100.

Besides a different timing of streamflow droughts in the year, the four threshold scenarios also resulted in different drought characteristics (e.g. deficit volume). Comparing drought characteristics between historical and future periods shows the changes that can be expected in the future. However, the four scenarios resulted in different future changes in drought characteristics (Table 3). The number of droughts will decrease in both catchments when the HVT is used. The number of droughts will increase in the Wolverine catchment when the transient threshold is used. In the Nigardsbreen catchment, the TVT-C indicates a decrease in the number of droughts and the TVT-D only results in a small increase in the number of droughts. The average duration will only increase in the HVT-D scenario (except RCP4.5 for Nigardsbreen); in the other threshold scenarios, the average duration is projected to decrease. The HVT-D and TVT-D result in a projected increase in deficit volumes, except for TVT-D in the Wolverine catchment. However, deficit volumes are projected to increase more drastically when HVT-D is used. The HVT-C causes in general a decrease in deficit volume, while the TVT-C causes an increase in the deficit volume. Average intensities are in general projected to increase for all scenarios, with one exception for both Nigardsbreen and Wolverine (see Table 3). For most threshold scenarios, the RCP8.5 will give a larger change in the drought characteristic than the RCP4.5 scenario compared to the historical period.

4.5 Effect of thresholds on analysing future drought processes

Using the four different methodological scenarios we can analyse streamflow drought processes differently. We separated the four scenarios into two comparisons: the glacier dy-namics effect and the influence of the threshold approach on analysing drought processes. To study the glacier dynamics effect, the transient threshold was used for both glacier area conceptualisations (Fig. 9). No historical variable threshold was used here to exclude the effect of changing peak flow discharges compared to the historical period. The thresholds in Fig. 9a and b are therefore based on the 30 previous years of discharge (TVT). In the constant glacier area conceptualisation, a drought occurs in streamflow in the beginning of September, while for the dynamic glacier area several streamflow droughts occur between June and September (Fig. 9). The long-term climatic changes cause the glacier to retreat in the future in the dynamic glacier conceptualisation. This glacier retreat can have an indirect effect on the occurrence of streamflow droughts because of less melt due to a smaller glacier. Streamflow droughts occurring in the summer period of 2092 in the Nigardsbreen catchment for the dynamic glacier area show this process (Fig. 9b). Streamflow droughts are caused by short-term (seasonal) anomalies in P (deficits) and T (lower) and additionally due to a retreating glacier resulting in less discharge from the glacier (Fig. 9). In the constant glacier area conceptualisation, the effect of long-term climate changes on glacier size is neglected and streamflow droughts are caused by short-term climate variability. In Fig. 9a, the drought in September is caused by below-normal temperatures, resulting in a deficit in Q_g and a drought in the total streamflow (Q). Furthermore, Fig. 9a shows that glacier melt in summer is buffering against the propagation of precipitation deficits. This effect gets lost with retreating glaciers and any remaining buffering against precipitation deficits needs to come from other stores, e.g. the snowpack and groundwater.

Table 3. Change in drought characteristics in the future compared to the historical period. The percentages show the increase or decrease of the respective drought characteristic with respect to the historical period for each catchment and each glacier area conceptualisation.

	Period	Scenario	Number		Avg. duration (d)		Avg. deficit (mm)		Avg. intensity (mm d^{-1})	
			RCP4.5	RCP8.5	RCP4.5	RCP8.5	RCP4.5	RCP8.5	RCP4.5	RCP8.5
Nigardsbreen	**Hist**	**HVT + C**	477		7.86		8.77		0.96	
	Fut	HVT + C	−80 %	−97 %	−39 %	−46 %	−61 %	−81 %	−35 %	−60 %
	Fut	TVT + C	−10 %	−47 %	−14 %	−26 %	25 %	−3 %	58 %	48 %
	Hist	**HVT + D**	467		8.06		9.34		1.04	
	Fut	HVT + D	−37 %	−58 %	−4 %	12 %	166 %	309 %	137 %	191 %
	Fut	TVT + D	9 %	3 %	−15 %	−18 %	38 %	66 %	63 %	97 %
Wolverine	**Hist**	**HVT + C**	400		10.21		26.68		3.21	
	Fut	HVT + C	−66 %	−81 %	−35 %	−39 %	9 %	−14 %	53 %	23 %
	Fut	TVT + C	23 %	21 %	−21 %	−38 %	79 %	88 %	106 %	142 %
	Hist	**HVT + D**	354		10.1		34.1		4.39	
	Fut	HVT + D	−21 %	−31 %	25 %	66 %	431 %	674 %	133 %	152 %
	Fut	TVT + D	72 %	91 %	−12 %	−12 %	−20 %	−13 %	−30 %	−21 %

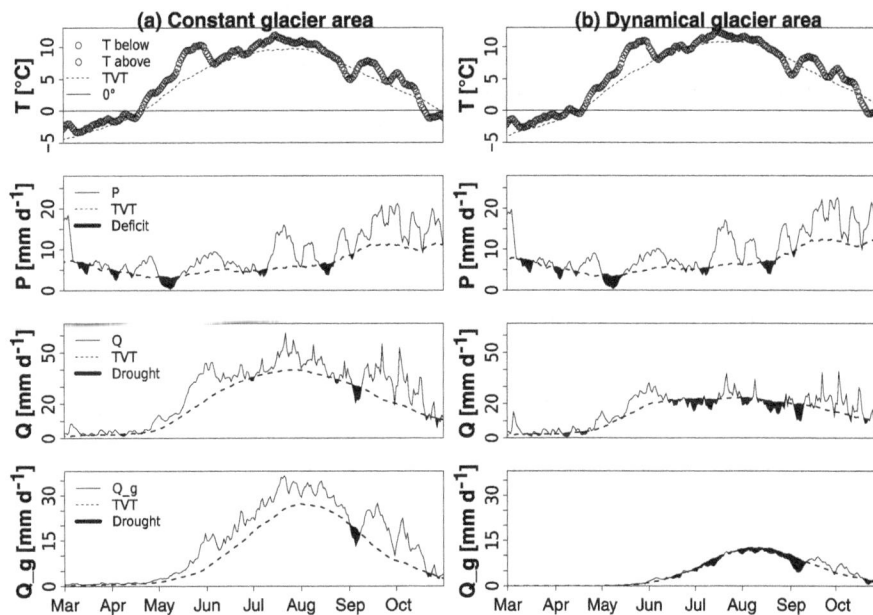

Figure 9. Example of streamflow droughts and causing factors (T, P and Q_g) for the different glacier area conceptualisations. Multi-model mean temperature, precipitation and discharge time series are presented for the Nigardsbreen catchment for March–October 2092, based on climate scenario RCP8.5. For the time series of P and T, a 7-day moving average was used. Droughts are analysed with the transient threshold. Note that T and P are slightly different in the panels **(a)** and **(b)** due to different lapse rates obtained during the calibration.

For comparison of the effect of the two threshold approaches on analysing drought processes, a dynamic glacier area conceptualisation was used for both thresholds (Fig. 10). The different thresholds clearly result in the identification of contrasting streamflow droughts in the Wolverine catchment in 2091. The HVT shows a long drought from July until October (shortly interrupted in September), while the TVT shows many streamflow droughts during the whole year (Fig. 10). The glacier has disappeared in 2091 in the Wolverine catchment, which caused a change in the regime. The HVT is based on the historical regime and the "drought" that can be seen is essentially the mismatch between the old and new regimes. Therefore, this drought occurs every year at the same moment, since the HVT is not changing and there is no glacier any more to produce a discharge peak in the summer. This "drought" does not represent extreme or exceptional discharge values and relating it to anomalies in P and T is not possible. T anomalies are mostly above the HVT temperature threshold, due to a warming climate, and can therefore not directly be used as explanation for droughts. Also, the deficits in P can not explain the large drought in the discharge. However, in the TVT approach, the threshold has adapted to the reduced summer discharge, like the thresholds of P and T have adapted (Fig. 10b). This causes temperatures to fluctuate around the threshold and these anomalies can be used to analyse the causing factors of drought in Q. Also, the deficits in P can be related to the droughts that are occurring in the streamflow. The TVT approach therefore could be used to study which drought processes and drought types (Van Loon and Van Lanen, 2012) will become important in the future.

5   Discussion

In this study, we aimed to systematically test the role of glacier changes and threshold approaches in simulating and analysing future streamflow droughts in glacierised catchments. The results indicate different effects of both methodological choices on drought characteristics and the analysis of drought processes, which is of major importance for further studies analysing climate change effects on streamflow droughts in cold climates. The study also showed that the methodological choices highlight different aspects of future streamflow droughts, and it is therefore essential for further studies to determine which aspect of drought one wants to study and choose the methods accordingly.

As glaciers have been shrinking and likely will further shrink in the future (e.g. Vaughan et al., 2013), there is wide consensus that glacier change needs to be accounted for in hydrological modelling. However, we have shown in this study that modelling with a constant glacier area can be interesting to analyse seasonal drought processes in the future, without taking into account the long-term changes of the glacier area. Analysing drought processes usually includes looking at anomalies in precipitation and temperature and their propagation through the hydrological cycle. Most drought processes occur within the season (Van Loon and Van Lanen, 2012) but some drought types can be classified as multi-season drought. An example is the snowmelt drought, which can be caused by high temperatures or low precipitation in winter, resulting in less snow supply to the snowpack, causing a drought in the snowmelt peak in summer due to less snow available for melt (Van Loon et al., 2015). In glacierised catchments, the time between the meteorological drivers and the resulting drought in streamflow can be even longer due to the long response time of glaciers (Bahr et al., 1998; Roe and O'Neal, 2009). A reduced winter mass balance would not directly result in a streamflow drought in

the glacier melt peak if temperatures are above or close to normal in summer. However, after several negative mass balance years and consequent glacier retreat, less glacier area and volume will be available for meltwater generation, possibly resulting in a drought when temperatures are close to or below normal in summer. Thus, the long-term effects of dynamical glaciers can influence droughts. Separating the effects of short-term climate variability and a changing glacier area and volume on droughts by using a constant and dynamic glacier area can therefore give useful insights on these intertwined processes.

Another option regarding the glacier modelling could be the full removal of the glacier. In theory, the comparison of simulated discharge without glaciers, with constant glaciers and with dynamic glaciers can give interesting information about the role of glaciers in causing or preventing streamflow droughts. For example, apart from distinguishing between the anomalies in glacier melt and glacier dynamics as causing factors of streamflow drought, also anomalies in snowmelt and precipitation deficits in relation to streamflow droughts could be better assessed. However, model parameters are calibrated to discharges and glacier mass balances of glacierised catchments and therefore reflect the typical sensitivities and relations among fluxes for glacierised catchments. Hence, these parameters cannot be directly used to simulate a non-glacierised catchment. We therefore did not include this option explicitly in our study. Nevertheless, in our dynamic glacier conceptualisation, we simulate a glacier disappearance for the Wolverine catchment from around 2060 onwards, while still using the same parameters. A solution, however, with time-varying parameters for simulation of long time periods and retreated glaciers does not yet exist (see, e.g. Merz et al., 2011; Thirel et al., 2015; Heuvelmans et al., 2004; Paul et al., 2007; Farinotti et al., 2012).

The dynamic glacier area representation used in this study is a simplification and therefore has its limitations. The $\triangle h$ parameterisation in HBV-light can, for example, not be used to simulate glacier advance compared to the defined glacier profile (see also Huss et al., 2008, 2010). Moreover, Huss et al. (2008) mention that this parameterisation is not able to reproduce the timescales for transfer of mass from the accumulation area to the ablation area. The change in volume is distributed over the glacier area to simulate an elevation change at the end of each year. Response time effects on drought can therefore not be directly analysed. However, the constant and dynamic glacier area conceptualisations are able to show the effect of short-term climate variability and long-term glacier area changes on streamflow droughts. Another drawback, in this HBV-light model version, is that elevations do not change after melting of glaciated model units. The surface lowering may in reality result in a positive feedback of melt due to higher temperatures and potentially less precipitation. Furthermore, this model version does not allow to use a seasonally varying discharge as a benchmark in the calibration (instead of the mean discharge; see Eq. 1),

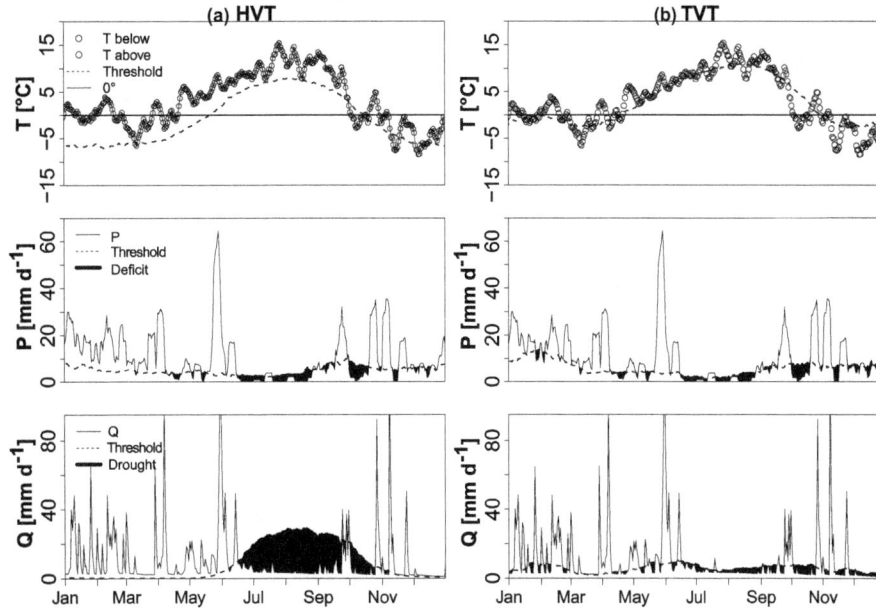

Figure 10. Example of streamflow drought and causing factors (T and P) for the different threshold methods. Panel **(a)** shows the HVT and **(b)** the TVT. Temperature, precipitation and discharge time series are presented for the Wolverine catchment for 2091 based on climate scenario RCP4.5 and the dynamic glacier area conceptualisation. For the P and T time series, a 7-day moving average was used.

which would be preferred when the regime shows a strong seasonality (Schaefli and Gupta, 2007). However, our objective function is not based on the whole discharge time series but only on the seasonal and peak discharges and the glacier mass balances, thereby partly taking the problem of calibrating on the mean discharge into account.

Despite these limitations, the implementation of the dynamic glacier area in the HBV model is an important improvement for the hydrological modelling in glacierised catchments. Many of the global hydrological models that have so far been applied to estimate changes in streamflow drought have not included glacier dynamics or any glacier component at all (e.g. Zhang et al., 2013). Compared to catchment-scale hydrological models which use approaches where glacier area is adjusted in larger jumps, without the coupling between melt and ice volume (e.g. Juen et al., 2007), the dynamic glacier area method used here is more applicable for the transient drought threshold approach because of the gradually changing discharge regime due to the gradually changing glacier. Using more advanced models to simulate glacier retreat may result in slightly different numbers in the timing of glacier retreat and changes in the discharge regime, but it would not change the results of this study regarding the use of the methodological options for drought analysis.

In our study, the glacier disappearance simulated by 2060 for the Wolverine catchment might be an unrealistically extreme result for most of the glacierised catchments in the world (Zemp et al., 2006; Rees and Collins, 2006; Radić et al., 2014; Bliss et al., 2014; Huss and Hock, 2015). The

use of a calibrated conceptual glaciohydrological model in our study which uses a simplification of glacier processes and does not take into account, e.g. a varying lapse rate (Gardner and Sharp, 2009), firn on the glacier, reduced albedo due to melt and explicit englacial and subglacial drainage, might have influenced the glacier melt and thereby also the rate of glacier disappearance. Also, the absence of a snow redistribution routine in our model, in which snow from higher elevation zones can be redistributed to the glacier (Seibert et al., 2017), might have influenced the rate of glacier retreat. The snow towers that appeared in our model, because snow was not redistributed (see also Freudiger et al., 2017), were checked for their possible error on the discharge simulations. The amount of SWE stored (or released in some elevation zones in the future) in the snow towers compared to the total discharge was however small (negligible up to a few percent). We therefore considered the effect of snow towers on our drought analysis to be small. Also, the assumption that parameters stay constant over time, while the catchment and climate are changing (Merz et al., 2011) (in this case changing glaciers) is causing some uncertainty.

We should also keep in mind that the future glacier area evolution has a large uncertainty caused by climate model uncertainties as shown in this study for the Nigardsbreen catchment (Fig. 5). The historical glacier area changes for Wolverine agree with the observed glacier area at the end of the historical period, but for Nigardsbreen a smaller glacier area than observed is simulated. This could be caused by the simplified modelling of glacier processes, the construction of the glacier profile and/or the climate forcing. We compared the annual average glacier melt contribution in the Nigardsbreen catchment with Engelhardt et al. (2014) and found comparable results (around 20 %). Nevertheless, both uncertainties, in the model and forcing, mainly influence the timing of changes in both catchments but not the processes that we studied and compared in the different scenarios, which is the main focus of this study.

Moreover, the two case study catchments in this study, with a different glacier area evolution and resulting changing discharge regime, showed the range of possible effects the methodological choices can have on future streamflow and drought projections. The glacier disappearance in the Wolverine catchment is a highly relevant and clear example in the discussion about drought definitions and thresholds in future projections. It also illustrates that the hydrological regime becomes more variable when the catchment changes from highly glacierised to non-glacierised (Fountain and Tangborn, 1985). This is important for streamflow drought analysis, since streamflow droughts will be more variable and mainly dependent on variability in precipitation, and it is therefore not appropriate to use a historical threshold that is based on other hydrological processes (stable glacier-dominated regime).

The other choice, which threshold approach to use, mainly relates to the question of the definition of a drought. For streamflow drought projections, a comparison with a historical period is always needed in order to assess the changes and to be able to understand them. However, one can raise the question if the threshold needs to be the same in the two periods (HVT approach). The results showed that, due to the regime shift, the HVT indicates severe droughts every year in summer. If we would have applied pooling (Fleig et al., 2006), the differences in drought characteristics between the threshold methods due to the regime shift would have been even more pronounced. Because this "regime shift drought" occurs each year, it will become the normal situation and it is clear that this mismatch of regimes can not be regarded as a drought. Therefore, the transient threshold is a better option to study droughts in glacierised catchments where discharge regimes change. Moreover, the advantage of TVT is that it can be used to analyse future drought processes which will be an important aspect for future water management. This study agrees with the findings of Wanders et al. (2015) that different threshold approaches can have substantial effects on future streamflow drought characteristics. Furthermore, the results confirm the findings of Van Huijgevoort et al. (2014) and Wanders et al. (2015) that in cold climates where regime shifts are expected the TVT is a better identifier of droughts than HVT. This is especially the case in glacierised basins as shown in this study, which are rapidly changing due to glacier retreat.

However, using the TVT, changes between historical and future situations cannot be assessed, because the benchmark itself is changing. Most studies (e.g. Forzieri et al., 2014) looking at future droughts in low flow periods have used

a historical threshold to define future droughts and conclude that low flows will increase, and therefore less droughts will occur. Here, the normal situation is changed (higher low flows), which is identified using the HVT. This information about changing normals is lost when only drought characteristics are analysed using the TVT. It is therefore important to complement the TVT drought characteristics with an analysis of the changes in the regime to put the drought results into perspective. This could be done, for example, by looking at the changes in the TVT itself or comparing the TVT with the HVT and by checking annual discharges (Figs. 7 and 6). In this study, the annual discharges of the Wolverine catchment are decreasing in the future, whereas the signal for Nigardsbreen is less clear. Apart from a changing seasonality, these annual discharges give information on how the total water availability will change.

Both threshold approaches thus take another viewpoint of drought. With the HVT, we look at future droughts from a viewpoint now, and with the transient threshold we change our viewpoint to the future and we then look at droughts. Since future droughts will also have impacts in the future, the latter viewpoint is more logical to study future droughts. However, the TVT also has some uncertainties. The main uncertainty concerns the adaptation that is assumed when using the transient threshold. The transient threshold changes every year and not always in the same direction and with the same magnitude. This would mean that society and ecosystems need to be flexible in the adaptation and the question is how adaptable we are to these regime changes and if we can assume that the same level of adaptation can be reached in both climate change scenarios (RCP4.5 and RCP8.5). Vidal et al. (2012), for example, discuss in their study about future droughts in France, in which the baseline of a standardised drought index is adapted each month, the feasibility of this time step and compare it with adaptation timescales for irrigated crops (seasonal or annual) and forestry (decadal). Nevertheless, several studies argue the use of "constant normals" as being representative for both the current and future climate and indicate ways to derive changing normals (e.g. Livezey et al., 2007; Arguez and Vose, 2011; Vidal et al., 2012).

Another aspect of the discussion about the definition of a drought is the use of a variable threshold to identify droughts. In contrast to other studies, which specifically look at low flow periods to analyse droughts (see, e.g. Hisdal et al., 2001; Fleig et al., 2006; Feyen and Dankers, 2009; Forzieri et al., 2014), for example, by using a constant instead of daily varying threshold, we include streamflow deficiencies in the high flow season as well in our streamflow drought definition. This is also done in many other studies that use a variable threshold level method (e.g. Van Loon et al., 2015; Fundel et al., 2013) or standardised drought indices (e.g. Shukla and Wood, 2008; Vidal et al., 2010), or in global-scale future drought studies (e.g. Van Huijgevoort et al., 2014; Prudhomme et al., 2014; Wanders et al., 2015), because it does fit with the definition of drought as below-normal water availability (Tallaksen and Van Lanen, 2004). However, the spatial and temporal scales in these studies can be different from our scales. Consequently, not all our identified streamflow droughts will lead to impacts. Nonetheless, in general, these droughts in terms of streamflow deficiencies might be important for, and could impact, downstream water users. It would be interesting to apply the methods and outcomes of this study to other glacierised catchments around the world, in particular those which are drier and therefore more dependent on glacial meltwater (e.g. Gascoin et al., 2011) and where climate change will likely have impacts on water availability and droughts.

6   Conclusions

This study systematically elucidated the effect of glacier dynamics and threshold approach on future streamflow drought characterisation and the analysis of the governing hydrological processes. The discharges and streamflow droughts of two case study catchments, Nigardsbreen (Norway) and Wolverine (Alaska), with a currently high percentage of glacier cover were studied. Streamflow was modelled with the HBV-light model for a historical period and into the future. This model accounts for the glacier retreat but also allows to keep glaciers constant, a feature that enabled this study to carry out a comparison of four potential views on future streamflow droughts. Assuming a constant glacier area and a threshold approach, whereby droughts are defined based on the historical hydrological regime, results in almost no droughts in the future, due to an increase in glacier melt. When the same historical threshold approach is applied to discharge simulated with glacier change, results show severe "regime shift droughts" in summer due to retreat, or even complete disappearance (Wolverine), of the glacier. If future droughts are studied from a future perspective, by using a transient threshold that changes with the changing hydrological regime, differences in drought characteristics between historical and future periods, and glacier dynamics options are smaller. Drought characteristics greatly differ among the four scenarios and these choices will therefore strongly influence future drought projections. We found the four options to be able to answer different questions about future streamflow drought in glacierised catchments: the transient threshold for analysing drought processes in the future, the historical threshold approach to assess changes between historical and future periods, the constant glacier area conceptualisation to analyse the effect of short-term climate variability and the dynamic glacier area to model realistic future discharges in glacierised catchments.

Most important for further future streamflow drought studies is to define what a future drought is and subsequently choose the right method. In addition to the definition of future droughts, questions that also need to be addressed in further studies are the relationships between the statistical description of droughts (the threshold based on a percentile of the flow duration curve) and the impacts and experiences of droughts by ecosystems and society. Are all droughts detected in the high flow season also experienced as droughts or, for example, only droughts with high deficits or long durations? Streamflow droughts upstream would mainly impact energy production and river ecology. However, if, for example, enough reservoir capacity is present for the energy production, a deficit in a part of the melt peak might be compensated by higher discharges from the glacier during the rest of the melt season and no impact is felt. In this study, an upstream perspective was used, but many people who depend on the water from glaciers live more downstream (e.g. water dependency in the Himalayas). Streamflow droughts in the high flow season upstream in glacierised catchments are related to droughts in the low flow season downstream, with potentially even larger impacts. Further research should investigate this relation and the impacts of drought downstream in these regions.

Table A1. Glacier and snow routine parameter values. All parameters were calibrated except CFR and CWH, indicated with *. For each catchment, two parameter sets were obtained: one for the dynamical glacier conceptualisation (D) and one for the static glacier area conceptualisation (C).

Parameter	Description	Nigardsbreen – C	Nigardsbreen – D	Wolverine – C	Wolverine – D
T_{calt} (°C/100 m)	T lapse rate	0.65	0.55	0.54	0.46
P_{calt} (%/100 m)	P lapse rate	13.40	15.43	15.98	12.70
TT (°C)	Threshold temperature	−0.17	−0.32	0.04	0.12
CFMAX (mm d °C^{-1})	Degree-day factor	2.34	3.17	2.67	1.94
SFCF (–)	Snowfall correction factor	1.00	0.95	1.69	1.88
CFR* (–)	Refreezing coefficient	0.05	0.05	0.05	0.05
CWH* (–)	Water holding capacity of snow	0.1	0.1	0.1	0.1
$CF_{glacier}$ (–)	Glacier melt correction factor	1.32	1.18	1.80	1.72
CF_{slope} (–)	Slope melt correction factor	2.67	1.54	1.65	2.57
KG_{min} (1 d^{-1})	Minimum outflow coefficient glacier storage	0.20	0.20	0.20	0.20
dKG (1 d^{-1})	Maximum minus minimum glacier storage outflow coefficient	0.50	0.39	0.50	0.50
AG (mm)	Calibration parameter	0.003	1.25	9.95	0.0003

Appendix A: Model parameters glacier and snow routine

In Table A1, the calibrated parameter values that were used in the glacier and snow routine of the HBV-light model are presented. A different parameter set was obtained for the dynamic and constant glacier area conceptualisations. The refreezing coefficient (CFR), which determines the amount of refreezing liquid water in the within the snowpack when temperatures are below the threshold temperature, and the water holding capacity of snow (CWH), which determines how much meltwater and rainfall are retained within the snowpack, were assigned a constant value and not calibrated. KG_{min}, dKG and AG are the parameters for the glacial water storage–outflow relationship (Stahl et al., 2008). The degree-day factor (CFMAX) is multiplied with $CF_{glacier}$ to simulate glacier melt and it is multiplied (divided) by CF_{slope} to calculate melt of snow and ice for south-facing slopes (north-facing slopes). No correction is used for east- and west-facing slopes.

maps are available via Matthias Huss and for some Norwegian glaciers via NVE. The climate model data are available from the Co-ordinated Regional Climate Downscaling Experiment (CORDEX) (http://www.cordex.org/). Glacier outlines can be obtained from GLIMS and NVE (Nigardsbreen) (http://www.glims.org/RGI/rgi50_dl.html and https://www.nve.no/hydrologi/bre/bredata/). The ASTER DEM can be downloaded from http://reverb.echo.nasa.gov/reverb/ and ERA-Interim data from http://apps.ecmwf.int/datasets/.

Author contributions. AVL and MVT conceived and designed the study. MVT carried out the modelling and analyses with feedback from AVL and MV. NW and MV provided the R script for TVT calculation and the HBV model with glacier routine, respectively. AVL and AT supervised the MSc thesis work that formed the basis of this paper. MVT, AVL and AT discussed the structure and content of the manuscript. MVT wrote the manuscript. All co-authors edited and revised the manuscript and approved the final version (AVL, NW, KS, MV, AT).

Competing interests. The authors declare that they have no conflict of interest.

Data availability. Data for the Nigardsbreen catchment are available via the Norwegian Water Resources and Energy Directorate (NVE) and for the Wolverine catchment via the US Geological Survey (USGS). Streamflow data and mass balances for Wolverine are also available online (https://waterdata.usgs.gov/nwis and https://alaska.usgs.gov/products/data.php?dataid=79). Ice thickness

Acknowledgements. We would like to thank Shad O'Neel and Louiss Sass from USGS for providing the mass balance data and climate observations for the Wolverine catchment. We thank Wai Kwok Wong, Liss Andreassen and Kjetil Melvold from NVE for providing the ice thickness data, the mass balance data, the metadata and the climate and discharge observations for the

Nigardsbreen catchment. We also thank Annemiek Stegehuis for extracting the CORDEX climate model data for the two catchments and Matthias Huss for providing the Wolverine ice thickness map. Furthermore, we thank Nick Barrand (University of Birmingham) for his support when setting up this research. We thank the University of Birmingham for covering the article processing charge. Finally, we thank the three reviewers, the editor and Lena M. Tallaksen for their helpful comments. AVL is funded by NWO Rubicon project no. 2004/08338/ALW and NW is funded by NWO Rubicon 825.15.003. This paper was developed within the framework of the UNESCO-IHP VIII FRIEND programme (EURO-FRIEND – Low flow and Drought group).

Edited by: Matthias Bernhardt

References

Akhtar, M., Ahmad, N., and Booij, M. J.: The impact of climate change on the water resources of Hindukush–Karakorum–Himalaya region under different glacier coverage scenarios, J. Hydrol., 355, 148–163, 2008.

Andreassen, L. M., Winsvold, S. H., Paul, F., and Hausberg, J.: Inventory of Norwegian Glaciers, NVE, Oslo, 2012.

Andreassen, L., Huss, M., Melvold, K., Elvehøy, H., and Winsvold, S.: Ice thickness measurements and volume estimates for glaciers in Norway, J. Glaciol., 61, 763–775, https://doi.org/10.3189/2015jog14j161, 2015.

Andreassen, L. M., Elvehøy, H., Kjøllmoen, B., and Engeset, R. V.: Reanalysis of long-term series of glaciological and geodetic mass balance for 10 Norwegian glaciers, The Cryosphere, 10, 535–552, https://doi.org/10.5194/tc-10-535-2016, 2016.

Arguez, A. and Vose, R. S.: The definition of the standard WMO climate normal: The key to deriving alternative climate normals, B. Am. Meteorol. Soc., 92, 699–704, https://doi.org/10.1175/2010bams2955.1, 2011.

Arnell, N. W.: The effect of climate change on hydrological regimes in Europe: a continental perspective, Global Environ. Chang., 9, 5–23, https://doi.org/10.1016/s0959-3780(98)00015-6, 1999.

Bahr, D. B., Meier, M. F., and Peckham, S. D.: The physical basis of glacier volume-area scaling, J. Geophys. Res.-Sol. Ea., 102, 20355–20362, https://doi.org/10.1029/97jb01696, 1997.

Bahr, D. B., Pfeffer, W. T., Sassolas, C., and Meier, M. F.: Response time of glaciers as a function of size and mass balance: 1. Theory, J. Geophys. Res.-Sol. Ea., 103, 9777–9782, https://doi.org/10.1029/98jb00507, 1998.

Bard, A., Renard, B., Lang, M., Giuntoli, I., Korck, J., Koboltschnig, G., Janža, M., d'Amico, M., and Volken, D.: Trends in the hydrologic regime of Alpine rivers, J. Hydrol., 529, 1823–1837, https://doi.org/10.1016/j.jhydrol.2015.07.052, 2015.

Barnett, T. P., Adam, J. C., and Lettenmaier, D. P.: Potential impacts of a warming climate on water availability in snow-dominated regions, Nature, 438, 303–309, https://doi.org/10.1038/nature04141, 2005.

Bates, B., Kundzewicz, Z. W., Wu, S., and Palutikof, J.: Climate Change and Water: Technical Paper vi, Intergovernmental Panel on Climate Change (IPCC), Geneva, IPCC secretariat, 2008.

Bergström, S. and Singh, V.: The HBV model, Computer Models of Watershed Hydrology, Water Resources Publications, Highlands Ranch, Colorado, USA, 443–476, 1995.

Bliss, A., Hock, R., and Radić, V.: Global response of glacier runoff to twenty-first century climate change, J. Geophys. Res.-Earth, 119, 717–730, https://doi.org/10.1002/2013jf002931, 2014.

Brouwer, C. and Heibloem, M.: Irrigation water management: irrigation water needs, Training Manual, Rome, Italy, FAO, 3, 1986.

Dee, D. P., Uppala, S. M., Simmons, A. J., Berrisford, P., Poli, P., Kobayashi, S., Andrae, U., Balmaseda, M. A., Balsamo, G., Bauer, P., Bechtold, P., Beljaars, A. C. M., van de Berg, L., Bidlot, J., Bormann, N., Delsol, C., Dragani R., Fuentes, M., Geer, A. J., Haimberger, L., Healy, S. B., Hersbach, H., Hólm, E. V., Isaksen, L., Kållberg, P., Köhler, M., Matricardi, M., McNally, A. P., Monge-Sanz, B. M., Morcrette, J.-J., Park, B.-K., Peubey, C., de Rosnay, P., Tavolato, C., Thépaut, J.-N., and Vitart, F.: The ERA-Interim reanalysis: Configuration and performance of the data assimilation system, Q. J. Roy. Meteor. Soc., 137, 553–597, 2011.

Déry, S. J., Stahl, K., Moore, R., Whitfield, P., Menounos, B., and Burford, J. E.: Detection of runoff timing changes in pluvial, nival, and glacial rivers of western Canada, Water Resour. Res., 45, W04426, https://doi.org/10.1029/2008wr006975, 2009.

Duethmann, D., Bolch, T., Farinotti, D., Kriegel, D., Vorogushyn, S., Merz, B., Pieczonka, T., Jiang, T., Su, B., and Güntner, A.: Attribution of streamflow trends in snow and glacier melt-dominated catchments of the Tarim River, Central Asia, Water Resour. Res., 51, 4727–4750, https://doi.org/10.1002/2014wr016716, 2015.

Engelhardt, M., Schuler, T. V., and Andreassen, L. M.: Contribution of snow and glacier melt to discharge for highly glacierised catchments in Norway, Hydrol. Earth Syst. Sci., 18, 511–523, https://doi.org/10.5194/hess-18-511-2014, 2014.

Farinotti, D., Usselmann, S., Huss, M., Bauder, A., and Funk, M.: Runoff evolution in the Swiss Alps: projections for selected high-alpine catchments based on ENSEMBLES scenarios, Hydrol. Process., 26, 1909–1924, https://doi.org/10.1002/hyp.8276, 2012.

Feyen, L. and Dankers, R.: Impact of global warming on streamflow drought in Europe, J. Geophys. Res.-Atmos., 114, D17116, https://doi.org/10.1029/2008jd011438, 2009.

Finger, D., Heinrich, G., Gobiet, A., and Bauder, A.: Projections of future water resources and their uncertainty in a glacierized catchment in the Swiss Alps and the subsequent effects on hydropower production during the 21st century, Water Resour. Res., 48, W02521, https://doi.org/10.1029/2011wr010733, 2012.

Fleig, A. K., Tallaksen, L. M., Hisdal, H., and Demuth, S.: A global evaluation of streamflow drought characteristics, Hydrol. Earth Syst. Sci., 10, 535–552, https://doi.org/10.5194/hess-10-535-2006, 2006.

Forzieri, G., Feyen, L., Rojas, R., Flörke, M., Wimmer, F., and Bianchi, A.: Ensemble projections of future streamflow droughts in Europe, Hydrol. Earth Syst. Sci., 18, 85–108, https://doi.org/10.5194/hess-18-85-2014, 2014.

Fountain, A. G. and Tangborn, W. V.: The effect of glaciers on streamflow variations, Water Resour. Res., 21, 579–586, https://doi.org/10.1029/WR021i004p00579, 1985.

Freudiger, D., Kohn, I., Seibert, J., Stahl, K., and Weiler, M.: Snow redistribution for the hydrological modeling of alpine catchments, WIRES Water, 4, e1232, https://doi.org/10.1002/wat2.1232, 2017.

Fundel, F., Jörg-Hess, S., and Zappa, M.: Monthly hydrometeorological ensemble prediction of streamflow droughts and corresponding drought indices, Hydrol. Earth Syst. Sci., 17, 395–407, https://doi.org/10.5194/hess-17-395-2013, 2013.

Gardner, A. S. and Sharp, M.: Sensitivity of net mass-balance estimates to near-surface temperature lapse rates when employing the degree-day method to estimate glacier melt, Ann. Glaciol., 50, 80–86, https://doi.org/10.3189/172756409787769663, 2009.

Gascoin, S., Kinnard, C., Ponce, R., Lhermitte, S., MacDonell, S., and Rabatel, A.: Glacier contribution to streamflow in two headwaters of the Huasco River, Dry Andes of Chile, The Cryosphere, 5, 1099–1113, https://doi.org/10.5194/tc-5-1099-2011, 2011.

Giorgi, F., Jones, C., and Asrar, G. R.: Addressing climate information needs at the regional level: the CORDEX framework, WMO Bull., 58, 175–183, 2009.

Gupta, H. V., Kling, H., Yilmaz, K. K., and Martinez, G. F.: Decomposition of the mean squared error and NSE performance criteria: Implications for improving hydrological modelling, J. Hydrol., 377, 80–91, https://doi.org/10.1016/j.jhydrol.2009.08.003, 2009.

Haylock, M., Hofstra, N., Klein Tank, A., Klok, E., Jones, P., and New, M.: A European daily high-resolution gridded data set of surface temperature and precipitation for 1950–2006, J. Geophys. Res.-Atmos., 113, D20119, https://doi.org/10.1029/2008jd010201, 2008.

Heuvelmans, G., Muys, B., and Feyen, J.: Evaluation of hydrological model parameter transferability for simulating the impact of land use on catchment hydrology, Phys. Chem. Earth, 29, 739–747, https://doi.org/10.1016/j.pce.2004.05.002, 2004.

Hisdal, H., Tallaksen, L., Peters, E., Stahl, K., and Zaidman, M.: Drought event definition, ARIDE Technical Rep., Oslo, Norway, University of Oslo, 6, 2000.

Hisdal, H., Stahl, K., Tallaksen, L. M., and Demuth, S.: Have streamflow droughts in Europe become more severe or frequent?, Int. J. Climatol., 21, 317–333, https://doi.org/10.1002/joc.619, 2001.

Horton, P., Schaefli, B., Mezghani, A., Hingray, B., and Musy, A.: Assessment of climate-change impacts on alpine discharge regimes with climate model uncertainty, Hydrol. Process., 20, 2091–2109, https://doi.org/10.1002/hyp.6197, 2006.

Huss, M.: Present and future contribution of glacier storage change to runoff from macroscale drainage basins in Europe, Water Resour. Res., 47, W07511, https://doi.org/10.1029/2010wr010299, 2011.

Huss, M. and Farinotti, D.: Distributed ice thickness and volume of all glaciers around the globe, J. Geophys. Res.-Earth, 117, F04010, https://doi.org/10.1029/2012jf002523, 2012.

Huss, M. and Hock, R.: A new model for global glacier change and sea-level rise, Front. Earth Sci., 3, 54, https://doi.org/10.3389/feart.2015.00054, 2015.

Huss, M., Farinotti, D., Bauder, A., and Funk, M.: Modelling runoff from highly glacierized alpine drainage basins in a changing climate, Hydrol. Process., 22, 3888–3902, https://doi.org/10.1002/hyp.7055, 2008.

Huss, M., Jouvet, G., Farinotti, D., and Bauder, A.: Future high-mountain hydrology: a new parameterization of glacier retreat, Hydrol. Earth Syst. Sci., 14, 815–829, https://doi.org/10.5194/hess-14-815-2010, 2010.

Immerzeel, W. W., Van Beek, L. P., and Bierkens, M. F.: Climate change will affect the Asian water towers, Science, 328, 1382–1385, https://doi.org/10.1126/science.1183188, 2010.

Immerzeel, W. W., Van Beek, L., Konz, M., Shrestha, A., and Bierkens, M.: Hydrological response to climate change in a glacierized catchment in the Himalayas, Climatic Change, 110, 721–736, 2012.

Jacob, D., Petersen, J., Eggert, B., Alias, A., Christensen, O. B., Bouwer, L. M., Braun, A., Colette, A., Déqué, M., Georgievski, G., Georgopoulou, E., Gobiet, A., Menut, L., Nikulin, G., Haensler, A., Hempelmann N., Jones, C., Keuler, K., Kovats, S., Kröner, N., Kotlarski, S., Kriegsmann, A., Martin, E., van Meijgaard, E., Moseley, C., Pfeifer, S., Preuschmann, S., Radermacher, C., Radtke, K., Rechid, D., Rounsevell, M., Samuelsson, P., Somot, S., Soussana, J. F., Teichmann, C., Valentini, R., Vautard, R., Weber, B., and Yiou, P.: EURO-CORDEX: new high-resolution climate change projections for European impact research, Reg. Environ. Change, 14, 563–578, 2014.

Jansson, P., Hock, R., and Schneider, T.: The concept of glacier storage: a review, J. Hydrol., 282, 116–129, https://doi.org/10.1016/s0022-1694(03)00258-0, 2003.

Jeelani, G., Feddema, J. J., Veen, C. J., and Stearns, L.: Role of snow and glacier melt in controlling river hydrology in Liddar watershed (western Himalaya) under current and future climate, Water Resour. Res., 48, W12508, https://doi.org/10.1029/2011wr011590, 2012.

Jonsdottir, H., Eliasson, J., and Madsen, H.: Assessment of serious water shortage in the Icelandic water resource system, Phys. Chem. Earth, 30, 420–425, https://doi.org/10.1016/j.pce.2005.06.007, 2005.

Juen, I., Kaser, G., and Georges, C.: Modelling observed and future runoff from a glacierized tropical catchment (Cordillera Blanca, Perú), Global Planet. Change, 59, 37–48, https://doi.org/10.1016/j.gloplacha.2006.11.038, 2007.

Kaser, G., Großhauser, M., and Marzeion, B.: Contribution potential of glaciers to water availability in different climate regimes, P. Natl. Acad. Sci. USA, 107, 20223–20227, https://doi.org/10.1073/pnas.1008162107, 2010.

Klok, E., Jasper, K., Roelofsma, K., Gurtz, J., and Badoux, A.: Distributed hydrological modelling of a heavily glaciated Alpine river basin, Hydrolog. Sci. J., 46, 553–570, https://doi.org/10.1080/02626660109492850, 2001.

Koboltschnig, G. R., Schöner, W., Zappa, M., and Holzmann, H.: Contribution of glacier melt to stream runoff: if the climatically extreme summer of 2003 had happened in 1979?, Ann. Glaciol., 46, 303–308, https://doi.org/10.3189/172756407782871260, 2007.

Konz, M. and Seibert, J.: On the value of glacier mass balances for hydrological model calibration, J. Hydrol., 385, 238–246, https://doi.org/10.1016/j.jhydrol.2010.02.025, 2010.

Lehner, B., Döll, P., Alcamo, J., Henrichs, T., and Kaspar, F.: Estimating the impact of global change on flood and drought risks in Europe: a continental, integrated analysis, Climatic Change, 75, 273–299, https://doi.org/10.1007/s10584-006-6338-4, 2006.

Li, H., Beldring, S., Xu, C.-Y., Huss, M., Melvold, K., and Jain, S. K.: Integrating a glacier retreat model into a hydrological model–Case studies of three glacierised catchments in Norway and Himalayan region, J. Hydrol., 527, 656–667, https://doi.org/10.1016/j.jhydrol.2015.05.017, 2015.

Livezey, R. E., Vinnikov, K. Y., Timofeyeva, M. M., Tinker, R., and van den Dool, H. M.: Estimation and extrapolation of climate normals and climatic trends, J. Appl. Meteorol. Clim., 46, 1759–1776, https://doi.org/10.1175/2007jamc1666.1, 2007.

Lutz, A. F., Immerzeel, W., Kraaijenbrink, P., Shrestha, A. B., and Bierkens, M. F.: Climate change impacts on the Upper Indus hydrology: sources, shifts and extremes, PLoS One, 11, e0165630, https://doi.org/10.1371/journal.pone.0165630, 2016.

Mayr, E., Hagg, W., Mayer, C., and Braun, L.: Calibrating a spatially distributed conceptual hydrological model using runoff, annual mass balance and winter mass balance, J. Hydrol., 478, 40–49, https://doi.org/10.1016/j.jhydrol.2012.11.035, 2013.

Merz, R., Parajka, J., and Blöschl, G.: Time stability of catchment model parameters: Implications for climate impact analyses, Water Resour. Res., 47, W02531, https://doi.org/10.1029/2010wr009505, 2011.

Miller, J. D., Immerzeel, W. W., and Rees, G.: Climate change impacts on glacier hydrology and river discharge in the Hindu Kush–Himalayas: a synthesis of the scientific basis, Mt. Res. Dev., 32, 461–467, https://doi.org/10.1659/mrd-journal-d-12-00027.1, 2012.

Naz, B. S., Frans, C. D., Clarke, G. K. C., Burns, P., and Lettenmaier, D. P.: Modeling the effect of glacier recession on streamflow response using a coupled glacio-hydrological model, Hydrol. Earth Syst. Sci., 18, 787–802, https://doi.org/10.5194/hess-18-787-2014, 2014.

O'Neel, S.: Supplemental Online Material of Assessing Streamflow Sensitivity to Variations in Glacier Mass Balance, 2014.

O'Neel, S., Hood, E., Arendt, A., and Sass, L.: Assessing streamflow sensitivity to variations in glacier mass balance, Clim. Change, 123, 329–341, https://doi.org/10.1007/s10584-013-1042-7, 2014.

O'Neel, S. R., Sass, L. C., McNeil, C., and McGrath, D.: USGS Alaska benchmark glacier mass balance data – Phase 1; Gulkana and Wolverine glaciers, US Geological Survey Data Release, https://doi.org/10.5066/F7HD7SRF, 2016.

Paul, F., Kääb, A., and Haeberli, W.: Recent glacier changes in the Alps observed by satellite: Consequences for future monitoring strategies, Global Planet. Change, 56, 111–122, https://doi.org/10.1016/j.gloplacha.2006.07.007, 2007.

Pfeffer, W. T., Arendt, A. A., Bliss, A., Bolch, T., Cogley, J. G., Gardner, A. S., Hagen, J.-O., Hock, R., Kaser, G., Kienholz, C., Miles, E. S., Moholdt, G., Mölg, N., Paul, F., Radić, V., Rastner, P., Raup, B. H., Rich, J., and Sharp, M. J.: The Randolph Consortium: a globally complete inventory of glaciers, J. Glaciol., 60, 537–552, https://doi.org/10.3189/2014jog13j176, 2014.

Prudhomme, C., Giuntoli, I., Robinson, E. L., Clark, D. B., Arnell, N. W., Dankers, R., Fekete, B. M., Franssen, W., Gerten, D., Gosling, S. N., Hagemann, S., Hannah, D. M., Kim, H., Masaki, Y., Satoh, Y., Stacke, T., Wada, Y., and Wisser, D.: Hydrological droughts in the 21st century, hotspots and uncertainties from a global multimodel ensemble experiment, P. Natl. Acad. Sci. USA, 111, 3262–3267, https://doi.org/10.1073/pnas.1222473110, 2014.

Radić, V., Bliss, A., Beedlow, A. C., Hock, R., Miles, E., and Cogley, J. G.: Regional and global projections of twenty-first century glacier mass changes in response to climate scenarios from global climate models, Clim. Dynam., 42, 37–58, https://doi.org/10.1007/s00382-013-1719-7, 2014.

Ragettli, S., Immerzeel, W. W., and Pellicciotti, F.: Contrasting climate change impact on river flows from high-altitude catchments in the Himalayan and Andes Mountains, P. Natl. Acad. Sci. USA, 113, 9222–9227, https://doi.org/10.1073/pnas.1606526113, 2016.

Rees, H. G. and Collins, D. N.: Regional differences in response of flow in glacier-fed Himalayan rivers to climatic warming, Hydrol. Process., 20, 2157–2169, https://doi.org/10.1002/hyp.6209, 2006.

Roe, G. H. and O'Neal, M. A.: The response of glaciers to intrinsic climate variability: observations and models of late-Holocene variations in the Pacific Northwest, J. Glaciol., 55, 839–854, https://doi.org/10.3189/002214309790152438, 2009.

Salzmann, N., Machguth, H., and Linsbauer, A.: The Swiss Alpine glaciers' response to the global "2 °C air temperature target", Environ. Res. Lett., 7, 044001, https://doi.org/10.1088/1748-9326/7/4/044001, 2012.

Schaefli, B. and Gupta, H. V.: Do Nash values have value?, Hydrol. Process., 21, 2075–2080, https://doi.org/10.1002/hyp.6825, 2007.

Schaefli, B., Hingray, B., Niggli, M., and Musy, A.: A conceptual glacio-hydrological model for high mountainous catchments, Hydrol. Earth Syst. Sci., 9, 95–109, https://doi.org/10.5194/hess-9-95-2005, 2005.

Seibert, J.: Multi-criteria calibration of a conceptual runoff model using a genetic algorithm, Hydrol. Earth Syst. Sci., 4, 215–224, https://doi.org/10.5194/hess-4-215-2000, 2000.

Seibert, J. and Vis, M. J. P.: Teaching hydrological modeling with a user-friendly catchment-runoff-model software package, Hydrol. Earth Syst. Sci., 16, 3315–3325, https://doi.org/10.5194/hess-16-3315-2012, 2012.

Seibert, J., Vis, M. J. P., Kohn, I., Weiler, M., and Stahl, K.: Technical Note: Representing glacier dynamics in a semi-distributed hydrological model, Hydrol. Earth Syst. Sci. Discuss., https://doi.org/10.5194/hess-2017-158, in review, 2017.

Sheffield, J. and Wood, E. F.: Drought: Past Problems and Future Scenarios, Earthscan, London and Washington, 2012.

Shukla, S. and Wood, A. W.: Use of a standardized runoff index for characterizing hydrologic drought, Geophys. Res. Lett., 35, L02405, https://doi.org/10.1029/2007gl032487, 2008.

Stahl, K., Moore, R., Shea, J., Hutchinson, D., and Cannon, A.: Coupled modelling of glacier and streamflow response to future climate scenarios, Water Resour. Res., 44, W02422, https://doi.org/10.1029/2007wr005956, 2008.

Sun, M., Li, Z., Yao, X., Zhang, M., and Jin, S.: Modeling the hydrological response to climate change in a glacierized high mountain region, northwest China, J. Glaciol., 61, 127–136, https://doi.org/10.3189/2015jog14j033, 2015.

Tallaksen, L. M. and Van Lanen, H. A.: Hydrological Drought: Processes and Estimation Methods for Streamflow and Groundwater, vol. 48, Elsevier, Amsterdam, 2004.

Tecklenburg, C., Francke, T., Kormann, C., and Bronstert, A.: Modeling of water balance response to an extreme future scenario in the Ötztal catchment, Austria, Adv. Geosci., 32, 63–68, https://doi.org/10.5194/adgeo-32-63-2012, 2012.

Teutschbein, C. and Seibert, J.: Bias correction of regional climate model simulations for hydrological climate-change impact studies: Review and evaluation of different methods, J. Hydrol., 456, 12–29, https://doi.org/10.1016/j.jhydrol.2012.05.052, 2012.

Thirel, G., Andréassian, V., and Perrin, C.: On the need to test hydrological models under changing conditions, Hydrolog. Sci. J., 60, 1165–1173, https://doi.org/10.1080/02626667.2015.1050027, 2015.

USGS Waterdata: http://maps.waterdata.usgs.gov/mapper/index.html (last access: 12 April 2016), 2016.

Van Beusekom, A. E., O'Nell, S. R., March, R. S., Sass, L. C., and Cox, L. H.: Re-analysis of Alaskan benchmark glacier mass-balance data using the index method, Tech. rep., US Geological Survey, Reston, Virginia, 2010.

Van Huijgevoort, M., Van Lanen, H., Teuling, A., and Uijlenhoet, R.: Identification of changes in hydrological drought characteristics from a multi-GCM driven ensemble constrained by observed discharge, J. Hydrol., 512, 421–434, https://doi.org/10.1016/j.jhydrol.2014.02.060, 2014.

Van Loon, A.: On the propagation of drought: how climate and catchment characteristics influence hydrological drought development and recovery, PhD thesis, Wageningen University, Wageningen, 2013.

Van Loon, A. F.: Hydrological drought explained, WIRES Water, 2, 359–392, https://doi.org/10.1002/wat2.1085, 2015.

Van Loon, A. F. and Van Lanen, H. A. J.: A process-based typology of hydrological drought, Hydrol. Earth Syst. Sci., 16, 1915–1946, https://doi.org/10.5194/hess-16-1915-2012, 2012.

Van Loon, A., Tijdeman, E., Wanders, N., Van Lanen, H., Teuling, A., and Uijlenhoet, R.: How climate seasonality modifies drought duration and deficit, J. Geophys. Res.-Atmos., 119, 4640–4656, https://doi.org/10.1002/2013jd020383, 2014.

Van Loon, A. F., Ploum, S. W., Parajka, J., Fleig, A. K., Garnier, E., Laaha, G., and Van Lanen, H. A. J.: Hydrological drought types in cold climates: quantitative analysis of causing factors and qualitative survey of impacts, Hydrol. Earth Syst. Sci., 19, 1993–2016, https://doi.org/10.5194/hess-19-1993-2015, 2015.

Van Loon, A. F., Gleeson, T., Clark, J., Van Dijk, A. I. J. M., Stahl, K., Hannaford, J., Di Baldassarre, G., Teuling, A. J., Tallaksen, L. M., Uijlenhoet, R., Hannah, D. M., Sheffield, J., Svoboda, M., Verbeiren, B., Wagener, T., Rangecroft, S., Wanders, N., and Van Lanen, H. A. J.: Drought in the Anthropocene, Nat. Geosci., 9, 89–91, https://doi.org/10.1038/ngeo2646, 2016.

van Vliet, M. T., Sheffield, J., Wiberg, D., and Wood, E. F.: Impacts of recent drought and warm years on water resources and electricity supply worldwide, Environ. Res. Lett., 11, 124021, https://doi.org/10.1088/1748-9326/11/12/124021, 2016.

Vaughan, D., Comiso, J., Allison, I., Carrasco, J., Kaser, G., Kwok, R., Mote, P., Murray, T., Paul, F., Ren, J., Rignot, E., Solomina, O., Steffen, K., and Zhang, T.: Observation: Cryosphere. In: Climate Change 2013: The Physical Science Basis. Contribution of Working Group I to the Fifth Assessment Report of the Intergovernmental Panel on Climate Change, Cambidge University Press, Cambridge, UK and New York, USA, 2013.

Verbunt, M., Gurtz, J., Jasper, K., Lang, H., Warmerdam, P., and Zappa, M.: The hydrological role of snow and glaciers in alpine river basins and their distributed modeling, J. Hydrol., 282, 36–55, https://doi.org/10.1016/s0022-1694(03)00251-8, 2003.

Vidal, J.-P., Martin, E., Franchistéguy, L., Habets, F., Soubeyroux, J.-M., Blanchard, M., and Baillon, M.: Multilevel and multiscale drought reanalysis over France with the Safran-Isba-Modcou hydrometeorological suite, Hydrol. Earth Syst. Sci., 14, 459–478, https://doi.org/10.5194/hess-14-459-2010, 2010.

Vidal, J.-P., Martin, E., Kitova, N., Najac, J., and Soubeyroux, J.-M.: Evolution of spatio-temporal drought characteristics: validation, projections and effect of adaptation scenarios, Hydrol. Earth Syst. Sci., 16, 2935–2955, https://doi.org/10.5194/hess-16-2935-2012, 2012.

Viviroli, D., Archer, D. R., Buytaert, W., Fowler, H. J., Greenwood, G. B., Hamlet, A. F., Huang, Y., Koboltschnig, G., Litaor, M. I., López-Moreno, J. I., Lorentz, S., Schädler, B., Schreier, H., Schwaiger, K., Vuille, M., and Woods, R.: Climate change and mountain water resources: overview and recommendations for research, management and policy, Hydrol. Earth Syst. Sci., 15, 471–504, https://doi.org/10.5194/hess-15-471-2011, 2011.

Wanders, N. and Van Lanen, H. A. J.: Future discharge drought across climate regions around the world modelled with a synthetic hydrological modelling approach forced by three general circulation models, Nat. Hazards Earth Syst. Sci., 15, 487–504, https://doi.org/10.5194/nhess-15-487-2015, 2015.

Wanders, N. and Wada, Y.: Human and climate impacts on the 21st century hydrological drought, J. Hydrol., 526, 208–220, https://doi.org/10.1016/j.jhydrol.2014.10.047, 2015.

Wanders, N., Wada, Y., and Van Lanen, H. A. J.: Global hydrological droughts in the 21st century under a changing hydrological regime, Earth Syst. Dynam., 6, 1–15, https://doi.org/10.5194/esd-6-1-2015, 2015.

Winsvold, S. H., Andreassen, L. M., and Kienholz, C.: Glacier area and length changes in Norway from repeat inventories, The Cryosphere, 8, 1885–1903, https://doi.org/10.5194/tc-8-1885-2014, 2014.

Wong, W. K., Beldring, S., Engen-Skaugen, T., Haddeland, I., and Hisdal, H.: Climate change effects on spatiotemporal patterns of hydroclimatological summer droughts in Norway, J. Hydrometeorol., 12, 1205–1220, 2011.

Xu, C.-Y., and Singh, V.: Evaluation and generalization of temperature-based methods for calculating evaporation, Hydrol. Process., 15, 305–319, https://doi.org/10.1002/hyp.119, 2001.

Zappa, M. and Kan, C.: Extreme heat and runoff extremes in the Swiss Alps, Nat. Hazards Earth Syst. Sci., 7, 375–389, https://doi.org/10.5194/nhess-7-375-2007, 2007.

Zemp, M., Haeberli, W., Hoelzle, M., and Paul, F.: Alpine glaciers to disappear within decades?, Geophys. Res. Lett., 33, L13504, https://doi.org/10.1029/2006gl026319, 2006.

Zhang, L., Su, F., Yang, D., Hao, Z., and Tong, K.: Discharge regime and simulation for the upstream of major rivers over Tibetan Plateau, J. Geophys. Res.-Atmos., 118, 8500–8518, https://doi.org/10.1002/jgrd.50665, 2013.

6

Variability in snow cover phenology in China from 1952 to 2010

Chang-Qing Ke[1,2,6], **Xiu-Cang Li**[3,4], **Hongjie Xie**[5], **Dong-Hui Ma**[1,6], **Xun Liu**[1,2], **and Cheng Kou**[1,2]

[1]Jiangsu Provincial Key Laboratory of Geographic Information Science and Technology, Nanjing University, Nanjing 210023, China
[2]Key Laboratory for Satellite Mapping Technology and Applications of State Administration of Surveying, Mapping and Geoinformation of China, Nanjing University, Nanjing 210023, China
[3]National Climate Center, China Meteorological Administration, Beijing 100081, China
[4]Collaborative Innovation Center on Forecast and Evaluation of Meteorological Disasters Faculty of Geography and Remote Sensing, Nanjing University of Information Science & Technology, Nanjing 210044, China
[5]Department of Geological Sciences, University of Texas at San Antonio, Texas 78249, USA
[6]Collaborative Innovation Center of South China Sea Studies, Nanjing 210023, China

Correspondence to: Chang-Qing Ke (kecq@nju.edu.cn)

Abstract. Daily snow observation data from 672 stations in China, particularly the 296 stations with over 10 mean snow cover days (SCDs) in a year during the period of 1952–2010, are used in this study. We first examine spatiotemporal variations and trends of SCDs, snow cover onset date (SCOD), and snow cover end date (SCED). We then investigate the relationships of SCDs with number of days with temperature below 0 °C (TBZD), mean air temperature (MAT), and Arctic Oscillation (AO) index. The results indicate that years with a positive anomaly of SCDs for the entire country include 1955, 1957, 1964, and 2010, and years with a negative anomaly of SCDs include 1953, 1965, 1999, 2002, and 2009. The reduced TBZD and increased MAT are the main reasons for the overall late SCOD and early SCED since 1952. This explains why only 12 % of the stations show significant shortening of SCDs, while 75 % of the stations show no significant change in the SCDs trends. Our analyses indicate that the distribution pattern and trends of SCDs in China are very complex and are not controlled by any single climate variable examined (i.e. TBZD, MAT, or AO), but a combination of multiple variables. It is found that the AO has the maximum impact on the shortening trends of SCDs in the Shandong peninsula, Changbai Mountains, Xiaoxingganling, and north Xinjiang, while the combined TBZD and MAT have the maximum impact on the shortening trends of SCDs in the Loess Plateau, Tibetan Plateau, and Northeast Plain.

1 Introduction

Snow has a profound impact on the surficial and atmospheric thermal conditions, and is very sensitive to climatic and environmental changes, because of its high reflectivity, low thermal conductivity, and hydrological effects via snowmelt (Barnett et al., 1989; Groisman et al., 1994). The extent of snow cover in the Northern Hemisphere has decreased significantly over the past decades because of global warming (Robinson and Dewey, 1990; Brown and Robinson, 2011). Snow cover showed the largest decrease in the spring, and the decrease rate increased for higher latitudes in response to larger albedo feedback (Déry and Brown, 2007). In North America, snow depth in central Canada showed the greatest decrease (Dyer and Mote, 2006), and snowpack in the Rocky Mountains in the United States declined (Pederson et al., 2013). However, in situ data showed a significant increase in snow accumulation in winter but a shorter snowmelt season over Eurasia (Bulygina et al., 2009). Decrease in snowpack has also been found in the European Alps in the last 20 years of the twentieth century (Scherrer et al., 2004), but a very long time series of snowpack suggests large decadal variability and overall weak long-term trends only (Scherrer et al., 2013). Meteorological data indicated that the snow cover over northwest China exhibited a weak upward trend in snow depth (Qin et al., 2006), with large spatiotemporal variations (Ke et al., 2009; Ma and Qin, 2012). Simulation experiments using climate models indicated that, with continuing global warming, the snow cover in China would show more variations in space and time than ever before (Shi et al., 2011; Ji and Kang, 2013). Spatiotemporal variations of snow cover are also manifested as snowstorms or blizzards, particularly excessive snowfall over a short time duration (Bolsenga and Norton, 1992; Liang et al., 2008; Gao, 2009; Wang et al., 2013; Llasat et al., 2014).

Total snow cover days in a year (SCDs hereafter) is an important index that represents the environmental features of climate (Ye and Ellison, 2003; Scherrer et al., 2004), and is directly related to the radiation and heat balance of the Earth–atmosphere system. The SCDs vary in space and time and contribute to climate change over short timescales (Zhang, 2005), especially in the Northern Hemisphere. Bulygina et al. (2009) investigated the linear trends of SCDs observed at 820 stations from 1966 to 2007, and indicated that the duration of snow cover decreased in the northern regions of European Russia and in the mountainous regions of southern Siberia, while it increased in Yakutia and the Far East. Peng et al. (2013) analysed trends in the snow cover onset date (SCOD) and snow cover end date (SCED) in relation to temperature over the past 27 years (1980–2006) from over 636 meteorological stations in the Northern Hemisphere. They found that the SCED remained stable over North America, whereas there was an early SCED over Eurasia. Satellite-derived snow data indicated that the average snow season duration over the Northern Hemisphere decreased at a rate of 5.3 days per decade between 1972/1973 and 2007/2008 (Choi et al., 2010). Their results also showed that a major change in the trend of snow duration occurred in the late 1980s, especially in western Europe, central and East Asia, and mountainous regions in western United States.

There are large spatiotemporal differences in the SCDs in China (Wang and Li, 2012). Analysis of 40 meteorological stations from 1971 to 2010 indicated that the SCDs had a significant decreasing trend in the western and south-eastern Tibetan Plateau, with the largest decline observed in Nielamu, reaching 9.2 days per decade (Tang et al., 2012). Data analysis also indicated that the SCDs had a linear decreasing trend at most stations in the Hetao region and its vicinity (Xi et al., 2009). However, analysis of meteorological station data in Xinjiang showed that the SCDs had a slight increasing trend, occurring mainly in 1960–1980 (Q. Wang et al., 2009). Li et al. (2009) analysed meteorological data from 80 stations in Heilongjiang province, Northeast China. Their results showed that the snow cover duration shortened, because of both the late SCOD (by 1.9 days per decade) and early SCED (by 1.6 days per decade), which took place mainly in the lower altitude plains.

The SCDs are sensitive to local winter temperature and precipitation, latitude (Hantel et al., 2000; C. Wang et al., 2009; Serquet et al., 2011; Morán-Tejeda et al., 2013), and altitudinal gradient and terrain roughness (Lehning et al., 2011; Ke and Liu, 2014). Essentially, the variation in SCDs

is mainly attributed to large-scale atmospheric circulation or climatic forcing (Beniston, 1997; Scherrer and Appenzeller, 2006; Ma and Qin, 2012; Birsan and Dumitrescu, 2014), such as monsoons, the El Niño–Southern Oscillation, the North Atlantic Oscillation, and the Arctic Oscillation (AO). Xu et al. (2010) investigated the relationship between the SCDs and monsoon index in the Tibetan Plateau and their results indicated that there were great spatial differences. As an index of the dominant pattern of non-seasonal sea-level pressure variations, the AO shows a large impact on the winter weather patterns of the Northern Hemisphere (Thompson and Wallace, 1998; Thompson et al., 2000; Gong et al., 2001; Wu and Wang, 2002; Jeong and Ho, 2005). The inter-annual variation of winter extreme cold days in the northern part of eastern China is closely linked to the AO (Chen et al., 2013). Certainly, the AO plays an important role in the variation of SCDs. An increase in the SCDs before 1990 and a decrease after 1990 have been reported in the Tibetan Plateau, and snow duration has positive correlations with the winter AO index (You et al., 2011), and a significant correlation between the AO and snowfall over the Tibetan Plateau on an inter-decadal timescale was also reported by Lü et al. (2008).

The focus of this study is the variability in the snow cover phenology in China. A longer time series of daily observations of snow cover is used for these spatial and temporal analyses. We first characterise the spatial patterns of change in the SCDs, SCOD, and SCED in different regions of China; we then examine the sensitivity of SCDs to the number of days with temperature below $0\,°C$ (TBZD), the mean air temperature (MAT), and the Arctic Oscillation (AO) index during the snow season (between SCOD and SCED).

2 Data and methods

2.1 Data

We use daily snow cover and temperature data in China from the 1 September 1951 to the 31 August 2010, provided by the National Meteorological Information Centre of China Meteorological Administration (CMA). According to the Specifications for Surface Meteorological Observations (China Meteorological Administration, 2003), an SCD is defined as a day when the snow cover in the area meets the following requirement: at least half of the observation field is covered by snow. For any day with at least half of the observation field covered by snow, snow depth is recorded as a rounded-up integer. For example, a normal SCD is recorded if the snow depth is equal to or more than $1.0\,cm$ (measured with a ruler), or a thin SCD if the snow depth is less than $1.0\,cm$. A snow year is defined as the time period from 1 September of the previous year to 31 August of the current year. For instance, September, October, and November 2009 are treated as the autumn season of snow year 2010, December 2009 and January and February 2010 as the winter season of snow year

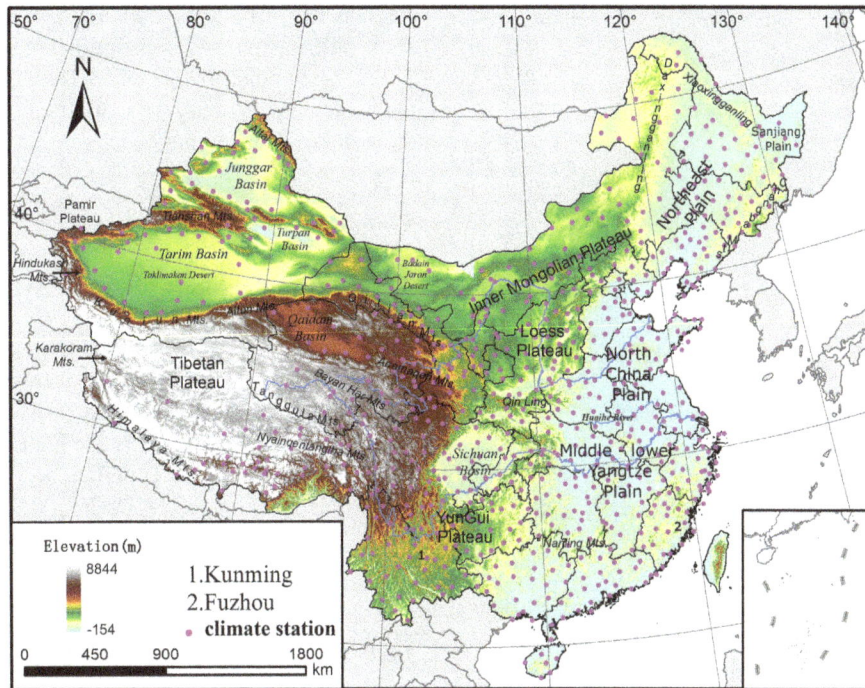

Figure 1. Locations of weather stations and major basins, mountains, and plains mentioned in the paper, overlying the digital elevation model for China.

2010, and March, April, and May 2010 as the spring season of snow year 2010.

Station density is high in eastern China, where the observational data for most stations are complete, with relatively long histories (as long as 59 years), while station density is low in western China, and the observation history is relatively short, although two of the three major snow regions are located in western China. If all stations with short time series are eliminated, the spatial representativeness of the data set would be a problem. Therefore, a time series of at least 30 years is included in this study.

Because of topography and climate conditions, the discontinuous nature of snowfall is obvious in western China, especially in the Tibetan Plateau, with patchy snow cover, and there are many thin SCD records (Ke and Li, 1998). However, in order to enhance data reliability, according to the previous studies (An et al., 2009; Wang and Li, 2012), thin SCDs in the original data set are not taken into account in this paper.

Totally, there are 722 stations in the original data set. Since station relocation and changes in the ambient environment could cause inconsistencies in the recorded data, we implement strict quality controls (such as inspection for logic, consistency, and uniformity) on the observational data sets in order to reduce errors (Ren et al., 2005). The standard normal homogeneity test (Alexandersson and Moberg, 1997) at the 95 % confidence level is applied to the SCDs and temperature series data in order to identify possible breakpoints.

Time series gap filling is performed after all inhomogeneities are eliminated, using nearest neighbour interpolation. After being processed as mentioned above, the 672 stations with annual mean SCDs greater than 1 (day) are finally selected for subsequent investigation (Fig. 1).

The observation period for each station is different, varying between 59 years (1951/1952–2009/2010) and 30 years (1980/1981–2009/2010). Overall, 588 stations have observation records between 50 and 59 years, 47 stations between 40 and 49 years, and 37 stations between 30 and 39 years (Fig. 2). Most of the stations with observation records of less than 50 years are located in remote or high-elevation areas. All 672 stations are used to analyse the spatiotemporal distribution of SCDs in China, while only 296 stations with more than 10 annual mean SCDs are used to study the changes of the relationships of SCOD, SCED, and SCDs with TBZD, MAT, and the AO index.

The daily AO index constructed by projecting the daily (00Z) 1000 mb height anomalies poleward of 20° N, from http://www.cpc.ncep.noaa.gov/products/precip/CWlink/daily_ao_index/ao.shtml, is used. A positive (negative) AO index corresponds to low (high) pressure anomalies throughout the polar region and high (low) pressure anomalies across the subtropical and midlatitudes (Peings et al., 2013). We average the daily AO indexes during the snow season of each station as the AO index of the snow year. A time series of AO indexes from 1952 to 2010, for each of the 296 stations, is then constructed.

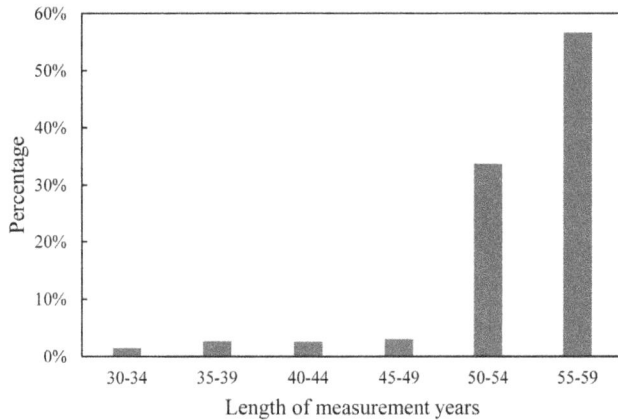

Figure 2. Percentage of weather stations with different measurement lengths.

A digital elevation model from the Shuttle Radar Topographic Mission (SRTM, http://srtm.csi.cgiar.org) of the National Aeronautics and Space Administration (NASA) with a resolution of 90 m and the administration map of China are used as the base map.

2.2   Methods

We apply a Mann–Kendall (MK) test to analyse the trends of SCDs, SCOD, and SCED. The MK test is an effective tool to extract the trends of time series, and is widely applied to the analysis of climate series (Marty, 2008). The MK test is characterised as being more objective, since it is a non-parametric test. A positive standardised MK statistic value indicates an upward or increasing trend, while a negative value demonstrates a downward or decreasing trend. Confidence levels of 90 and 95 % are taken as thresholds to classify the significance of positive and negative trends of SCDs, SCOD, and SCED.

At the same time, if SCDs, SCOD, or SCED at one climate station has a significant MK trend (above 90 %), their linear regression analyses are performed against time, respectively. The slopes of the regressions represent the changing trends and are expressed in days per decade. The statistical significance of the slope for each of the linear regressions is assessed by the Student's t test (two-tailed test of the Student t distribution), and confidence levels of 90 and 95 % are considered.

Correlation analysis is used to examine the SCDs relationships with the TBZD, MAT, and the AO index, and the Pearson product-moment correlation coefficients (PPMCCs) have been calculated. The PPMCC is a widely used estimator for describing the spatial dependence of rainfall processes, and it indicates the strength of the linear covariance between two variables (Habib et al., 2001; Ciach and Krajewski, 2006). The statistical significance of the correlation coefficients is calculated using the Student's t test, and confidence levels above 90 % are considered significant in our analysis.

The spatial distribution of SCDs, SCOD, and SCED, and their calculated results, are spatially interpolated by applying the ordinary Kriging method.

3   Results

3.1   Cross-validation of the spatial interpolations

All mean errors are near zero, all average standard errors are close to the corresponding root mean squared errors, and all root mean squared standardised errors are close to 1 (Table 1). Prediction errors are unbiased and valid, except for slightly overestimated coefficients of variation (CVs) and slightly underestimated SCDs in 2002. Overall, the interpolation results have small errors and are acceptable.

3.2   Spatiotemporal variations of SCDs

3.2.1   Spatial distribution of SCDs

The analysis of observations from 672 stations indicates that there are three major stable snow regions with more than 60 annual mean SCDs (Li, 1990): Northeast China, north Xinjiang, and the Tibetan Plateau, with Northeast China being the largest of the three (Fig. 3a). In the Daxinganling, Xiaoxingganling, and Changbai Mountains of Northeast China, there are more than 90 annual mean SCDs, corresponding to a relatively long snow season. The longest annual mean SCDs, 163 days, is at Arxan Station (in the Daxinganling Mountains) in Inner Mongolia. In north Xinjiang, the SCDs are relatively long in the Tianshan and Altun Mountains, followed by the Junggar Basin. The annual mean SCDs in the Himalayas, Nyainqentanglha, Tanggula Mountains, Bayan Har Mountains, Anemaqen Mountains, and Qilian Mountains of the Tibetan Plateau are relatively long, although most of these regions have fewer than 60 annual SCDs. The Tibetan Plateau has a high elevation, a cold climate, and many glaciers, but its mean SCDs are not as large as those of the other two stable snow regions.

Areas with SCDs of 10–60 per year are called unstable snow regions with annual periodicity (definite snow cover every winter) (Li, 1990). It includes the peripheral parts of the three major stable snow regions, Loess Plateau, Northeast Plain, North China Plain, Shandong peninsula, and regions north of the Qinling–Huaihe line (along the Qinling Mountains and Huaihe River to the east). Areas with SCDs of 1–10 per year are called unstable snow regions without annual periodicity (the mountainous regions are excluded) (Li, 1990). It includes the Qaidam Basin, the Badain Jaran desert, the peripheral parts of Sichuan Basin, the northeast part of the Yungui Plateau, and the middle and lower Yangtze River Plain. Areas with occasional snow and mean annual SCDs of less than 1.0 (day) are distributed north of the Sichuan

Table 1. Prediction errors of cross-validation for the spatial interpolation with the ordinary Kriging method (the unit is day for snow cover days (SCDs), snow cover onset day (SCOD), and snow cover end day (SCED); there is no unit for the coefficient of variation (CV)).

Item (figures)	Mean error	Average standard error	Root mean squared error	Root mean squared standardised error
Mean SCD$_S$ (Fig. 3a)	−0.0230	11.0558	13.7311	1.1097
CV (Fig. 3b)	0.0017	0.7364	0.5510	0.7579
SCD$_S$ in 1957 (Fig. 5a)	−0.0015	11.1561	13.4662	1.1898
SCD$_S$ in 2002 (Fig. 5b)	0.0306	6.6185	8.5887	1.2522
SCD$_S$ in 2008 (Fig. 5c)	0.0477	7.3167	8.1968	1.0969
SCED in 1957 (Fig. 5d)	−0.0449	15.0528	18.9860	1.1921
SCED in 1997 (Fig. 5e)	0.0696	15.5722	17.7793	1.1040
SCOD in 2006 (Fig. 5f)	0.0482	15.4503	16.1757	1.0449
SCOD (Fig. 8a)	0.0293	11.2458	13.9078	1.1712
SCED (Fig. 8b)	−0.0222	15.2265	18.3095	1.1308

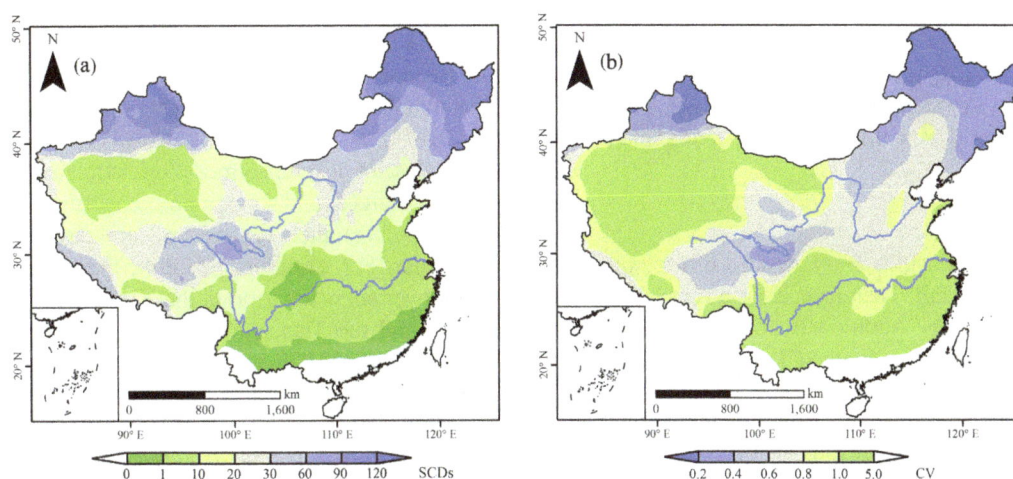

Figure 3. Annual mean snow cover days (SCDs) from 1980/1981 to 2009/2010 (**a**), and their coefficients of variation (CVs) (**b**).

Basin and in the belt along Kunming, Nanling Mountains, and Fuzhou (approximate latitude of 25° N). Because of the latitude or local climate and terrain, there is no snow in the Taklimakan Desert, Turpan Basin, the Yangtze River Valley in the Sichuan Basin, the southern parts of Yunnan, Guangxi, Guangdong, and Fujian, and on the island of Hainan.

The spatial distribution pattern of SCDs based on climate data with longer time series is similar to previous studies (Li and Mi, 1983; Li, 1990; Liu et al., 2012; C. Wang et al., 2009; Wang and Li, 2012). Snow distribution is closely linked to latitude and elevation, and is generally consistent with the climate zones (Lehning et al., 2011; Ke and Liu, 2014). There are relatively more SCDs in Northeast China and north Xinjiang, and fewer SCDs to the south (Fig. 3a). In the Tibetan Plateau, located in south-western China, the elevation is higher than eastern areas at the same latitude, and the SCDs are greater than in eastern China (Tang et al., 2012). The amount of precipitation also plays a critical role in determining the SCDs (Hantel et al., 2000). In the north-eastern coastal areas of China, which are affected consider-

ably by the ocean, there is much precipitation. In north Xinjiang, which has a typical continental (inland) climate, the precipitation is less than in Northeast China, and there are more SCDs in the north of Northeast China than in north Xinjiang (Dong et al., 2004; Q. Wang et al., 2009). Moreover, the local topography has a relatively large impact on the SCDs (Lehning et al., 2011). The Tarim Basin is located inland, with relatively little precipitation, thus snowfall there is extremely rare except in the surrounding mountains (Li, 1993). The Sichuan Basin is surrounded by high mountains, therefore situated in the precipitation shadow in winter, resulting in fewer SCDs (Li and Mi, 1983; Li, 1990).

The three major stable snow regions, Northeast China, north Xinjiang, and the eastern Tibetan Plateau, have smaller CVs in the SCDs (Fig. 3b). Nevertheless, the SCDs in arid or semi-arid regions, such as South Xinjiang, the northern and south-western Tibetan Plateau, and central and western Inner Mongolia, show large fluctuations because there is little precipitation during the cold seasons, and certainly little snowfall and large CVs of SCDs. In particular, the Taklimakan

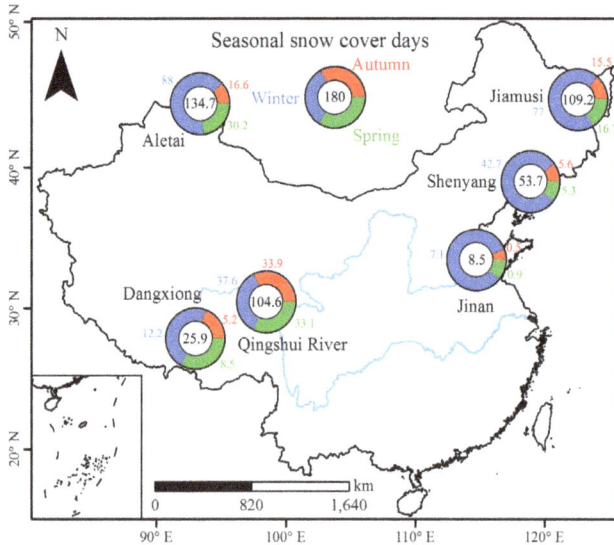

Figure 4. Seasonal variation of SCDs; the number in the centre denotes annual mean SCDs, the blue colour in the circle the SCDs represents the winter season, the green colour spring, and the red colour autumn.

Desert in the Tarim Basin is an extremely arid region, with only occasional snowfall. Therefore, it has a very large range of fluctuations of SCDs. Additionally, the middle and lower Yangtze River Plain also has large SCDs fluctuations because of warm-temperate or subtropic climate with a short winter and little snowfall. Generally, the fewer the SCDs, the larger the CV (C. Wang et al., 2009). This is consistent with other climate variables, such as precipitation (Yang et al., 2015).

3.2.2 Temporal variations of SCDs

Seasonal variation of SCDs is primarily controlled by temperature and precipitation (Hantel et al., 2000; Scherrer et al., 2004; Liu et al., 2012). In north Xinjiang and Northeast China, snow is primarily concentrated in the winter (Fig. 4). In these regions, the SCDs exhibit a single-peak distribution. In the Tibetan Plateau, however, the seasonal variation of SCDs is slightly different, i.e. more snow in the spring and autumn combined than in the winter. The mean temperature and precipitation at Dangxiong station (30°29' N, 91°06' E; 4200.0 m) in winter are -7.7 °C and 7.9 mm, respectively, and those at Qingshuihe station (33°48' N, 97°08' E; 4415.4 m) are -15.8 °C and 16.3 mm, respectively. It is too cold and dry to produce enough snow in the Tibetan Plateau (Hu and Liang, 2014).

The temporal variation of SCDs shows very large differences from 1 year to another. We define a year with a positive (negative) anomaly of SCDs in the following way: for a given year, if 70 % of the stations have a positive (negative) anomaly and 30 % of the stations have SCDs larger (smaller) than the mean $\pm$ 1 standard deviation (1 SD), it is regarded as

a year with a positive (negative) anomaly of SCDs. The years with a positive anomaly of SCDs in China are 1955, 1957, 1964, and 2010 (Table 2). Moreover, the stations with SCDs larger than the mean + 2 SD account for 25 and 26 % of all stations in 1955 and 1957, respectively, and these 2 years are considered as years with an extremely positive anomaly of SCDs. In 1957, there was an almost nationwide positive anomaly of SCDs except for north Xinjiang (Fig. 5a). This 1957 event had a great impact on agriculture, natural ecology, and social-economic systems, and resulted in a heavy snow-caused disaster (Hao et al., 2002).

Years with a negative anomaly of SCDs include 1953, 1965, 1999, 2002, and 2009 (Table 2). If there is too little snowfall in a specific year, a drought is possible. Drought resulting from little snowfall in the cold season is a slow process and can sometimes cause serious damages. For example, East China displayed an apparent negative anomaly of SCDs in 2002 (Fig. 5b), and had very little snowfall, leading to an extreme winter drought in Northeast China, where snowfall is the primary form of winter precipitation (Fang et al., 2014).

Because of different atmospheric circulation backgrounds, vapour sources, and topographic conditions in different regions of China, there are great differences in the SCDs, even in 1 year. For example, in 2008, there were more SCDs and longer snow duration in the Yangtze River Basin, North China, and the Tianshan Mountains in Xinjiang (Fig. 5c), especially in the Yangtze River Basin, where large snowfall was normally not observed. However, four episodes of severe and persistent snow, extreme low temperatures, and freezing weather occurred in 2008 and led to a large-scale snowstorm in this region (Gao, 2009). As reported by the Ministry of Civil Affairs of China, the 2008 snowstorm killed 107 people and caused losses of USD 15.45 billion. Both the SCDs and scale of economic damage broke records from the past 5 decades (Wang et al., 2008). On the contrary, there was no snowstorm in north Xinjiang, the Tibetan Plateau, and Pan-Bohai Bay region in 2008. Moreover, Northeast China had an apparent negative anomaly of SCDs (Fig. 5c).

There are great differences in the temporal variations of SCDs, even in the three major stable snow regions. If we redefine a year with a positive (negative) anomaly of SCDs using a much higher standard (i.e. 80 % of stations have a positive (negative) anomaly and 40 % of stations have an SCDs larger (smaller) than the mean $\pm$ 1 SD), it is found that 1957, 1973, and 2010 are years with a positive anomaly of SCDs in Northeast China, while 1959, 1963, 1967, 1998, 2002, and 2008 are years with a negative anomaly of SCDs (Table 3, Fig. 5a–c). Years with a positive anomaly of SCDs in north Xinjiang include 1960, 1977, 1980, 1988, 1994, and 2010, and years with a negative anomaly of SCDs include 1974, 1995, and 2008 (Table 3, Fig. 5c). North Xinjiang is one of the regions prone to extreme snow events, where frequent heavy snowfall greatly affects the development of animal husbandry (Hao et al., 2002).

Table 2. Percentage (%) of stations with anomalies (P for positive and N for negative) of snow cover days (SCDs) in a year, snow cover onset date (SCOD), and snow cover end date (SCED). Percentage (%) of stations with anomalies of SCDs, SCOD, and SCED larger (smaller) than the mean ± 1 or 2 standard deviations (SD), with the bold number denoting years with a positive (negative) anomaly of SCDs, and late (early) years for SCOD or SCED in China. All the percentages are calculated based on 672 stations.

Year	SCDs P	1 SD	2 SD	−2 SD	−1 SD	N	SCOD P	1 SD	2 SD	−2 SD	−1 SD	N	SCED P	1 SD	2 SD	−2 SD	−1 SD	N
1952	31	2	0	13	33	69	69	40	21	2	9	31	55	17	2	12	17	45
1953	28	7	0	**3**	**36**	**72**	40	8	2	2	18	60	37	8	1	10	18	63
1954	57	31	12	0	8	43	35	8	4	1	18	65	56	11	0	0	10	44
1955	**79**	**45**	**25**	1	5	21	37	9	4	1	22	63	77	21	2	1	6	23
1956	46	10	0	0	4	54	69	20	2	0	9	31	61	24	1	2	12	39
1957	**85**	**62**	**26**	0	3	15	26	6	1	0	15	74	**84**	**35**	**5**	1	4	16
1958	48	15	4	0	14	52	46	17	0	0	18	54	52	17	3	4	18	48
1959	28	7	1	4	23	72	53	26	8	1	18	47	59	23	3	1	5	41
1960	37	13	3	0	16	63	49	11	2	0	10	51	59	24	6	4	18	41
1961	36	7	1	1	18	64	25	9	2	1	27	75	30	6	1	7	26	70
1962	41	11	3	0	10	59	44	13	4	2	10	56	58	18	3	0	11	42
1963	25	5	2	2	27	75	34	14	5	1	23	66	51	14	0	8	17	49
1964	**76**	**36**	**11**	0	1	24	31	3	1	4	24	69	64	18	1	0	5	36
1965	26	8	0	**1**	**32**	**74**	59	18	5	1	8	41	55	14	2	3	17	45
1966	28	6	1	0	13	72	46	21	6	0	13	54	67	12	1	2	5	33
1967	31	5	0	3	23	69	40	11	3	2	15	60	43	5	0	3	12	57
1968	61	29	12	3	8	39	35	8	1	0	13	65	34	13	0	4	26	66
1969	42	18	5	4	21	58	45	13	1	3	20	55	67	20	1	1	7	33
1970	46	15	1	2	11	54	38	10	3	2	24	62	62	19	3	0	7	38
1971	53	12	1	1	9	47	38	15	4	1	17	62	53	9	1	1	8	47
1972	55	23	11	0	8	45	37	9	2	1	21	63	46	16	4	1	9	54
1973	50	19	2	1	7	50	35	10	1	2	23	65	43	9	1	1	8	57
1974	33	8	0	3	23	67	53	29	6	1	11	47	52	12	1	1	10	48
1975	41	10	4	1	15	59	26	7	2	1	21	74	43	15	3	2	16	57
1976	35	11	3	1	23	65	60	25	12	0	5	40	**77**	**31**	**5**	1	3	23
1977	45	20	3	0	9	55	28	5	1	0	25	72	57	14	3	2	12	43
1978	60	22	8	0	2	40	43	13	2	2	13	57	55	10	1	0	8	45
1979	41	8	1	0	7	59	43	11	1	0	20	57	**79**	**32**	**2**	0	4	21
1980	39	12	1	0	5	61	41	9	1	1	16	59	82	27	2	0	4	18
1981	42	13	2	0	13	58	45	20	4	2	18	55	44	13	1	2	15	56
1982	40	12	1	1	15	60	23	9	2	**0**	30	77	58	23	6	6	16	42
1983	50	19	6	0	12	50	44	14	1	1	11	56	67	26	2	1	9	33
1984	26	9	1	1	28	74	68	32	16	0	5	32	48	8	1	2	13	52
1985	66	24	3	0	3	34	32	8	1	1	24	68	46	8	2	1	8	54
1986	50	14	2	0	12	50	32	5	1	1	19	68	63	18	4	3	10	38
1987	67	23	4	0	4	33	40	7	1	2	15	60	60	23	3	1	8	40
1988	56	17	1	0	2	44	24	6	1	3	26	76	69	23	0	1	7	31
1989	47	18	4	0	11	53	71	29	7	1	6	29	41	6	1	3	18	59
1990	56	19	2	0	7	44	52	9	1	0	9	48	49	12	1	2	10	51
1991	34	4	0	2	9	66	60	21	3	0	4	40	72	26	3	1	4	28
1992	50	13	4	1	7	50	54	18	5	0	4	46	50	13	1	5	19	50
1993	58	19	2	1	4	42	43	9	1	0	17	57	49	18	2	2	21	51
1994	58	19	2	0	4	42	28	6	2	1	22	72	39	11	0	3	18	61
1995	36	10	3	3	15	64	57	24	3	1	15	43	49	8	1	7	18	51
1996	26	8	2	2	22	74	**71**	**30**	**4**	0	5	29	55	11	1	2	15	45
1997	37	3	0	1	18	63	44	13	3	2	12	56	18	4	2	**9**	**49**	**82**
1998	34	8	2	4	18	66	37	11	3	1	20	63	30	9	1	7	25	70
1999	25	4	1	**1**	**35**	**75**	61	23	12	1	7	39	51	11	2	5	15	49
2000	64	17	4	0	5	36	59	18	2	0	9	41	39	7	0	5	22	61
2001	67	29	8	0	5	33	39	16	2	1	22	61	42	17	1	3	15	58
2002	17	2	0	**5**	**32**	**83**	59	22	4	1	4	41	31	6	0	12	30	69
2003	57	29	4	1	8	43	36	6	1	0	21	64	50	9	2	6	18	50
2004	35	3	1	0	16	65	42	11	2	1	26	58	32	7	1	13	33	68
2005	60	18	1	0	4	40	48	15	2	0	11	52	33	4	0	2	19	67
2006	48	11	3	0	8	52	**70**	**33**	**7**	0	5	30	57	16	0	1	10	43
2007	30	6	1	0	22	70	69	25	5	1	6	31	29	3	1	7	26	71
2008	43	19	5	3	20	57	68	27	7	0	8	32	41	10	1	4	24	59
2009	24	6	0	**1**	**31**	**76**	73	23	9	0	5	27	27	4	0	3	25	73
2010	**75**	**42**	**11**	0	10	25	42	11	2	1	18	58	72	20	1	1	7	28

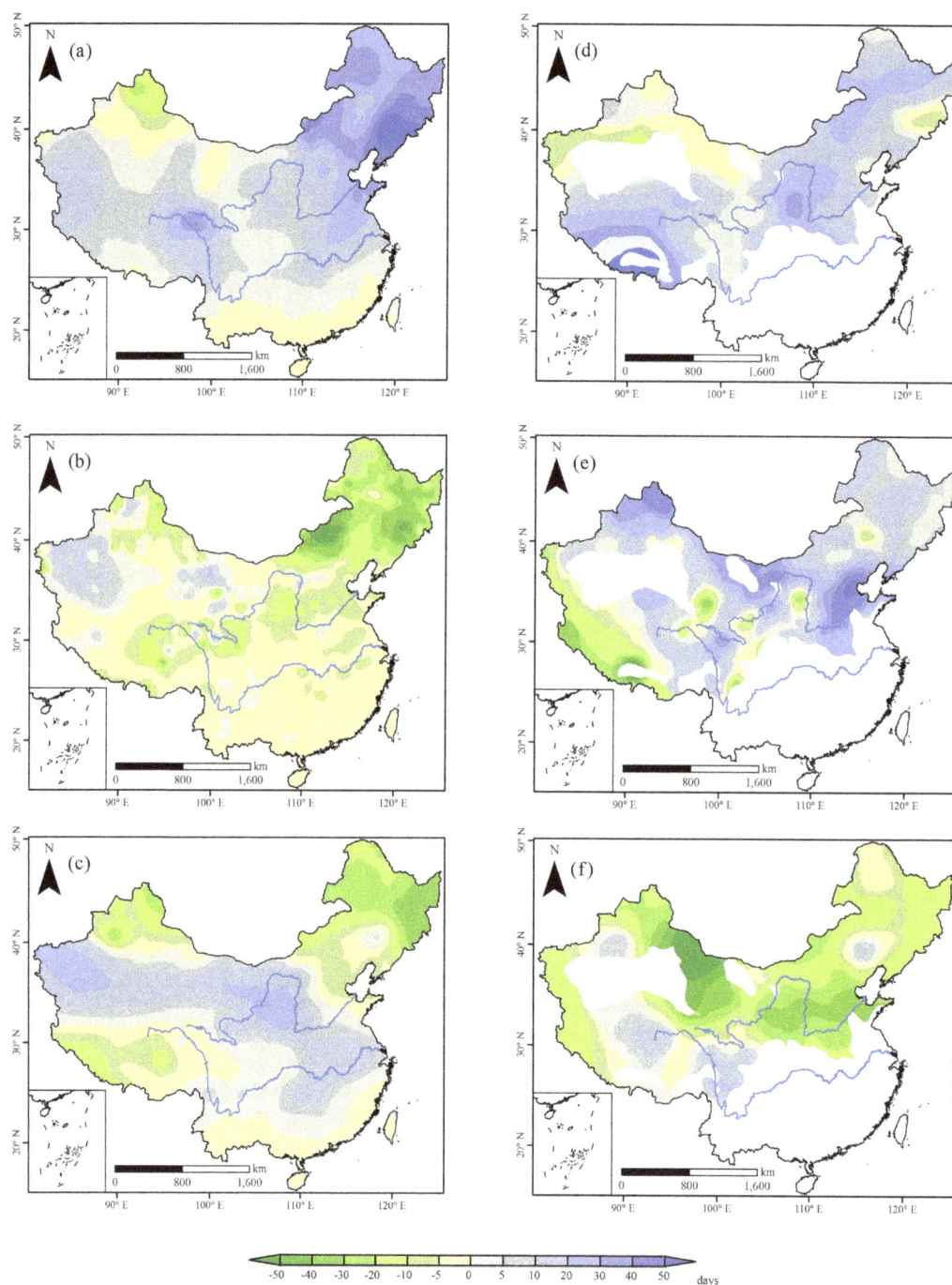

Figure 5. SCDs anomalies in 1957 (**a**), 2002 (**b**) and 2008 (**c**), anomaly of snow cover onset date (SCOD) in 2006 (**d**), and anomalies of snow cover end date (SCED) in 1957 (**e**) and 1997 (**f**).

Years with a positive anomaly of SCDs in the Tibetan Plateau include 1983 and 1990, whereas years with a negative anomaly of SCDs include 1965, 1969, and 2010 (Table 3). The climate in the Tibetan Plateau is affected by the Indian monsoon from the south, westerlies from the west, and the East Asian monsoon from the east (Yao et al., 2012).

Therefore, there is a spatial difference in the SCDs within the Tibetan Plateau, and a difference in the spatiotemporal distribution of snowstorms (Wang et al., 2013). Our results differ from the conclusions drawn by Dong et al. (2001), as they only used data from 26 stations, covering only a short period (1967–1996).

Table 3. The same as Table 2, but only for the years with a positive (negative) anomaly of SCDs and only for the three major stable snow regions: Northeast China (78 stations), north Xinjiang (21 stations), and the Tibetan Plateau (63 stations).

Year	Northeast China						North Xinjiang						Tibetan Plateau					
	P	1 SD	2 SD	−2 SD	−1 SD	N	P	1 SD	2 SD	−2 SD	−1 SD	N	P	1 SD	2 SD	−2 SD	−1 SD	N
1957	**98**	**72**	**16**	0	0	2	22	0	0	2	33	78	74	52	13	0	4	26
1959	2	0	0	**15**	**73**	98	88	38	0	0	0	12	37	11	3	0	6	63
1960	39	14	1	0	26	61	**100**	**88**	**29**	0	0	0	23	0	0	3	30	77
1963	11	0	0	**6**	**41**	89	26	0	0	5	26	74	20	0	0	0	28	80
1965	66	24	0	1	16	34	21	0	0	0	37	79	12	4	0	**4**	**50**	**88**
1967	16	0	0	**14**	**59**	**84**	78	22	0	0	6	22	23	6	0	0	15	77
1969	21	1	0	15	43	79	78	28	0	0	6	22	4	0	0	**6**	**53**	**96**
1973	**89**	**60**	**4**	0	0	11	42	0	0	5	11	58	36	11	2	0	21	64
1974	55	18	0	3	21	45	5	0	0	**21**	**58**	**95**	38	3	0	2	14	62
1977	73	32	4	0	5	27	**95**	**74**	**0**	0	5	5	36	19	7	0	7	64
1980	65	18	1	0	8	35	**95**	**63**	**5**	0	0	5	45	10	2	0	3	55
1983	62	23	3	0	3	38	26	0	0	0	21	74	**95**	**60**	**19**	0	0	5
1988	70	23	0	0	3	30	**100**	**68**	**11**	0	0	0	52	22	5	0	2	48
1990	40	0	0	0	11	60	32	5	0	0	21	68	**81**	**41**	**3**	0	0	19
1994	94	29	1	0	0	6	**95**	**53**	**0**	0	0	5	46	14	2	0	11	54
1995	33	1	0	3	15	67	5	0	0	**21**	**74**	**95**	75	42	11	0	0	25
1998	4	0	0	**14**	**64**	**96**	63	5	0	5	11	37	82	39	12	0	0	18
2002	4	0	0	**19**	**63**	**96**	26	0	0	5	21	74	22	2	0	0	15	78
2008	7	0	0	**11**	**48**	**93**	5	0	0	**5**	**47**	**95**	59	6	0	2	14	41
2010	**92**	**69**	**17**	0	3	8	**100**	**67**	**11**	0	0	0	15	6	0	**2**	**50**	**85**

Table 4. Significance of trends according to the Mann–Kendall test of SCDs, SCOD, and SCED, significance of relationships among SCDs, SCOD, SCED, respectively, with TBZD, significance of relationship between SCDs and MAT, and significance of relationship between SCDs and AO (296 stations in total). All of them have two significance levels, 90 and 95 %.

		SCDs			SCOD			SCED		
		95 %	90 %	I*	95 %	90 %	I*	95 %	90 %	I*
Trend	Positive	19	37	125	178	196	74	1	3	37
	Negative	26	35	99	5	8	18	72	103	153
TBZD	Positive	124	154	126	0	1	50	72	99	170
	Negative	1	1	15	61	87	158	0	2	25
MAT	Positive	0	2	22						
	Negative	114	148	124						
AO	Positive	31	45	90						
	Negative	33	48	113						

Note: I* represents insignificant trends or relations.

3.2.3 SCD trends

Changing trends of annual SCDs are examined, as shown in Fig. 6a, and summarised in Table 4. Among the 296 stations, there are 35 stations (12 %) with a significant negative trend, and 37 stations (13 %) with a significant positive trend (both at the 90 % level), while 75 % of stations show no significant trends. The SCDs exhibit a significant downward trend in the Xiaoxingganling, the Changbai Mountains, the Shandong peninsula, the Qilian Mountains, the North Tianshan Mountains, and the peripheral zones in the south and eastern Tibetan Plateau (Fig. 6a). For example, the SCDs decreased by 50 days from 1955 to 2010 at the Kuandian station in Northeast China, 28 days from 1954 to 2010 at the Hongliuhe station in Xinjiang, and 10 days from 1958 to 2010 at the Gangcha station on the Tibetan Plateau (Fig. 7a–c).

The SCDs in the Bayan Har Mountains, the Anemaqen Mountains, the Inner Mongolia Plateau, and the Northeast Plain, exhibit a significant upward trend (Fig. 6a). For example, at the Shiqu station on the eastern border of the Tibetan Plateau, the SCDs increased 26 days from 1960 to 2010 (Fig. 7d). The coexistence of negative and positive trends in the change of SCDs was also reported by Bulygina et al. (2009) and Wang and Li (2012).

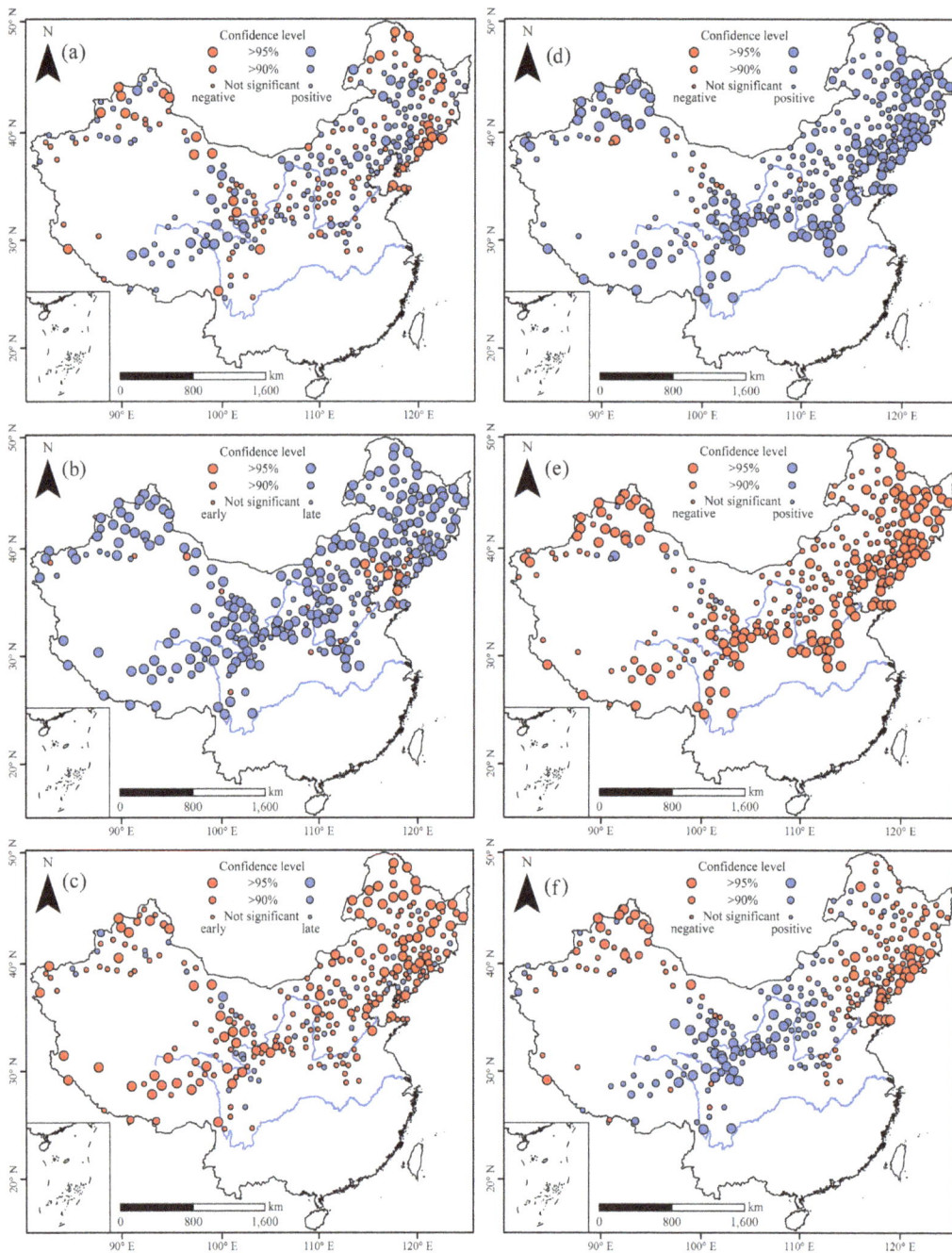

Figure 6. Significance of trends according to the Mann–Kendall test of SCDs (**a**), SCOD (**b**), and SCED (**c**) from the 296 stations with more than 10 annual mean SCDs, significance of relationship between the SCDs and days with temperature below 0 °C (TBZD) (**d**), significance of relationship between the SCDs and mean air temperature (MAT) (**e**), and significance of relationship between the SCDs and Arctic Oscillation (AO) index (**f**).

3.3 Spatiotemporal variations of SCOD

3.3.1 SCOD variations

The SCOD is closely related to both latitude and elevation (Fig. 8a). For example, snowfall begins in September on the

Tibetan Plateau, in early or middle October on the Daxinganling, and in middle or late October on the Altai Mountains in Xinjiang. The SCOD also varies from one year to another (Table 2). Using the definition of a year with a positive (negative) anomaly of SCDs, as introduced before (i.e. 70 % stations with positive (negative) SCOD anomaly and 30 % sta-

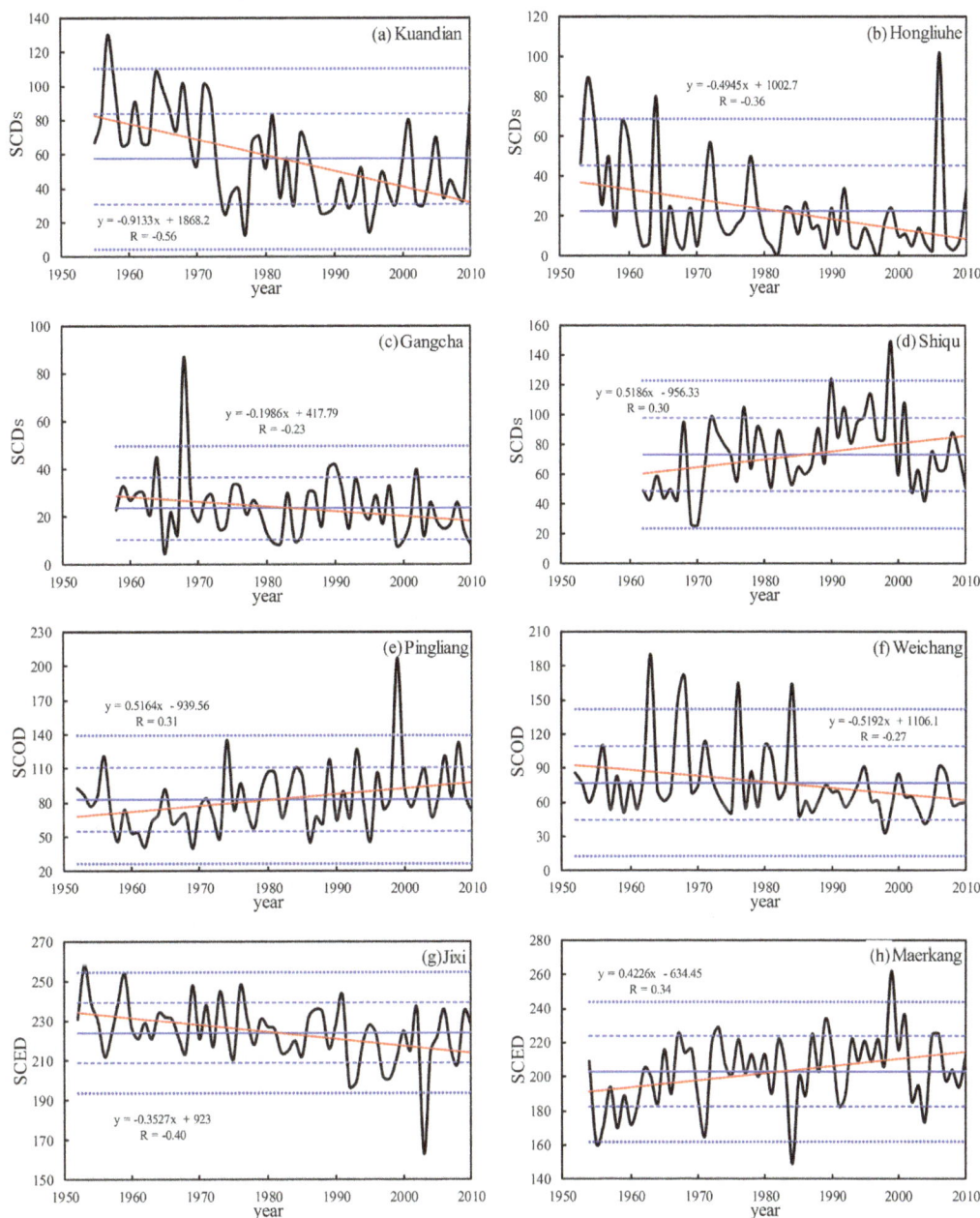

Figure 7. Variations in SCDs at Kuandian (40°43′ N, 124°47′ E; 260.1 m) **(a)**, Hongliuhe (41°32′ N, 94°40′ E; 1573.8 m) **(b)**, Gangcha (37°20′ N, 100°08′ E; 3301.5 m) **(c)**, and Shiqu (32°59′ N, 98°06′ E; 4533.0 m) **(d)**; SCOD at Pingliang (35° 33′ N, 106°40′ E; 1412.0 m) **(e)** and Weichang (41°56′ N, 117°45′ E; 842.8 m) **(f)**; and SCED at Jixi (45°18′ N, 130°56′ E; 280.8 m) **(g)**, and Maerkang (31°54′ N, 102°54′ E; 2664.4 m) **(h)**. (The unit on the y axis in the panels **(e)**, **(f)**, **(g)**, and **(h)** denotes the Julian day using 1 September as reference).

tions with SCOD larger (smaller) than the mean ±1 SD), we consider a given year as a late (early) SCOD year. Two years, 1996 and 2006, can be considered as late SCOD years on a large scale (Table 2), especially in 2006, in East China and the Tibetan Plateau (Fig. 5d). Only 1 year, 1982, can be considered as an early SCOD year.

3.3.2 SCOD trends

There are 196 stations (66 %) with a significant trend of late SCOD, and eight stations (3 %) with a significant trend of early SCOD (both at the 90 % level), while 31 % of the stations show no significant trends (Table 4). The SCOD in the major snow regions in China exhibits a significant trend towards late SCOD (Fig. 6b). These significantly late trends

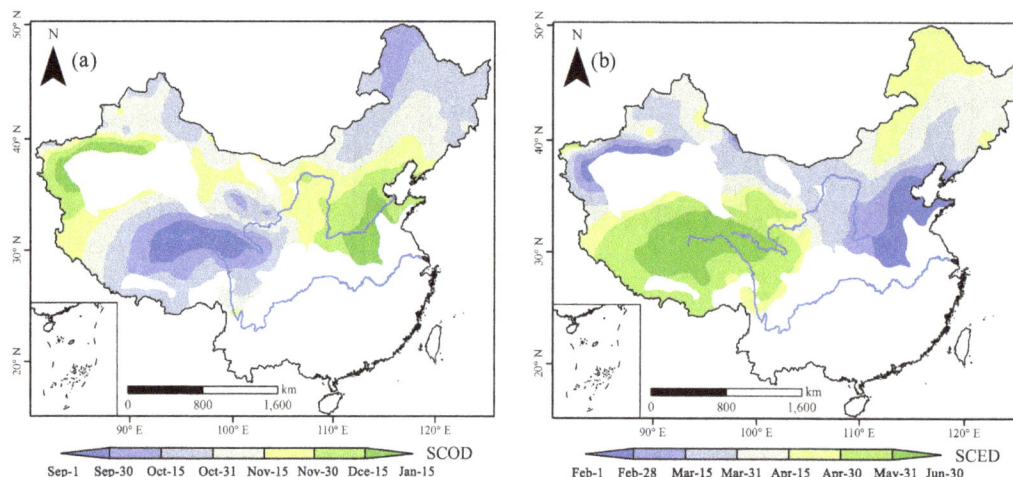

Figure 8. Spatial distribution of SCOD (**a**) and SCED (**b**) based on the stations with an average of more than 10 SCDs.

dominate the major snow regions in China. In particular, the late SCOD in Northeast China is consistent with a previous study (Li et al., 2009). Only the SCOD in the east Liaoning Bay region exhibits a significant trend towards early SCOD. For example, the SCOD at the Pingliang station in Gansu province shows a late rate of 5.2 days per decade from 1952 to 2010, but the SCOD at the Weichang station in Hebei province shows an early rate of 5.2 days per decade from 1952 to 2010 (Fig. 7e–f).

3.4 Spatiotemporal variations of SCED

3.4.1 SCED variations

The pattern of SCED is similar to that of SCOD (Fig. 8b), i.e. places with early snowfall normally show late snowmelt, while places with late snowfall normally show early snowmelt. Like the SCOD, temporal variations of SCED are large (Table 2). Using the same standard for defining the SCOD anomaly, we judge a given year as a late (early) SCED year. Three years, 1957, 1976 and 1979, can be considered as late SCED years on a large scale (Table 2). It is evident that 1957 was a typical year whose SCED was late, which was also the reason for the great SCDs (Fig. 5a and e). The SCED in 1997 was early for almost all of China except for the Tibetan Plateau, western Tianshan Mountains, and western Liaoning (Fig. 5f).

3.4.2 SCED trends

For the SCED, there are 103 stations (35 %) with a significantly early trend (at the 90 % level), while 64 % of stations show no significant trends (Table 4). The major snow regions in China all show early SCED, significant for Northeast China, north Xinjiang, and the Tibetan Plateau (Fig. 6c). The tendency of late SCED is limited, with only three stations

(1 %) showing a significant trend. For example, the SCED at the Jixi station in Northeast China shows an early rate of 3.5 days per decade from 1952 to 2010, while the SCED at the Maerkang station in Sichuan province shows a late rate of 4.2 days per decade from 1954 to 2010 (Fig. 7g–h).

4 Discussion

In the context of global warming, 196 stations (66 %) show significantly late SCOD, and 103 stations (35 %) show significantly early SCED, all at the 90 % confidence level. It is not necessary for one station to show both significantly late SCOD and early SCED. This explains why only 12 % of stations show a significantly negative SCDs trend, while 75 % of stations show no significant change in the trends of SCDs. The latter is inconsistent with the overall shortening of the snow period in the Northern Hemisphere reported by Choi et al. (2010). One reason could be the different time periods used in the two studies, 1972–2007 in Choi et al. (2010) as compared to 1952–2010 in this study. Below, we discuss the possible connections between the spatiotemporal variations of snow cover and the warming climate and changing AO.

4.1 Relationship with TBZD

The number of days with temperature below 0 °C (TBZD) plays an important role in the SCDs. There are 280 stations (95 % of 296 stations) showing positive correlations between TBZD and SCDs, with 154 of them (52 %) having significantly positive correlations (Table 4, Fig. 6d). For example, there is a significantly positive correlation between SCDs and TBZD at the Chengshantou station (Fig. 9a). Therefore, generally speaking, the smaller the TBZD, the shorter the SCDs.

For the SCOD, there are 245 stations with negative correlations with TBZD, accounting for 83 % of 296 stations,

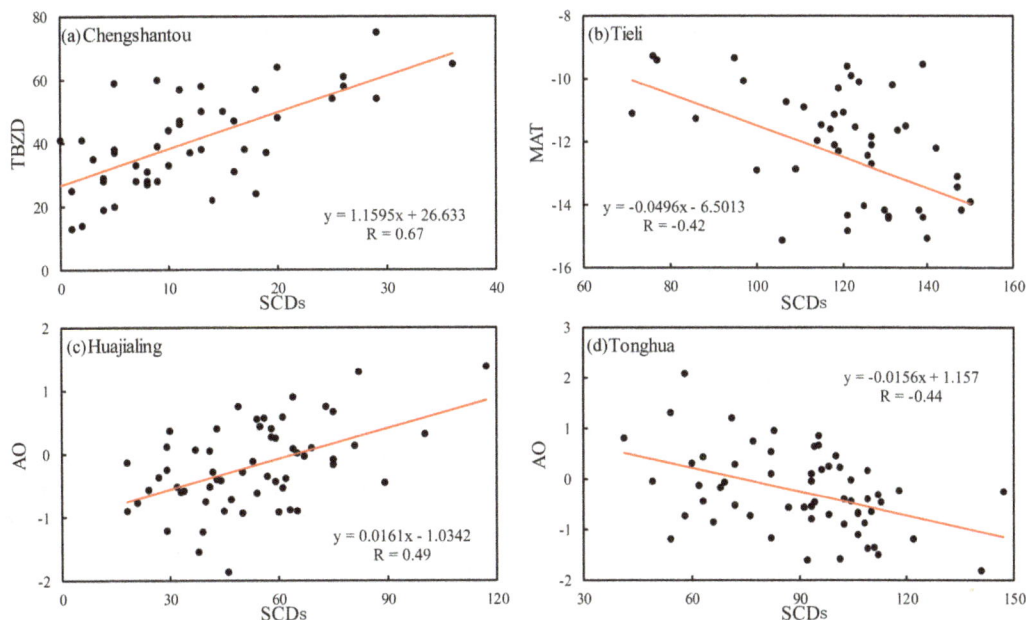

Figure 9. SCDs relationships with TBZD at Chengshantou (37°24′ N, 122°41′ E; 47.7 m) (**a**), MAT at Tieli (46°59′ N, 128°01′ E; 210.5 m) (**b**), and AO index at Huajialing (35°23′ N, 105°00′ E; 2450.6 m) (**c**) and Tonghua (41°41′ N, 125°54′ E; 402.9 m) (**d**).

whereas only 51 stations (17 %) show positive correlations (Table 4). This means that for smaller TBZD, the SCOD is later. For the SCED, there are 269 stations with positive correlations, accounting for 91 % of 296 stations, whereas only 27 stations (9 %) have negative correlations. This means that for smaller TBZD, the SCED is earlier.

Very similar results are found for the MAT (Table 4, Fig. 6e), and Fig. 9b shows an example (the Tieli station).

4.2   Relationship with AO

Although the AO index has shown a strong positive trend in the past decades (Thompson et al., 2000), its impact on the SCDs in China is spatially distinctive. Positive correlations (46 % of 296 stations) are found in the eastern Tibetan Plateau and the Loess Plateau (Table 4, Fig. 6f), and Fig. 9c shows an example (the Huajialing station). Negative correlations (54 % of 296 stations) exist in north Xinjiang, Northeast China, and the Shandong peninsula, and Fig. 9d shows an example (the Tonghua station).

5   Conclusion

This study examines the snow cover change based on 672 stations in 1952–2010 in China. Specifically, the 296 stations with more than 10 annual mean SCDs are used to study the changing trends of SCDs, SCOD, and SCED, and SCD relationships with TBZD, MAT, and AO index during snow seasons. Some important results are summarised below.

Northeast China, north Xinjiang, and the Tibetan Plateau are the three major snow regions. The overall inter-annual variability of SCDs is large in China. The years with a positive anomaly of SCDs in China include 1955, 1957, 1964, and 2010, while the years with a negative anomaly of SCDs are 1953, 1965, 1999, 2002, and 2009. Only 12 % of stations show a significantly negative SCDs trend, while 75 % of stations show no significant SCDs trends. Our analyses indicate that the distribution pattern and trends of SCDs in China are very complex and are not controlled by any single climate variable examined (i.e. TBZD, MAT, or AO), but by a combination of multiple variables.

It is found that significantly late SCOD occurs in nearly the whole of China except for the east Liaoning Bay region; significantly early SCED occurs in nearly all major snow regions in China. Both the SCOD and SCED are closely related to the TBZD and MAT, and are mostly controlled by local latitude and elevation. Owing to global warming since the 1950s, the reduced TBZD and increased MAT are the main reasons for overall late SCOD and early SCED, although it is not necessary for one station to experience both significantly late SCOD and early SCED. This explains why only 12 % of stations show significantly negative trends in SCDs, while 75 % of stations show no significant SCDs trends.

Long-duration, consistent records of snow cover and depth are rare in China because of many challenges associated with taking accurate and representative measurements, especially in western China; the station density and metric choice also vary with time and locality. Therefore, more accurate and reliable observation data are needed to further analyse the spatiotemporal distribution and features of snow cover phenology. Atmospheric circulation causes variability in the snow cover phenology, and its effect requires deeper investigations.

Acknowledgements. This work is financially supported by the Program for National Nature Science Foundation of China (no. 41371391), and the Program for the Specialized Research Fund for the Doctoral Program of Higher Education of China (no. 20120091110017). This work is also partially supported by the Collaborative Innovation Center of Novel Software Technology and Industrialization. We would like to thank the National Climate Center of China (NCC) in Beijing for providing valuable climate data sets. We thank the three anonymous reviewers and the editor for valuable comments and suggestions that greatly improved the quality of this paper.

Edited by: H.-J. Hendricks Franssen

References

Alexandersson, H. and Moberg, A.: Homogenization of Swedish temperature data Part 1: homogeneity test for linear trends, Int. J. Climatol., 17, 25–34, 1997.

An, D., Li, D., Yuan, Y., and Hui, Y.: Contrast between snow cover data of different definitions, J. Glaciol. Geocrol., 31, 1019–1027, 2009.

Barnett, T. P., Dumenil, L., and Latif, M.: The effect of Eurasian snow cover on regional and global climate variations, J. Atmos. Sci., 46, 661–685, 1989.

Beniston, M: Variations of snow depth and duration in the Swiss Alps over the last 50 years: Links to changes in large-scale climatic forcings, Clim. Change, 36, 281–300, 1997.

Birsan, M. V. and Dumitrescu, A.: Snow variability in Romania in connection to large-scale atmospheric circulation, Int. J. Climatol., 34, 134–144, 2014.

Bolsenga, S. J. and Norton, D. C.: Maximum snowfall at long-term stations in the U.S./Canadian Great Lakes, Nat. Hazards, 5, 221–232, 1992.

Brown, R. D. and Robinson, D. A.: Northern Hemisphere spring snow cover variability and change over 1922–2010 including an assessment of uncertainty, The Cryosphere, 5, 219–229, doi:10.5194/tc-5-219-2011, 2011.

Bulygina, O. N., Razuvaev, V. N., and Korshunova, N. N.: Changes in snow cover over Northern Eurasia in the last few decades, Environ. Res. Lett., 4, 045026, doi:10.1088/1748-9326/4/4/045026, 2009.

Chen, S., Chen, W., and Wei, K.: Recent trends in winter temperature extremes in eastern China and their relationship with the Arctic Oscillation and ENSO, Adv. Atmos. Sci., 30, 1712–1724, 2013.

China Meteorological Administration: Specifications for Surface Meteorological Observations, Beijing, China Meteorological Press, 1–62, 2003.

Choi, G., Robinson, D. A., and Kang, S.: Changing Northern Hemisphere snow seasons, J. Climate, 23, 5305–5310, 2010.

Ciach, G. J. and Krajewski, W. F.: Analysis and modeling of spatial correlation structure in small-scale rainfall in Central Oklahoma, Adv. Water Resour., 29, 1450–1463, 2006.

Déry, S. J. and Brown, R. D.: Recent Northern Hemisphere snow cover extent trends and implications for the snow-albedo feedback, Geophys. Res. Lett., 34, L22504, doi:10.1029/2007GL031474, 2007.

Dong, A., Guo, H., Wang, L., and Liang, T.: A CEOF analysis on variation about yearly snow days in Northern Xinjiang in recent 40 years, Plateau Meteorol., 23, 936–940, 2004.

Dong, W., Wei, Z., and Fan, J.: Climatic character analysis of snow disasters in east Qinghai-Xizang Plateau livestock farm, Plateau Meteorol., 20, 402–406, 2001.

Dyer, J. L. and Mote, T. L.: Spatial variability and trends in observed snow depth over North America, Geophys. Res. Lett., 33, L16503, doi:10.1029/2006GL027258, 2006.

Fang, S., Qi, Y., Han, G., Zhou, G., and Cammarano, D.: Meteorological drought trend in winter and spring from 1961 to 2010 and its possible impacts on wheat in wheat planting area of China, Sci. Agr. Sin., 47, 1754–1763, 2014.

Gao, H.: China's snow disaster in 2008, who is the principal player?, Int. J. Climatol., 29, 2191–2196, 2009.

Gong, D. Y., Wang, S. W., and Zhu, J. H.: East Asian winter monsoon and Arctic oscillation, Geophys. Res. Lett., 28, 2073–2076, 2001.

Groisman, P. Y., Karl, T. R., and Knight, R. W.: Observed impact of snow cover on the heat-balance and the rise of continental spring temperatures, Science, 263, 198–200, 1994.

Habib, E., Krajewski, W. F., and Ciach, G. J.: Estimation of rainfall interstation correlation, J. Hydrometeorol., 2, 621–629, 2001.

Hantel, M., Ehrendorfer, M., and Haslinger, A.: Climate sensitivity of snow cover duration in Austria, Int. J. Climatol., 20, 615–640, 2000.

Hao, L., Wang, J., Man, S., and Yang, C.: Spatio-temporal change of snow disaster and analysis of vulnerability of animal husbandry in China, J. Nat. Disaster, 11, 42–48, 2002.

Hu, H. and Liang, L.: Temporal and spatial variations of snowfall in the east of Qinghai-Tibet Plateau in the last 50 years, Acta Geogr. Sin., 69, 1002–1012, 2014.

Jeong, J. H. and Ho, C. H.: Changes in occurrence of cold surges over East Asia in association with Arctic oscillation, Geophys. Res. Lett., 32, L14704, doi:10.1029/2005GL023024, 2005.

Ji, Z. and Kang, S.: Projection of snow cover changes over China under RCP scenarios Clim. Dyn., 41, 589–600, 2013.

Ke, C. Q. and Li, P. J.: Spatial and temporal characteristics of snow cover over the Tibetan plateau, Acta Geogr. Sin., 53, 209–215, 1998.

Ke, C. Q. and Liu, X.: MODIS-observed spatial and temporal variation in snow cover in Xinjiang, China, Clim. Res., 59, 15–26, 2014.

Ke, C. Q., Yu, T., Yu, K., Tang, G. D., and King, L.: Snowfall trends and variability in Qinghai, China, Theor. Appl. Climatol., 98, 251–258, 2009.

Lehning, M., Grünewald, T., and Schirmer, M.: Mountain snow distribution governed by an altitudinal gradient and terrain roughness, Geophys. Res. Lett., 38, L19504, doi:10.1029/2011GL048927, 2011.

Li, D., Liu, Y., Yu, H. and Li, Y.: Spatial-temporal variation of the snow cover in Heilongjiang Province in 1951-2006, J. Glaciol. Geocrol., 31, 1011–1018, 2009.

Li, P. J.: Dynamic characteristic of snow cover in western China, Acta Meteorol. Sin., 48, 505–515, 1993.

Li, P. J.: A preliminary study of snow mass variations over past 30 years in China, Acta Geogr. Sin., 48, 433–437, 1990.

Li, P. J. and Mi, D.: Distribution of snow cover in China, J. Glaciol. Geocrol., 5, 9–18, 1983.

Liang, T. G., Huang, X. D., Wu, C. X., Liu, X. Y., Li, W. L., Guo, Z. G., and Ren, J. Z.: An application of MODIS data to snow cover monitoring in a pastoral area: A case study in Northern Xinjiang, China, Remote Sens. Environ., 112, 1514–1526, 2008.

Liu, Y., Ren, G., and Yu, H.: Climatology of Snow in China, Sci. Geogr. Sin., 32, 1176–1185, 2012.

Llasat, M. C., Turco, M., Quintana-Seguí, P., and Llasat-Botija, M.: The snow storm of 8 March 2010 in Catalonia (Spain): a paradigmatic wet-snow event with a high societal impact, Nat. Hazards Earth Syst. Sci., 14, 427–441, doi:10.5194/nhess-14-427-2014, 2014.

Lü, J. M., Ju, J. H., Kim, S. J., Ren, J. Z., and Zhu, Y. X.: Arctic Oscillation and the autumn/winter snow depth over the Tibetan Plateau, J. Geophys. Res., 113, D14117, 2008.

Ma, L. and Qin, D.: Temporal-spatial characteristics of observed key parameters of snow cover in China during 1957–2009, Sci. Cold Arid Reg., 4, 384–393, 2012.

Marty, C.: Regime shift of snow days in Switzerland, Geophys. Res. Lett., 35, L12501, doi:10.1029/2008GL033998, 2008.

Morán-Tejeda, E., López-Moreno, J. I., and Beniston, M.: The changing roles of temperature and precipitation on snowpack variability in Switzerland as a function of altitude, Geophys. Res. Lett., 40, 2131–2136, 2013.

Pederson, G. T., Betancourt, J. L., and Gregory, J. M.: Regional patterns and proximal causes of the recent snowpack decline in the Rocky Mountains, U.S., Geophys. Res. Lett., 40, 1811–1816, 2013.

Peings, Y., Brun, B., Mauvais, V., and Douville, H.: How stationary is the relationship between Siberian snow and Arctic Oscillation over the 20th century, Geophys. Res. Lett., 40, 183–188, 2013.

Peng, S., Piao, S., Ciais, P., Friedlingstein, P., Zhou, L., and Wang, T.: Change in snow phenology and its potential feedback to temperature in the Northern Hemisphere over the last three decades, Environ. Res. Lett., 8, 014008, doi:10.1088/1748-9326/8/1/014008, 2013.

Qin, D., Liu, S., and Li, P.: Snow cover distribution, variability, and response to climate change in western China, J. Climate, 19, 1820–1833, 2006.

Ren, G. Y., Guo, J., Xu, M. Z., Chu, Z. Y., Zhang, L., Zou, X. K., Li, Q. X., and Liu, X. N.: Climate changes of China's mainland over the past half century, Acta. Meteorol. Sin., 63, 942–956, 2005.

Robinson, D. A. and Dewey, K. F.: Recent secular variations in the extent of northern hemisphere snow cover, Geophys. Res. Lett., 17, 1557–1560, 1990.

Scherrer, S. C., Appenzeller, C., and Laternser, M.: Trends in Swiss Alpine snow days: The role of local- and large-scale climate variability, Geophys. Res. Lett., 31, L13215, doi:10.1029/2004GL020255, 2004.

Scherrer, S. C., and Appenzeller, C.: Swiss Alpine snow pack variability: major patterns and links to local climate and large-scale flow, Clim. Res., 32, 187–199, 2006.

Scherrer, S. C., Wüthrich, C., Croci-Maspoli, M., Weingartner, R., and Appenzeller, C.: Snow variability in the Swiss Alps 1864-2009, Int. J. Clim., 33, 3162–3173, doi:10.1002/joc.3653, 2013.

Serquet, G., Marty, C., Dulex, J.-P., and Rebetez, M.: Seasonal trends and temperature dependence of the snowfall/precipitation-day ratio in Switzerland, Geophys. Res. Lett., 38, L07703, doi:10.1029/2011GL046976, 2011.

Shi, Y., Gao, X., Wu, J., and Giorgi, F.: Changes in snow cover over China in the 21st century as simulated by a high resolution regional climate model, Environ. Res. Lett., 6, 045401, doi:10.1088/1748-9326/6/4/045401, 2011.

Tang, X., Yan, X., Ni, M., and Lu, Y.: Changes of the snow cover days on Tibet Plateau in last 40 years, Acta. Geogr. Sin., 67, 951–959, 2012.

Thompson, D. W. J. and Wallace, J. M.: The Arctic oscillation signature in the wintertime geopotential height and temperature fields, Geophys. Res. Lett., 25, 1297–1300, 1998.

Thompson, D. W. J., Wallace, J. M., and Hegerl, G. C.: Annular modes in the extratropical circulation, part II: Trends, J. Climate, 13, 1018–1036, 2000.

Wang, C. and Li, D.: Spatial-temporal variations of the snow cover days and the maximum depth of snow cover in China during recent 50 years, J. Glaciol. Geocrol., 34, 247–256, 2012.

Wang, C., Wang, Z., and Cui, Y.: Snow cover of China during the last 40 years: Spatial distribution and interannual variation, J. Glaciol. Geocrol., 31, 301-310, 2009.

Wang, L., Gao, G., Zhang, Q., Sun, J. M., Wang, Z. Y., Zhao, Y., Zhao, S. S., Chen, X. Y., Chen, Y., Wang, Y. M., Chen, L. J., and Gao, H.: Characteristics of the extreme low-temperature, heavy snowstorm and freezing disasters in January 2008 in China, Meteorol. Mon., 34, 95–100, 2008.

Wang, Q., Zhang, C., Liu, J., and Liu, W.: The changing tendency on the depth and days of snow cover in Northern Xinjiang, Adv. Clim. Change Res., 5, 39–43, 2009.

Wang, W., Liang, T., Huang, X., Feng, Q., Xie, H., Liu, X., Chen, M., and Wang, X.: Early warning of snow-caused disasters in pastoral areas on the Tibetan Plateau, Nat. Hazards Earth Syst. Sci., 13, 1411–1425, doi:10.5194/nhess-13-1411-2013, 2013.

Wu, B. Y. and Wang, J.: Winter Arctic oscillation, Siberian high and East Asian winter monsoon, Geophys. Res. Lett., 29, 1897, doi:10.1029/2002GL015373, 2002.

Xi, Y., Li, D., and Wang, W.: Study of the temporal-spatial characteristics of snow covers days in Hetao and its vicinity, J. Glaciol. Geocrol., 31, 446–456, 2009.

Xu, L., Li, D., and Hu, Z.: Relationship between the snow cover day and monsoon index in Tibetan Plateau, Plateau Meterol., 29, 1093–1101, 2010.

Yang, H., Yang, D., Hu, Q., and Lv, H.: Spatial variability of the trends in climatic variables across China during 1961–2010, Theor. Appl. Climatol., 120, 773–783, 2015.

Yao, T., Thompson, L., Yang, W., Yu, W., Gao, Y., Guo, X., Yang, X., Duan, K., Zhao, H., Xu, B., Pu, J., Lu, A., Xiang, Y., Kattel, D. B., and Joswiak, D.: Different glacier status with atmospheric circulations in Tibetan Plateau and surroundings, Nature Clim. Change, 2, 663–667, 2012.

Ye, H. and Ellison, M.: Changes in transitional snowfall season length in northern Eurasia, Geophys. Res. Lett., 30, 1252,
  doi:10.1029/2003GL016873, 2003.
You, Q., Kang, S., Ren, G., Fraedrich, K., Pepin, N., Yan, Y., and

Ma, L.: Observed changes in snow depth and number of snow
  days in the eastern and central Tibetan Plateau, Clim. Res., 46,
  171–183, 2011.
Zhang, T.: Influence of the seasonal snow cover on the ground thermal regime: An overview, Rev. Geophys., 43, 1–23, 2005.

7

Global re-analysis datasets to improve hydrological assessment and snow water equivalent estimation in a sub-Arctic watershed

David R. Casson[1,2], **Micha Werner**[1,2], **Albrecht Weerts**[2,3], **and Dimitri Solomatine**[1,4]

[1]IHE Delft Institute of Water Education, Hydroinformatics Chair Group, P.O. Box 3015, 2601 DA, Delft, the Netherlands
[2]Deltares, Operational Water Management, P.O. Box 177, 2600 MH, Delft, the Netherlands
[3]Wageningen University and Research, Hydrology and Quantitative Water Management group, P.O. Box 47, 6700 AA, Wageningen, the Netherlands
[4]Delft University of Technology, Water Resources Section, P.O. Box 5048, 2600 GA, Delft, the Netherlands

Correspondence: David R. Casson (dave.casson@deltares.nl)

Abstract. Hydrological modelling in the Canadian sub-Arctic is hindered by sparse meteorological and snowpack data. The snow water equivalent (SWE) of the winter snowpack is a key predictor and driver of spring flow, but the use of SWE data in hydrological applications is limited due to high uncertainty. Global re-analysis datasets that pro-vide gridded meteorological and SWE data may be well suited to improve hydrological assessment and snowpack simulation. To investigate representation of hydrological pro-cesses and SWE for application in hydropower operations, global re-analysis datasets covering 1979–2014 from the European Union FP7 eartH2Observe project are applied to global and local conceptual hydrological models. The re-cently developed Multi-Source Weighted-Ensemble Precip-itation (MSWEP) and the WATCH Forcing Data applied to ERA-Interim data (WFDEI) are used to simulate snowpack accumulation, spring snowmelt volume and annual stream-flow. The GlobSnow-2 SWE product funded by the Euro-pean Space Agency with daily coverage from 1979 to 2014 is evaluated against in situ SWE measurement over the local watershed. Results demonstrate the successful application of global datasets for streamflow prediction, snowpack accumu-lation and snowmelt timing in a snowmelt-driven sub-Arctic watershed. The study was unable to demonstrate statisti-cally significant correlations ($p < 0.05$) among the measured snowpack, global hydrological model and GlobSnow-2 SWE compared to snowmelt runoff volume or peak discharge. The GlobSnow-2 product is found to under-predict late-season snowpacks over the study area and shows a premature de-cline of SWE prior to the true onset of the snowmelt. Of the datasets tested, the MSWEP precipitation results in annual SWE estimates that are better predictors of snowmelt volume and peak discharge than the WFDEI or GlobSnow-2. This study demonstrates the operational and scientific utility of the global re-analysis datasets in the sub-Arctic, although knowl-edge gaps remain in global satellite-based datasets for snow-pack representation, for example the relationship between passive-microwave-measured SWE to snowmelt runoff vol-ume.

1 Introduction

Snowpack accumulation and melt are the main drivers of hydrology and peak flow events in high-latitude ($> 60° N$) watersheds. The snow water equivalent (SWE) stored in the winter snowpack is the key contributor and predictor of spring and summer streamflow (Liu et al., 2015). In situ mea-surement of SWE can provide valuable information for op-erational water managers, but data collection is challenging in remote high-latitude watersheds, and uncertainty in max-imum annual SWE remains a key constraint in hydrological forecasting (Larue et al., 2017). In northern Canada, uncer-tainty in SWE measurement and a lack of developed hydro-logical modelling tools result in high uncertainty in the pre-diction of snowmelt-driven flood events, leading to infras-tructure risk and hindering operational water management. Climate change is also shifting the hydrology regime at high latitudes, with global circulation models and observational trends indicating a reduction in spring snowpack duration, although the trend in SWE is less clear (Brown and Mote, 2009; Rees et al., 2014). This will increase risk to hydro-electric facilities, mining operations and local communities as rapid spring snowmelt, rain-on-snow events and variable precipitation patterns that cause flooding become more severe (AMAP, 2012; McCabe et al., 2007; National Research Council, 2007).

SWE measurements from ground and remote-sensing sources have high uncertainty for hydrological application. Although field measurement of SWE can be accurate at point locations, these provide only limited spatial and temporal coverage. Precipitation gauge measurements to quantify snowfall at high latitudes have high uncertainty due to the scarcity of meteorological stations, short duration of meteorological measurement records and systematic measurement error (Devine and Mekis, 2008; Mekis and Vincent, 2011; Sugiura et al., 2006). Remote sensing is used to monitor snow on a global scale and measurement of snow depth with passive microwave has the advantage of frequent revisit times, long-term data records and a large spatial extent of data collection (Nolin, 2011). GlobSnow-2 provides a long-term (1979–2014) daily record of SWE over the Northern Hemisphere (Luojus et al., 2014). However, passive microwave measurement of SWE is limited for the measurement of deep or wetted snowpacks, relies on estimates of density, and tends to underestimate SWE in tundra environments (Rees et al., 2007).

Global re-analysis data products, which integrate multiple data sources, are well suited to provide meteorological data at high latitudes due to complete spatial and extended temporal coverage. Research into the reliability of re-analysis products at high latitudes is, however, limited due to a lack of reliable precipitation and SWE data (Mudryk et al. 2015; Wong et al., 2017).

In this study a locally distributed conceptual hydrological model using a simplified snow accumulation and melt routine is forced with eartH2Observe meteorological data to simulate SWE and catchment discharge.

Meteorological datasets generated as part of the eartH2Observe project have been used to force global hydrological models (Schellekens et al., 2017). These global hydrological models can be used to improve understanding of water resources in regions like the sub-Arctic, where information is lacking and the models have large uncertainties in part due to simplifications of physical processes (Bierkens and Van Beek, 2009; van Dijk et al., 2014). This study examines the application of global re-analysis data products for hydrological modelling and representation of SWE in the Snare Watershed in the Canadian sub-Arctic. The available datasets hold great potential to allow accurate discharge modelling for sub-Arctic watersheds and development of more advanced modelling systems. This has practical relevance for operational water management at high latitudes

and provides a basis for hydrological forecasting and data assimilation to further improve model performance.

The three main goals of this paper are as follows:

1. determine the skill of a locally distributed conceptual hydrological model for a snowmelt-driven, high-latitude watershed forced with long-term meteorological re-analysis data developed in the eartH2Observe project;

2. assess the representation of SWE in both the local- and global-scale models and compare to the GlobSnow-2 daily SWE product as well as available long-term records of snowpack surveys;

3. determine the predictive capacity of SWE measurement from in situ snowpack surveys, GlobSnow-2 SWE as well as local and global hydrological models for snowmelt volume and peak discharge rates.

2 Study area and context

The Snare Watershed is located in the northern extent of the Mackenzie River basin in Canadian sub-Arctic. The watershed covers an area of roughly $14\,000\,\mathrm{km}^2$ above a cascade of four hydropower stations as depicted in Fig. 1. The Snare Watershed is typical of many watersheds across northern Canada where temporal and spatial coverage of meteorological data is very sparse, but where historic discharge gauging records are available.

The Snare Watershed has low topographic relief and is characterized by low rolling hills of exposed bedrock with depressions from glacier-scouring forming wetlands, shallow lakes and streams (ECG, 2008). The southern extent of the watershed is boreal forest, while the northern extent is above the treeline and is covered mostly by shrub and sedge tundra (Government of Canada, 2013). Annual precipitation is generally low and in the range of 200 to 500 mm and temperatures are below $0\,^\circ\mathrm{C}$ for extended periods in the winter months (ECG, 2008).

Several meteorological stations have been installed in the Snare Watershed; however, precipitation records are very short, with a maximum duration of 3 years. Gauge measurement of snowfall is known to have systematic underestimation, and large bias correction factors (80 %–120 %) are required for snowfall at high latitudes, though factors in the boreal and tundra region of the Snare Watershed may be closer to only 20 % (Mekis and Vincent, 2011; Yang et al., 2005). Snowpacks accumulated from winter snowfall are highly spatially variable in depth and SWE, with lower accumulation over lake and plateau areas (Rees et al., 2014). Snowfall measurements at high latitudes are particularly difficult to verify due to the sublimation effects on precipitation totals (Mekis and Hogg, 1998).

Sublimation, the direct conversion of snow particles to vapour, is a major factor in removing snow from tundra areas

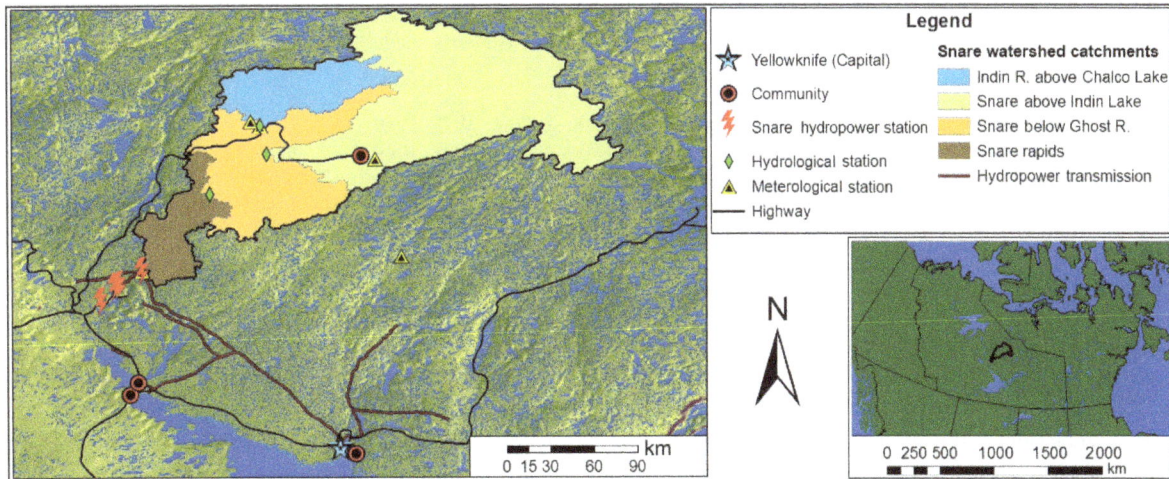

Figure 1. Snare Watershed location in Northwest Territories, Canada.

(Marsh et al., 1995) and along with wind redistribution is a key driver of spatial variability and quantity of SWE. Sublimation estimates in the sub-Arctic boreal forest and tundra regions vary considerably in a general range from 10 % to 50 % of total snowfall (Dery and Yau, 2002; Liston et al., 2002; Marsh et al., 1995; Pomeroy et al., 1997, 1999). Direct measurement of sublimation is very difficult, so values are more often determined through water balance assessment (Liston and Sturm, 2004).

Improved modelling of streamflow and SWE has a direct benefit for the operation of active hydropower facilities in the Snare Watershed. Current approaches for hydropower operations in the Snare Watershed use ground SWE measurements and matching with historical discharge records with similar flow characteristics to anticipate discharge. The system planner uses anticipated streamflow to determine whether to hold or spill water, and whether it is necessary to order diesel should hydroelectric generation fall short and need to be offset using generators. This forecasting approach is limited as it cannot incorporate additional information such as changing temperature regimes, antecedent water storage and meteorological forecasts. In this study the operational context of the Snare Hydro System is used to demonstrate that global datasets are not only useful for broad-scale assessment, but can be applied for accurate discharge modelling and development of a hydrological forecasting system.

3 Methodology

3.1 Hydrological models

3.1.1 The wflow-HBV model

The wflow-HBV is based on the conceptual HBV-96 algorithm and is developed as a distributed hydrological modelling platform using the PCRaster python framework (Karssenberg et al., 2010; OpenStreams, 2016). The wflow-HBV includes a simplified snow accumulation and melt routine based on the degree-day method and kinematic wave approximation for routing (Bergström, 1992). The snow routine does consider snowpack melt and refreezing, but not moisture loss from the snowpack (sublimation) and wind redistribution. Several attempts have been made to improve on the snowmelt modelling of the HBV model, but it has been found that inclusion of more advanced routines and additional input data have had only limited improvement of results (Lindstrom et al., 1997). The wflow-HBV model is highly parameterized and requires a structured approach to calibration to achieve suitable streamflow and physical process representation.

A Python-based framework for optimization, pyOpt, was implemented for calibration of the wflow-HBV model (Perez et al., 2012). Single-objective, constrained parameter optimization of the Nash–Sutcliffe efficiency (NSE) was performed using the Augmented Lagrangian Harmony Search Optimizer (Geem et al., 2001). Constraints on specific model parameters based on land cover type and introduction of lakes and reservoirs were used to improve physical process representation. Historical discharge data were separated into calibration, validation and testing periods. The difference between validation and testing periods is that validation results are seen and evaluated by the modeller in an iterative calibration process, while testing data are not used until the final model parameter values are set. A calibration period as shown in Table 1 was selected to correspond with available discharge data and representative peak flow events in each catchment, and to allow sufficient additional discharge data for validation and testing of the model.

Table 1. Calibration, validation and testing periods.

Catchment	Calibration	Validation	Testing
Catchment 1: Indin River above Chalco Lake	2000–2009	1978–1999	2010–2014
Catchment 2: Snare River above Indin Lake	2000–2004	1998–1999, 2005–2010	2010–2014
Catchment 3: Snare River above Ghost River	2000–2009	1984–1999	2010–2014

3.1.2 Global hydrological models

A set of global hydrological and land-surface models were considered in this study and presented in Table 2. Model state variables such as SWE for selected models and forcing datasets can be obtained from the eartH2Observe project Water Cycle Integrator (WCI) (EartH2Observe, 2017).

3.2 Data

3.2.1 Meteorological data

Meteorological stations are sparse in the study area, as they are across northern Canada (Mekis and Vincent, 2011). Local meteorological stations data collected from Government of Canada Historical Climate Data records were reviewed to determine consistency and completeness (ENR, 2016; Simpson, 2016). With the exception of the Yellowknife station, precipitation records for both rainfall and snowfall were, however, found to be incomplete or of short duration. Temperature records for several nearby stations shown in Fig. 1 were found to be complete and suitable for comparison validation.

Global re-analysis datasets generated as part of the eartH2Observe project were used as forcing data for the wflow-HBV model. The primary precipitation forcing dataset used is the Multi-Source Weighted-Ensemble Precipitation (MSWEP), available at a daily timestep from 1979 to 2015 at a resolution of $0.25° \times 0.25°$. MSWEP was created through combination of gauge, satellite and re-analysis data and includes a long-term bias correction procedure based on discharge observations (H. E. Beck et al., 2017). Precipitation and temperature data from the WATCH Forcing Data applied to ERA-Interim reanalysis data (WFDEI) were used at a daily timestep 1979–2012 at a resolution of $0.5° \times 0.5°$ (Weedon et al., 2014). Potential evapotranspiration (PET) for this study was selected as Penman–Monteith calculated at a daily timestep at a $0.25° \times 0.25°$ resolution based on eartH2Observe Water Resource Re-analysis 2 (WRR2) data (Allen et al., 1998).

Available ground-based weather station data sources and long-term climate normals were used to validate the re-analysis datasets from eartH2Observe. Mean annual precipitation for the eartH2Observe datasets are comparable at the nearest gauge with long-term records at Yellowknife. Undercatch-corrected annual mean precipitation totals for Yellowknife were 377.7 mm, with MSWEP and WFDEI to-

talling 356.3 and 370.7 mm respectively (ENR, 2016). A comparison of monthly precipitation to undercatch-corrected local datasets shows slightly better correlation and performance for MSWEP ($y = 0.93x$, $R^2 = 0.27$) than WFDEI ($y = 0.88x$, $R^2 = 0.25$). Daily mean temperature data for several local stations were well correlated with WFDEI (Lower Carp Lake, $R^2 = 0.98$; Indin River, $R^2 = 0.97$) and showed low biases.

3.2.2 Discharge data

Discharge in the Snare Watershed follows a distinct and highly seasonal pattern which is typical of the sub-Arctic (Kokelj, 2003). Low winter flows are followed by a large peak discharge due to snowmelt. In some years, rainfall in the late fall will cause a notable secondary peak before flow recession in the end of the year. Discharge is available both as a historic time series from as early as 1978 and in near-real time provided by the Water Survey of Canada (ENR, 2016, 2017) for the three hydrological stations presented in Fig. 1. Although the period of record is different for each of the three stations, the annual water yields are well correlated between the three catchments, helping to validate the rating curves and reported discharge rates.

3.2.3 In situ SWE data

Measurement of SWE can be performed in situ with accurate snow depth and density at point locations. However, the resulting datasets have limited spatial and temporal coverage (Derkson et al., 2008). The in situ measurements, or snowpack surveys, are often collected near the end of the snow accumulation season to provide advance information for anticipated snowmelt volume. A long-term record (1978–2016) of end-of-winter snowpack surveys is available at locations distributed across the Snare Watershed (GNWT, 2017). Snowpack survey measurements contain inherent uncertainty related to site selection, sampling protocols and interpolation methods used to create spatial estimates. Despite these limitations, snowpack survey data are considered the most reliable SWE available in the study area.

3.2.4 GlobSnow-2 SWE data

GlobSnow-2 SWE, hereafter referred to as GlobSnow, is a long-term (1979–present) daily record of SWE covering the non-mountainous areas of the Northern Hemisphere (Luo-

Table 2. Global model and process summary.

Model	Evaporation	Snow	Lakes/reservoirs	Routing	Reference
HTESSEL	Penman–Monteith	Energy balance	No	CaMa-Flood	Dutra et al. (2009)
JULES	Penman–Monteith	Energy balance	No	No	Clark et al. (2011)
PCR-GLOBWB	Hamon (Tier 1) or imposed as forcing	Temperature-based Melt factor	Yes	Travel-time approach	Bierkens and Van Beek (2009)
W3RA	Penman–Monteith	Degree day	No	Cascading linear reservoirs	van Dijk et al. (2014)
WaterGAP3	Priestley–Taylor	Degree day	Yes	Manning–Strickler	Flörke et al. (2013)

jus et al., 2014). GlobSnow uses a Bayesian non-linear iterative assimilation approach with passive microwave measurements and ground-based weather station measurements to create a 25 km by 25 km gridded SWE product (Takala, 2011). GlobSnow has limitations and uncertainty consistent with the measurement of SWE from passive microwave measurements, leading to underestimation in tundra environments due to several contributing factors (Rees et al., 2007). Passive microwave algorithms provide limited measurement of melting snow as the presence of even small amounts of water in the snowpack results in an emissivity similar to land with no snow cover (Nolin, 2011). In GlobSnow, a microwave-derived dry snow mask is first used to determine snow-covered area and SWE retrievals are only retained for those areas determined to have snow cover. When snow is wet, the snow-masking procedure underestimates the snow-covered area.

GlobSnow algorithm performance has been tested in Canada by comparing retrievals to in situ measurements for a variety of Canadian land covers (Snauffer et al., 2016). The overall RMSE for comparison with Canadian data is 40 mm, although algorithm retrieval is poor for boreal forest snow where the SWE is greater than 150 mm (Takala et al., 2011). Sparsity of weather station snow depth measurements in boreal regions results in stronger weighting of microwave-based retrievals in the GlobSnow algorithm, contributing to underestimation of SWE due to the volume scatter from dry snowpacks exceeding 150 mm.

3.3 Snowmelt volume

Snowmelt volume and peak discharge were calculated and extracted from the measured discharge data at the Catchment 3 outlet. No local or global model data were used in these calculations. Snowmelt volume was approximated using the local minimum method from the hydrograph stream flow separation program (HYSEP) implemented in MATLAB (Burkley, 2012). This is a mathematical technique that mimics manual methods for stream flow separation as opposed to an explicit representation of the physical processes (Sloto and Crouse, 1996). Secondary hydrograph peaks that occurred after the freshet peak and are driven by late-season rainfall events were removed in the snowmelt volume calculation. The separation of rainfall-driven flow increases was

performed using a simple exponential regression to estimate the regression curve from the spring melt hydrograph (Toebes et al., 1969). The method applied in this study results in an annual mean contribution of SWE to total stream flow of 63 %, with a standard deviation of 10 %. These values of snowmelt contribution to streamflow are consistent with literature estimates (30 %–80 %) from more detailed catchment studies (DDC, 2014; McNamara et al., 1998; Schelker et al., 2013; Stieglitz et al., 1999), if a little on the high side.

3.4 Prediction of snowmelt volume and peak discharge from maximum annual SWE

Prediction of spring streamflow is largely dependent on the accuracy of SWE estimates prior to snowmelt (Sospedra-Alfonso et al., 2016). Rank correlation analysis is used to compare maximum annual SWE to the corresponding spring snowmelt volume and peak discharge. Use of maximum annual SWE allows comparison between local and global model datasets, GlobSnow and in situ measurements. Spearman's rho is used as a non-parametric measure of the monotonicity (i.e. whether the trend is entirely increasing or decreasing) between datasets as calculated in Eq. (1) (Yue et al., 2002).

$$r_{\mathrm{s}} = 1 - \frac{6 \sum d_i^2}{n(n^2 - 1)} \ \text{where} \ d_i = \mathrm{rg}(X_i) - \mathrm{rg}(Y_i), \qquad (1)$$

where r_{s} is Spearman's rho and $\mathrm{rg}(X_i)$ is the rank of observation X_i in a sample of size n. Spearman's rho test includes a two-sided p value for significance. The period of record for all rank correlation analysis was 1985 to 2012.

4 Results

4.1 Discharge simulations

Graphical results for the testing period of the wflow-HBV model presented in Fig. 2 show good or acceptable overall model representation of discharge. From the graphical assessment, it appears that model results could be improved with slightly greater attenuation of streamflow. Modelled discharge in 2014 is anomalous with over-prediction of the discharge volume due to snowmelt contribution to streamflow. Analyses of the in situ data show that low snowpack SWE

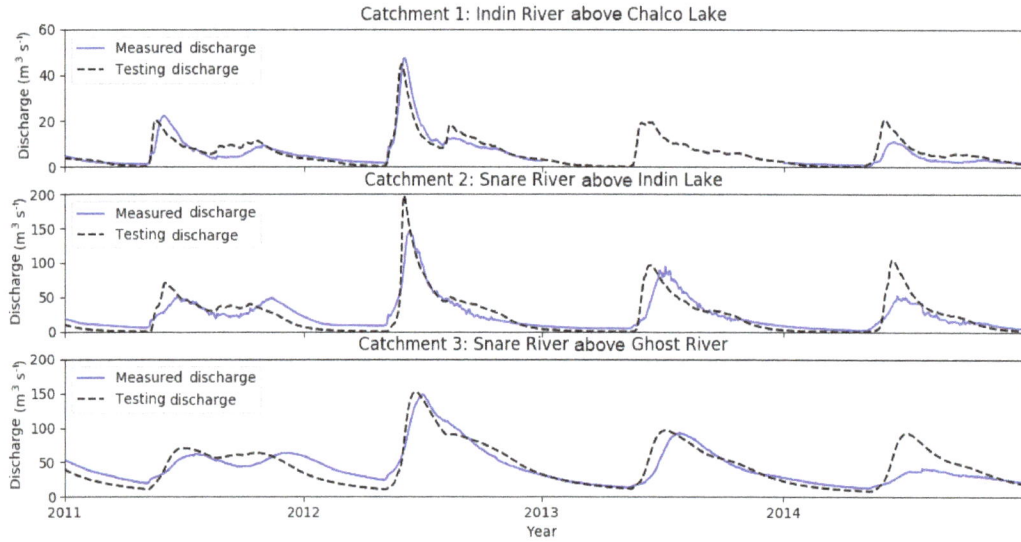

Figure 2. The wflow-HBV discharge results for the testing period.

Table 3. The wflow-HBV discharge statistical results.

Variable	Catchment 1			Catchment 2			Catchment 3		
	Calibration	Validation	Testing	Calibration	Validation	Testing	Calibration	Validation	Testing
Duration (yr)	22	9	4	9	3	4	20	6	4
Error statistics									
NSE	0.84	0.68	0.80	0.88	0.68	0.59	0.83	0.70	0.67
KGE	0.88	0.65	0.88	0.91	0.83	0.70	0.90	0.74	0.81
PBIAS (%)	−2.6	−15.1	6.0	−5.0	−6.5	0.5	−3.3	−15.2	3.4
RSR	0.44	0.77	0.46	0.32	0.56	0.51	0.39	0.64	0.51

Note: NSE = Nash Sutcliffe efficiency, KGE = Kling–Gupta efficiency, PBIAS = percent bias, RSR = root mean squared error observations standard deviation ratio.

was recorded in snowpack surveys collected in 2014, though this is not reflected in the MSWEP forcing data.

Results only from the testing period are shown graphically in Fig. 2, while the performance statistics over the calibration, validation and testing periods are shown in Table 3. These statistics would generally be classified as good or very good calibration under the model evaluation guidelines defined by Moriasi et al. (2007). NSE values can be in the range of $-\infty$ to 1 where 1 indicates the ideal with no difference between simulated and observed values. (Nash and Sutcliffe, 1970). Percent bias (PBIAS) gives a measure of the tendency of the simulated results to be larger or less than the observed values. RMSE-observations standard deviation ratio (RSR) has the benefit of a normalization and scaling factor, which facilities comparison (Moriasi et al., 2007). Evaluation using KGE is similar to NSE, with an ideal optimized value of 1 (Gupta et al., 2009).

4.2 Snow water equivalent

The accumulated SWE over the Snare Watershed has been measured by in situ snowpack surveys and can be used to evaluate GlobSnow-2 and hydrological models. Figure 3 shows the quantity and timing of SWE accumulation and melt patterns over the period of record. Each snowpack survey point is the spatial mean of a set of snowpack survey stations collected in the same field program. The line graphs represent the spatial mean of daily mean, maximum and minimum SWE estimates from GlobSnow-2, hydrological and land surface models.

The comparison of the GlobSnow data with the in situ SWE measurements in Fig. 4, where the blue crosses are the observations taken in early spring while the red asterisks are the observations from late spring, shows GlobSnow tends to overestimate SWE in the early season and underestimate in the late season. Error is also correlated to the magnitude of the GlobSnow measurement (right-hand figure). The assumption of a constant density of $0.24\,\mathrm{g\,cm^{-3}}$ in the Glob-

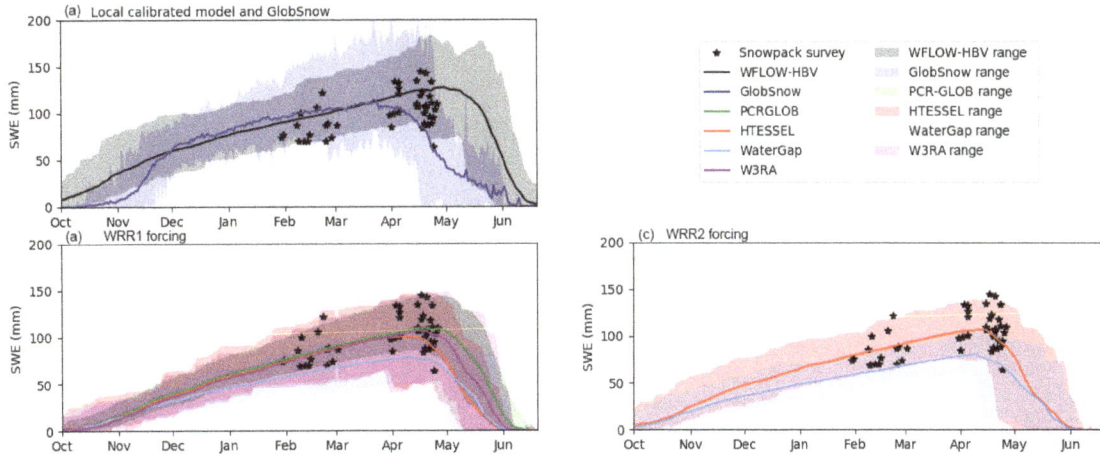

Figure 3. Daily mean, maximum and minimum daily SWE compared to ground measurements (1980–2012).

Snow retrieval algorithm contributes to this trend. The mean density in the Snare Watershed snow surveys is $0.21 \, \text{g cm}^{-3}$, with a standard deviation of $0.06 \, \text{g cm}^{-3}$ (GNWT, 2016). The assumption of constant density would lead to overestimation of SWE for freshly fallen snow and underestimation for mature snowpacks.

The high overall RMSE (45.1 %) and PBIAS (18.3 %), showing under-prediction by GlobSnow, are consistent with a recent validation study of GlobSnow over Canadian boreal forest and tundra environments (Larue et al., 2017; Takala, 2011). In this study, a key contributing factor to the high RMSE is that comparison is made with late season measurements where GlobSnow SWE retrievals have premature decline. The spatial distribution of RMSE and PBIAS in Fig. 5 indicates better performance over the northern tundra areas compared to southern areas where boreal forest land cover dominates. The checkered pattern of the error statistics is due to the 25 km by 25 km resolution of the GlobSnow product. Observations were interpolated to the 25 km grid using inverse distance weighting.

4.3 Prediction of snowmelt volume and peak discharge

Maximum annual SWE is a key predictor of spring and summer streamflow rates. Rank correlation analysis provides evaluation of the predictive power of measured and modelled SWE for snowmelt volume and peak discharge rates. Table 4 shows results for Spearman's rho (r_s) and two-sided p test (p), correlating the maximum SWE found in each of the dataset–model combinations considered, and the observed snowmelt volume and peak discharge. The last column provides the correlation to the SWE obtained from ground-based measurements.

The selection of forcing data has a clear effect on correlation of model maximum annual SWE to snowmelt volume, peak discharge and in situ data. MSWEP forcing precipitation showed superior performance to WFDEI irrespec-

tive of the model used. The local wflow-HBV model forced with MSWEP is the best and only statistically significant ($p < 0.05$) predictor of snowmelt volume and peak discharge. This can be attributed to the calibration of the local model, while global models are generally uncalibrated. GlobSnow has poor correlation to snowmelt volume, peak discharge and in situ data, which is consistent with expected limitations from SWE measurement with passive microwave measuring deep and late-season snowpacks.

The period used for rank correlation analysis was 1985–2012, meaning the wflow-HBV model was calibrated over 18.5 % (5 years) to 37.0 % (10 years) of the rank correlation analysis time period. The higher Spearman coefficient performance of the wflow-HBV model in rank correlation analysis may be partly attributed to improved process representation of snow accumulation and removal processes, including interception and precipitation biases. The quantification of the improvement in inter-annual variability and rank correlation due to correlation has not been investigated in this study. The dominant driver of the rank correlation analysis is the choice of forcing meteorological data.

5 Discussion

5.1 Global re-analysis datasets for predicting streamflow, snowpack accumulation and melt

Global re-analysis datasets applied in this study provide considerable advantages in hydrological assessment in a high-latitude watershed compared to what can be achieved with in situ data. Local meteorological datasets are simply too short, inconsistent and spatially disperse to be applied in long-term modelling. The use of hydrological models allows the estimation of hydrological state variables such as snowpack accumulation and streamflow using both local and global conceptual hydrological models.

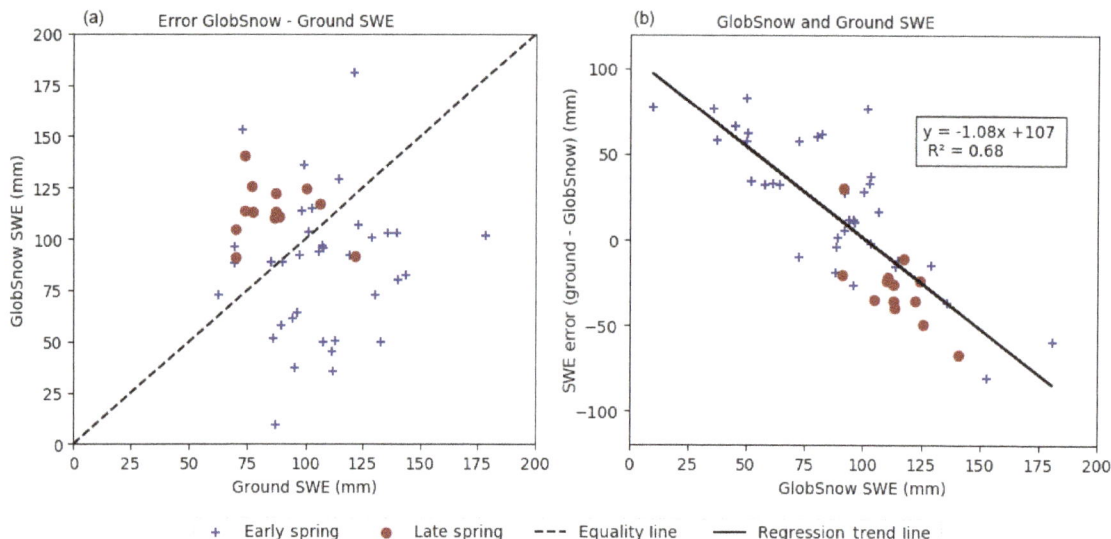

Figure 4. GlobSnow and Ground SWE Measurement Comparison.

Table 4. SWE, snowmelt and peak discharge rank correlation analysis.

Model	Forcing dataset	Snowmelt volume		Peak discharge		Ground SWE measurement	
		r_S	p	r_S	p	r_S	p
wflow-HBV	MSWEP	0.52	0.004	0.54	0.003	0.53	0.004
HTESSEL	MSWEP	0.47	0.011	0.48	0.010	0.55	0.002
JULES	MSWEP	0.47	0.012	0.48	0.010	0.62	0.000
WaterGap	MSWEP	0.34	0.076	0.36	0.063	0.67	0.000
HTESSEL	WFDEI	0.25	0.193	0.25	0.201	0.04	0.834
JULES	WFDEI	0.23	0.243	0.23	0.250	0.01	0.976
WaterGap	WFDEI	0.17	0.382	0.13	0.509	0.15	0.440
W3RA	WFDEI	0.15	0.451	0.10	0.601	0.16	0.409
PCR-GLOB	WFDEI	0.14	0.465	0.12	0.532	0.15	0.438
GlobSnow	Passive microwave/snow gauge data	0.14	0.484	0.18	0.360	0.18	0.371

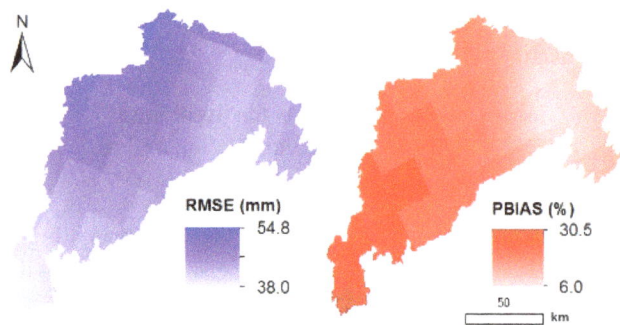

Figure 5. Error from comparison of GlobSnow SWE and interpolated in situ SWE (1980–2012).

The local watershed model in this study, forced with global re-analysis datasets and calibrated to available streamflow records is able to reliably and accurately model streamflow based on calibration, validation and testing of statistical results. The wflow-HBV model is conceptual and has limited representation of physical snow processes; however, the modelled maximum annual SWE was found to be a better predictor of snowmelt volume and peak discharge than snowpack survey data as the Spearman coefficient is higher and p value is lower ($p < 0.05$).

Assimilation of snowpack survey data for model state update has the potential to improve SWE estimates and optimally use available information. Data assimilation requires estimates of both model state and observational uncertainty, quantification of which would improve understanding to the relative reliability and applicability of data sources (Liu et al., 2012).

In global hydrological models, which are not calibrated to streamflow data, MSWEP has better performance over the Snare Watershed in predicting snowmelt volume and peak discharge compared to WFDEI. The selection of forcing data in this study has a greater effect than the choice of conceptual hydrological model, owing to the control over precipitation volumes. Studies of streamflow in calibrated versions of the global hydrological models have also found superior performance using MSWEP (H. E. Beck et al, 2017).

Limitations of hydrological models in high-latitude watersheds include a lack of important physical processes such as permafrost interactions, ice effects on rivers and lake outlets and complex processes in the snowpack. Calibration of highly parameterized models such as wflow-HBV masks underlying physical processes and does not explicitly represent them. This limits applicability for certain types of assessment such as permafrost thaw with climate change, which will alter runoff processes (Duan et al., 2017). Incorporating additional remote sensing data, including land and lake cover, can improve the spatial representation of physical processes and allow assessment based on land use changes.

5.2 SWE measurement for operation and planning purposes

SWE is used by operational water managers to predict the inflow volumes from snowmelt and to anticipate peak discharges. The results of this study demonstrate, however, that SWE measurement for application in hydrological forecasting is still problematic in the Snare Watershed. Consideration of multiple data sources and methodological improvement of data collection can be used to update model states.

In situ measurement of SWE from snowpack surveys provides an end-of-season snapshot measurement and, due to the long data record in the Snare Watershed, allows comparison with previous years. Field data collection could be improved with strategies that consider topographical and vegetative characteristics of the watershed to improve and standardize site selection (Rees et al., 2014). The recognition that while inter-annual variability of snowpack is high, distribution patterns are relatively consistent would improve SWE measurement due to typifying station measurements based on topographic relief.

Snowpack SWE in the conceptual hydrological models forced by MSWEP and WFDEI global have comparable magnitudes to snowpack survey measurements. Given that conceptual models do not include sublimation, which is known to remove a large quantity of snowpack SWE, the MSWEP and WDFEI global re-analysis datasets tend to underestimate actual snowfall. This is difficult to verify as precipitation gauge measurements at high latitudes are known to have large under-catch. Sublimation of snowpack SWE is also very difficult to measure and verify, particularly from remote sensing data (Petropoulos, 2013).

GlobSnow is well suited to providing accessible, timely SWE data as supplementary information for water managers and for assimilation into operational modelling systems. Snow data assimilation for hydrological forecasting is an emerging field and can be applied to operational water management systems (Huang et al., 2017; Montero et al., 2016). However, SWE products based on passive microwave measurements such as GlobSnow under-predict SWE of tundra and boreal environments present across northern Canada (Larue et al., 2017; Takala, 2011). Improvement of retrieval algorithms and the assimilation of in situ estimates can reduce error, though overcoming inherent the limitations of measuring deep (> 150 mm) or wetted snowpack will require novel approaches. Our results suggest that the assumption of a constant density used in GlobSnow is a source of error in the early and late periods of accumulation, and advancing over this assumption could help improve the SWE estimates from products such as GlobSnow.

5.3 Global re-analysis datasets for local application

To be of use in operational managers and planners, the global re-analysis datasets and hydrological models presented in this study must provide reliable data to inform decision making and decrease uncertainty. In the context of the Snare Watershed and snowmelt-driven hydropower operations, the snowpack SWE is the predominant source of uncertainty. Current operation of the Snare Hydro System relies on local expert knowledge, historical records and surrogate hydrographs. These methods will be challenged by changes to local hydrology, snow duration and snowmelt quantity with climate change.

The use of global re-analysis datasets helps with short-term planning by allowing the development of more reliable and accurate hydrological models, which form the basis of forecasting systems. Hydrological models developed with local data alone will have greater calibration parameter uncertainty and less rigorous validation. The calibrated wflow-HBV model was integrated into the Delft-FEWS operational forecasting platform (Werner et al., 2013). The use of this established framework and forecasting tool can improve operator confidence around water release and operation within water license limits.

This study demonstrates that SWE estimation for prediction of snowmelt volume and peak discharge is a persistent challenge. Choice of forcing data has a large effect compared to selection of model, and while global hydrological models can replicate the magnitude of end of season SWE, the difficultly is in accurately predicting inter-annual variability. SWE estimation from passive microwave measurements was found to be a poor predictor, which is consistent with a recent validation study of GlobSnow over eastern Canada that concludes the product accuracy to currently be insufficient for hydrologic simulations (Larue et al., 2017). SWE measurement from passive microwave has poor agreement with spring discharge volume, possibly due to algorithm errors at high SWE values (Rawlins et al., 2006). A locally calibrated hydrological-model-generated snowpack SWE that the more predictive of snowmelt volume and peak discharge than uncalibrated global models.

The manual collection of end of-winter snowpack survey data is justified, as the study shows that ground data are a comparatively reliable predictor of snowmelt contribution to streamflow and peak discharge. Improved field measurement techniques that exploit snow distribution across local topography could help further improve the quality, frequency and predictive ability of ground measurement data. These data could be optimally merged with model data using data assimilation methods (Sun et al., 2016).

The methods described in this study improve representation of the hydrological processes and forecasting application could allow a better operational strategy to be implemented. Global datasets, and in particular meteorological re-analysis data, are useful not only for broad scale assessment, but can be applied for accurate discharge modelling and development of a hydrological forecasting system. This has practical relevance for operational water management in the sub-Arctic.

6   Conclusions

This study demonstrates that considerable gains in hydrological assessment and model performance for high-latitude watersheds can be achieved with global re-analysis datasets and conceptual hydrological models. The findings of this study are relevant to operational water management in high-latitude catchments with sparse meteorological data and to current scientific research in the estimation of SWE with global remote sensing and re-analysis data. The methods described in this study can be readily applied in the Canadian sub-Arctic where watersheds do not have comprehensive meteorological data or operational hydrological models.

Results of the application of global re-analysis datasets to a locally distributed conceptual model (wflow-HBV) show that the spring snowmelt discharge can be predicted well in terms of timing and magnitude over a 30-year period. Model performance for discharge and select physical processes is improved through constrained parameter optimization, but it is also clear from the results that the calibrated HBV model parameters may compensate for cryosphere processes such as sublimation that are lacking in the model.

This study highlighted the limitations of SWE derived from global re-analysis datasets and conceptual hydrological models to predict the volume of snowmelt and peak discharge rates. Comparison of global re-analysis datasets in the eartH2Observe project shows improved performance in MSWEP precipitation forcing compared to WFDEI for snowpack representation. MSWEP forcing data produced

more realistic inter-annual snowpack SWE, which was better able to predict snowmelt volume and peak spring discharge. This finding was consistent for five global hydrological models assessed over the local study area, demonstrating the importance of precipitation forcing data relative to model structure. Data products available in near-real time such as MSWEP-NRT, which is a variant of the historic MSWEP dataset, can be similarly applied to model forcing in remote regions. Using Delft-FEWS, scheduled model runs can be used to keep model states current and generate regularly scheduled hydrological forecasts (H. Beck et al., 2017).

SWE estimation for prediction of snowmelt volume and peak discharge is a persistent challenge. SWE products based on passive microwave measurements such as GlobSnow under-predict SWE in boreal and tundra environments, particularly in the late winter season prior to snowmelt. Improvement of retrieval algorithms and the assimilation of in situ estimates can reduce error, though overcoming inherent limitations measuring deep ($> 150\,\text{mm}$) or wetted snowpacks will require novel approaches. Our results suggest the assumption of a constant density used in GlobSnow is a source of error in the early and late periods of accumulation, and not making this assumption could help improve the SWE estimates from products such as GlobSnow.

This study has demonstrated the utility of global re-analysis datasets for hydrological assessment in the data-sparse Canadian sub-Arctic. In the operational context of the Snare Hydro System, the length and breadth of hydrological assessment presented here is much greater than could be achieved with local meteorological data. Further research can focus on the optimal merging of observed and modelled snow data to improve predictability of snowmelt volume and peak discharge. The continued development of these datasets and modelling frameworks is promising, helping to improve the understanding of water resources in data-sparse northern regions in the face of climate change.

Author contributions. DRC, MW, AW and DS contributed to the design, methodology, and implementation of the research. MW, AW and DRS supervised the research. DRC and AW designed and implemented the model. DRC performed the model calibration, data processing and analysis. DRC wrote the paper with contribution from MW.

Competing interests. The authors declare that they have no conflict of interest.

Special issue statement. This article is part of the special issue "Integration of Earth observations and models for global water resource assessment". It is not associated with a conference.

Acknowledgements. This research received funding from the European Union Seventh Framework Programme (FP7/2007-2013) under grant agreement no. 603608, "Global Earth Observation for integrated water resource assessment": eartH2Observe. This study would not have been possible without access to open-source datasets generated from the eartH2Observe project and the development of open-source software packages including wflow and Delft-FEWS. Chris Derksen and Emanuel Dutra are acknowledged for their advice throughout the study and helpful comments on the paper. The first author acknowledges the Master programme "Water Science and Engineering" (specialisation in Hydroinformatics) of IHE Delft Institute for Water Education, in the framework of which this study has been carried out.

Edited by: Gianpaolo Balsamo

References

Allen, R. G., Pereira, L. S., Raes, D., and Smith, M: Crop evapotranspiration-Guidelines for computing crop water requirements, FAO Irrigation and drainage paper, 56, 6541, 1998.

AMAP: Arctic Climate Issues 2011: Changes in Arctic Snow, Water, Ice and Permafrost, SWIPA 2011, Gaustadalléen 21, 0349 Oslo, Norway, 2012.

Beck, H., van Dijk, A., Leviizzani, V., Schellekens, J., Miralles, G., Martrens, B., de Roo, A., Pappenberger, F., Huffman, G., and Wood, E.: MSWEP: 3-hourly 0.1? fully global precipitation (1979–present) by merging gauge, satellite, and weather model data [Abstract], Geophysical Research Abstracts, 19, 2017.

Beck, H. E., van Dijk, A. I. J. M., Levizzani, V., Schellekens, J., Miralles, D. G., Martens, B., and de Roo, A.: MSWEP: 3-hourly 0.25° global gridded precipitation (1979–2015) by merging gauge, satellite, and reanalysis data, Hydrol. Earth Syst. Sci., 21, 589–615, https://doi.org/10.5194/hess-21-589-2017, 2017.

Bergström, S.: The HBV Model – its structure and applications, NORRKÖPING, Sweden, 1992.

Bierkens, M. F. P. and Van Beek, L.: The Global Hydrological Model PCR-GLOBWB, P.O. Box 80115, 3508 TC, Utrecht, the Netherlands, 2009.

Brown, R. D. and Mote, P. W.: The Response of Northern Hemisphere Snow Cover to a Changing Climate, J. Climate, 22, 2124–2145, 2009.

Burkley, J.: Hydrograph Separation using HYDSEP, File Exchange, MathWorks, available at: http://nl.mathworks.com/matlabcentral/fileexchange/36387-hydrograph-separation-using-hydsep/content/f_hysep.m (last access: 30 April 2017), 2012.

Clark, D. B., Mercado, L. M., Sitch, S., Jones, C. D., Gedney, N., Best, M. J., Pryor, M., Rooney, G. G., Essery, R. L. H., Blyth, E., Boucher, O., Harding, R. J., Huntingford, C., and Cox, P. M.: The Joint UK Land Environment Simulator (JULES), model description – Part 2: Carbon fluxes and vegetation dynamics,

Geosci. Model Dev., 4, 701–722, https://doi.org/10.5194/gmd-4-701-2011, 2011.

DDC: Hydrology Baseline Report, ANNEX X: APPENDIX A: ANNOTATED BIBLIOGRAPHY, Golder Associates Ltd., Yellowknife, NWT, Canada, 2014.

Derkson, C., Sturm, M., Liston, G. E., Holmgren, J., Huntington, H., Silis, A., and Solie, D.: Northwest Territories and Nunavut Snow Characteristics from a Subarctic Traverse: Implications for Passive Microwave Remote Sensing, J. Hydrometeorol., 10, 448–462, 2008.

Dery, S. and Yau, M.: Large-scale mass balance effects of blowing snow and surface sublimation, J. Geophys. Res.-Atmos., 107, 213–227, https://doi.org/10.1029/2001jd001251, 2002.

Devine, K. and Mekis, E.: Field accuracy of Canadian rain measurements, Atmos. Ocean, 46, 213–227, https://doi.org/10.3137/ao.460202, 2008.

Duan, L., Man, X., Kurylyk, B., and Cai, T.: Increasing Winter Baseflow in Response to Permafrost Thaw and Precipitation Regime Shifts in Northeastern China, Water, 9, 25, https://doi.org/10.3390/w9010025, 2017.

Dutra, E., Balsamo, G., Viterbo, P., Mirand, P., Beljaars, A., Schar, C., and Elder, K.: New snow scheme in HTESSEL: description and offline validation, ECMWF607, available at: https://www.ecmwf.int/sites/default/files/elibrary/2009/9167-new-snow-scheme-htessel-description-and-offline-validation.pdf (last access: 30 April 2017), 2009.

eartH2Observe: earth2observe downscaling tools, available at: https://github.com/earth2observe/downscaling-tools/releases, last access: 25 November 2016.

eartH2Observe: eartH2Observe Water Cycle Integrator (WCI): https://wci.earth2observe.eu/, last access: 3 March 2017.

ECG: Ecological regions of the Northwest Territories, Government of Northwest Territories, Yellowknife, NT, Canada, 2008.

ENR: Historical Climate Data – Online Database, available at: http://climate.weather.gc.ca/index_e.html, last access: 28 October 2016.

ENR: Real-Time Hydrometric Data, available at: https://wateroffice.ec.gc.ca/mainmenu/real_time_data_index_e.html#wb-cont, last access: 1 April 2017.

Flörke, M., Kynast, E., Bärlund, I., Eisner, S., Wimmer, F., and Alcamo, J.: Domestic and industrial water uses of the past 60 years as a mirror of socio-economic development: A global simulation study, Global Environ. Chang., 23, 144–156, 2013.

Geem, Z., Kim, J., and Loganathan, G.: A New Heuristic Optimization Algorithm: Harmony Search, Simulation, 76, 60–68, 2001.

GNWT: Snow Survey – Spreadsheet Summary, E. a. N. R. Government of the Northwest Territories, Water Resources Division. Yellowknife, Northwest Territories, Canada, available at: http://www.enr.gov.nt.ca/programs/snow-surveys, last access: 1 April 2017.

Government of Canada: National Ecological Framework, available at: http://sis.agr.gc.ca/cansis/nsdb/ecostrat/index.html, last access: 29 January 2013.

Gupta, H. V., Kling, H., Koray, Y., and Martinez, G.: Decomposition of the mean squared error and NSE performance criteria: Implications for improving hydrological modelling, J. Hydrol., 377, 80–91, 2009.

Huang, C., Newman, A. J., Clark, M. P., Wood, A. W., and Zheng, X.: Evaluation of snow data assimilation using the ensemble Kalman filter for seasonal streamflow prediction in the western United States, Hydrol. Earth Syst. Sci., 21, 635–650, https://doi.org/10.5194/hess-21-635-2017, 2017.

Karssenberg, D., Schmitz, O., Salamon, P., de Jong, K., and Bierkens, M.: A software framework for construction of process-based stochastic spatio-temporal models and data assimilation, Environ. Modell. Softw., 25, 489–502, https://doi.org/10.1016/j.envsoft.2009.10.004, 2010.

Kokelj, S.: Hydrologic Overview of the North and South Slave Regions Water Resources Division Yellowknife, Northwest Territories, Canada, 2003.

Larue, F., Royer, A., De Sève, D., Langlois, A., Roy, A., and Brucker, L.: Validation of GlobSnow-2 snow water equivalent over Eastern Canada, Remote Sens. Environ., 194, 264–277, https://doi.org/10.1016/j.rse.2017.03.027, 2017.

Lindstrom, G., Johansson, B., Persson, M., Gardelin, M., and Bergstrom, S.: Development and test of the distributed HBV-96 hydrological model, J. Hydrol., 201, 272–288, 1997.

Liston, G., McFadden, J., Sturm, M., and Pielke, R.: Modelled changes in arctic tundra snow, energy and moisture fuxes due to increased shrubs, Glob. Change Biol., 8, 17–32, 2002.

Liston, G. and Sturm, M.: The role of winter sublimation in the Arctic moisture budget, Nord. Hydrol., 35, 325–334, 2004.

Liu, Y., Weerts, A. H., Clark, M., Hendricks Franssen, H.-J., Kumar, S., Moradkhani, H., Seo, D.-J., Schwanenberg, D., Smith, P., van Dijk, A. I. J. M., van Velzen, N., He, M., Lee, H., Noh, S. J., Rakovec, O., and Restrepo, P.: Advancing data assimilation in operational hydrologic forecasting: progresses, challenges, and emerging opportunities, Hydrol. Earth Syst. Sci., 16, 3863–3887, https://doi.org/10.5194/hess-16-3863-2012, 2012.

Liu, Y., Peters-Lidard, C. D., Kumar, S. V., Arsenault, K. R., and Mocko, D. M.: Blending satellite-based snow depth products with in situ observations for streamflow predictions in the Upper Colorado River Basin, Water Resour. Res., 51, 1182–1202, https://doi.org/10.1002/2014WR016606, 2015.

Luojus, K., Pullianen, J., Takala, M., J. Lemmetyinen, Kangwa, M., M. Eskelinen (FMI), (SYKE), S. M., R. Solberg, A.-B. S. N., G. Bippus, E. R., T. Nagler (ENVEO), (EC), C. D., (GAMMA), A. W., Wunderle, S., (UniBe), F. H., Fontana, F., and (MeteoSwiss), N. F.: GlobSnow-2 Final Report, available at: http://www.globsnow.info/docs/GlobSnow_2_Final_Report_release.pdf (last access: 30 April 2017), European Space Agency, 2014.

Marsh, P., Quinton, B., and Pomeroy, J.: HYDROLOGICAL PROCESSES AND RUNOFF AT THE ARCTIC TREELINE IN NORTHWESTERN CANADA, National Hydrology Research Institute Saskatoon, Sask., 1995.

McCabe, G. J., Clark, M., and Hay, L. E.: Rain on Snow Events in the Western United States, American Meterological Society, 2007.

McNamara, J., Kane, D., and Hinzman, L.: An analysis of streamflow hydrology in the Kuparuk River Basin, Arctic Alaska: a nested watershed approach J. Hydrol., 206, 39–57, 1998.

Mekis, E. and Hogg, W.: Rehabilitation and analysis of Canadian daily precipitation time series, Atmos. Ocean, 37, 53–85, https://doi.org/10.1080/07055900.1999.9649621, 1998.

Mekis, E. and Vincent, L.: An Overview of the Second Generation Adjusted Daily Precipitation Dataset for Trend Analysis in Canada, Atmos. Ocean, 49, 163–177, 2011.

Montero, R. A., Schwanenberg, D., Krahe, P., Lisniak, D., Sensoy, A., Sorman, A. A., and Akkol, B.: Moving horizon estimation for assimilating H-SAF remote sensing data into the HBV hydrological model, Adv. Water Resour., 92, 248–257, https://doi.org/10.1016/j.advwatres.2016.04.011, 2016.

Moriasi, D. N., Arnold, J. G., Van Liew, M. W., Bingner, R. L., Harmel, R. D., and Veith, T. L.: Model Evaluation Guidelines for Systematic Quantification of Accuracy in Watershed Simulations, T. ASABE, 50, 885–900, 2007.

Mudryk, L. R., Derksen, C., Kushner, P. J., and Brown, R.: Characterization of Northern Hemisphere Snow Water Equivalent Datasets, 1981–2010, J. Climate, 28, 8037–8051, 2015.

Nash, J. E. and Sutcliffe, J. V.: River flow forecasting through conceptual models part I – A discussion of principles, J. Hydrol., 10, 282–290, https://doi.org/10.1016/0022-1694(70)90255-6, 1970.

National Ecological Framework: available at: http://sis.agr.gc.ca/cansis/nsdb/ecostrat/index.html, last access: 29 January 2013.

National Research Council: Colorado River Basin Water Management: Evaluating and Adjusting to Hydroclimatic Variability, 978-0-309-10524-8, 2007.

Nolin, A. W.: Recent advances in remote sensing of seasonal snow, J. Glaciol., 56, 1141–1150, 2011.

OpenStreams: wflow stable release documentation, available at: http://wflow.readthedocs.io/en/stable/, last access: 12 September 2016.

Perez, R., Jansen, P., and Joaquim, R.: pyOpt: A Python-Based Object-Oriented Framework for Nonlinear Constrained Optimization, Struct. Multidiscip. O., 45, 101–118 2012.

Petropoulos, G.: Remote Sensing of Energy Fluxes and Soil Moisture Content: Challenges and Future Outlook, CRC Press, 2013.

PML RGS THREDDS Data Server: available at: https://wci.earth2observe.eu/thredds/catalog-earth2observe.html, last access: 25 November 2015.

Pomeroy, J., Marsh, P., and Gray, D.: Application of a Distributed Blowing Snow Model to the Arctic, Hydrol. Process., 11, 1451–1464, 1997.

Pomeroy, J. W., Hedstrom, N. R., Parviainen, J., and Granger, R. J.: The Snow Mass Balance of Wolf Creek , Yukon : Effects of Snow Sublimation and Redistribution, National Water Research Institute, 15–30, 1999.

Rawlins, M., Fahnestock, M., Frolking, S., and Vorosmarty, C.: On the Evaluation of Snow Water Equivalent Estimates over the Terrestrial Arctic Drainage Basin, 63nd EASTERN SNOW CONFERENCE, Newark, Delaware, USA, 2006.

Rees, A., English, M., Derksen, C., and Silis, A.: The Distribution and Properties and Role of Snow Cover in the Open Tundra 64th EASTERN SNOW CONFERENCE St. John's, Newfoundland, Canada, 2007.

Rees, A., English, M., Derkson, C., Toose, P., and Sillis, A.: Observations of late winter Canadian Tundra snow sover properties, Hydrol. Process., 28, 3962–3977, https://doi.org/10.1002/hyp.9931, 2014.

Schelker, J., Kuglerová, L., Eklöf, K., Bishop, K., and Laudon, H.: Hydrological effects of clear-cutting in a boreal forest – Snowpack dynamics, snowmelt and streamflow responses, J. Hydrol.,

484, 105–114, https://doi.org/10.1016/j.jhydrol.2013.01.015, 2013.

Schellekens, J., Dutra, E., Martínez-de la Torre, A., Balsamo, G., van Dijk, A., Sperna Weiland, F., Minvielle, M., Calvet, J.-C., Decharme, B., Eisner, S., Fink, G., Flörke, M., Peßenteiner, S., van Beek, R., Polcher, J., Beck, H., Orth, R., Calton, B., Burke, S., Dorigo, W., and Weedon, G. P.: A global water resources ensemble of hydrological models: the eartH2Observe Tier-1 dataset, Earth Syst. Sci. Data, 9, 389–413, https://doi.org/10.5194/essd-9-389-2017, 2017.

Simpson, G.: genURLs.R, available at: https://gist.github.com/gavinsimpson/8c13e3c5f905fd67cf85, last access: 28 October 2016.

Sloto, R. and Crouse, M.: HYSEP: A Computer Program for Streamflow Hydrograph Separation and Analysis, Lemoyne, Pennsylvania, USA, 1996.

Snauffer, A., Hsieh, W., and Cannon, A.: Comparison of gridded snow water equivalent products with in situ measurements in British Columbia, Canada, J. Hydrol., 541, 714–726, https://doi.org/10.1016/j.jhydrol.2016.07.027, 2016.

Sospedra-Alfonso, R., Merryfield, W. J., and Kharin, V. V.: Representation of Snow in the Canadian Seasonal to Interannual Prediction System. Part II: Potential Predictability and Hindcast Skill, J. Hydrometeorol., 17, 2511–2535, https://doi.org/10.1175/jhm-d-16-0027.1, 2016.

Stieglitz, M., Hobbie, J., Giblin, A., and Kling, G.: Hydrologic modelling of an Arctic Tundra Basin: Toward pan-Arctic predictions, J. Geophys. Res., 104, 507–527, 1999.

Sugiura, K., Ohata, T., and Yang, D.: Catch Characteristics of Precipitation Gauges in High-Latitude Regions with High Winds, J. Hydrometeorol., 7, 984–994, https://doi.org/10.1175/jhm542.1, 2006.

Sun, L., Seidou, O., Nistor, I., and Liu, K.: Review of the Kalman-type hydrological data assimilation, Hydrolog. Sci. J., 61, 2348–2366, https://doi.org/10.1080/02626667.2015.1127376, 2016.

Takala, M., Luojus, K., Pulliainen, J., Derksen, C., Lemmetyinen, J., Kärnä, J.-P., Koskinen, J., and Bojkov, B.: Estimating northern hemisphere snow water equivalent for climate research through assimilation of space-borne radiometer data and ground-based measurements, Remote Sens. Environ., 115, 3517–3529, https://doi.org/10.1016/j.rse.2011.08.014, 2011.

Toebes, C., Morrissey, W. B., Shorter, R., and Hendy, M.: Base Flow Recession Curves, Wellington, New Zealand, 1969.

van Dijk, A. I. J. M., Renzullo, L. J., Wada, Y., and Tregoning, P.: A global water cycle reanalysis (2003–2012) merging satellite gravimetry and altimetry observations with a hydrological multi-model ensemble, Hydrol. Earth Syst. Sci., 18, 2955–2973, https://doi.org/10.5194/hess-18-2955-2014, 2014.

Weedon, G. P., Balsamo, G., Bellouin, N., Gomes, S., Best, M. J., and Viterbo, P.: The WFDEI meteorological forcing data set: WATCH Forcing Data methodology applied to ERA-Interim reanalysis data, Water Resour. Res., 50, 7505–7514, https://doi.org/10.1002/2014WR015638, 2014.

Werner, M., Schellekens, J., Gijsbers, P., van Dijk, M., van den Akker, O., and Heynert, K.: The Delft-FEWS flow forecasting system, Environ. Modell. Softw., 40, 65–77, https://doi.org/10.1016/j.envsoft.2012.07.010, 2013.

wflow stable release documentation: available at: http://wflow.readthedocs.io/en/stable/, last access: 25 January 2016.

Wong, J. S., Razavi, S., Bonsal, B. R., Wheater, H. S., and Asong, Z. E.: Inter-comparison of daily precipitation products for large-scale hydro-climatic applications over Canada, Hydrol. Earth Syst. Sci., 21, 2163–2185, https://doi.org/10.5194/hess-21-2163-2017, 2017.

WSC: Historical Hydrometric Data Search, available at: https://wateroffice.ec.gc.ca/mainmenu/historical_data_index_e.html, last access: 2 November 2016.

Yang, D., Kane, D., Zhang, Z., Legates, D., and Goodison, B.: Bias corrections of long-term (1973–2004) daily precipitation data over the northern regions, Geophys. Res. Lett., 32, L19501, https://doi.org/10.1029/2005GL024057, 2005.

Yue, S., Pilon, P., and Cavadia, G.: Power of the Mann-Kendall and Spearman's rho tests for detecting monotonic trends in hydrological series, J. Hydrol., 259, 254–271, 2002.

8

Forest impacts on snow accumulation and ablation across an elevation gradient in a temperate montane environment

Travis R. Roth and Anne W. Nolin

Water Resource Sciences, Oregon State University, Corvallis, OR 97331, USA

Correspondence to: Travis R. Roth (rothtra@science.oregonstate.edu)

Abstract. Forest cover modifies snow accumulation and ablation rates via canopy interception and changes in sub-canopy energy balance processes. However, the ways in which snowpacks are affected by forest canopy processes vary depending on climatic, topographic and forest characteristics. Here we present results from a 4-year study of snow–forest interactions in the Oregon Cascades. We continuously monitored snow and meteorological variables at paired forested and open sites at three elevations representing the Low, Mid, and High seasonal snow zones in the study region. On a monthly to bi-weekly basis, we surveyed snow depth and snow water equivalent across 900 m transects connecting the forested and open pairs of sites. Our results show that relative to nearby open areas, the dense, relatively warm forests at Low and Mid sites impede snow accumulation via canopy snow interception and increase sub-canopy snowpack energy inputs via longwave radiation. Compared with the Forest sites, snowpacks are deeper and last longer in the Open site at the Low and Mid sites (4–26 and 11–33 days, respectively). However, we see the opposite relationship at the relatively colder High sites, with the Forest site maintaining snow longer into the spring by 15–29 days relative to the nearby Open site. Canopy interception efficiency (C_{IE}) values at the Low and Mid Forest sites averaged 79 and 76 % of the total event snowfall, whereas C_{IE} was 31 % at the lower density High Forest site. At all elevations, longwave radiation in forested environments appears to be the primary energy component due to the maritime climate and forest presence, accounting for 93, 92, and 47 % of total energy inputs to the snowpack at the Low, Mid, and High Forest sites, respectively. Higher wind speeds in the High Open site significantly increase turbulent energy exchanges and snow sublimation. Lower wind speeds in the High Forest site create preferential snowfall deposition. These results show the im-

portance of understanding the effects of forest cover on sub-canopy snowpack evolution and highlight the need for improved forest cover model representation to accurately predict water resources in maritime forests.

1 Introduction

Snowpacks the world over are changing. Increasing global temperatures and accompanied climatic changes are altering snowpack characteristics and shifting melt timing earlier (McCabe and Clark, 2005; Mote, 2006; Mussleman et al., 2017). The timing, intensity, and duration of snowmelt depend on climatic and physiographic variables. In the topographically diverse western US the distribution of snow cover is governed by regional climate, elevation, vegetation presence/absence, and forest structure (Elder et al., 1998; Harpold et al., 2013). Forests overlap with mountains across this region and modify snow accumulation and ablation rates through canopy interception and a recasting of the sub-canopy energy balance (Hedstrom and Pomeroy, 1998; López-Moreno and Stähli, 2008; Varhola et al., 2010). Recently, a considerable amount of effort has been expended in research into the snow–forest processes that control the distribution of snow in mountainous regions (Stähli and Gustafsson, 2006; Jost et al., 2007; López-Moreno and Latron, 2008; Musselman et al., 2008; Ellis et al., 2013; Moeser et al., 2015). While these studies have focused on cold, predominately continental snowpacks, few have investigated snow–forest process interaction in warm maritime environments where snow is especially sensitive to changes in energy balance (Storck et al., 2002; Lundquist et al., 2013).

Maritime snowpacks accumulate and reside at temperatures near the melting point. Such snowpacks do not fit the simple accumulation–ablation model of a monotonic increase until peak snow water equivalent (SWE) followed by a monotonic decrease to snow disappearance. Such temperature sensitive snowpacks may experience disproportionate effects of climate warming and changing forest cover (Nolin and Daly, 2006; Dickerson-Lange et al., 2015). Ramifications of these impacts have far reaching eco-hydrological impacts across the snowmelt dependent western US, highlighting the continued need for research into snow–forest process interactions in maritime montane settings (Mote, 2006; Harpold et al., 2015; Vose et al., 2016).

In the Pacific Northwest, United States (PNW), mountain environments are a disparate composite of forest cover driven by forest harvest, regrowth, and natural disturbance. Forest disturbance can have significant impacts on snow processes, whose effects can range from immediate (Boon, 2009) to decadal (Lyon et al., 2008; Gleason and Nolin, 2016). At the stand scale, forests attenuate wind speeds, thereby suppressing turbulent mixing of the near-surface atmosphere (Liston and Sturm, 1998); modify the radiation received at the snow surface through shifts in shortwave and longwave contributions and reduced surface albedo (Sicart et al., 2004; O'Halloran et al., 2012; Gleason et al., 2013); and temporally shift seasonal- and event-scale accumulation and ablation patterns through canopy snowfall interception (Varhola et al., 2010). Natural and anthropogenic alterations in forest cover such as mountain pine beetle infestation, forest management practices, and forest fire affect snow processes by modifying forest structure, i.e., canopy cover and gap size (Boon, 2009; Bewley et al., 2010; Ellis et al., 2013) and snow albedo (Gleason et al., 2013; Gleason and Nolin, 2016). The frequency and intensity of forest fires have been increasing (Westerling et al., 2006; Miller et al., 2009; Spracklen et al., 2009), impacting accumulation and ablation rates (Gleason et al., 2013), and are anticipated to continue increasing (Moritz et al., 2012; Westerling et al., 2011), while prolonged droughts, and a future of increasing drought prevalence, have increased water stress, creating changes in forest characteristics across the western US (Allen, 2010; Choat, 2012; Dai, 2013). Disturbances of this type alter the snow–forest dynamic through a modification of the magnitudes of central process relationships, often resulting in unanticipated outcomes (Lundquist et al., 2013). The present reality and specter of continued future change to climate and forest cover underscores the increasing importance of characterizing vegetation impacts on snow accumulation and ablation within warm, topographically varied terrains.

Elevation (as a proxy for temperature) and forest canopy cover are important controls on peak snow accumulation (Geddes et al., 2005; Jost et al., 2007). Elevation drives snow accumulation and is the principle predictor of peak snow water equivalent (Gray, 1979; Elder et al., 1991; Sproles et al., 2013). The partitioning of precipitation between rainfall and snowfall is determined by atmospheric temperature and the elevation of the rain–snow transition can be described as a function of the temperature lapse rate. Forest canopies intercept snow, reducing sub-canopy accumulation (Schmidt and Gluns, 1991; Hedstom and Pomeroy, 1998; Musselman et al., 2008). The magnitude and rate of canopy interception are also affected by air temperature. Air temperature has been shown to have an inverse relationship with canopy interception (Andreadis et al., 2009) and a nonlinear correlation with event size (Hedstrom and Pomeroy, 1998); these relationships are often based on a few measurements and at a single point. Forests also reduce solar radiation reaching the snowpack surface (Link and Marks, 1999; Hardy et al., 2004) and increase longwave radiation at the snowpack surface (Lundquist et al., 2013), thus modifying net radiation (Sicart et al., 2004). Forest cover reduces wind speed, thereby reducing latent and sensible heat flux at the snowpack surface (Link and Marks, 1999; Boon, 2009). The direct effect of wind speed on canopy snow interception has not been explicitly studied, with most research focusing on wind redistribution of snow (Gary, 1974; Pomeroy et al., 1997; Liston and Sturm, 1998; Woods et al., 2006). Research demonstrates that forests reduce wind speed and can lead to increased snow accumulation in canopy gaps or forest clearcuts where wind speeds decline and snow is released from upwind canopy flow (Gary, 1974). These combined forest effects on sub-canopy energy and mass balance can accelerate or delay the onset and rate of snowmelt (Varhola et al., 2010). These studies highlight the key differences between forested and open areas, and the effects of elevation on snowpack evolution. With strong agreement that the western US will be facing warmer winters in the future and new understanding that snow in forested regions is more sensitive to increased temperatures than snow in non-forested regions (Lundquist et al., 2013), it is critical that we measure, characterize, and understand maritime snow–forest interactions. This study examines and evaluates the combined effects of forest cover, climate variability, and elevation on snow accumulation and ablation in a maritime montane environment. Specifically, we focus on the following research questions.

1. To what extent do forests modify snow accumulation and ablation in a maritime temperate forest?

2. How does canopy interception affect sub-canopy snowpack evolution across an elevation gradient?

3. How does forest cover affect the sub-canopy snow surface energy balance relative to adjacent open areas and what are the principal drivers of melt?

In subsequent sections, we describe the study area; present research methods for field measurements, energy balance calculations, and snow modeling; present our key findings; and conclude with a description of potential applications and future steps.

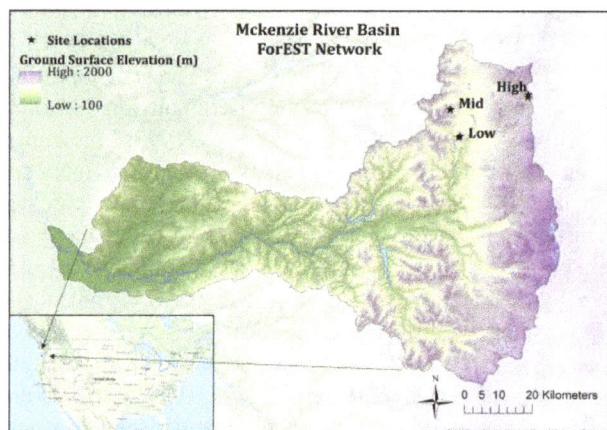

Figure 1. The Oregon ForEST network sites of the McKenzie River basin.

2 Methods

2.1 Description of the study area

The McKenzie River basin (MRB) is part of the greater Willamette River basin in western Oregon, USA (Fig. 1). It covers an area of 3041 km^2 and spans an elevation range from 150 m to over 3100 m at the crest of the Cascades Mountains that flank its eastern boundary. Orographic uplift results in average annual precipitation ranging from 1000 mm at lower elevations to over 3500 mm at the highest elevations in the basin (Jefferson et al., 2008). The rain–snow transition zone sits between 500 and 1200 m (Marks et al., 1998). The area above the transition zone accounts for 12 % of the total area with the Willamette River basin, yet contributes 60–80 % of summer baseflow to the Willamette River (Brooks et al., 2012). The MRB elevation between 1000 and 2000 m is especially important as it comprises 42 % of the total area within the MRB and snowmelt from this elevation band accounts for nearly 93 % of the total snow water storage (Sproles et al., 2013). Warm snowpack conditions facilitate frequent melt events during the winter months of December, January and February (DJF), commonly masking the distinction between accumulation and ablation periods. Nolin and Daly (2006) showed that snowpack in this region has an acute sensitivity to temperature, with the low elevation snow zones of the Oregon Cascades classified as the most "at-risk" snow within the region. The Natural Resources Conservation Service (NRCS) has been monitoring seasonal snowpack within the MRB since the early 1980s by a point-based snow telemetry (SNOTEL) network. Placement of SNOTEL stations was designed to be representative of water producing regions of a watershed and yet network stations were ultimately placed in protected, accessible locations (Molotch and Bales, 2006). However, the limited configuration was not designed to understand forest–snow processes or with future climate change in mind, and therefore

a statistically unbiased approach to site selection that is spatially representative is needed for any substantial snow observation network (Molotch and Bales, 2006). This underscores the need for intelligent and statistically relevant snow monitoring sites that go beyond the existing network. Section 2.2 outlines the snow monitoring network we deployed in water year (WY) 2012 that meets these stated needs.

2.2 The Oregon ForEST network

The Oregon Forest Elevation Snow Transect (ForEST) network extends from the rain–snow transition zone through the seasonal snow zone in the Oregon Cascades with paired forested and open sites at three elevations, Low (1150 m), Mid (1325 m) and High (1465 m) (Fig. 1). The ForEST network was designed to efficiently represent the range of peak SWE within the basin. Using a binary regression tree (BRT) approach, we identified elevation, vegetation type and vegetation density as the key predictor variables and we used them to classify the basin and locate our network sites (Molotch and Bales, 2006; Gleason et al., 2017). At each of three elevation zones, we established Open (low forest density) and Forest (high forest density) site pairs in adjacent areas, while controlling for slope and aspect. Open sites consisted of < 20 % canopy cover, while corresponding Forest sites had > 60 % canopy cover based on the 2001 National Vegetation Cover Database (Homer et al., 2007), and were subsequently verified by in situ measurements.

At each of the six sites within the ForEST network tower-based instruments continuously measured snow depth, incoming and reflected shortwave radiation, air temperature, relative humidity, wind speed, wind direction, and soil temperature and soil moisture (Table S1 in the Supplement). Sensor measurement frequency was 15 s with output values averaged over a 10 min period. The suite of sensors allowed the calculation of the snow surface energy balance through either direct measurement, e.g., solar radiation, or empirical equations, e.g., turbulent fluxes or longwave radiation. The snow-climate monitoring stations were deployed and active for the duration of the snow season at all sites, typically from mid to late November through May, with minimal disruptions due to battery or mechanical failures. We present results from the Low and Mid sites for WY 2012–WY 2015 and results from the High sites which were added to the network for WY 2014 and WY 2015. Additionally, SWE and snow depth measurements were collected along 900 m transects ("snow courses") extending from the Forest to Open sites in the low, mid, and high elevation zones. SWE measurement locations were restricted to > 50 m from the forest edge to eliminate canopy edge effects. These snow course surveys were conducted on a monthly basis during the accumulation period, and then bi-weekly during the ablation phase until the snow disappearance date (SDD). SWE was measured using a snow tube (Federal sampler) and snow depth was measured using a steel probe pole. Within each vegetation cover type, e.g., Open or Forest sites, SWE measurements were made at 100 m intervals with snow depth measurements every 5 m. Snow course data used in this analysis are from WY 2012 to WY 2015 for all ForEST network sites. To estimate SDD for each site we calculated the snowpack ablation rate using median snow depths from the last two snow courses of the season and linearly extrapolating to the date of zero snow depth. SDD represents the date when the primary seasonal snowpack disappears and does not take into account late season periods of accumulation/ablation. We excluded data from the historically low WY 2015 due to a near absence of winter snow.

2.3 Canopy interception efficiency

Forest structure characteristics at each site were quantified using ground-based conventional forest inventory methods. At transect locations coinciding with SWE measurements, individual tree characteristics were measured within each quadrat and averaged for that particular site, i.e., diameter at breast height (DBH), crown radius, tree height, and tree species (Table 1). Forest density was performed using a plotless density estimator approach described in Elzinga et al. (1998). The forest canopy at each site was further characterized using skyward looking hemispherical photographs acquired using a Nikon Coolpix 990 digital camera equipped with a FC-E8 fisheye converter, which has a 180° field-of-view (Inoue et al., 2004). The hemispherical photographs were assessed with the Gap Light Analyzer 2.0 to measure leaf area index (LAI) and canopy closure (CC), which is the complement of the sky view fraction (Frazer et al., 1999).

During the snow accumulation period forest canopy plays a large role in reducing snowpack by intercepting incoming snowfall, prohibiting a significant portion from accumulating on the forest floor. A forest canopy is the integrated sum of the forest overlaying the ground surface; this includes needles, leaves, branches, and trunks. The canopy structure is the primary control on canopy interception, followed by event-specific variables, i.e., event size, air temperature, and wind speed (Varhola et al., 2010). Canopy snow interception is inherently difficult to accurately quantify due to the temporally sensitive impacts of local climate on the canopy itself and the limited measurement capabilities to directly measure canopy interception (Martin et al., 2013; Friesen et al., 2014). From measured snowfall at each climate station within the ForEST network we calculated percent canopy interception efficiency (C_{IE}) for daily snowfall events. A snowfall event is defined as the daily increase in measured snow depth in the Open sites greater than 3 cm. Ryan et al. (2008) showed that acoustic snow depth measurement error for the Campbell Scientific SR50a is ±2 cm under normal field conditions. Therefore, to reduce the influence of depth measurement error on our snow event classification, we used a ≥ 3 cm threshold for our analysis. C_{IE} is calculated as

$$C_{IE} = \left[\frac{O_S - F_S}{O_S} \right] \times 100, \tag{1}$$

where O_S and F_S are the measured snowfall (cm) in the Open and Forest sites, respectively. C_{IE} was calculated for individual events and for seasonal averages at each Forest site.

2.4 Snow surface energy balance

A snow surface energy balance was calculated at a daily time step using aggregated 10 min meteorological measurements from each site. Each energy balance component was either directly measured or calculated using empirically derived equations valid for a maritime snowpack. Total energy into the snowpack equals the combined incoming and outgoing energies experienced at the surface of the snowpack. The governing equation for the snow surface energy balance is

$$\Delta Q = Q_S + Q_L + Q_E + Q_H + Q_C \tag{2}$$

where ΔQ is the change in total energy present at the snow surface (W m^{-2}); Q_S is total solar radiation (W m^{-2}); Q_L is total longwave radiation (W m^{-2}); Q_E is latent heat (W m^{-2}); Q_H is sensible heat (W m^{-2}); and Q_C is conductive energy (W m^{-2}).

A critical component within the snow surface energy balance calculations is the determination of the snow surface temperature, T_{snow} (Andreas, 1986). T_{snow} controls directional energy flows by regulating temperature and vapor flux gradients between the atmosphere and the snowpack, which control the sensible and latent heat transfer, respectively. T_{snow} is also the primary control of longwave radiation emitted from the snowpack. However, T_{snow} is difficult to directly measure and is therefore estimated as a function of the dew-point (frost-point) temperature, T_{dew}, as demonstrated by Raleigh et al. (2013). Using T_{dew} to estimate daily averages of T_{snow} reduces bias and is a reasonable first-order approximation at standard height measurements (Raleigh et al., 2013).

2.4.1 Solar radiation

Incoming and reflected solar radiation were each measured using an upward facing and downward facing LI-200s™ pyranometer (LI-COR). The pyranometers have a spectral range of 400–1100 nm and a field-of-view of 180°. Net solar radiation is calculated as

$$Q_S = S_{in} \times (1 - \alpha) \tag{3}$$

where S_{in} equals the measured incoming shortwave radiation (W m^{-2}). Albedo, α, was calculated as the ratio of reflected and incoming measured solar radiation. When periods of newly fallen snow obscured the upward facing solar pyranometer, i.e., when $\frac{S_{out}}{S_{in}} > 1$, a value of $\alpha = 0.9$ was used.

Table 1. Site forest characteristics with the associated SD for each measurement.

Site	DBH (cm)	Height (m)	Crown diameter (m)	Forest density per $10\,\mathrm{m}^2$	S_{VF} (%)	Study duration average C_{IE} (%)
Low Forest	52.1 ± 20.0	33.7 ± 10.4	9.4 ± 0.8	19.4 ± 2.0	10.9	79
Low Open	17.3 ± 4.3	8.9 ± 2.6	3.7 ± 0.4	15.7 ± 5.1	68.7	–
Mid Forest	36.5 ± 17.4	21.2 ± 10.1	6.7 ± 0.8	20.7 ± 1.5	10.1	76
Mid Open	19.0 ± 8.3	11.8 ± 3.4	4.0 ± 0.2	15.8 ± 6.7	61.6	–
High Forest	21.4 ± 4.1	14.2 ± 3.7	2.8 ± 0.1	19.0 ± 12.8	35.1	39
High Open*	29.4 ± 10.3	9.9 ± 3.4	0.4 ± 0.6	13.1 ± 3.9	88.1	–

* Includes fire related standing dead trees.

Similarly, when $\frac{S_{\mathrm{out}}}{S_{\mathrm{in}}} < 0.3$, a value of $\alpha = 0.3$ was used to adequately simulate the lower bound of forest floor albedo during the ablation period (Melloh et al., 2002).

2.4.2 Longwave radiation

Longwave radiation is rarely directly measured in the seasonal snow zone due to the high cost in both absolute, e.g., instrument cost, and relative terms, e.g., energy requirements. Longwave radiation balance was calculated as

$$Q_{\mathrm{L}} = L\uparrow + L\downarrow \tag{4}$$

where $L\downarrow$ is the calculated longwave radiation received by the snowpack surface and $L\uparrow$ is the calculated longwave radiation emitted by the snow surface. Longwave radiation emitted at the snow surface is approximated by

$$L\uparrow = \varepsilon_{\mathrm{snow}}\,\sigma\,T_{\mathrm{snow}}^4 \tag{5}$$

where $\varepsilon_{\mathrm{snow}}$ is the snow surface emissivity and is set at 0.96 (Link and Marks, 1999).

A variety of empirically derived formulas exist for calculating incoming longwave radiation under clear (L_{clear}) and cloudy skies at various sites throughout the world (Brutsaert, 1975; Sicart et al., 2004; Flerchinger et al., 2009). All derivations are variations of the general form of the Stefan–Boltzmann equation that relates clear sky incoming longwave radiation to atmospheric emissivity ($\varepsilon_{\mathrm{clear}}$), the Stefan–Boltzman constant (σ), and air temperature T_{air} (K).

$$L_{\mathrm{clear}} = \varepsilon_{\mathrm{clear}}\,\sigma\,T_{\mathrm{air}}^4 \tag{6}$$

Many of these parameterizations are site specific or do not incorporate a cloud cover component or account for longwave radiation emitted from the canopy (Hatfield et al., 1983; Alados-Alboledas et al., 1995). The presence and type of cloud cover affects how longwave radiation is absorbed and transmitted through the atmospheric air column, significantly affecting emissivity and subsequently the magnitude of incoming longwave radiation (Sicart et al., 2004; Lundquist

et al., 2013). Incorporating a sky view factor (S_{VF}) into the longwave radiation calculations allowed us to partition the incoming longwave into atmospheric and forest canopy contributions.

Following Flerchinger et al. (2009) we performed a comparative analysis of various longwave radiation algorithms and measured net longwave radiation. Table S2 shows two clear sky algorithms and three cloud correction algorithms used in the comparison, totalling six combinations in all, with the "best-fit" algorithm determined by root mean squared error (RMSE). We measured longwave radiation using a Huskeflux NR1 net radiometer during spring 2013 for a 2-week period in a forested site within the MRB (Gleason et al., 2013) and for a 10-day period in an adjacent open area, excluding a 4-day period of rain. The NR1 measures four separate components of the surface radiation balance, separately measuring incoming and reflected solar radiation and both incoming and outgoing far infra-red radiation. The pyrogeometers have a built-in Pt100 temperature sensor for calculation of both the sky and surface temperature. Additionally, they are heated, with temperature compensation, to avoid moisture build-up on the thermopile sensors. The predicted incoming longwave radiation results of each method were then compared to the NR1 measured incoming longwave radiation using RMSE (Table S3). We found that the best approximation for incoming longwave energy was the clear sky algorithm of Dilley and O'Brien (1998) combined with the cloud adjustment of Crawford and Duchon (1999). The combined Crawford–Dilley method was therefore used in all longwave calculations going forward and is calculated as

$$L\downarrow = (S_{\mathrm{VF}})\,\varepsilon_{\mathrm{adj}}\sigma\,T_{\mathrm{air}}^4 + (1 - S_{\mathrm{VF}})\,\varepsilon_{\mathrm{snow}}\sigma\,(T_{\mathrm{C}}^4) \tag{7}$$

where SVF is the sky view factor and represents the fraction of viewable sky from the perspective of the ground surface; $\varepsilon_{\mathrm{adj}}$ is the adjusted atmospheric emissivity; and T_{C} is the temperature of the forest canopy (K). T_{C} is highly variable and typically not directly measured. The literature suggests a range of temperature of an increase of 4–30 K from measured air temperature (Derby and Gates, 1966; Pomeroy et al., 2003; Essery et al., 2008). We assumed canopy temperature to be equal to $T_{air} + 4$ K based on Boon (2009). Adjusted emissivity accounts for changes in atmospheric emissivity due to cloud cover and is found by adjusting the clear sky emissivity (ε_{clear}) by some estimation of cloud cover. The Dilley and O'Brien (1998) clear sky algorithm is as

$$L_{clear} = 59.38 + 113.7 \cdot \left(\frac{T_{air}}{273.16}\right)^6 + 96.96\sqrt{\frac{\omega}{25}}, \tag{8}$$

$$\omega = \frac{465 \frac{e_0}{100}}{T_{air}}. \tag{9}$$

The Crawford and Duchon (1999) cloud correction adjusted algorithm requires ε_{clear}, which we computed from Eq. (8) and is in the following form:

$$\varepsilon_{adj} = (1-r) + r \cdot \varepsilon_{clear}, \tag{10}$$

where r is the solar ratio, an approximation of cloud cover, and is equal to the ratio of measured incoming solar radiation and potential solar radiation (Lhomme et al., 2007).

2.4.3 Turbulent heat flux

The turbulent fluxes of latent and sensible heat are calculated using indirect methods. Latent heat exchange was calculated using the method found by Kustas et al. (1994):

$$Q_E = \left(\rho_a 0.622 \frac{L_v}{P_a}\right) C_e U_Z (e_a - e_0), \tag{11}$$

where ρ_a is the density of air (kg m^{-3}), L_v is the latent heat of vaporization or sublimation (J kg^{-1}), P_a is the total atmospheric pressure (Pa), C_e is the bulk transfer coefficient for vapor exchange, $U(z)$ is the wind speed at height Z (m) above the snow surface (m s^{-1}), e_a is the atmospheric vapor pressure at height Z above the snow surface (Pa), and e_0 is the vapor pressure at the snow surface (Pa). This calculation favors the bulk aerodynamic approach adapted from Brutsaert (1982), as direct measurement is limited and successful implementation difficult in remote environments (Moore, 1983; Marks and Dozier, 1992; Marks et al., 1998). C_{en} is the bulk transfer coefficient for vapor exchange under neutral stability and is calculated as

$$C_{en} = k^2 \left[\ln\left(\frac{Z}{Z_0}\right)\right]^{-2} \tag{12}$$

where k is von Karman's constant 0.4 ($-$) and Z is the height of the measurement above the snow surface (m) and was 3 m above the snow-free ground surface for the Low and Mid sites and 4.5 m for the High sites. Additionally, the surface roughness length Z_0 is a primary control on the bulk transfer coefficient, Eq. (12). The roughness length is affected by snow properties and is generally found to have values ranging from 0.001 to 0.005 m (Moore, 1983; Morris, 1989). This

value represents the mean height of snow surface obstacles that impede air movement over the snow surface. In our analysis we used a median value, 0.003 m, due to the variable nature of the seasonal snowpack.

The bulk aerodynamic approach is guided by stability conditions in the air above the snow surface. The stability of the air column is determined by application of the dimensionless bulk Richardson number (Ri_B) which relates the density gradient to the velocity gradient, in this case the energy of buoyancy forces to the energy created by shear stress forces. Ri_B is calculated as

$$Ri_B = \frac{g Z (T_{air} - T_{snow})}{0.5 (T_{air} + T_{snow}) U(z)^2} \tag{13}$$

where g is the acceleration due to gravity, 9.81 m s^{-2}. As Eq. (13) shows, the stability of the atmosphere is temperature dependent. Under stable conditions where the relatively warm air column settles the snow surface will cool and become dense, impeding turbulent mixing. Conversely, when the air column is relatively colder than the snow surface, free convection of the air column exists where the air warms and expands, causing increased mixing and unstable conditions. Positive values of Ri_B indicate stable conditions, whereas negative values indicate instability. Corrections for atmospheric stability effects are inconsistent within the literature and therefore remain an area of continued study (Anderson, 1976; Oke, 1987; Kustas et al., 1994; Andreas, 2002). In this study we employ Eqs. (14a) and (14b) as the general stability correction equations (Oke, 1987):

$$\text{Unstable:} \quad \frac{C_e}{C_{en}} = (1 - 16Ri_B)^{0.75}; \tag{14a}$$

$$\text{Stable:} \quad \frac{C_e}{C_{en}} = (1 - 5Ri_B)^2. \tag{14b}$$

Sensible heat exchange, much like latent heat exchange, is controlled by temperature, wind speed, roughness length, and atmospheric stability conditions. Sensible heat flux was calculated as

$$Q_H = \rho_a C_p C_h u_a (T_{air} - T_{snow}) \tag{15}$$

where c_p is the specific heat of dry air (J kg^{-1} K^{-1}) and C_h is the bulk transfer coefficient for sensible heat. Here we assumed that $C_e = C_h$ and $C_{en} = C_{hn}$.

3 Results

3.1 Snow surveys

Values for 1 April SWE, as calculated from the NRCS SNOTEL stations, range from 9 % (WY 2015) to 139 % (WY 2012) of the 30-year median reference period (1981–2010). Snow surveys conducted at the Low and Mid elevation sites

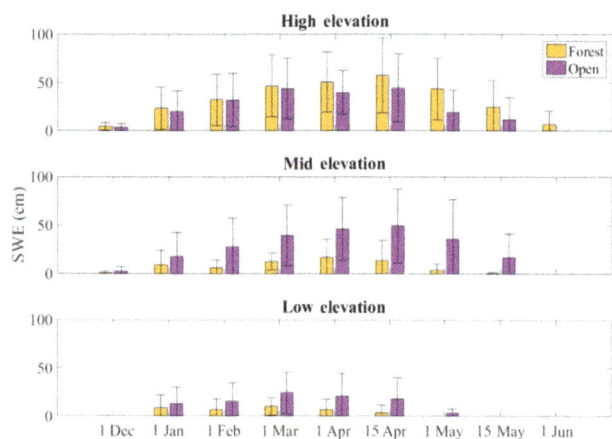

Figure 2. Average snow water equivalent (SWE) for the Open and Forest sites within the ForEST network, WY 2012–2014.

for WY 2012–14 show SWE at the Open site to be consistently greater and snow cover lasting longer into the spring than at the adjacent Forest site (Fig. 2). During the average snow year of WY 2013 (93 % of the 30-year median) the Low and Mid sites showed substantial differences between Open and Forest SWE throughout the accumulation and ablation seasons, whereas at the High sites SWE amounts were similar in Open and Forest. Conversely, snow lasted longer into the spring at the High Forest site relative to the High Open site. Because 1 April SWE may not accurately represent annual peak SWE at low and mid elevations within the PNW, we use the date of peak SWE in the following analysis. Therefore, peak SWE at the Low Open site was 209, 215, 225, and 242 % of the Forest site peak SWE, respectively, for WY 2012–WY 2015. Peak SWE at the Mid Open site was 200, 280, 328, and 302 % of the Forest site peak SWE, respectively, for WY 2012–WY 2015. However, SWE at the High Forest site is consistently higher than at the High Open site, 111, 103, 125, and 110 % for WY 2012–WY 2015, respectively.

Excluding the historically low snowpack of WY 2015 (Sproles et al., 2017), the 3-year average snow depth ablation rates in the Forest sites at Low and Mid elevation were 1.3 and 1.2 cm d^{-1}, while the Open sites were 4.1 and 3.1 cm d^{-1}, respectively (Table 2). Melt rates at the High site were greater at both sites than their lower elevation counterparts, with a rate of 4.7 cm d^{-1} at the High Open site and a rate of 3.2 cm d^{-1} for the High Forest site. At Mid Open snow persistence exceeds that of the Mid Forest site by 11–33 days. This is a similar finding to the low elevation sites, where snow lasted longer at Low Open by 4–26 days compared with the Low Forest site. Conversely, the High Forest site maintains snow longer into the spring by 15–29 days when compared to the High Open site.

3.2 Forest characteristics and canopy interception efficiency

Results show that C_{IE} in the Low and Mid Forest sites, for WY 2012–WY 2015, was 79 and 76 % of the total event snowfall, whereas C_{IE} was 31 % at the High Forest site (Table 2). C_{IE} showed no significant threshold behavior between event size and C_{IE}, although there is an inverse relationship between duration and C_{IE} at the Low and Mid sites. Events that lasted for a single day had an average canopy interception efficiency of 87 % with a reduction in average C_{IE} with increasing event length, from 73 % for a 2-day event and 57 % for a 3-day event to 51 % for any event lasting longer than 4 days. Due to the low snow years of WY 2014 and WY 2015 the High site had only four events that lasted longer than 1 day, and therefore no relationship with event duration could be identified. Using event-based C_{IE} for all snowfall events we calculated how much snow was removed by the canopy at each elevation and compared that with each event snowfall amount (Fig. 3). The low elevation site has a high correlation between C_{IE} and event size for all qualifying events ($R^2 = 0.86$) and an estimated overall snow removal efficiency of 58 %. The Mid elevation site has a lower correlation ($R^2 = 0.64$) between C_{IE} and event size and an overall snow removal efficiency of 42 %. The linear relationship of the Low and Mid sites is similar to what Storck et al. (2002) found for a single Douglas fir (*Pseudotsuga menziesii*) over a 2-year study in Oregon, that 60 % of event snowfall was intercepted by the canopy. This relationship does not hold at the high elevation site, with an overall snow removal efficiency of only 4 %. Further analysis using the Spearman rank correlation non-parametric measure shows similar results. The Spearman rank correlation coefficient (r_s) is 0.89 for both the low and mid elevation sites. This correlation does not persist at the high elevation sites ($r_s = -0.05$). We note an apparent threshold behavior where events less than 15 cm have a stronger linear relationship between event size and C_{IE} (Fig. 3) and the canopy was more effective at snow removal for events in that range compared with events greater than 15 cm. For events < 15 cm, canopy removal rates increase to 88 % for the Low site and 89 % for the Mid site, and interestingly, a weak correlation emerges, R^2 of 0.27, with 50 % removal for the High site.

3.3 Energy balance

To better understand the energy balance effect of forest canopies on snow accumulation and ablation, we calculated the mean daily energy balance components for the low and mid elevation sites for WY 2012–WY 2015 and for WY 2014 and WY 2015 for both high elevation sites (Fig. 4). Net radiation is the major component at all sites, while the turbulent fluxes and sensible and latent heat are only significant at the High Open site. Turbulent fluxes at all other sites are only episodically important and do not account for any significant

Table 2. Summary snow statistics for WY 2012–WY 2014 – Oregon ForEST network.

Site	WY2012			WY2013			WY2014		
	Peak SWE (cm)	C_{IE} (%)	Ablation rate (depth cm day^{-1})	Peak SWE (cm)	C_{IE} (%)	Ablation rate (depth cm day^{-1})	Peak SWE (cm)	C_{IE} (%)	Ablation rate (depth cm day^{-1})
Low Forest	23	70	1.6	24	75	1.9	8	92	0.4
Low Open	48	–	4.0	51	–	4.3	18	–	1.3
Mid Forest	45	70	1.0	26	75	1.3	12	83	1.1
Mid Open	89	–	3.8	73	–	2.5	38	–	4.5
High Forest	100	–	4.1	73	–	2.4	59	39	3.1
High Open	90	–	5.4	71	–	2.9	42	–	5.9

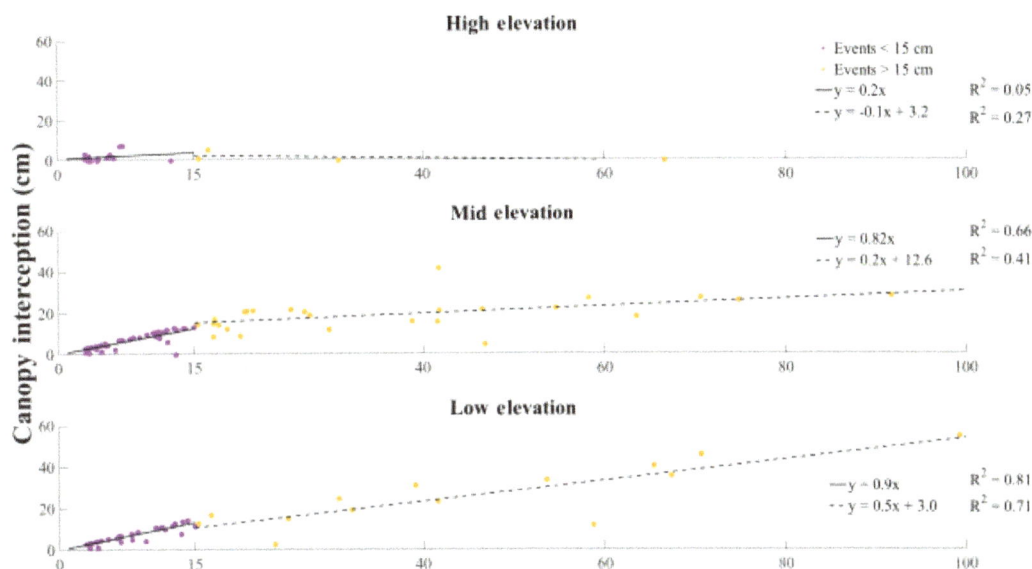

Figure 3. Canopy interception depth vs. event snowfall within the ForEST network.

amount of energy at the monthly or annual timescales. On an annual basis, shortwave radiation is the primary component of the energy balance at all Open sites, whereas longwave radiation dominates at all Forest sites. There is a strong dominance of shortwave (longwave) energy at the Low and Mid Open (Forest) sites, where it accounts for 89 and 71 % (93 and 92 %) of the average annual net energy balance, respectively. At the High sites this trend persists, although the magnitudes change. Within the High Forest site, shortwave radiation accounts for the majority of energy received at the snow surface, but the annual total is reduced by 53 %, with net longwave radiation accounting for 47 %. Conversely, at the High Open site solar radiation accounts for 71 % of the annual total, while longwave radiation is reduced to 7 %. The turbulent fluxes account for the remaining 22 %.

The stable atmospheric conditions at all sites, except the High Open site, reduce the turbulent fluxes to consistently insignificant values at the daily timescale, with only a few days over the course of the study period where these fluxes

persist (Fig. 4). Not surprising then is the importance of the radiative fluxes for the net energy balance at all sites outside of the High Open site. Longwave radiation dominates at the Low and Mid Forest sites regardless of elevation or year (Figs. S1–S4 in the Supplement). Snowpack melt response to the increased longwave radiation in the forest from lasting events can be substantial. For example, at the Mid Forest site during an 8-day mid-January period, longwave radiation at the snow surface increased 71 W m^{-2} (225 % increase), while snowmelt response was immediate and significant, attributed to a reduction of 32 cm (37 %) in snowpack depth (Fig. 5). During the same period, longwave radiation increased 56 W m^{-2} (342 % increase) at the Mid Open site, while snowpack was reduced by 6 cm (5 %). Throughout WY 2013 longwave radiation inputs are shown to have a strong inverse correlation with snowpack depth at the Mid Forest site (Fig. 5). This is not the case at the Mid Open site, where snowmelt is driven by shortwave radiation, with few accumulation season melt events at all, with snowpack set-

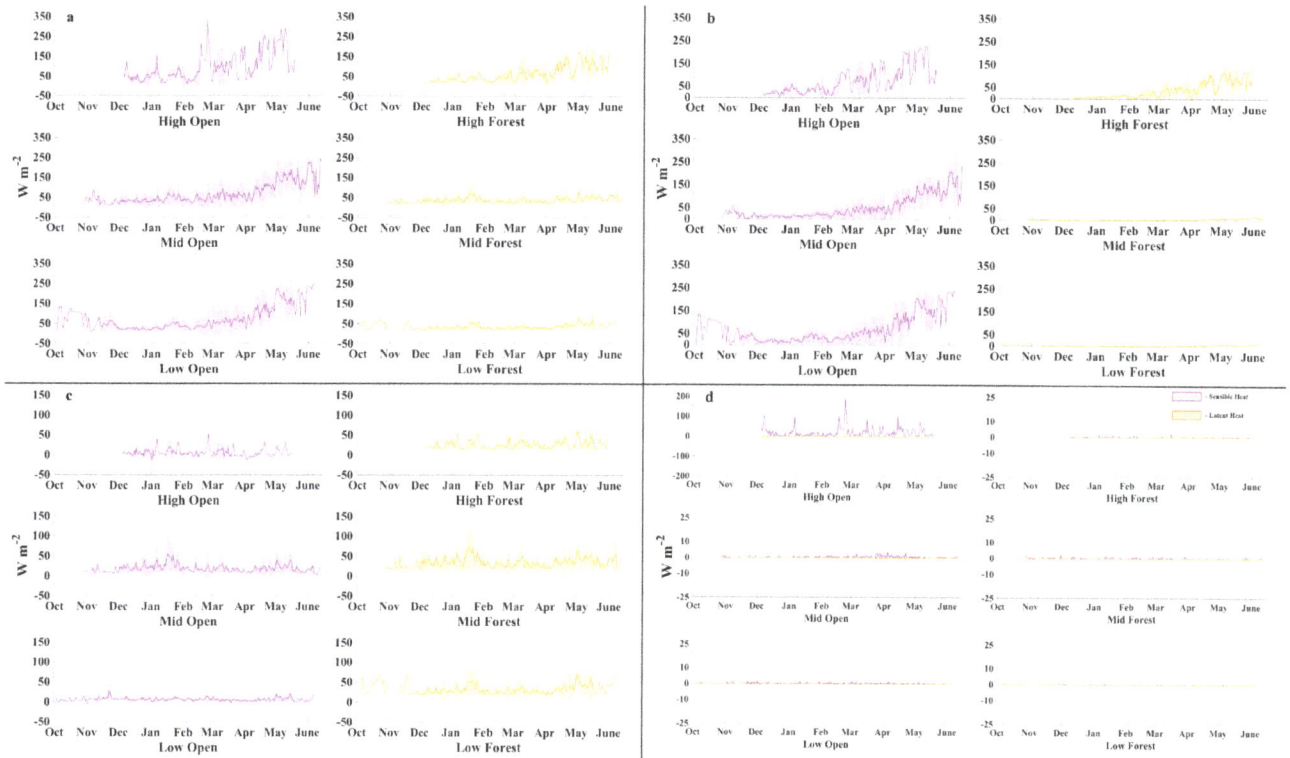

Figure 4. Calculated daily mean energy balance in $W\,m^{-2}$ (solid line) and the range of values (shaded area) for **(a)** net energy at the snow surface; **(b)** net solar radiation; **(c)** net longwave radiation; and **(d)** net turbulent energy at the snow surface for each site within the ForEST network, WY 2012–WY 2015.

tling attributed to the major snow reduction event in late December. A similar analysis at the High sites shows shortwave radiation driving the snowmelt response to mid-season melt events (Fig. S4). WY 2015 was a historically low year for the Pacific Northwest (Sproles et al., 2017); however, over a 4-day period in early January 2015 a large melt event occurred where the High Forest experienced a 37 % reduction in snow depth and the High Open snow depth reduced by 50 % (Fig. S4). Longwave radiation increased 94 % at the Forest site, attributed to 71 % of the total energy budget during the event. Conversely, the Open site longwave radiation increased 366 %, yet accounted for only 26 % of the total net energy budget, with shortwave radiation at 49 % and the net turbulent flux contributing the rest.

Air temperature is a first-order control in longwave radiation calculation and therefore it is expected that the lower and thus warmer sites will experience a larger percentage of net radiation in the form of longwave radiation. Average monthly air temperatures show that the High Forest site is 1.9 and 1.8 °C cooler during the winter months (DJF) than the Low and Mid Forest sites, respectively (Fig. 6). Colder temperatures reduce the longwave radiation received at the snow surface during the winter months as longwave radiation is nonlinearly controlled by air temperature (Eq. 7). The reduced longwave input and lower forest density at the High

Forest site is reflected in the radiation budget where the net longwave energy component is 25 % less than the net longwave energy at the Low and Mid Forest sites.

Wind speeds at all sites except at the High Open site are relatively weak and inconsistent, resulting in little turbulent mixing. Sustained (annual average) wind speeds at the High Open site are over 5 times greater than at any other site, with peak daily maxima more than 9 times greater (Fig. 7). At the High Open site high wind speeds occur frequently, while all other sites experience low wind speeds and little variability. Mean winter wind speed for the High Open site is $3.6\,m\,s^{-1}$. Mean winter wind speeds for the Low and Mid Open sites are both $0.7\,m\,s^{-1}$. The high wind speeds cause instability and subsequent turbulent mixing, resulting in much larger turbulent fluxes at the High Open site. Conversely, when wind speeds are low, minimal, if any, mixing occurs, and a decoupling of the snow surface and the atmosphere can persist. Calculation of the Richardson number (Eq. 13) determines the stability of the atmosphere, and where values are greater than 0.2, this decoupling occurs. Although there is no consensus on what threshold this critical value should be, we use a threshold of 0.2 (Raleigh et al., 2013). Over the course of the study the Ri_B value within each cover type at the Low and Mid elevation sites and the High Forest site exceeds the critical value for the majority of days. For example, in WY

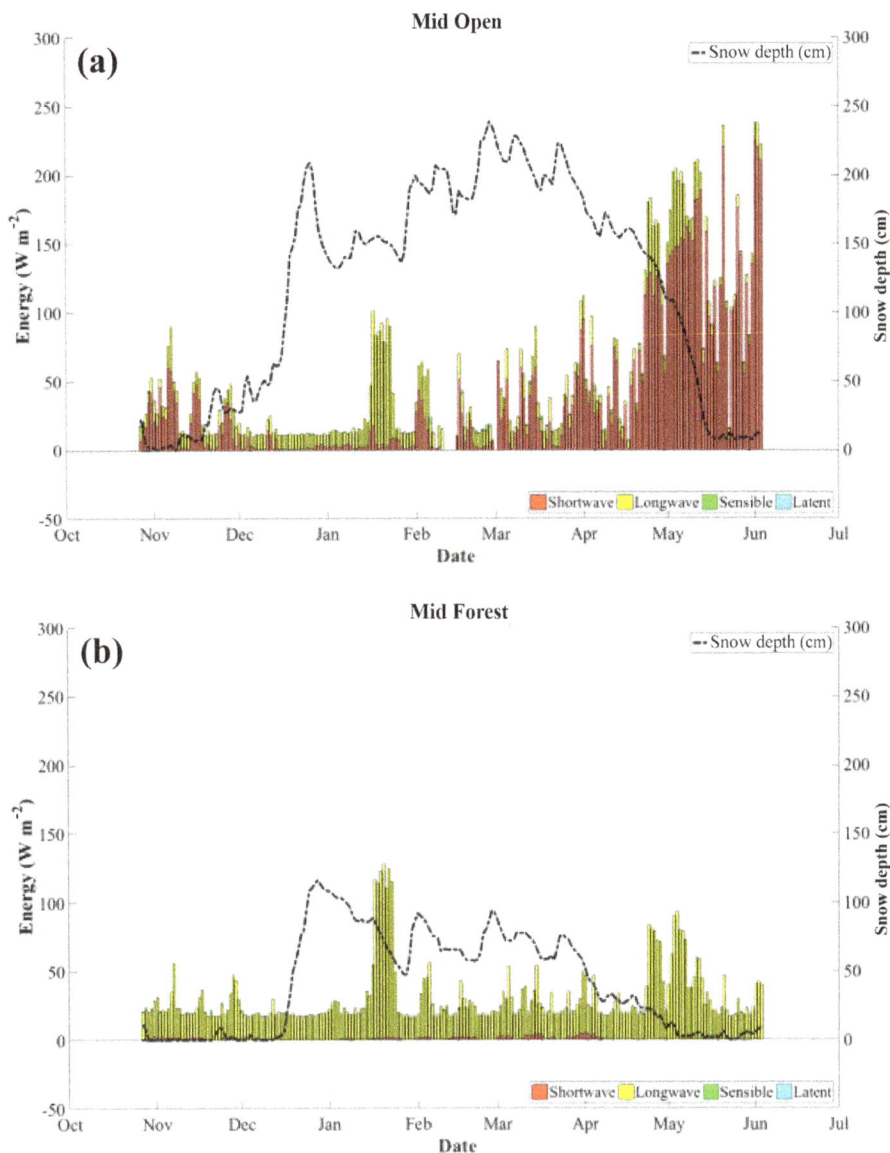

Figure 5. Calculated daily mean energy balance component magnitudes (bars) and the daily measured snow depth (dashed line) for Mid Open **(a)** and Mid Forest **(b)** during WY 2013.

2014 the critical value was exceeded 60 % of the time at both the Low sites, 76 and 71 % at the Mid Open and Mid Forest sites, 82 % of the time at the High Forest site, and only 10 % of the time at the High Open site.

Forest structure at the Low and Mid Forest sites is typified by average crown diameters of 9.4 and 6.7 m and average LAIs of 2.4 and 2.7, respectively. At the High Forest site average crown diameter and LAI were measured as 2.8 and 1.1 m, respectively. A multi-layered and randomly distributed forest canopy greatly impacts the amount of solar radiation reaching the forest floor through beam attenuation (Campbell, 1986). Forest canopies provide solar shading as the spring progresses and solar angle increases, intensifying the incoming solar radiation. At the Low and Mid Forest sites

where canopy interception is high, the impact of solar shading becomes less pronounced and snowpack SWE is not preserved late into the spring. With snowfall magnitude essentially the same at the mid and high elevations, we see that the snowpack lasts much longer into the spring at the High Forest site when forest shading has a meaningful effect on reducing solar inputs into the snowpack.

4   Discussion

In maritime snow zones where winter precipitation is often a mix of rain and snow, multiple mechanisms align to contradict the conventional wisdom that snow is retained longer in forests than in open areas (Link and Marks, 1999; Jost

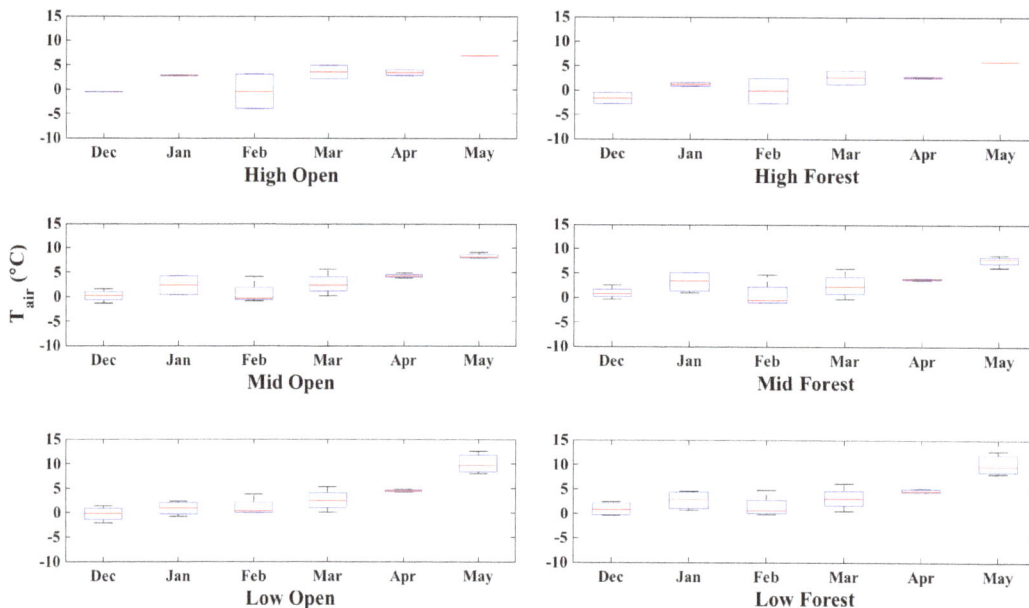

Figure 6. Boxplot of average monthly air temperature for each site within the ForEST network, WY 2012–WY 2015.

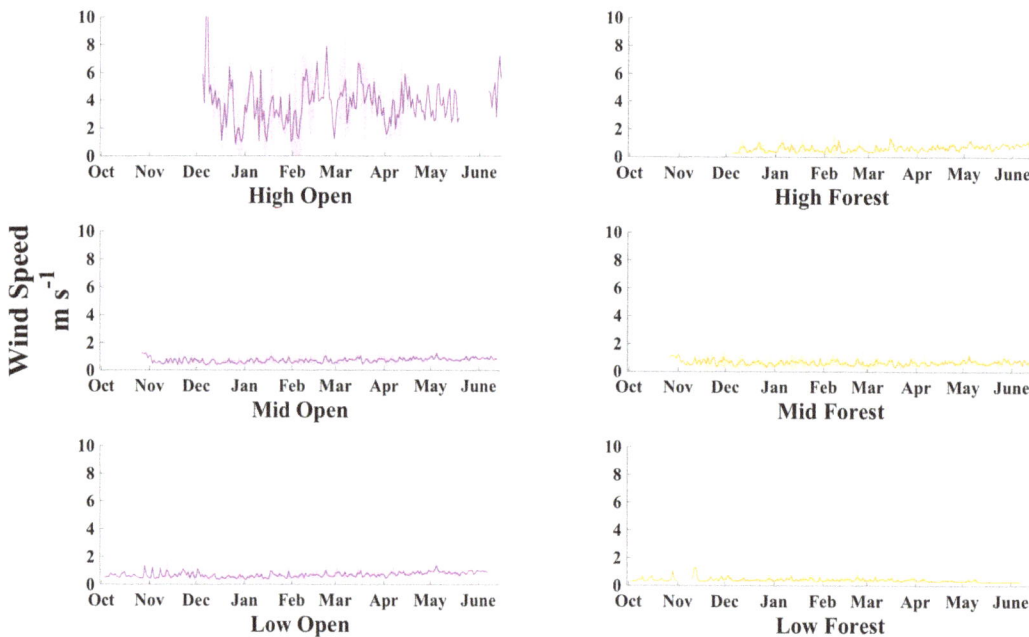

Figure 7. Daily average wind speed (heavy solid line) and the range of wind speeds (shaded area) for each site within the ForEST network, WY 2012–WY 2015.

et al., 2007; Musselman et al., 2008). Multi-layered forest cover and a relatively warm forest increase canopy interception efficiency, resulting in significant reductions in sub-canopy snow accumulation (Storck et al., 2002). While no significant relationship existed between daily air temperature and C_{IE} within our study ($p > 0.005$), a threshold behavior appears to exist where events under 15 cm seem to be highly correlated with C_{IE}. This suggests a nonlinear relationship for event-scale canopy interception in dense, relatively warm forests. The slopes of trend lines in Fig. 3 show that the dense forests at these Low and Mid Forest sites remove a considerable amount of snow from each event, significantly reducing sub-canopy accumulation. The high snow removal capacities of these forests suggest canopy density is a first-order process in snow accumulation.

While few studies in maritime forested environments on the energy balance exist, there is evidence of longwave radiation as the dominating term during rain on snow (ROS) events within forests (Berris and Harr, 1987; Mazurkiewicz et al., 2008; Garvelmann et al., 2014). Berris and Harr (1987) showed that longwave radiation accounted for 38–88 % of all ROS event snowmelt. Garvelmann et al. (2014) found that in two ROS events longwave radiation accounted for 55.1 and 38.8 % of the net energy balance, although this may be biased low due to the inability to accurately capture tree trunk temperature. Although Mazurkiewicz et al. (2008) did not differentiate between radiation terms, they found that net radiation was the largest contributor to melt. The highly nonlinear relationship between air temperature and incoming longwave radiation formulation is apparent in the net radiation budget analysis. Infrequent cloud-free days and the warm, dense forests of the study area combine to emit a significant amount of longwave radiation to the snow surface (Berris and Harr, 1987; Sicart et al., 2004; Garvelmann et al., 2014). This leads to a positive net snow surface energy balance and midwinter melt events, most pronounced at the warmer lower elevation sites. With prolonged exposure to longwave radiation emitted by the canopy and the high efficiency of warm forest canopy interception capabilities, low elevation maritime subcanopy snowpacks are relatively thin and do not persist long enough into the spring season to benefit from forest shading. This creates a radiative paradox where the longwave radiation emitted by dense and relatively warm forest cover exceeds the resulting reduction in shortwave radiation due to forest shading (Sicart et al., 2004; Lawler and Link, 2011; Lundquist et al., 2013). The higher elevation sites experience colder air temperatures, higher wind speeds, and lower forest density, which combine to decrease C_{IE} and the impact of longwave radiation on mid-winter melt events. Furthermore, relatively low ablation rates for the Low and Mid Forest sites suggest that forests do provide some radiative shading during the melt season. However, the benefit of solar shading can only be realized if a sufficient snow cover is present. Otherwise, the effects of reduced solar inputs become secondary and it is the accumulation rate, or more precisely, the efficiency of the canopy interception, that is the principle control on the date of snow disappearance.

Here, we considered that wind may have an impact on canopy snow unloading and subsequent increases in subcanopy snow accumulation. While a seasonal mean presents a general view of the wind environment at each Open site, it masks the variability of wind gusts that can drive snow redistribution. Using the 10 min mean wind speeds better depicts the wind characteristics that can affect wind redistribution of snow. Pomeroy and Gray (1990) suggest that for wet snow a snow transport wind threshold of 7–10 m s^{-1} measured at 10 m above the ground surface must be exceeded before any redistribution can occur. Using this threshold, the High Open site measured wind speeds met or exceeded the lower threshold 9.9 % of the entire record and 14.4 % if we

translate measured wind speed to $Z = 10$ m using a simple wind profile power law. This represents a substantial amount of the snow season and enough to suggest that wind redistribution is possible. More likely is the wind effect on deposition of snowfall. The influence of the forest on the reduction of wind speeds at the high elevation sites can lead to preferential deposition within the forest as the wind speeds attenuate. Once snow is deposited onto the ground the wet maritime snow makes it difficult to be redistributed as a result of saltation and suspension. However, the Open site experiences high enough sustained wind speeds to effectively redistribute and transport wet maritime snow from the High Open site into the adjacent High Forest site. Although the magnitude of this redistribution of snow from Open to Forest is unknown, it is reasonable to assume that it is not insignificant considering the sustained high winds of the High Open environment.

The effects of elevation position within a watershed and forest structure on snow persistence can have serious implications within a warming climate. Sproles et al. (2013) documented a 150 m increase in the elevation of the snow line for every 1 °C temperature increase and showed that projected temperature increases of about 2 °C would shift precipitation at 1500 m from snowfall to a rain–snow mix. If that were to occur, then forests at that elevation, e.g., the High Forest site, that now help maintain late spring snowpacks would likely behave more like the lower elevation forests in which snowmelt occurs earlier than in the Open areas, effectively offsetting any solar shading gains that the forest can provide in the present. Peak SWE and spring runoff would be reduced at these higher elevations. These high elevation forests could lose their dry season "moisture subsidy" and suffer increased moisture stress, with wide-ranging implications for forest and water resource managers.

5 Conclusions

This paper highlights the complex snow–forest process relationships and suggests that forest cover is a principal control on snow persistence due to reduced accumulation from canopy interception and earlier/faster melt due to increased longwave radiation. High density, relatively warm forests have high canopy interception efficiency that controls subcanopy snowpack evolution and mediates the amount of springtime solar shading of the snowpack. The cooler and less dense High Forest site has a reduced interception efficiency and acts as a snow deposition reservoir for the nearby windy High Open site. Net radiation drives the snow surface energy balance, with the partitioning between longwave and shortwave a function of forest complexity. Our study demonstrates the sensitivity of Pacific Northwest snowpack development to temperature and forest cover. Nolin and Daly (2006) demonstrated that much of the Oregon Cascade snowpack is at risk, the ForEST network included, by looking at temperature only. Similarly, Sproles et al., 2013 showed that the lower boundary of the snow zone has little resilience to a warming world. Our paper demonstrates that understanding the snowpack energy budget is key to understanding how forests influence snow accumulation and melt. By quantifying the mechanisms of how vegetation affects sub-canopy snowpack energy balance, the results of this study provide the basis for understanding the sensitivity of maritime snowpacks to a changing climate. As climate continues to warm, we anticipate reduced snow accumulation at elevations where snowfall shifts to a rain–snow mix, and amplified sub-canopy melt rates due to longwave radiative heating in warmer forests, thereby reducing overall forest snow retention. However, higher elevation colder sites with a less dense forest can mitigate that to some extent by retaining the snowpack longer through lower relative forest longwave emission and lower canopy interception. A key finding within this study is that throughout the study duration, one that saw high inter-annual snowfall variability, a definitive pattern emerged within the energy budget and snowpack dynamics across the network. The energy budget format that we present here goes beyond the temperature only approach while getting at the causal effects and mechanisms of the challenge of vegetation–snowpack interactions for a warming climate.

While these results are focused on the Oregon Cascades, they have broader implications for other relatively warm forested snow environments with elevation gradients, such as parts of the California Sierra Nevada, the Japanese and European Alps, and the Pyrenees (Lundquist et al., 2013). These results will aid in improving parameterizations of snow–forest interactions in physically based snow hydrology models and land surface models. Additionally, as climate change alters regional snow deposition patterns across the western US, our findings are applicable to land and water managers, seeking to improve forest snowpack retention, enhance forest health, and improve streamflow forecasting. This study demonstrates the value of plot-scale snow–forest process studies for improving our understanding of the forest effects on snowpack evolution. Future work will focus on a multi-scale approach that incorporates remote sensing and snow hydrology modeling to identify forest structure metrics that are well suited to accurately modeling snow–forest interactions. Such an approach will allow the snow community to quantify the improvement of snow–forest interactions across spatial scales and enhance model prediction for landscape and regional applications.

Competing interests. The authors declare that they have no conflict of interest.

Acknowledgements. This research was made possible by funding provided by the National Science Foundation (EAR 1039192) and from a NASA Earth Science Student Fellowship (16-EARTH16F-0426). We thank Willamette National Forest for providing access permits for the ForEST network. Additional material support was provided by the Western Ecology Division office of the Environmental Protection Agency, with special thanks to Ron Waschmann. We thank the many student interns who assisted in snow surveys and site maintenance.

Edited by: Jan Seibert

References

Alados-Alboledas, L., Vida, J., and Olmo, F. J.: The estimation of thermal atmospheric radiation under cloudy skies, Int. J. Climate, 15, 107–116, 1995.

Allen, C. D., Macalady, A. K., Chenchouni, H., Bachelet, D., McDowell, N., Vennetier, M., Kitzberger, T., Rigling, A., Breshears, D. D., Hogg, E. H., Gonzalez, P., Fensham, R., Zhang, Z., Castro, J., Demidova, N., Lim, J. H., Allard G., Running, S. W., Semerci, A., and Cobb, N.: A global overview of drought and heat-induced tree mortality reveals emerging climate change risks for forests, Forest Ecol. Manag., 259, 660–684, 2010.

Anderson, E. A.: A point energy balance and mass balance model of snow cover, Silver Spring, Md., US DOC NOAA, Tech Rpt., NWS 19, 1976.

Andreadis, K. M., Storck, P., and Lettenmaier, D. P.: Modeling snow accumulation and ablation processes in forested environments, Water Resour. Res., 45, 1–13, https://doi.org/10.1029/2008WR007042, 2009.

Andreas, E. L.: A new method of measuring the snow-surface temperature, Cold Reg. Sci. Technol., 12, 139–156, https://doi.org/10.1016/0165-232X(86)90029-7, 1986.

Andreas, E. L.: Parameterizing scalar transfer over snow and ice: a review, J. Hydrometeorol., 3, 417–432, https://doi.org/10.1175/1525-7541(2002)003<0417:PSTOSA>2.0.CO;2, 2002.

Berris, S. N. and Harr, R. D.: Comparative snow accumulation and melt during rainfall in forested and clear-cut plots in the western Cascades of Oregon, Water Resour. Res., 23, 135–142, https://doi.org/10.1029/WR023i001p00135, 1987.

Bewley, D., Alila, Y., and Varhola, A.: Variability of snow water equivalent and snow energetics across a large catchment subject to Mountain Pine Beetle infestation and rapid salvage logging, J. Hydrol., 388, 464–479, 2010.

Boon, S.: Snow ablation energy balance in a dead forest stand, Hydrol. Process., 23, 2600–2610, https://doi.org/10.1002/hyp.7246, 2009.

Brooks, J. R., Wigington, P. J., Phillips, D. L., Comeleo, R., and Coulombe, R.: Willamette River Basin surface water isoscape ($\delta^{18}O$ and δ^2H): temporal changes of source water within the river, Ecosphere, 3, 39, https://doi.org/10.1890/es11-00338.1, 2012.

Brutsaert, W.: On a derivable formula for long-wave radiation from clear skies, Water Resour. Res., 11, 742–744, 1975.

Brutsaert, W.: Evaporating Into the Atmosphere: Theory, History, and Applications, D. Reidel Publ. Co., Dordrecht, 1982.

Campbell, G. S.: Extinction coefficients for radiation in plant canopies calculated using an ellipsoidal inclination angle distribution, Agr. Forest Meteorol., 36, 317–321, 1986.

Choat, B., Jansen, S., Brodribb, T. J., Cochard, H., Delzon, S., Bhaskar, R., Bucci, S. J., Field, T. S., Gleason, S. M., Hacke, U. G., Jacobsen, A. L., Lens, F., Maherali, H., Martinez-Vilalta, J., Mayr, S., Mencuccini, M., Mitchell, P. J., Nardini, A., Pittermann, J., Pratt, R. B., Sperry, J. S., Westoby, M., Wright, I. J., and Zanne, A. E.: Global convergence in the vulnerability of forests to drought, Nature, 491, 752–755, 2012.

Crawford, T. M. and Duchon, C. E.: An improved parameterization for estimating effective atmospheric emissivity for use in calculating daytime downwelling longwave radiation, J. Appl. Meteorol., 48, 474–480, 1999.

Dai, A.: Increasing drought under global warming in observations and models, Nat. Clim. Change, 3, 52–58, https://doi.org/10.1038/nclimate1633, 2013.

Derby, R. W. and Gates, D. M.: The temperature of tree trunks – calculated and observed, Am. J. Bot., 53, 580–587, 1966.

Dickerson-Lange, S. E., Lutz, J. A., Martin, K. A., Raleigh, M. S., Gersonde, R., and Lundquist, J. D: Evaluating observational methods to quantify snow duration under diverse forest canopies, Water Resour. Res., 15, 1203–1224, https://doi.org/10.1002/2014WR015744, 2015.

Dilley, A. C. and O'Brien, D. M.: Estimating downward clear sky long-wave irradiance at the surface from screen temperature and precipitable water, Q. J. Roy. Meteor. Soc., 124, 1391–1401, 1998.

Elder, K., Dozier, J., and Michaelsen, J.: Snow accumulation and distribution in an alpine watershed, Water Resour. Res., 27, 154–1552, 1991.

Elder, K., Rosenthal, W., and Davis, R. E.: Estimating the spatial distribution of snow water equivalence in a montane watershed, Hydrol. Process., 12, 1793–1808, 1998.

Ellis, C. R., Pomeroy, J. W., and Link, T. E.: Modeling increases in snowmelt yield and desynchronization resulting from forest gap-thinning treatments in a northern mountain headwater basin, Water Resour. Res., 49, 936–949, https://doi.org/10.1002/wrcr.20089, 2013.

Elzinga, C. L., Salzer, D. W., and Willoughby, J. W.: Measuring and Monitoring Plant Populations, USDI Bureau of Land Management Technical Reference 1730-1, National Business Center, Denver, CO, 1998.

Essery, R., Pomeroy, J. W., Ellis, C., and Link, T. E.: Modeling longwave radiation to snow beneath forest canopies using hemispherical photography or linear regression, Hydrol. Process., 22, 2788–2800, https://doi.org/10.1002/hyp.6930, 2008.

Flerchinger, G. N., Xaio, W., Marks, D., Sauer, T. J., and Yu, Q.: Comparison of algorithms for incoming atmospheric long-wave radiation, Water Resour. Res., 45, W03423, https://doi.org/10.1029/2008WR007394, 2009.

Frazer, G., Canham, C., and Lertzman, K.: Gap Light Analyzer (GLA): Imaging Software to Extract Canopy Structure and Gap Light Transmission Indices from True-Colour Fisheye Photographs: User's Manual and Program Documentation, Simon Fraser University, Burnaby, BC, https://doi.org/10.1016/S0168-1923(01)00274-X, 1999.

Friesen, J., J. Lundquist, J., and Van Stan, J. T.: Evolution of forest precipitation water storage measurement methods, Hydrol. Process., 29, 2504–2520, https://doi.org/10.1002/hyp.10376, 2014.

Gary, H. L.: Snow accumulation and snowmelt as influenced by a small clearing in a lodgepole pine forest, Water Resour. Res., 10, 348–353, https://doi.org/10.1029/WR010i002p00348, 1974.

Garvelmann, J., Pohl, S., and Weiler, M.: Variability of observed energy fluxes during rain-on-snow and clear sky snowmelt in a mid-latitude mountain environment, J. Hydrometeorol., 15, 1220–1236, 2014.

Geddes, C. A., Brown, D. G., and Farge, D. B.: Topography and Vegetation as Predictors of Snow Water Equivalent Across the Alpine Treeline Ecotone at Lee Ridge, Glacier National Park, Montana, USA, Arct. Antarct. Alp. Res., 37, 197–205, 2005.

Gleason, K. E. and Nolin, A. W.: Charred forests accelerate snow albedo decay: parameterizing the post-fire radiative forcing on snow for three years following fire, Hydrol. Process., 197–205, https://doi.org/10.1002/hyp.10897, 2016.

Gleason, K. E., Nolin, A. W., and Roth, T. R.: Charred forests increase snowmelt: effects of burned woody debris and incoming solar radiation on snow ablation, Geophys. Res. Lett., 40, 4654–4661, https://doi.org/10.1002/grl.50896, 2013.

Gleason, K. E., Nolin, A. W., and Roth, T. R.: Developing a representative snow-monitoring network in a forested mountain watershed, Hydrol. Earth Syst. Sci., 21, 1137–1147, https://doi.org/10.5194/hess-21-1137-2017, 2017.

Gray, D. M.: Snow accumulation and distribution, in: Proceedings, Modelling of Snow Cover Runoff, edited by: Colbeck, S. C., and Ray, M., US Army Cold Regions Research and Engineering Laboratory, Hanover, NH, 3–33, 1979.

Hardy, J. P., Marks, D., Link, T., and Koenig, G.: Variability of the below canopy thermal structure over snow, Eos Trans. AGU, Fall Meet Suppl., 85(47): F448, 2004.

Harpold, A. A., Biederman, J. A., Condon, K., Merino, M., Korgaonkar, Y., Nan, T., Sloat, L. L., Ross, M., and Brooks, P. D.: Changes in snow accumulation and ablation following the Las Conchas Forest Fire, New Mexico, USA, Ecohydrology, 7, 440–452, https://doi.org/10.1002/eco.1363, 2013.

Harpold, A. A., Molotch, N. P., Musselman, K. N., Bales, R. C., Kirchner, P. B., Litvak, M., and Brooks, P. D.: Soil moisture response to snowmelt timing in mixed-conifer subalpine forests, Hydrol. Process., 29, 2782–2798, https://doi.org/10.1002/hyp.10400, 2015.

Hatfield, J. L., Reginato, R. J., and Idso, S. B.: Comparison of longwave radiation calculation methods over the United States, Water Resour. Res., 19, 285–288, 1983.

Hedstrom, N. R. and Pomeroy, J. W.: Measurements and modeling of snow interception in the boreal forest, Hydrol. Process., 12, 1611–1625, https://doi.org/10.1002/(SICI)1099-1085(199808/09)12:10/11<1611::AID-HYP684>3.0.CO;2-4, 1998.

Homer, C., Dewitz, J., Fry, J., Coan, M., Hossain, N., Larson, C., Herold, N., McKerrow, A., VanDriel, J. N., and Wickham, J.: Completion of the 2001 National Land Cover Database for the Conterminous United States, Photogramm. Eng. Rem. S., 73, 337–341, 2007.

Inoue, A., Yamamoto, K., Mizoue, N., and Kawahara, Y.: Calibrating view angle and lens distortion of the Nikon fish-eye converter FC-E8, J. Forest Res., 9, 17, https://doi.org/10.1007/s10310-003-0073-8, 2004.

Jefferson, A., Nolin, A., Lewis, S., and Tague, C.: Hydrogeologic controls on streamflow sensitivity to climate variation, Hydrol. Process., 22, 4371–4385, https://doi.org/10.1002/hyp.7041, 2008.

Jost, G., Weiler, M., Gluns, D. R., and Alila, Y.: The influence of forest and topography on snow accumulation and melt at the watershed-scale, J. Hydrol., 347, 101–115, https://doi.org/10.1016/j.jhydrol.2007.09.006, 2007.

Kustas, W. P., Rango, A., and Uijlenhoet, R.: A simple energy budget algorithm for the snowmelt runoff model, Water Resour. Res., 30, 1515–1527, 1994.

Lawler, R. R. and Link, T. E.: Quantification of incoming allwave radiation in discontinuous forest canopies with application to snowmelt prediction, Hydrol. Process., 25, 3322–3331, https://doi.org/10.1002/hyp.8150, 2011.

Lhomme, J. P., Vacher, J. J., and Rocheteau, A.: Estimating downward long-wave radiation on the Andean Altiplano, Agr. Forest Meteorol., 145, 139–1482, 2007.

Link, T. E. and Marks, D.: Distributed simulation of snowcover mass- and energy-balance in the boreal forest, Hydrol. Process., 13, 2439–2452, 1999.

Liston, G. E. and Sturm, M.: A snow-transport model for complex terrain, J. Glaciol., 44, 498–516, 1998.

López-Moreno, J. I. and Latron, J.: Influence of canopy density on snow distribution in a temperate mountain range, Hydrol. Process., 22, 117–126, 2008.

López-Moreno, J. I. and Stähli, M.: Statistical analysis of the snow cover variability in a subalpine watershed: assessing the role of topography and forest interactions, J. Hydrol., 348, 379–394, 2008.

Lundquist, J. E., Dickerson-Lange, S. E., Lutz, J. A., and Cristea, N. C.: Lower forest density enhances snow retention in regions with warmer winters: a global framework developed from plot-scale observations and modeling, Water Resour. Res., 49, 6356–6370, https://doi.org/10.1002/wrcr.20504, 2013.

Lyon, S. W., Troch, P. A., Broxton, P. D., Molotch, N. P., and Brooks, P. D.: Monitoring the timing of snowmelt and the initiation of streamflow using a distributed network of temperature/light sensors, Ecohydrology, 1, 215–224, https://doi.org/10.1002/eco.18, 2008.

Marks, D. and Dozier, J.: Climate and energy exchange at the snow surface in the alpine region of the Sierra Nevada, 2, Snow cover energy balance, Water Resour. Res., 28, 3043–3054, https://doi.org/10.1029/92WR01483, 1992.

Marks, D., Kimball, J., Tingey, D., and Link, T. E.: The sensitivity of snowmelt processes to climate conditions and forest cover during rain-on-snow: a case study of the 1996 Pacific Northwest flood, Hydrol. Process., 12, 1569–1587, https://doi.org/10.1002/(SICI)1099-1085(199808/09)12:10/11, 1998.

Martin, K. A., Van Stan, J. T., Dickerson-Lange, S. E., Lutz, J. A., Berman, J. W., Gersonde, R., and Lundquist, J. D.: Development and testing of a snow interceptometer to quantify canopy water storage and interception processes in the rain/snow transition zone of the North Cascades, Washington, USA, Water Resour. Res., 49, 3243–3256, https://doi.org/10.1002/wrcr.20271, 2013.

Mazurkiewicz, A. B., Callery, D. G., and McDonnell, J. J.: Assessing the controls of the snow energy balance and water available for runoff in a rain-on-snow environment, J. Hydrol., 354, 1–14, https://doi.org/10.1016/j.jhydrol.2007.12.027, 2008.

McCabe, G. J. and Clark, M. P.: Trends and variability in snowmelt runoff in the western United States, J. Hydrometeorol., 6, 476–482, 2005.

Melloh, R. A., Hardy, J. P., Bailey, R. N., and Hall, T.: An efficient snow albedo model for the open and subcanopy, Hydrol. Process., 16, 3571–3584, https://doi.org/10.1002/hyp.1229, 2002.

Miller, J. D., Safford, H. D., Crimmins, M. A., and Thode, A. E.: Quantitative evidence for increasing forest fire severity in the Sierra Nevada and southern Cascade Mountains, California and Nevada, USA, Ecosystems, 12, 16–32, 2009.

Moeser, D., Stähli, M., and Jonas, T.: Improved snow interception modeling using canopy parameters derived from airborne LiDAR data, Water Resour. Res., 51, 5041–5059, https://doi.org/10.1002/2014WR016724, 2015.

Molotch, N. P. and Bales, R. C.: SNOTEL representativeness in the Rio Grande headwaters on the basis of physiographics and remotely sensed snow cover persistence, Hydrol. Process., 20, 723–739, 2006.

Moore, R. D.: On the use of bulk aerodynamic formulae over melting snow, Nord. Hydrol., 14, 193–206, 1983.

Moritz, M. A., Parisien, M. A., Batllori, E., Krawchuk, M. A., Van Dorn, J., Ganz, D. J., and Hayhoe, K.: Climate change and disruptions to global fire activity, Ecosphere, 3, 49, https://doi.org/10.1890/ES11-00345.1, 2012.

Morris, E.: Turbulent transfer over snow and ice, J. Hydrol., 105, 205–223, 1989.

Mote, P. W.: Climate-driven variability and trends in mountain snowpack in western North America, J. Climate, 19, 6209–6220, 2006.

Musselman, K., Molotch, N. P., and Brooks, P. D.: Effects of vegetation on snow accumulation and ablation in a mid-latitude sub-alpine forest, Hydrol. Process., 22, 2767–2776, https://doi.org/10.1002/hyp.7050, 2008.

Musselman, K., Clark, M. P., Liu, C., Ikeda, K., and Rasmussen, R.: Slower snowmelt in a warmer world, Nat. Clim. Change, 7, 214–220, https://doi.org/10.1038/NCLIMATE3225, 2017.

Nolin, A. W. and Daly, C.: Mapping "at risk" snow in the Pacific Northwest, J. Hydrometeorol., 7, 1164–1172, https://doi.org/10.1175/JHM543.1, 2006.

O'Halloran, T. L., Law, B. E., Goulden, M. L., Wang, Z., Barr, J. G., Schaaf, C., Brown, M., Fuentes, J. D., Göckede, M., and Black, A.: Radiative forcing of natural forest disturbances, Glob. Change Biol., 18, 555–565, 2012.

Oke, T. R.: Boundary Layer Climates, Routledge, 2nd edition, Methuen, London, vol. 8, 262–303, 1987.

Pomeroy, J. W. and Gray, D. M.: Saltation of snow, Water Resour. Res., 26–27, 1583–1594, 1990.

Pomeroy, J. W., Marsh, P., and Gray, D. M.: Application of a distributed blowing snow model to the Arctic, Hydrol. Process., 11, 1451–1464, 1997.

Pomeroy, J. W., Gray, D. M., Hedstrom, N. R., and Janowicz, J. R.: Prediction of seasonal snow accumulation in cold climate forests, Hydrol. Process., 16, 3543–3558, https://doi.org/10.1002/hyp.1228, 2003.

Raleigh, M. S., Landry, C. C., Hayashi, M., Quinton, W. L., and Lundquist, J. D.: Approximating snow surface temperature from standard temperature and humidity data: new possibilities for snow model and remote sensing evaluation, Water Resour. Res., 49, 8053–8069, https://doi.org/10.1002/2013WR013958, 2013.

Roth, T. R. and Nolin, A. W.: Willamette Water 2100 Forest Elevational Snow Transect (ForEST) Project: dataset available at https://ir.library.oregonstate.edu/xmlui/handle/1957/59984 (last access: 19 July 2017), https://doi.org/10.7267/N900001K, 2016.

Ryan, W. A., Doesken, N. J., and Fassnacht, S. R.: Evaluation of ultrasonic snow depth sensors for US snow measurements, J. Atmos. Ocean. Tech., 25, 667–684, https://doi.org/10.1175/2007JTECHA947.1, 2008.

Schmidt, R. A. and Gluns, D. R.: Snowfall interception on branches of three conifer species, Can. J. Forest Res., 21, 1262–1269, 1991.

Sicart, J. E., Essery, R. L. H., Pomeroy, J. W., Hardy, J., Link, T. E., and Marks, D.: A sensitivity study of daytime net radiation during snowmelt to forest canopy and atmospheric conditions, J. Hydrometeorol., 5, 774–784, https://doi.org/10.1175/1525-7541(2004)005<0774:ASSODN>2.0.CO;2, 2004.

Spracklen, D. V., Mickley, L. J., Logan, J. A., Hudman, R. C., Yevich, R., Flannigan, M. D., and Westerling, A. L.: Impacts of climate change from 2000 to 2050 on wildfire activity and carbonaceous aerosol concentrations in the western United States, J. Geophys. Res.-Atmos., 114, D20301, https://doi.org/10.1029/2008JD010966, 2009.

Sproles, E. A., Nolin, A. W., Rittger, K., and Painter, T. H.: Climate change impacts on maritime mountain snowpack in the Oregon Cascades, Hydrol. Earth Syst. Sci., 17, 2581–2597, https://doi.org/10.5194/hess-17-2581-2013, 2013.

Sproles, E. A., Roth, T. R., and Nolin, A. W.: Future snow? A spatial-probabilistic assessment of the extraordinarily low snowpacks of 2014 and 2015 in the Oregon Cascades, The Cryosphere, 11, 331–341, https://doi.org/10.5194/tc-11-331-2017, 2017.

Stähli, M. and Gustafsson, D.: Long-term investigations of the snow cover in a subalpine semi-forested catchment, Hydrol. Process., 20, 411–428, 2006.

Storck, P., Lettenmaier, D. P., and Bolton, S.: Measurement of snow interception and canopy effects on snow accumulation and melt in a mountainous maritime climate, Oregon, United States, Water Resour. Res., 38, 1223, https://doi.org/10.1029/2002WR001281, 2002.

Varhola, A., Coops, N. C., Weiller, M., and Moore, R. D.: Forest canopy effects on snow accumulation and ablation: an integrative review of empirical result, J. Hydrol., 392, 219–233, https://doi.org/10.1016/j.jhydrol.2010.08.009, 2010.

Vose, J. M., Miniat, C. F., Luce, C. H., Asbjornsen, H., Caldwell, P. V., Campbell, J. L., Grant, G. E., Isaak, D. J., Loheide II, S. P., and Sun, G.: Ecohydrological implications of drought for forests in the United States, For. Ecol. Manage., 380, 335–345, https://doi.org/10.1016/j.foreco.2016.03.025, 2016.

Westerling, A. L., Hildago., H. G., Cayan, D. R., and Swetnam, T. W.: Warming and earlier spring increase in western US forest wildfire activity, Science, 313, 940, https://doi.org/10.1126/science.1128834, 2006.

Westerling, A. L., Turner, M. G., Smithwick, E. A. H., Romme, W. H., and Ryan, M. G.: Continued warming could transform Great Yellowstone fire regimes by mid-21st century, P. Natl. Acad. Sci. USA, 108, 32, 13165–13170, 2011.

Woods, S., Ahl, R., Sappington, J., and McCaughey, W.: Snow accumulation in thinned lodgepole pine stands, Montana, USA, Forest Ecol. Manag. 235, 202–211, 2006.

Now you see it, now you don't: a case study of ephemeral snowpacks and soil moisture response in the Great Basin, USA

Rose Petersky[1] and Adrian Harpold[1,2,3]

[1]Graduate Program of Hydrologic Sciences, University of Nevada,
1664 N Virginia St., Reno, NV 89557, USA
[2]Natural Resources Environmental Science Department, University of Nevada,
1664 N Virginia St., Reno, NV 89557, USA
[3]Global Water Center, University of Nevada, 1664 N Virginia St., Reno, NV 89557, USA

Correspondence: Adrian Harpold (aharpold@cabnr.unr.edu)

Abstract. Ephemeral snowpacks, or those that persist for < 60 continuous days, are challenging to observe and model because snow accumulation and ablation occur during the same season. This has left ephemeral snow understudied, despite its widespread extent. Using 328 site years from the Great Basin, we show that ephemeral snowmelt causes a 70-days-earlier soil moisture response than seasonal snowmelt. In addition, deep soil moisture response was more variable in areas with seasonal snowmelt. To understand Great Basin snow distribution, we used MODIS and Snow Data Assimilation System (SNODAS) data to map snow extent. Estimates of maximum continuous snow cover duration from SNODAS consistently overestimated MODIS observations by > 25 days in the lowest (< 1500 m) and highest ($>$ 2500 m) elevations. During this time period snowpack was highly variable. The maximum seasonal snow cover during water years 2005–2014 was 64 % in 2010 and at a minimum of 24 % in 2014. We found that elevation had a strong control on snow ephemerality, and nearly all snowpacks over 2500 m were seasonal except those on south-facing slopes. Additionally, we used SNODAS-derived estimates of solid and liquid precipitation, melt, sublimation, and blowing snow sublimation to define snow ephemerality mechanisms. In warm years, the Great Basin shifts to ephemerally dominant as the rain–snow transition increases in elevation. Given that snow ephemerality is expected to increase as a consequence of climate change, physics-based modeling is needed that can account for the complex energetics of shallow snowpacks in complex terrain. These modeling efforts will need to be supported by field observations of mass and energy and linked to finer remote sensing snow products in order to track ephemeral snow dynamics.

1 Introduction

Seasonal snowmelt supplies water to one-sixth of the world's population, which supports one-fourth of the global economy (Barnett et al., 2005; Sturm et al., 2017). Seasonal snowpack provides predictable melt timing and volumes in the spring, which influences streamflow timing, surface water, and groundwater availability (Berghuijs et al., 2014; Jasechko et al., 2014; Stewart et al., 2005). Reliable spring snowmelt also provides a strong control on vegetation phenology and productivity in many ecosystems (Parida and Buermann, 2014; Trujillo et al., 2012). Despite the importance of seasonal snow to water supplies, much of the world's snow is ephemeral (or intermittent), which means it melts and sublimates throughout the snow cover season instead of having one consistent period of snowmelt. Even small shifts from seasonal to ephemeral snowpacks due to regional warming could disrupt snowmelt rates and timing. A shift from seasonal to ephemeral snowpacks will also have negative implications for the winter tourism that requires continuous snow cover, as well as water management and hydropower that relies on the predictability of snowmelt from mountain reservoirs (Schmucki et al., 2017; Sturm et al., 2017). The hydrological impacts of ephemeral snowpacks have received little study.

Snowmelt influences a variety of terrestrial hydrological processes and states, particularly soil moisture dynamics in areas with low summer precipitation (Harpold and Molotch, 2015; Seyfried et al., 2009). Snowmelt-derived soil moisture is a primary control on streamflow generation and timing and ecosystem productivity in many semi-arid systems (Jefferson, 2011; McNamara et al., 2005; Schwinning and Sala, 2004; Stielstra et al., 2015; Trujillo et al., 2012). Although few studies have isolated their hydrological importance, ephemeral snowpacks modify the intensity and duration of precipitation inputs to soil by storing and releasing water in a less predictable way than seasonal snow. For example, McNamara et al. (2005) described five predictable phases of soil moisture evolution in semi-arid watersheds with seasonally dominant snowmelt: (1) a summer dry period; (2) a transitional fall wetting period; (3) a winter wet, low-flux period; (4) a spring wet, high-flux period; and (5) a transitional late-spring drying period. Soil moisture response to ephemeral snowmelt is likely to sit between the predictable timing and rates of seasonal snow and the stochastic nature of rainfall, but few observations across this gradient exist. Despite the hydrological and ecological importance of ephemeral snow, there are no widely accepted methodologies to classify, map, and model snow ephemerality.

One commonly used snowpack classification system by Sturm et al. (1995) divides snowpack into six categories and defines ephemeral snowpacks as those persisting for less than 60 consecutive days, are less than 50 cm depth, and have less than three different snow layers (Sturm et al., 1995). While it is arbitrary, using the 60-day threshold allows for comparisons between the extent of ephemeral snow to previous studies and among different areas. The Sturm et al. (1995) classification system is also incorporated into physical snowpack models, such as SnowModel (Liston and Elder, 2006), to separate seasonal and ephemeral snowpacks into different modeling domains. Models often make this separation because the energetics of ephemeral snowpacks are much more sensitive to basal melt from ground heat flux. Additionally, cold content varies more rapidly through time in shallow ephemeral snowpacks. Most physics-based models (e.g., Liston and Elder, 2006) are optimized for seasonal snow and produce less accurate results over ephemeral snow (Kelleners et al., 2010; Kormos et al., 2014).

Ground-based and remote sensing observations have their own strengths and weaknesses for observing ephemeral snowpacks and soil moisture response. Most ground-based snow measurement stations (e.g., the Natural Resources Conservation Service Snow Telemetry, NRCS SNOTEL) in the Great Basin, and the western United States, are built to observe seasonal snow (Fig. 1). This is because sites are typ-

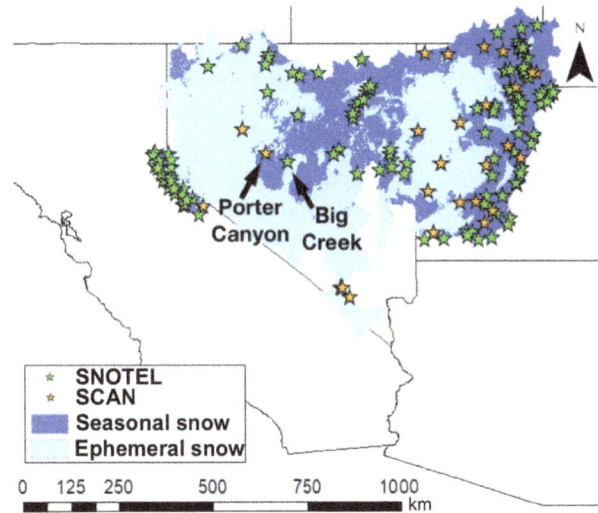

Figure 1. Locations of SNOTEL and SCAN stations in the Great Basin, USA, in ephemeral and seasonal snow as defined by < 60 or ≥ 60 days of maximum consecutive snow duration, respectively.

ically placed in topographically sheltered forest gaps that retain snow longer than nearby terrain. This improves the skill of streamflow forecasting, the primary goal of the SNOTEL network, but means that most SNOTEL sites only have ephemeral snow cover in exceptionally dry or warm years (Serreze et al., 1999). Only 2 of the 131 SNOTEL stations in the Great Basin experienced an ephemeral snow season on average (Fig. 1) each water year from 2005 to 2014. The scarcity of ground-based ephemeral snow and soil moisture data has changed slightly in recent years with additional measurements at the NRCS Soil Climate Analysis Network (SCAN) (Fig. 1) and increased deployment in research watersheds (Anderton et al., 2002; Jost et al., 2007). On average, 26 out of 39 SCAN stations in the Great Basin experienced ephemeral snow cover each year (Fig. 1). However, the lack of field observations from ephemeral snowpacks with co-located soil moisture has limited previous investigations (e.g., Sturm et al., 2010).

Spectral remote sensing collects observations over all cloud-free areas but has its own sets of advantages and challenges for observing ephemeral snow. One issue is that there are multiple methods to define the start and end of the observed snow-covered period. Often, it is defined as the date of the first and last remotely sensed observations of snow cover (e.g., Choi et al., 2010; Kimball et al., 2004; Nitta et al., 2014). Because this approach does not account for intermittent snow-free periods, it tends to overestimate snow duration and miss important ephemeral dynamics (Thompson and Lees, 2014). Snow persistence thresholds can be used to define snow ephemerality, but no standard persistence threshold exists (e.g., Gao et al., 2011; Karlsen et al., 2007). Given the intermittent nature of ephemeral snow, observations must be daily or finer to capture its dynamics (Wang et al., 2014). Consequently, products like Landsat that has a 16-day overpass and Sentinel that has 5–10-day overpass do poorly at estimating snow seasonality compared to products like the MODIS that have twice daily overpass, but they offer untapped potential for merged products with higher spatial and temporal resolution. Moreover, high cloud cover reduces observation frequency and limits the ability to observe ephemeral snow events. Like with ground-based snow research, some remote-sensing-based studies exclude ephemeral events altogether (e.g., Sugg et al., 2014). Only a limited number of algorithms have been developed to handle ephemeral snow specifically. For example, the algorithm developed by Thompson and Lees (2014) uses daily MOD10A1 data and accounts for snow absences in the middle of the snow season, but their study was challenging to verify and applied only in a small area of Australia. Given the current lack of ground-based observations (Fig. 1), there is great potential to use finer-scale satellite products and employ more refined methods targeted at areas with ephemeral snow.

There are a variety of underlying processes that cause ephemeral snowpack and challenge snow models. Based on previous classification systems, we define three mechanisms causing ephemeral snowpacks: (1) rainfall limiting the accumulation of snowpack, (2) snowpack ablation from melt or sublimation, and (3) wind scour removing snowpacks. All of these mechanisms have a variety of underlying atmospheric and snowpack processes that challenge prediction with snow models. At rain–snow transition elevations, even small temperature variations and other atmospheric variables can alter the mixture of rainfall and snowfall (Harpold et al., 2017b; Jefferson, 2011; Klos et al., 2014). Complete snow water equivalent (SWE) removal from melt or sublimation is also another common cause of snow ephemerality (Clow, 2010; Leathers et al., 2004; Mote et al., 2005; Sospedra-Alfonso and Merryfield, 2017). Typically, physics-based models overestimate modeled SWE in ephemeral snowpack, due to neglect or underestimation of ground heat flux and the challenges of tracking cold content in shallow snowpacks (Cline, 1997; Hawkins and Ellis, 2007; Kelleners et al., 2010; Kormos et al., 2014; Tyler et al., 2008; Şensoy et al., 2006; Slater et al., 2017). Models parameterize energy fluxes differently, which can lead to differences in model estimates of sublimation and melt (Essery et al., 2009; Sospedra-Alfonso et al., 2016; Schmucki et al., 2014). Removal of snowpack from wind scour is an important control on snow accumulation in alpine regions but is often neglected in models altogether (e.g., Mernild et al., 2017; Pomeroy, 1991; Winstral et al., 2013). Widespread evidence exists that wind redistribution of snow can cause ephemeral snowpacks that are consistent from year to year because of topography and dominant wind directions (Hood et al., 1999). The three mechanisms causing ephemeral snow (i.e., rain–snow transition, ablation by sublimation and melt, and wind scour) have fundamentally different underlying causes, with variable and poorly quantified sensitivities to climate and land cover variability.

The goal of this paper is to use the Great Basin as a case study to estimate the distribution and mechanisms causing ephemeral snow to better constrain their impact on soil moisture and hydrological response. We adapt the classification from Sturm et al. (1995) to map snow and soil moisture response across the Great Basin, compare remotely sensed and modeled estimates of ephemeral snow, and develop our own metrics to further classify snow seasonality. The Great Basin is ideal for this investigation because it spans dramatic gradients of elevation and hydroclimatology with large areas of both seasonal and ephemeral snow. This prototypical area depends disproportionately on mountain snowpack for water supplies, contains few ground-based observations, and there is relatively little winter cloud cover to limit spectral remote sensing techniques. Three research questions guide our analyses of ephemeral snowpacks in the Great Basin. (1) What are the implications for soil moisture from seasonal to ephemeral snowmelt? (2) How does topography affect snow seasonality? And (3) what mechanisms cause ephemeral snowpacks and how does that vary with climate? We find that ephemeral snow originates from melt and shifts to lower-elevation rain–snow transitions during warm winters, which leads to a fundamentally different soil moisture response than from seasonal snowmelt.

2 Study area

The Great Basin is the closed basin between the Wasatch and southern mountain ranges in Utah and the eastern slope of the Sierra Nevada mountain range in California. The region is known for having "internal drainage", which means that none of the waterways travel to the ocean (Svejcar, 2015). The climate is semi-arid and the ecosystem is shrub-dominated (Svejcar, 2015; West, 1983). We defined the Great Basin region based on the hydrologic unit code (HUC) Region 16 adapted from Seaber et al. (1987) by the United States Geological Survey (USGS) (Fig. A1). Precipitation in the Great Basin varies widely between < 10 cm in many of the lower elevations and > 100 cm in many of the high-elevation mountains (Fig. A2). Overall, the Great Basin has a mean winter (defined as 1 December to 1 April) precipitation of 12 cm and a mean winter temperature of 0.4 °C (Fig. A2; Abatzoglou, 2012).

3 Methods

In order to compare the effect of snow ephemerality on soil moisture patterns, we first investigated snow and soil moisture response for SNOTEL and SCAN stations within the Great Basin. To evaluate how soil moisture varies based on snowpack parameters during a drought year (water year

2015) and a non-drought year (water year 2016), we chose two SNOTEL stations – Porter Canyon (ID 2170, elevation 2191 m) and Big Creek Summit (ID 337, elevation 2647 m) – that differ in elevation but are in close proximity. We used average snow water equivalent (SWE) data from snow pillows to determine snow cover. We categorized each day as snow covered if continuous SWE was greater than 0.1 cm. We then designated site years as seasonal or ephemeral depending on if continuous snow cover was greater or less than 60 days, respectively. For these stations, we compared percent soil moisture, at 5 and 50 cm soil depth along with snow depth, and SWE. We then also acquired soil moisture and SWE data at 5 and 50 cm for all the SNOTEL and SCAN stations in the Great Basin in water years 2014–2016 and categorized site years from those stations as ephemeral or seasonal. We discarded years and stations containing more than 7 days of continuous missing data or soil moisture values that were 0 %. To compare the timing of snow and peak soil moisture, we then took the difference between the day of last snow and the day with peak median 10-day soil moisture for each year at each site. It should be noted that ablation on the snow pillow may be impacted by differences in ground heat flux and co-location issues with the soil moisture sensors. We also calculated the coefficient of variation (CV; 1 standard deviation divided by the mean) of soil moisture for each year at each station.

We mapped ephemeral snow across the Great Basin using two methods: spectral remote sensing with MODIS data and modeled Snow Data Assimilation System (SNODAS) data. We used Google Earth Engine to analyze the data, which is a cloud-based computing platform optimized for mapping large datasets (Gorelick et al., 2017). The MODIS dataset used was the 2010 MODIS/Terra Snow Cover Daily L3 Global 500 m Grid (MOD10A) and we used the normalized difference snow index (NDSI) with parameters outlined in Hall et al. (2006) to find fractional snow-covered data. The equation for calculating NDSI in MOD10 is

$$NDSI = \frac{Band\ 4 - Band\ 6}{Band\ 4 + Band\ 6}. \tag{1}$$

A pixel is then mapped as containing fractional snow using the NDSI value, as long as the reflectance in Band 2 is > 10 % (Hall et al., 2001). We classified all pixels with a snow fraction of 30–100 as *snow*, pixels with snow fractions between 0 and 30 as *no snow*, and pixels that had all other designations as *other*. We also used an algorithm derived from Thompson and Lees (2014) to minimize the impact of cloud cover in our MODIS data. The algorithm "grows" the boundaries of all areas containing snow and reclassifies pixels that were classified as *other* to *snow* if the corresponding pixels in the previous image were classified as *snow*. It also reclassifies pixels that were classified as *other* to *no snow* if the corresponding pixels in the previous image were *no snow*.

To determine the number of ephemeral and seasonal snow events, we used a Google Earth Engine function to note the

day of the water year when snow appeared (when a pixel went from being classified as *no snow* in the previous day to classified as *snow* in the current day) and when snow disappeared (a pixel went from being classified as *snow* in the previous day to being classified as *no snow* in the current day), and we determined the length of snow cover by subtracting the day of snow appearance from the day of snow disappearance. If the length of snow cover was < 60 days, then the snow event was classified as ephemeral. Otherwise, if the length of snow cover was ≥ 60 days, the snow event was categorized as seasonal. In addition to these metrics, we derived a snow seasonality metric (SSM) to quantify a MODIS pixel's tendency to have ephemeral or seasonal snow, rather than a binary metric like < 60 days. The SSM is depicted in Eq. (2) and it works by classifying every day where there was seasonal snow present as 1 and every day where there was ephemeral snow present as −1, and then averaging all −1 and +1 values. This created a −1 to 1 scale, where −1 signifies that all the snow-covered days in a given pixel within 1 water year were ephemeral and +1 signifies that they were all seasonal.

$$SSM = \frac{Days_{Seasonal} - Days_{Ephemeral}}{Days_{Total}} \tag{2}$$

Additionally, we discarded all instances where snow was absent for 1 day only from the overall record of snow disappearance and appearance because there were numerous artifacts from the MOD10A NDSI processing that lead to single-day snow disappearance during long stretches of snow cover. The 1-day snow events were also removed from the SNODAS algorithm to make both algorithms more consistent. For each water year from 2005 to 2014, we recorded the maximum total number of days where snow was present (to be referred to as the maximum snow duration).

To determine the relationship between elevation and snow seasonality, we took the average maximum snow duration across water years 2005–2014 and used elevation and aspect as measured by a digital elevation model (DEM) obtained from the Shuttle Radar Topography Mission resampled to the same resolution with bilinear sampling (Farr et al., 2007). To calculate northness, we used the following equation:

$$Northness = \cos\left(\frac{aspect \cdot \pi}{180}\right). \tag{3}$$

We then categorized each MODIS pixel based on five 500 m elevation bins from a range of 1000 to > 3000 m. Then, to remove bias based on the size of each bin, we used random sampling to make each bin contain the same number of points as the least full bin (13 548 points that were > 3000 m). Then we combined each resampled bin into one dataset and created heat maps to compare the elevation vs. the average maximum snow duration. We also use the same method to compare aspect to average maximum snow duration using eight 45° bins from a range of 0 to 360°. We randomly sampled 195 163 points from each bin (with the size

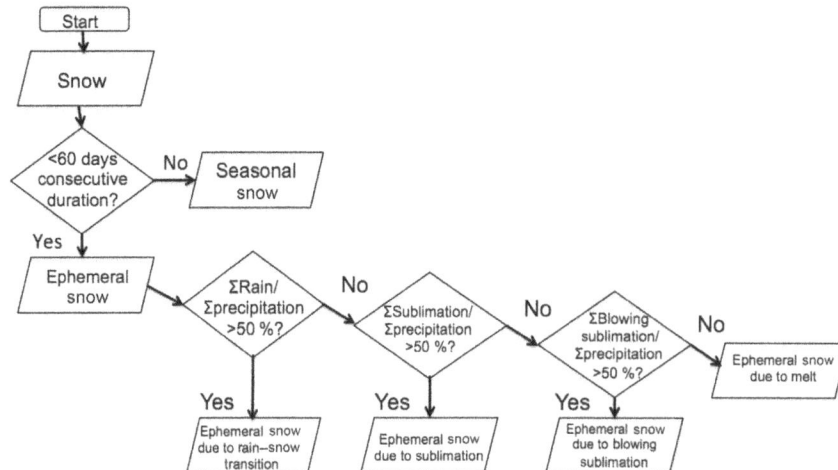

Figure 2. Diagram of the process for the ephemeral snow mechanism model. Seasonal snow outputs were rejected, and all other outputs were categorized.

of the bin ranging from 315 to 360°). After resampling, we combined all the bins together and split them into three elevation categories: low elevation (elevation < 1500 m), medium elevation (1500 ≥ elevation < 2500), and high elevation (elevation ≥ 2500 m). Then, we resampled again to 82 823 points per bin (the size of the high-elevation bin).

We used SNODAS data to differentiate the mechanisms that cause snow to become ephemeral. The four mechanisms were assigned if the net ablation (or rain) exceeded 50 % of the total winter precipitation (Fig. 2): (1) a mixture of rain and snow limiting snow accumulation (the rain–snow transition), (2) snowpack loss due to sublimation, (3) snowpack loss due to melt, and (4) snowpack loss due to wind scour. We determined the prevailing mechanism in each 1000 m SNODAS pixel in each year. We used Google Earth Engine to execute the modeled algorithm on each 1000 m SNODAS pixel in the Great Basin. We then chose 6 years (2009–2014) and created histograms of each mechanism by elevation for each year.

4   Results and discussion

4.1   Ephemeral snow and soil water inputs

In order to quantify differing soil moisture responses between seasonal and ephemeral snowpacks that have important ecohydrological implications for the Great Basin, we use the five phases in the McNamara et al. (2005) framework for soil moisture response to seasonal snowmelt. First, we qualitatively compare two nearby sites with differing snow regimes. Second, we make quantitative analyses using all of the soil moisture records available in snow-covered places of the Great Basin (Fig. 3).

We contrast soil moisture response at two adjacent SNO-TEL stations that differ in elevation by > 500 m (Fig. 1)

to illustrate differences between ephemeral and seasonal snowmelt. Soil moisture at 5 and 50 cm depth was used to represent shallow and deep responses during a drought year (water year 2015) and a typical year (water year 2016). Porter Canyon had ephemeral snow (28 days maximum duration) in 2015 and seasonal snow (116 days) in 2016 (Fig. 3a). Big Creek had seasonal snowpack both years, although much shallower snowpack in 2015 (Fig. 3b). When seasonal snowpack is present at both sites in 2016, soil moisture follows the phases outlined by McNamara et al. (2005) for a semi-arid, snowmelt-driven environment. Shallow and deep soil moisture was in a low-flux state during December–February (DJF) at Big Creek in 2016 (Fig. 3f). During March–May (MAM), soil moisture increased substantially and was in a high-flux state. Average shallow soil moisture in 2015 and 2016 was similar in the MAM period (24.4 % and 24.8 %, respectively) and DJF period (11.3 % and 19.8 %), suggesting that snow storage and melt negates differences in early season soil moisture between years with very different winter precipitation. Porter Canyon also showed a similar soil moisture increase in the MAM period after a stable low-flux pattern in the DJF period during water year 2016. Both sites also reach their near maximum annual soil moisture coincident with snow disappearance in 2016 (Harpold and Molotch, 2015), but Porter Canyon has snow disappearance in both years that preceded peak soil moisture by several months. The deeper 50 cm soil moisture had a smaller and shorter peak during 2015 at Porter Canyon as compared to 2016 and the Big Creek response.

Using similar records to those illustrated at these two sites, we use 328 site years (50 ephemeral and 278 seasonal site years) from all SNOTEL and SCAN sites in the Great Basin (Fig. 1) over water years 2014, 2015, and 2016 to illustrate the broader patterns of soil moisture response to ephemeral and seasonal snowmelt. We found that soil

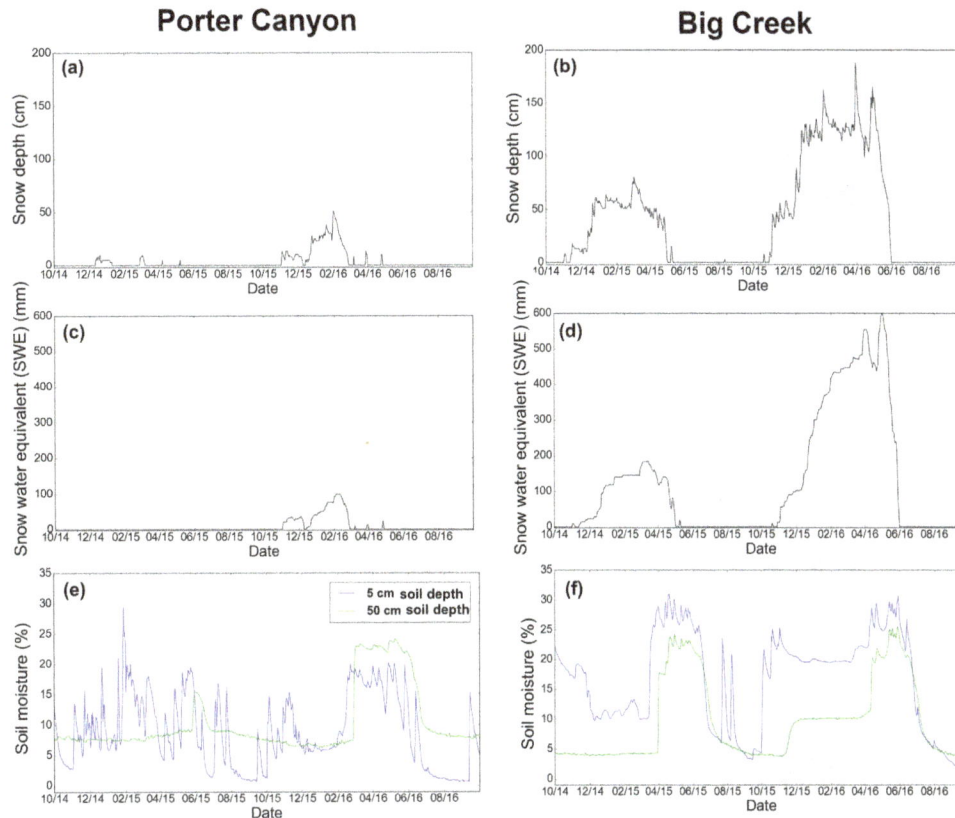

Figure 3. (a, b) Snow depth, **(c, d)** snow water equivalent, and **(e, f)** soil moisture measured at Porter Canyon and Big Creek Snow Telemetry (SNOTEL) stations for water years 2015–2016, which were a drought year and a typical year, respectively.

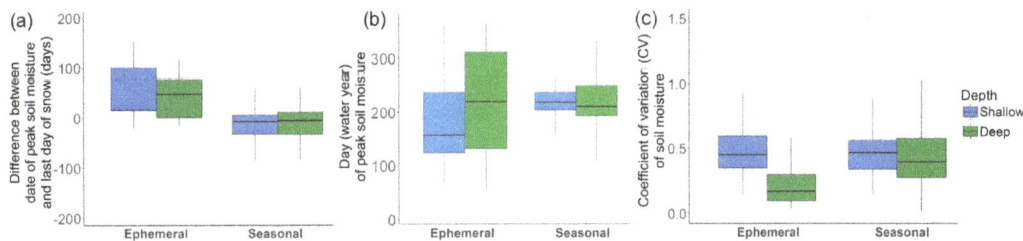

Figure 4. (a) The difference between date of peak soil moisture and last day of snow (days) for shallow (5 cm) and deep (50 cm) soil moisture during water years 2014–2016 in Great Basin SNOTEL stations with ephemeral snow (50 site years) and seasonal snow (278 site years). **(b)** Day of peak soil moisture for SNOTEL and SCAN stations for shallow (5 cm) and deep (50 cm) soil moisture during water years 2014–2016. **(c)** The coefficient of variation (CV) for shallow (5 cm) and deep (50 cm) soil moisture during water years 2014–2016

moisture following seasonal snowmelt reached a maximum 5 and 7 days prior to snow disappearance for shallow and deep soil moisture, respectively. This confirms previous findings that seasonal snowmelt drives coincident wetting and deeper water percolation (Harpold and Molotch, 2015; McNamara et al., 2005). In contrast, the median soil moisture peaked 79 and 48 days after of snow disappearance from ephemeral snowmelt for shallow and deep soil moisture, respectively (Fig. 4a). This is consistent with the peak shallow soil moisture occurring much earlier in the water year in shallow ephemeral snowmelt areas (Fig. 4b). The later deep

soil moisture response in ephemeral areas reflects the lack of response, or low coefficient of variation, as compared to seasonal snowmelt (Fig. 4c). The lower CV for deep ephemeral snowmelt (0.2) compared to deep seasonal snowmelt (0.4–0.5) is indicative of reduced deep percolation and less water becoming available to groundwater and streamflow.

The differences in soil moisture response between seasonal and ephemeral snowpacks across the Great Basin could have important consequences for vegetation phenology and runoff generation. For example, the timing of soil moisture is a strong control on the timing and amount of net ecosys-

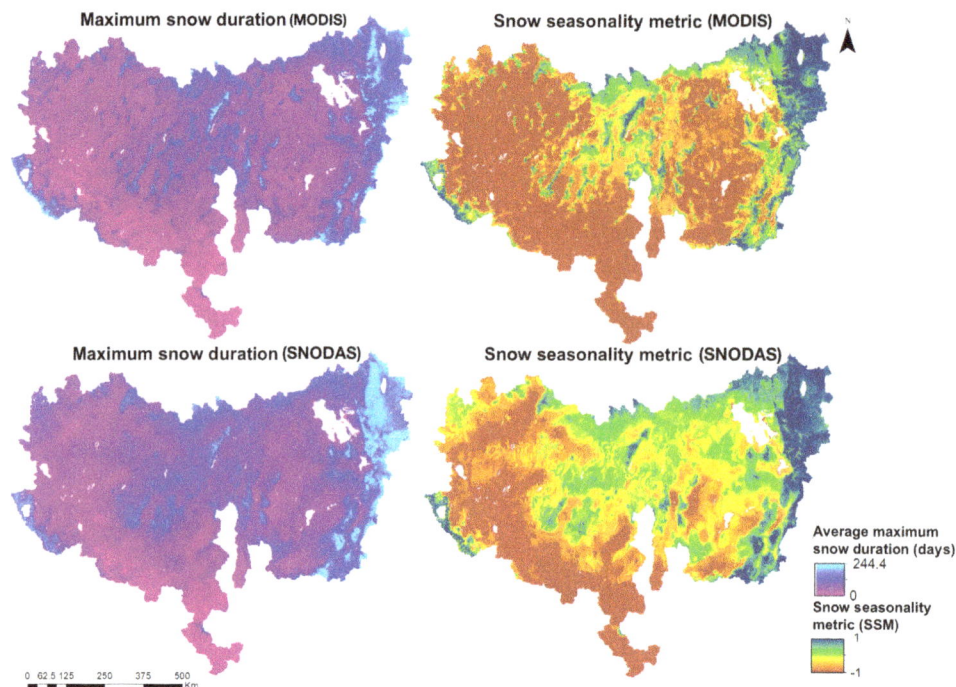

Figure 5. The average maximum consecutive snow duration (maximum snow duration) and snow seasonality metric (SSM) for the Great Basin measured using MODIS and Snow Data Assimilation System (SNODAS) data in the Great Basin, USA, for water years 2005–2014.

tem productivity (Inouye, 2008), with earlier snowmelt causing an earlier and longer growing season with reduced carbon uptake (Hu et al., 2010; Winchell et al., 2016). Harpold (2016) also showed that earlier snow disappearance generally led to more days of soil moisture below wilting point at SNOTEL sites. Our finding that soil moisture peaked earlier in ephemeral snowmelt than seasonal snowmelt is thus likely to be correlated with reduced vegetation productivity and increased late season water stress in many areas. In addition to stressing local vegetation, ephemeral snowmelt may reduce groundwater recharge and streamflow. For example, baseflow contributions to streamflow and overall water yield declined when snowmelt rates were smaller (Barnhart et al., 2016; Earman et al., 2006; Trujillo and Molotch, 2014), and overall water yields were lower in basins receiving more rain and less snow (Berghuijs et al., 2014). Changes in percolation patterns also affect the distribution of more shallow rooting plants versus deeper rooting plants that need long duration soil moisture pulses to grow and reproduce (Schwinning and Sala, 2004). These differences in how ephemeral versus seasonal snowmelt affects soil moisture provide a strong motivation to understand the distribution and causes of ephemeral snowpacks across the Great Basin.

4.2 Topographic controls on snow seasonality

In a typical year, much the Great Basin experiences ephemeral snow (Fig. 5) that can only be comprehensively observed with remote sensing platforms because of the lack

of standard ground stations (Fig. 1). Using MODIS imagery, there are two new metrics to estimate snow ephemerality with daily snow cover products: (1) the maximum consecutive snow duration and (2) the snow seasonality metric. The SSM describes both the consecutive snow season length and shoulder-season ephemerality. A SSM value < 1 means an area experiences at least one ephemeral snow event. The average SSM was −0.4 (Fig. 5), suggesting that on average the Great Basin was dominated by ephemeral snow extent. Maximum consecutive snow duration can be compared to the Sturm et al. (1995) 60-day threshold for ephemeral snow, as done in this case, but it is flexible enough to include a threshold of any day length. The average maximum consecutive snow duration in the Great Basin from MODIS data was 42.1 days (Fig. 5). We found higher estimates of the average maximum consecutive snow duration measured using SNODAS of 62.9 days but a similar average SSM of −0.4 (Fig. 5). While the maps of the two products tend to produce similar results (Fig. 5), the SNODAS spatial patterns often miss finer-scale topographic controls (e.g., Wasatch mountains in the far eastern Great Basin) and overestimates snow durations in the colder, lower elevations (e.g., basins below the Ruby Mountains in the central Great Basin). In general, SNODAS overestimates snow duration in areas with the longest and shortest snow durations, i.e., highest and lowest elevations (Fig. 6). In these critical water supply areas > 2500 m, where snow would persist for > 150 days according to MODIS, the SNODAS estimates were often biased by > 50 days (Fig. 6). We explore the challenges of coarse, physically based models, such as SNODAS, later in this paper.

We investigate elevation and aspect as proxies for snowpack mass and energy dynamics in order to expand our understanding of snow ephemerality. Elevation is a primary control on near-surface air temperature due to the adiabatic lapse rate (Bishop et al., 2011; Greuell and Smeets, 2001; Nolin and Daly, 2006). Prior research has found that there is a strong elevation dependence on snowmelt timing, runoff generation, snow water equivalent, and snow season length (Hunsaker et al., 2012; Jefferson, 2011; Jost et al., 2007; Molotch and Meromy, 2014). Elevation effects are likely due to a variety of factors, including temperature controls on the rain–snow transition, longwave radiation in cloudy areas, and sensible heat flux. Aspect is often a secondary control on snow distributions because it influences incoming shortwave radiation (Jost et al., 2007; Pomeroy et al., 2003) and wind patterns (Knowles et al., 2015; Leathers et al., 2004; Winstral et al., 2013). Shortwave radiation is the primary driver of ablation via melt and sublimation (Cline, 1997; Marks and Dozier, 1992).

Dividing the Great Basin into low elevations (< 1500 m), mid-elevations (1500–2500 m), and high elevations (> 2500 m) illustrated elevation's dominant role on snow cover duration (Fig. 7). Across the Great Basin, 96.2 % of the low-elevation area and 75.2 % of the mid-elevation area had a maximum consecutive snow duration of < 60 days. Conversely, only 10.5 % of high elevations had a maximum consecutive snow duration of < 60 days (Fig. 7). The results suggest that mid- and low elevations of the Great Basin are more likely to be ephemerally dominant. The heat maps also illustrate that elevation alone is not a strong predictor of maximum consecutive snow cover days (Fig. 7). We use three smaller mountain ecoregions (Fig. A1) to illustrate variability in elevation effects (Fig. 8). There were similar average maximum snow duration values in the Ruby Mountains (Fig. 8a), eastern Sierra Nevada (Fig. 8b), and western Wasatch–Uinta ecoregion (Fig. 8c) (107, 100, and 95 days, respectively). However, snow in the Ruby Mountains persisted longer than the Sierra Nevada and Wasatch–Uinta ecoregions. The Sierra Nevada ecoregion had a weaker relationship between snow persistence and elevation above 2500 m, while the Wasatch–Uinta ecoregion had a weaker relationship with elevation below 2500 m (Fig. 8). These differing relationships between maximum snow duration and elevation suggest other factors are affecting snow ephemerality.

Aspect is also an important control on snow seasonality in the Great Basin, but its importance is limited to mid- and high elevations. We find that there are shorter maximum snow durations in south-facing aspects at elevations > 1500 m (Fig. 9). At low elevations, the difference in average maximum snow duration between north- and south-facing slopes was 0.4 days, while for mid- and high elevations,

it was 2 and 5 days, respectively (Fig. 9). This is consistent with aspect strongly controlling solar radiation, which is the main energy input to the snowpack. This suggests that deeper, high-elevation snowpacks ablate in response to greater solar radiation and corresponding warmer temperature on south-facing hillslopes (Hinckley et al., 2014; Kormos et al., 2014). In contrast, lower-elevation areas appear to have maximum snow duration caused by factors other than aspect. This is consistent with the outsized importance of other energy fluxes and factors, like ground heat flux and rain–snow transition elevation, that are not captured by aspect and elevation (Figs. 7, 8 and 9).

4.3 Proximate mechanisms controlling snow ephemerality

We propose a three-mechanism classification scheme to help frame our understanding of snow ephemerality: (1) rain–snow transitions limit snow accumulation, (2) snowpack ablation from melt and sublimation, and (3) wind scour or redistribution. Probably the most explored and observed mechanism is the potential for rising rain–snow transition elevations to limit snow accumulation and duration (Bales et al., 2006; Klos et al., 2014; Knowles and Cayan, 2004; Mote, 2006). Reduction in snow duration can also be caused by the melt of snowpack (Mote, 2006) and losses from sublimation (Harpold et al., 2012; Hood et al., 1999); however, much less is known about the role and distribution of these processes outside of the seasonal snowpack zone. Finally, wind scour can reduce snowpacks by redistributing it to other areas or by increasing blowing wind sublimation (Knowles et al., 2015; Leathers et al., 2004).

We chose 6 years to evaluate the dominant mechanisms causing snowpack ephemerality using a new classification system (Fig. 2) based on SNODAS data that compared favorably to estimates from MODIS (Figs. 5 and 6). In that 6-year period, the year with the lowest average winter (1 December to 1 April) temperature using gridded meteorological (GRIDMET) 4 km resolution surface temperature estimates was 2013 at −0.9 °C, while the year with the highest average winter temperature was 2014 at 1.0 °C (Abatzoglou, 2012; Table 1). In water year 2013 and water year 2010, the two coldest years, seasonal snowpacks were dominant in most of the Great Basin and western United States (Figs. 10–11). In the coldest years of 2010 and 2013, the rain–snow transition and melt caused ephemerality to shift lower in elevation (Fig. 11). In the warmest year of 2014, seasonal snowpack was lowest at lower elevations throughout the western US mountain ranges (Fig. 10), including the Great Basin where the increase in ephemeral snowpacks at higher elevations was due primarily to a rain–snow mechanism (Figs. 10 and 11). Melt-caused snow ephemerality also increased in the warm 2014, but ephemeral snow remained sparse above 2500 m in all years. Overall, our findings are consistent with the importance of variability in rain–snow transition elevations limit-

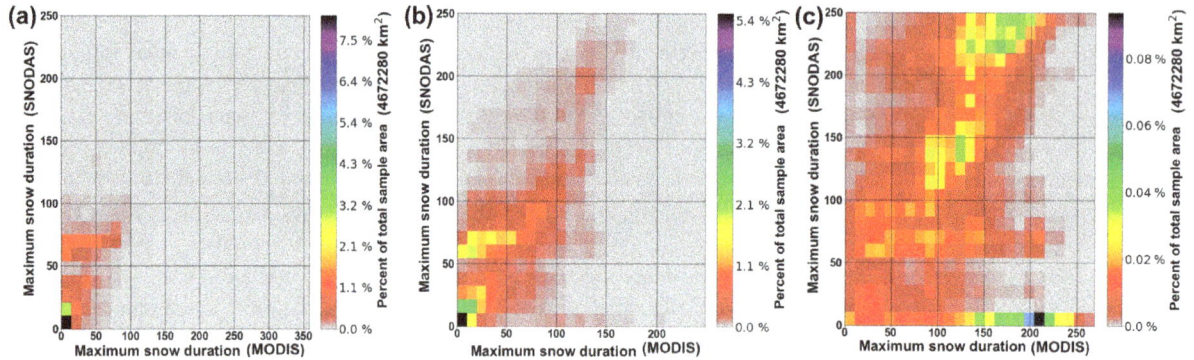

Figure 6. Maximum consecutive snow duration (maximum snow duration) measured using MODIS and Snow Data Assimilation System (SNODAS) data at **(a)** low elevations (0–1500 m), **(b)** medium elevations (1500–2000 m), and **(c)** high elevations (2000 m +).

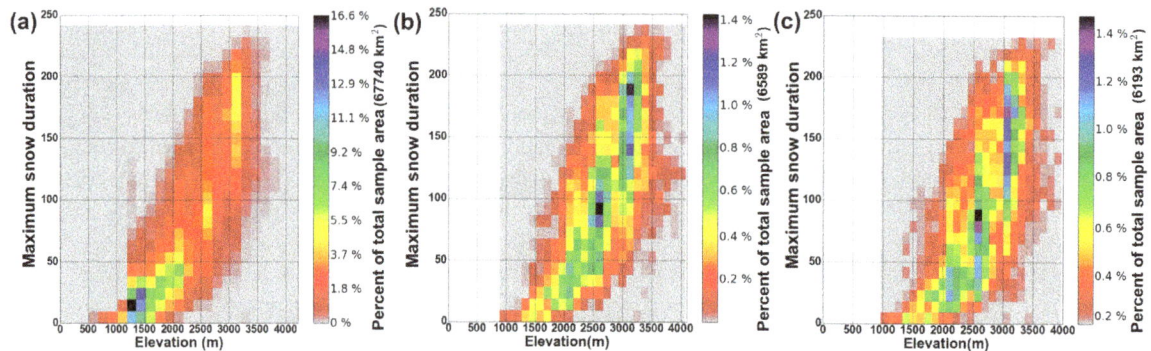

Figure 7. Heat maps of the relationship between elevation and average maximum consecutive snow duration (maximum snow duration) from MODIS at **(a)** all slopes, **(b)** north-facing slopes only, and **(c)** south-facing slopes only in the Great Basin, USA. North facing was defined as northness > 0.25 and south facing was defined as northness < −0.25. Color bar scale is different in panel **(a)**, reflecting the much larger area at low elevation.

ing snow accumulation and duration (Bales et al., 2006; Klos et al., 2014; Knowles and Cayan, 2004; Mote, 2006). Sublimation was only present as a limiting mechanism in 2010 and only for a small area (Fig. 10). Blowing snow sublimation was not the dominant cause of snow ephemerality in the Great Basin for any year; SNODAS struggles to represent wind redistribution of snow (Clow et al., 2012; Hedrick et al., 2015). Our approach to classify proximate causes of snow ephemerality has some limitations. Namely, it assigns only a single mechanism to each grid cell when there could be multiple mechanisms. Moreover, the method cannot consider changes in the mechanisms with time (e.g., melt tends to occur more in spring) because we applied annualized estimates of snow cover duration and concerns about the fidelity of the SNODAS model at short timescales.

The mechanisms causing snow ephemerality that can be inferred from the SNODAS model have important implications for water availability in the Great Basin, but there is less confidence in the model fidelity in these shallow snowpacks given their differences with the MODIS observations (Fig. 6). These limitations are present in all snowpack energy models because the models were developed for deeper

snowpacks where terms like ground heat flux and albedo depth relationships can be ignored or are insensitive (Cline, 1997; Harstveit, 1984; Liang et al., 1994; Tyler et al., 2008; Slater et al., 2017). In shallow snowpacks, these terms are more critical (Hawkins and Ellis, 2007; Şensoy et al., 2006; Slater et al., 2017; Tyler et al., 2008), and the lack of SWE means the internal energy state of the snowpack (i.e., cold content) is more easily varied by short-term climate forcing (e.g., warm, sunny days) (Liston, 1995). Ephemeral snowpacks also exist at lower elevations with warmer soils and increased ground heat flux (Slater et al., 2017; Tyler et al., 2008). Uncertainty in the rain–snow transition principally arises from predicting climate forcing and in particular temperature and humidity in places like the Great Basin (Harpold et al., 2017a). However, the underlying phase prediction method and related model decisions and climate forcing data can also be important for the quality of precipitation phase prediction (Harpold et al., 2017b). Further complicating rain–snow transition mechanisms is the storage or drainage of liquid water on existing snowpacks (Lundquist et al., 2008; Marks et al., 2001). Although SNODAS assimilates MODIS imagery into the model, it does not appear

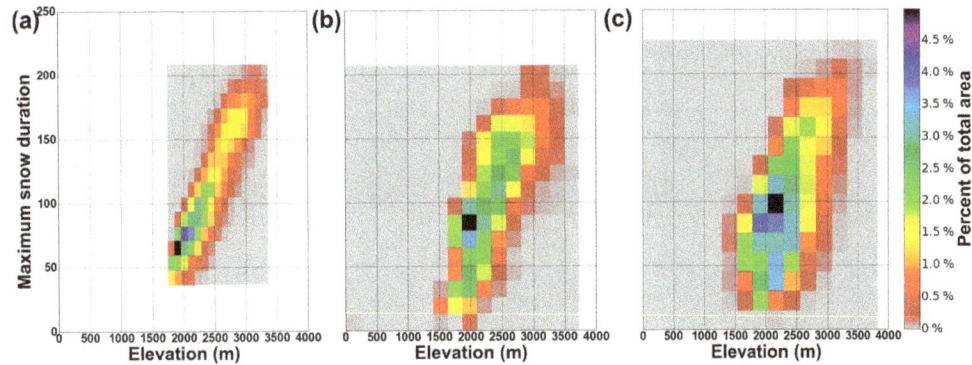

Figure 8. Heat maps showing the relationship between elevation and average maximum consecutive snow duration (maximum snow duration) for three seasonally dominant ecoregions in the Great Basin: **(a)** the Ruby Mountains, **(b)** the Sierra Nevada mountains, and **(c)** the Wasatch–Uinta Mountains.

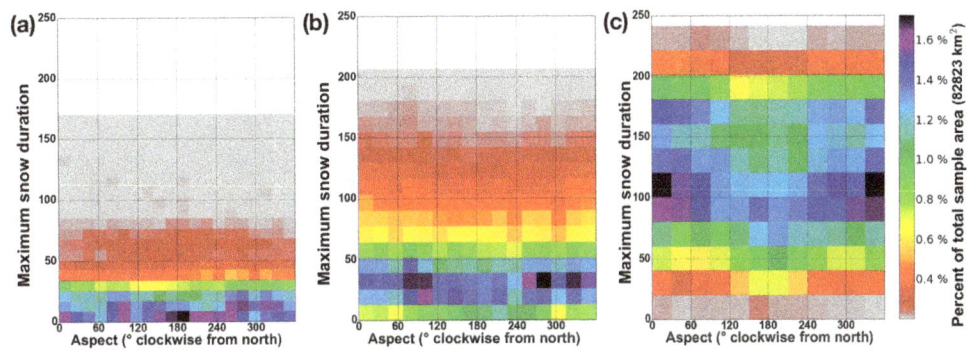

Figure 9. Heat maps of the relationship between aspect and average maximum consecutive snow duration (maximum snow duration) at **(a)** low elevations (0–1500 m), **(b)** medium elevations (1500–2500 m), and **(c)** high elevations (2500 m+).

to capture the finer elevation patterns we found using the MOD10A product (Figs. 5 and 6) and, in particular, seemed to overestimate consecutive days of snow cover. Part of the challenge at higher elevations is modeling blowing snow patterns over 1 km grid cells, which gives consistent lower accuracy of SNODAS above the tree line and in more windy areas (Clow et al., 2012; Hedrick et al., 2015). The Great Basin shows tremendous variability in snow ephemerality caused by interactions of topography, elevation, and prevailing wind (Figs. 10–11) and, thus, represents an area where improvements in the physically based modeling will be critical to predicting snow water resources under a variable and changing climate.

5   Conclusions

Mapping, measuring, and modeling ephemeral snow is challenging with current techniques, but it is vital for understanding future water resources and vegetation water use. Ephemeral snowpacks do not have distinct accumulation and ablation periods, which means the timing of soil moisture input varies and is more challenging to predict than seasonal snowmelt (e.g., McNamara et al., 2005). Consequently,

as snowpacks shift from seasonal to ephemeral, there are potential ecohydrological consequences such as changes to vegetation response, vegetation distribution, drainage, lateral water flow, and solute transport. Our work shows that, while topography and climate variability have strong controls on the distribution of ephemeral snowpacks (Figs. 7 and 10), those factors will not be sufficient for predicting snow ephemerality under varying climate. Instead, there is a need for physics-based models capable of capturing the three broad mechanisms identified by this study: (1) rain–snow transitions limit snow accumulation, (2) snowpack ablation from melt and sublimation, and (3) wind scour and redistribution. These classifications could help better identify local and regional sensitivity to increased snow ephemerality (Figs. 10 and 11). This work has also highlighted major weaknesses in the observational infrastructure, data analysis, and modeling techniques needed to support the growing importance of ephemeral snowpacks in the Great Basin. In light of these diverse needs, we conclude with a short summary of recommendations meant to guide future research directions.

– Improving and standardizing snow ephemerality metrics: our research suggests there is a snow duration threshold where snowpack and soil moisture pat-

Table 1. Average winter (1 December–1 April) temperature (°C) and average elevation (m) for both dominant mechanisms of snow ephemerality and seasonal snow from 2009 to 2014 in the Great Basin.

Water year	Average winter temperature (°C)	Mean elevation for rain–snow transition (m)	Mean elevation for melt (m)	Mean elevation for seasonal snow (m)
2009	0.1	1806	1751	1728
2010	−0.6	1811	1747	1761
2011	−0.2	1803	1766	1700
2012	0.4	1803	1745	1710
2013	−0.9	1816	1710	1754
2014	1.0	1790	1749	1732

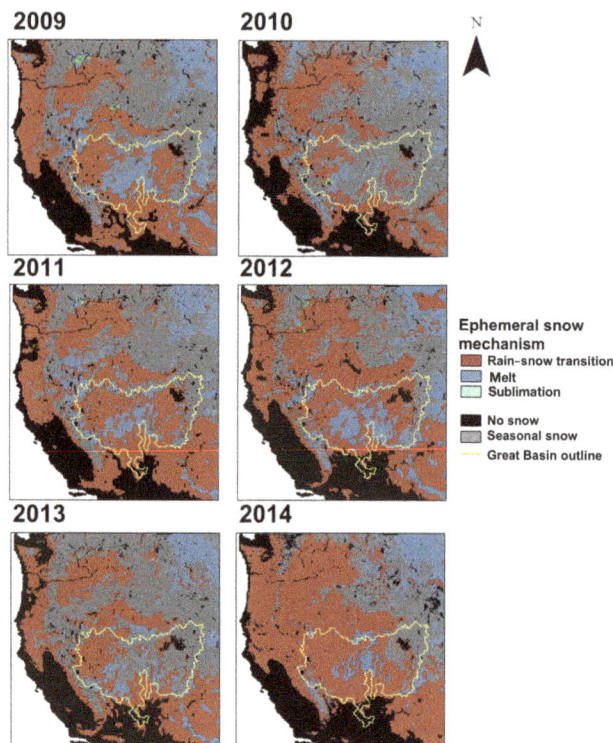

Figure 10. Dominant mechanisms for snow ephemerality from water years 2009–2014 in the western United States. Areas with seasonal snow (grey), no snow (black), and water bodies (black) are also depicted. The Great Basin region is outlined in yellow.

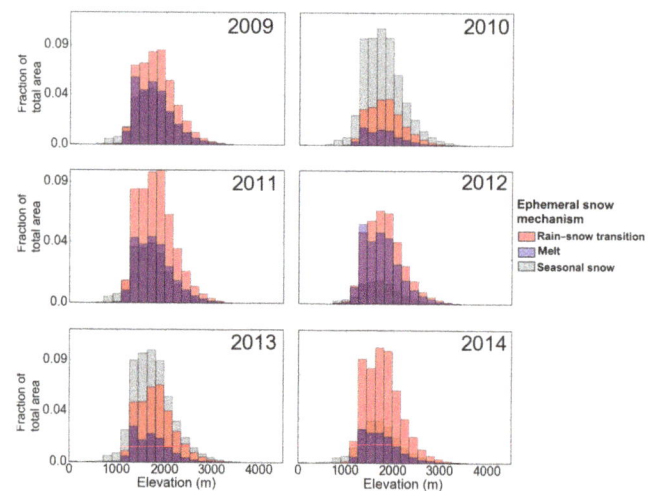

Figure 11. Histograms of the relationship between elevation and the dominant mechanisms for snow ephemerality in the Great Basin from water years 2009–2014.

terns begin to resemble seasonal instead of ephemeral snowmelt and perhaps a second threshold when they begin to resemble rain (Fig. 3). Yet evidence that this threshold is near the 60 days used in Sturm et al. (1995), or consistent across space, is lacking. Instead of using this arbitrary 60-day threshold, it is recommended that future research use the snow properties and soil moisture response of ephemeral snowpacks combined with a sensitivity analysis to create a snow duration threshold capable of differentiating seasonal and ephemeral soil moisture response (e.g., McNamara et al., 2005).

– Increasing snow and soil moisture observations in ephemeral areas: in the Great Basin, only 2 Snow Telemetry stations and 26 Soil Climate Analysis Network stations observe ephemeral snowpacks (Fig. 1). The lack of observations makes it more difficult to develop relationships between snowmelt and soil moisture. To help develop better criteria for categorizing snowpack as ephemeral, we need more snow and soil moisture observations in ephemeral areas. Also, observing both shallow and deep soil moisture can add significant hydrological inferences. We can then also use these observations to verify results derived from remote sensing and physically based models.

– Improved remote sensing algorithms: there is no consistent standard for defining the length of the snow-covered period. It is still common for papers to define the length of a snow-covered period by the first and last days of snow cover. This approach does not account for short-term snow disappearance between those days. Approaches that report the total number of snow-covered days miss information contained during shown snow-free periods. Additionally, there is no consistent algorithm for accounting for cloud cover and that may make these types of methods infeasible for some regions. More widespread use of the object-oriented techniques, like the one used in this study, is needed to evaluate their efficacy and accuracy across differing regions and snow regimes.

– Improved spatial resolution and fidelity of snow and climate data: the MOD10A data product has a spatial resolution of 500 m. The coarse resolution made it difficult to verify our ephemeral snow results with SNOTEL observations that use 3 m wide snow pillows. Topographic complexity leads to variations in climate on much finer resolutions than the 4000 m gridded meteorology data used for this analysis. Gridded snow and climate data should have a spatial resolution more consistent with the variability in snowpacks on the order of 10–100 m. While very fine resolution climate datasets are beginning to be produced, there is a large need to merge existing remote sensing snow observations into a data product that maximizes the current space and time resolutions across different remote sensing platforms (e.g., the spatial resolution of Sentinel 2 but the temporal resolution of MODIS).

– Improved physics-based modeling: identifying weaknesses in physically based models was not the objective of this study; however, it is clear this is a need for better prediction of snow ephemerality. Improving model parameterization of ground heat flux and ensuring the temporal model resolution is sufficient to capture rapid changes in cold content are two ways to improve these models. These improvements are contingent on new and better observations of mass and energy fluxes to support greater model fidelity in ephemeral snow.

Appendix A: Additional information about the study area and ephemeral snow algorithm

Figure A1 is an elevation map of the Great Basin, USA, showing key ecoregions and major cities. Figure A2 is a map of average winter (1 December–1 April) temperature, precipitation, and radiation across water years 2001–2015. Figure A3 shows how the measured number of ephemeral and seasonal snow events at SNOTEL sites corresponded to the number derived from the ephemeral snow algorithm. Figure A4 shows how the 30 % snow fraction was chosen using a sensitivity analysis.

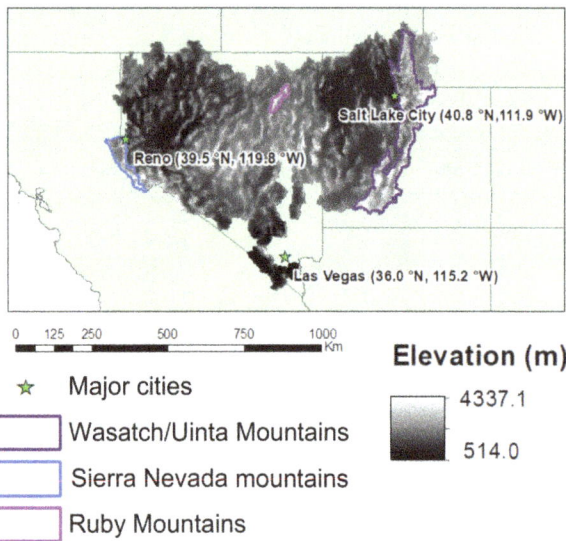

Figure A1. Map of the Great Basin region, USA, as defined by the USGS HUC Region 16 along with major cities and mountain ranges. The Sierra Nevada, Ruby, and Wasatch–Uinta mountain ranges are highlighted.

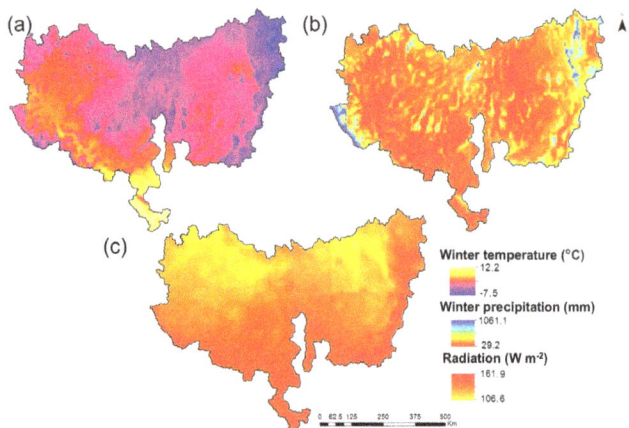

Figure A2. (a) Average winter temperature, **(b)** average winter precipitation, and **(c)** average winter radiation across water years 2001–2015 in the Great Basin.

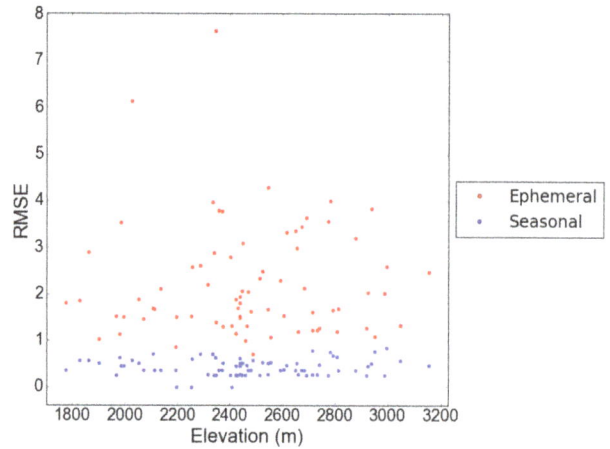

Figure A3. Root mean square errors between the number of observed ephemeral and seasonal snow events at Snow Telemetry (SNOTEL) stations and the number of ephemeral and seasonal snow events derived from the algorithm in Google Earth Engine in each 500 m MODIS pixel corresponding to that station. Measured snow water equivalent of 0.3 cm or greater was used to determine snow presence for SNOTEL sites.

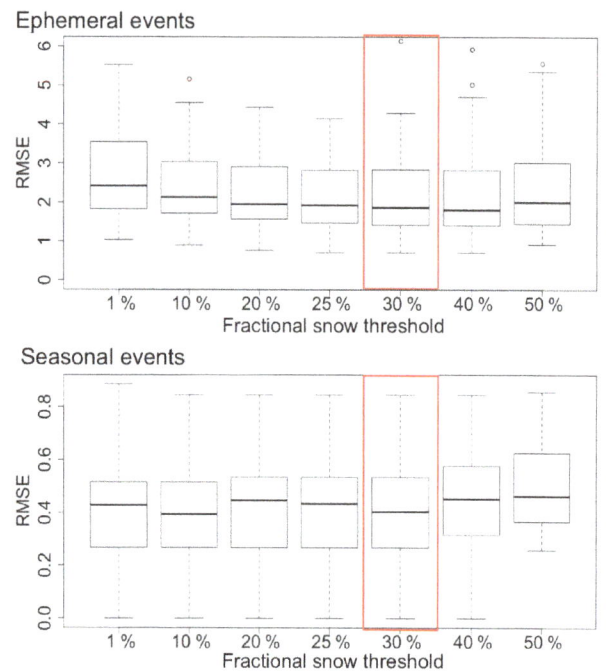

Figure A4. Box plots depicting the root mean square errors between the number of observed ephemeral and seasonal snow events at Snow Telemetry (SNOTEL) stations and the number of ephemeral and seasonal snow events derived from the algorithm in Google Earth Engine in each 500 m MODIS pixel corresponding to that station at snow fractions of 1–50 %. The chosen snow fraction was 30 % (highlighted in red).

Author contributions. AH conceived the idea. RP and AH designed the study. RP created the algorithms and performed the analysis. Both RP and AH interpreted the data and wrote the paper.

Competing interests. The author declares that there is no conflict of interest.

Special issue statement. This article is part of the special issue "Understanding and predicting Earth system and hydrological change in cold regions". It is not associated with a conference.

Acknowledgements. We acknowledge the NASA Space Grant Consortium and USDA NIFA NEV05293 for providing funding. Patrick Longley helped in creating the snow ephemerality metric. Charles Morton of the Desert Research Institute and the members of the Google Groups Earth Engine Forum helped with Google Earth Engine. We also thank Scott Tyler for his support.

Edited by: Sean Carey

References

Abatzoglou, J. T.: Development of gridded surface meteorological data for ecological applications and modelling, Int. J. Climatol., 33, 121–131, https://doi.org/10.1002/joc.3413, 2012.

Anderton, S. P., White, S. M., and Alvera, B.: Micro-scale spatial variability and the timing of snowmelt runoff in a high mountain catchment, J. Hydrol., 268, 158–176, https://doi.org/10.1016/S0022-1694(02)00179-8, 2002.

Bales, R. C., Molotch, N. P., Painter, T. H., Dettinger, M. D., Rice, R., and Dozier, J.: Mountain hydrology of the western United States, Water Resour. Res., 42, W08432, https://doi.org/10.1029/2005WR004387, 2006.

Barnett, T. P., Adam, J. C., and Lettenmaier, D. P.: Potential impacts of a warming climate on water availability in snow-dominated regions, Nature, 438, 303–309, 2005.

Barnhart, T. B., Molotch, N. P., Livneh, B., Harpold, A. A., Knowles, J. F., and Schneider, D.: Snowmelt rate dictates streamflow, Geophys. Res. Lett., 43, 8006–8016, 2016.

Berghuijs, W., Woods, R., and Hrachowitz, M.: A precipitation shift from snow towards rain leads to a decrease in streamflow, Nat. Clim. Change, 4, 583–586, 2014.

Bishop, M. P., Björnsson, H., Haeberli, W., Oerlemans, J., Shroder, J. F., and Tranter, M.: Encyclopedia of snow, ice and glaciers, Springer Science & Business Media, New York, NY, USA, 2011.

Choi, G., Robinson, D. A., and Kang, S.: Changing Northern Hemisphere Snow Seasons, J. Climate, 23, 5305–5310, https://doi.org/10.1175/2010JCLI3644.1, 2010.

Cline, D. W.: Effect of seasonality of snow accumulation and melt on snow surface energy exchanges at a continental alpine site, J. Appl. Meteorol., 36, 32–51, 1997.

Clow, D. W.: Changes in the Timing of Snowmelt and Streamflow in Colorado: A Response to Recent Warming, J. Climate, 23, 2293–2306, https://doi.org/10.1175/2009JCLI2951.1, 2010.

Clow, D. W., Nanus, L., Verdin, K. L., and Schmidt, J.: Evaluation of SNODAS snow depth and snow water equivalent estimates for the Colorado Rocky Mountains, USA, Hydrol. Process., 26, 2583–2591, 2012.

Earman, S., Campbell, A. R., Phillips, F. M., and Newman, B. D.: Isotopic exchange between snow and atmospheric water vapor: Estimation of the snowmelt component of groundwater recharge in the southwestern United States, J. Geophys. Res., 111, D09302, https://doi.org/10.1029/2005JD006470, 2006.

Essery, R., Rutter, N., Pomeroy, J., Baxter, R., Stähli, M., Gustafsson, D., Barr, A., Bartlett, P., and Elder, K.: SNOWMIP2: An evaluation of forest snow process simulations, B. Am. Meteorol. Soc., 90, 1120–1135, 2009.

Farr, T. G., Rosen, P. A., Caro, E., Crippen, R., Duren, R., Hensley, S., Kobrick, M., Paller, M., Rodriguez, E., Roth, L., Seal, D., Shaffer, S., Shimada, J., Umland, J., Werner, M., Oskin, M., Burbank, D., and Alsdorf, D.: The Shuttle Radar Topography Mission, Rev. Geophys., 45, RG2004, https://doi.org/10.1029/2005RG000183, 2007.

Gao, Y., Xie, H., and Yao, T.: Developing snow cover parameters maps from MODIS, AMSR-E, and blended snow products, Photogramm. Eng. Rem. S., 77, 351–361, 2011.

Gorelick, N., Hancher, M., Dixon, M., Ilyushchenko, S., Thau, D., and Moore, R.: Google Earth Engine: Planetary-scale geospatial analysis for everyone, Remote Sens. Environ., 202, 18–27, 2017.

Greuell, W. and Smeets, P.: Variations with elevation in the surface energy balance on the Pasterze (Austria), J. Geophys. Res., 106, 31717–31727, 2001.

Hall, D., Salomonson, V., and Riggs, G.: MODIS/Terra snow cover daily L3 global 500 m grid, Version 5.[Tile h09v04], National Snow and Ice Data Center, Boulder, Colorado, USA, 2006.

Hall, D. K., Riggs, G. A., Salomonson, V. V., Barton, J., Casey, K., Chien, J., DiGirolamo, N., Klein, A., Powell, H., and Tait, A.: Algorithm theoretical basis document (ATBD) for the MODIS snow and sea ice-mapping algorithms, NASA GSFC, Greenbelt, MD, USA, 2001.

Harpold, A., Brooks, P., Rajagopal, S., Heidbuchel, I., Jardine, A., and Stielstra, C.: Changes in snowpack accumulation and ablation in the intermountain west, Water Resour. Res., 48, W11501, https://doi.org/10.1029/2012WR011949, 2012.

Harpold, A. A.: Diverging sensitivity of soil water stress to changing snowmelt timing in the Western US, Adv. Water Resour., 92, 116–129, 2016.

Harpold, A. A. and Molotch, N. P.: Sensitivity of soil water availability to changing snowmelt timing in the western US, Geophys. Res. Lett., 42, 8011–8020, 2015.

Harpold, A. A., Rajagopal, S., Crews, J., Winchell, T., and Schumer, R.: Relative humidity has uneven effects on shifts from snow to rain over the western US, Geophys. Res. Lett., 44, 9742–9750, 2017a.

Harpold, A. A., Kaplan, M. L., Klos, P. Z., Link, T., McNamara, J. P., Rajagopal, S., Schumer, R., and Steele, C. M.: Rain or snow: hydrologic processes, observations, prediction, and research needs, Hydrol. Earth Syst. Sci., 21, 1–22, https://doi.org/10.5194/hess-21-1-2017, 2017b.

Harstveit, K.: Snowmelt modelling and energy exchange between the atmosphere and a melting snow cover, in: Proceedings of 18th International Conference for Alpine Meteorology, 1984, Opatija, Croatia (formerly Yugoslavia), 334–337, 1984.

Hawkins, T. W. and Ellis, A. W.: A case study of the energy budget of a snowpack in the arid, subtropical climate of the southwestern United States, Journal of the Arizona-Nevada Academy of Science, 39, 1–13, 2007.

Hedrick, A., Marshall, H.-P., Winstral, A., Elder, K., Yueh, S., and Cline, D.: Independent evaluation of the SNODAS snow depth product using regional-scale lidar-derived measurements, The Cryosphere, 9, 13–23, https://doi.org/10.5194/tc-9-13-2015, 2015.

Hinckley, E.-L. S., Ebel, B. A., Barnes, R. T., Anderson, R. S., Williams, M. W., and Anderson, S. P.: Aspect control of water movement on hillslopes near the rain–snow transition of the Colorado Front Range, Hydrol. Process., 28, 74–85, https://doi.org/10.1002/hyp.9549, 2014.

Hood, E., Williams, M., and Cline, D.: Sublimation from a seasonal snowpack at a continental, mid-latitude alpine site, Hydrol. Process., 13, 1781–1797, 1999.

Hu, J., Moore, D. J., Burns, S. P., and Monson, R. K.: Longer growing seasons lead to less carbon sequestration by a subalpine forest, Glob. Change Biol., 16, 771–783, 2010.

Hunsaker, C. T., Whitaker, T. W., and Bales, R. C.: Snowmelt runoff and water yield along elevation and temperature gradients in California's southern Sierra Nevada, JAWRA J. Am. Water Resour. As., 48, 667–678, 2012.

Inouye, D. W.: Effects of climate change on phenology, frost damage, and floral abundance of montane wildflowers, Ecology, 89, 353–362, 2008.

Jasechko, S., Birks, S. J., Gleeson, T., Wada, Y., Fawcett, P. J., Sharp, Z. D., McDonnell, J. J., and Welker, J. M.: The pronounced seasonality of global groundwater recharge, Water Resour. Res., 50, 8845–8867, 2014.

Jefferson, A. J.: Seasonal versus transient snow and the elevation dependence of climate sensitivity in maritime mountainous regions, Geophys. Res. Lett., 38, L16402, https://doi.org/10.1029/2011GL048346, 2011.

Jost, G., Weiler, M., Gluns, D. R., and Alila, Y.: The influence of forest and topography on snow accumulation and melt at the watershed-scale, J. Hydrol., 347, 101–115, https://doi.org/10.1016/j.jhydrol.2007.09.006, 2007.

Karlsen, S. R., Solheim, I., Beck, P. S., Høgda, K. A., Wielgolaski, F. E., and Tømmervik, H.: Variability of the start of the growing season in Fennoscandia, 1982–2002, Int. J. Biometeorol., 51, 513–524, 2007.

Kelleners, T., Chandler, D., McNamara, J. P., Gribb, M. M., and Seyfried, M.: Modeling runoff generation in a small snow-dominated mountainous catchment, Vadose Zone J., 9, 517–527, 2010.

Kimball, J., McDonald, K., Frolking, S., and Running, S.: Radar remote sensing of the spring thaw transition across a boreal landscape, Remote Sens. Environ., 89, 163–175, https://doi.org/10.1016/j.rse.2002.06.004, 2004.

Klos, P. Z., Link, T. E., and Abatzoglou, J. T.: Extent of the rain snow transition zone in the western US under historic and projected climate, Geophys. Res. Lett., 41, 4560–4568, 2014.

Knowles, J. F., Harpold, A. A., Cowie, R., Zeliff, M., Barnard, H. R., Burns, S. P., Blanken, P. D., Morse, J. F., and Williams, M. W.: The relative contributions of alpine and subalpine ecosystems to the water balance of a mountainous, headwater catchment, Hydrol. Process., 29, 4794–4808, 2015.

Knowles, N. and Cayan, D. R.: Elevational dependence of projected hydrologic changes in the San Francisco estuary and watershed, Climatic Change, 62, 319–336, 2004.

Kormos, P. R., Marks, D., McNamaraa, J. P., Marshall, H. P., Winstral, A., and Flores, A. N.: Snow distribution, melt and surface water inputs to the soil in the mountain rain–snow transition zone, J. Hydrol., 519, 190–204, https://doi.org/10.1016/j.jhydrol.2014.06.051, 2014.

Leathers, D. J., Graybeal, D., Mote, T., Grundstein, A., and Robinson, D.: The role of airmass types and surface energy fluxes in snow cover ablation in the central Appalachians, J. Appl. Meteorol., 43, 1887–1899, 2004.

Liang, X., Lettenmaier, D. P., Wood, E. F., and Burges, S. J.: A simple hydrologically based model of land surface water and energy fluxes for general circulation models, J. Geophys. Res., 99, 14415–14428, 1994.

Liston, G. E.: Local advection of momentum, heat, and moisture during the melt of patchy snow covers, J. Appl. Meteorol., 34, 1705–1715, 1995.

Liston, G. E. and Elder, K.: A Distributed Snow-Evolution Modeling System (SnowModel), J. Hydrometeorol., 7, 1259–1276, https://doi.org/10.1175/JHM548.1, 2006.

Lundquist, J. D., Neiman, P. J., Martner, B., White, A. B., Gottas, D. J., and Ralph, F. M.: Rain versus snow in the Sierra Nevada, California: Comparing Doppler profiling radar and surface observations of melting level, J. Hydrometeorol., 9, 194–211, 2008.

Marks, D. and Dozier, J.: Climate and energy exchange at the snow surface in the alpine region of the Sierra Nevada: 2. Snow cover energy balance, Water Resour. Res., 28, 3043–3054, 1992.

Marks, D., Link, T., Winstral, A., and Garen, D.: Simulating snowmelt processes during rain-on-snow over a semi-arid mountain basin, Ann. Glaciol., 32, 195–202, 2001.

McNamara, J. P., Chandler, D., Seyfried, M., and Achet, S.: Soil moisture states, lateral flow, and streamflow generation in a semi-arid, snowmelt-driven catchment, Hydrol. Process., 19, 4023–4038, https://doi.org/10.1002/hyp.5869, 2005.

Mernild, S. H., Liston, G. E., Hiemstra, C. A., Malmros, J. K., Yde, J. C., and McPhee, J.: The Andes Cordillera. Part I: snow distribution, properties, and trends (1979–2014), Int. J. Climatol., 37, 1680–1698, 2017.

Molotch, N. P. and Meromy, L.: Physiographic and climatic controls on snow cover persistence in the Sierra Nevada Mountains, Hydrol. Process., 28, 4573–4586, 2014.

Mote, P. W.: Climate-driven variability and trends in mountain snowpack in western North America, J. Climate, 19, 6209–6220, 2006.

Mote, P. W., Hamlet, A. F., Clark, M. P., and Lettenmaier, D. P.: Declining mountain snowpack in western North America, B. Am. Meteorol. Soc., 86, 39–50, 2005.

Nitta, T., Yoshimura, K., Takata, K., O'ishi, R., Sueyoshi, T., Kanae, S., Oki, T., Abe-Ouchi, A., and Liston, G. E.: Representing Variability in Subgrid Snow Cover and Snow Depth in a Global Land Model: Offline Validation, J. Climate, 27, 3318–3330, https://doi.org/10.1175/JCLI-D-13-00310.1, 2014.

Nolin, A. W. and Daly, C.: Mapping "at risk" snow in the Pacific Northwest, J. Hydrometeorol., 7, 1164–1171, 2006.

Parida, B. R. and Buermann, W.: Increasing summer drying in North American ecosystems in response to longer nonfrozen periods, Geophys. Res. Lett., 41, 5476–5483, 2014.

Petersky, R. and Harpold, A.: Now You See It Now You Don't: A Case Study of Ephemeral Snowpacks in the Great Basin U.S.A., ScholarWorks, available at: https://scholarworks.unr.edu/handle/11714/2952, last access: 11 September 2018.

Pomeroy, J.: Transport and sublimation of snow in wind-scoured alpine terrain, in: Snow, Hydrology and Forests in Alpine Areas, edited by: Bergman, H., Lang, H., Frey, W., Issler, D., and Salm, B., IAHS Press, 205, 131–140, 1991.

Pomeroy, J., Toth, B., Granger R., Hedstrom, N., and Essery, R.: Variation in surface energetics during snowmelt in a subarctic mountain catchment, J. Hydrometeorol., 4, 702–719, 2003.

Schmucki, E., Marty, C., Fierz, C., and Lehning, M.: Evaluation of modelled snow depth and snow water equivalent at three contrasting sites in Switzerland using SNOWPACK simulations driven by different meteorological data input, Cold Reg. Sci. Technol., 99, 27–37, https://doi.org/10.1016/j.coldregions.2013.12.004, 2014.

Schmucki, E., Marty, C., Fierz, C., Weingartner, R., and Lehning, M.: Impact of climate change in Switzerland on socioeconomic snow indices, Theor. Appl. Climatol., 127, 875–889, 2017.

Schwinning, S. and Sala, O. E.: Hierarchy of responses to resource pulses in arid and semi-arid ecosystems, Oecologia, 141, 211–220, 2004.

Seaber, P. R., Kapinos, F. P., and Knapp, G. L.: Hydrologic Unit Maps, US Government Printing Office, Denver, CO, USA, 1987.

Şensoy, A., Şorman, A., Tekeli, A., Şorman, A., and Garen, D.: Point-scale energy and mass balance snowpack simulations in the upper Karasu basin, Turkey, Hydrol. Process., 20, 899–922, 2006.

Serreze, M. C., Clark, M. P., Armstrong, R. L., McGinnis, D. A., and Pulwarty, R. S.: Characteristics of the western United States snowpack from snowpack telemetry (SNOTEL) data, Water Resour. Res., 35, 2145–2160, 1999.

Seyfried, M., Grant, L., Marks, D., Winstral, A., and McNamara, J.: Simulated soil water storage effects on streamflow generation in a mountainous snowmelt environment, Idaho, USA, Hydrol. Process., 23, 858–873, 2009.

Slater, A. G., Lawrence, D. M., and Koven, C. D.: Process-level model evaluation: a snow and heat transfer metric, The Cryosphere, 11, 989–996, https://doi.org/10.5194/tc-11-989-2017, 2017.

Sospedra-Alfonso, R. and Merryfield, W. J.: Influences of Temperature and Precipitation on Historical and Future Snowpack Variability over the Northern Hemisphere in the Second Generation Canadian Earth System Model, J. Climate, 30, 4633–4656, https://doi.org/10.1175/JCLI-D-16-0612.1, 2017.

Sospedra-Alfonso, R., Mudryk, L., Merryfield, W., and Derksen, C.: Representation of Snow in the Canadian Seasonal to Interannual Prediction System. Part I: Initialization, J. Hydrometeorol., 17, 1467–1488, https://doi.org/10.1175/JHM-D-14-0223.1, 2016.

Stewart, I. T., Cayan, D. R., and Dettinger, M. D.: Changes toward earlier streamflow timing across western North America, J. Climate, 18, 1136–1155, 2005.

Stielstra, C. M., Lohse, K. A., Chorover, J., McIntosh, J. C., Barron-Gafford, G. A., Perdrial, J. N., Litvak, M., Barnard, H. R., and Brooks, P. D.: Climatic and landscape influences on soil moisture are primary determinants of soil carbon fluxes in seasonally snow-covered forest ecosystems, Biogeochemistry, 123, 447–465, 2015.

Sturm, M., Holmgren, J., and Liston, G. E.: A seasonal snow cover classification system for local to global applications, J. Climate, 8, 1261–1283, 1995.

Sturm, M., Taras, B., Liston, G. E., Derksen, C., Jonas, T., and Lea, J.: Estimating Snow Water Equivalent Using Snow Depth Data and Climate Classes, J. Hydrometeorol., 11, 1380–1394, https://doi.org/10.1175/2010JHM1202.1, 2010.

Sturm, M., Goldstein, M. A., and Parr, C.: Water and life from snow: A trillion dollar science question, Water Resour. Res., 53, 3534–3544, https://doi.org/10.1002/2017WR020840, 2017.

Sugg, J. W., Perry, L. B., Hall, D. K., Riggs, G. A., and Badurek, C. A.: Satellite perspectives on the spatial patterns of new snowfall in the Southern Appalachian Mountains, Hydrol. Process., 28, 4602–4613, 2014.

Svejcar, T.: The Northern Great Basin: A Region of Continual Change, Rangelands, 37, 114–118, https://doi.org/10.1016/j.rala.2015.03.002, 2015.

Thompson, J. A. and Lees, B. G.: Applying object-based segmentation in the temporal domain to characterise snow seasonality, ISPRS J. Photogramm., 97, 98–110, 2014.

Trujillo, E. and Molotch, N. P.: Snowpack regimes of the western United States, Water Resour. Res., 50, 5611–5623, 2014.

Trujillo, E., Molotch, N. P., Goulden, M. L., Kelly, A. E., and Bales, R. C.: Elevation-dependent influence of snow accumulation on forest greening, Nat. Geosci., 5, 705–709, 2012.

Tyler, S. W., Burak, S. A., McNamara, J. P., Lamontagne, A., Selker, J. S., and Dozier, J.: Spatially distributed temperatures at the base of two mountain snowpacks measured with fiber-optic sensors, J. Glaciol., 54, 673–679, 2008.

Wang, Z., Schaaf, C. B., Strahler, A. H., Chopping, M. J., Roman, M. O., Shuai, Y., Woodcock, C. E., Hollinger, D. Y., and Fitzjarrald, D. R.: Evaluation of MODIS albedo product (MCD43A) over grassland, agriculture and forest surface types during dormant and snow-covered periods, Remote Sens. Environ., 140, 60–77, 2014.

West, N.: Great Basin-Colorado plateau sagebrush semi-desert, Temperate Deserts and Semi-Deserts, 5, 331–369, 1983.

Winchell, T. S., Barnard, D. M., Monson, R. K., Burns, S. P., and Molotch, N. P.: Earlier snowmelt reduces atmospheric carbon uptake in midlatitude subalpine forests, Geophys. Res. Lett., 43, 8160–8168, 2016.

Winstral, A., Marks, D., and Gurney, R.: Simulating wind-affected snow accumulations at catchment to basin scales, Adv. Water Resour., 55, 64–79, 2013.

10

Towards a tracer-based conceptualization of meltwater dynamics and streamflow response in a glacierized catchment

Daniele Penna[1], **Michael Engel**[2], **Giacomo Bertoldi**[3], **and Francesco Comiti**[2]

[1]Department of Agricultural, Food and Forestry Systems, University of Florence, via San Bonaventura 13, 50145 Florence, Italy

[2]Faculty of Science and Technology, Free University of Bozen-Bolzano, Piazza dell' Università 5, 39100 Bolzano, Italy

[3]Institute for Alpine Environment, EURAC – European Academy of Bolzano/Bozen, viale Druso 1, 39100 Bolzano, Italy

Correspondence to: Daniele Penna (daniele.penna@unifi.it)

Abstract. Multiple water sources and the physiographic heterogeneity of glacierized catchments hamper a complete conceptualization of runoff response to meltwater dynamics. In this study, we used environmental tracers (stable isotopes of water and electrical conductivity) to obtain new insight into the hydrology of glacierized catchments, using the Saldur River catchment, Italian Alps, as a pilot site. We analysed the controls on the spatial and temporal patterns of the tracer signature in the main stream, its selected tributaries, shallow groundwater, snowmelt and glacier melt over a 3-year period. We found that stream water electrical conductivity and isotopic composition showed consistent patterns in snowmelt-dominated periods, whereas the streamflow contribution of glacier melt altered the correlations between the two tracers. By applying two- and three-component mixing models, we quantified the seasonally variable proportion of groundwater, snowmelt and glacier melt at different locations along the stream. We provided four model scenarios based on different tracer signatures of the end-members; the highest contributions of snowmelt to streamflow occurred in late spring–early summer and ranged between 70 and 79 %, according to different scenarios, whereas the largest inputs by glacier melt were observed in mid-summer, and ranged between 57 and 69 %. In addition to the identification of the main sources of uncertainty, we demonstrated how a careful sampling design is critical in order to avoid underestimation of the meltwater component in streamflow. The results of this study supported the development of a conceptual model of streamflow response to meltwater dynamics in the Saldur catchment, which is likely valid for other glacierized catchments worldwide.

1 Introduction

Glacierized catchments are highly dynamic systems characterized by large complexity and heterogeneity due to the interplay of several geomorphic, ecological, climatic and hydrological processes. Particularly, the hydrology of glacierized catchments significantly impacts downstream settlements, ecosystems and larger catchments that are directly dependent on water deriving from snowmelt, glacier melt or high-elevation springs (Finger et al., 2013; Engelhardt et al., 2014). Water seasonally melting from snowpack and glacier bodies can constitute a larger contribution to annual streamflow than rain (Cable et al., 2011; Jost et al., 2012), and is widely used, especially in Alpine valleys, for irrigation and hydropower production (Schaefli et al., 2007; Beniston, 2012). It is therefore pivotal for an effective adoption of water resources strategies to understand the origin of water and to quantify the proportion of snowmelt and glacier melt in streamflow (Finger et al., 2013; Fan et al., 2015). To achieve this goal it is critical to gain a more detailed understanding of the hydrological functioning of glacierized catchments through the analysis of the spatial and temporal variability of water sources and the spatial and seasonal meltwater (snowmelt plus glacier melt) dynamics.

Hydrochemical tracers (e.g. temporary storage of winter–early spring precipitation in the snowpack and in the glacier body and their melting during the late spring and summer controls the variability in solute and isotopic compositions of stream water (Kendall and McDonnell, 1998). Therefore, hydrochemical tracers allow for an effective identification of water sources and their variability within the catchments and over different seasons, providing essential information about water partitioning and water dynamics and improving our understanding of complex hydrology and hydroclimatology of the catchment (Rock and Mayer, 2007; Fan et al., 2015; Xing et al., 2015). Particularly, a few works relied on stable isotopes of water (^{2}H and ^{18}O) used in combination with EC to evaluate the role played by meltwater in the hydrology of glacierized catchments. For instance, some of these investigations allowed for the separation of streamflow into subglacial-, englacial-, melt- and rainfall-derived components in the South Cascade Glacier, USA (Vaughn and Fountain, 2005), into components due to monsoon rainfall runoff, post-monsoon interflow, winter snowmelt and groundwater (the latter estimated up to 40 % during summer and monsoon periods) in the Ganga River, Himalaya (Maurya et al., 2011), and into snowmelt, ice melt and shallow groundwater components in Arctic catchments characterized by a gradient of glacierization (Blaen et al., 2014). Other researchers assessed the possibility to use isotopes and EC as complementary tracers, in addition to water temperature, to identify a permafrost-related component in spring water in a glacierized catchment in the Ortles-Cevedale massif, Italian Alps (Carturan et al., 2016).

Two recent studies used stable isotopes and EC over a 3-year period to assess water origin and streamflow contributors in the glacierized Saldur River catchment, Italian Alps. Penna et al. (2014) showed a preliminary analysis on the highly complex EC and isotopic signature of different waters sampled in the catchment, identifying distinct tracer signals in snowmelt and glacier melt. These two end-members dominated the streamflow throughout the late spring and summer, whereas liquid precipitation played a secondary role, limited to rare intense rainfall events. They also assessed, without quantifying it, the switch from snowmelt- to glacier melt-dominated periods, and estimated that the snowmelt fraction in groundwater ranged between 21 and 93 %. Engel et al. (2016) employed two- and three-component mixing models to quantify the relative contribution of snowmelt, glacier melt and groundwater to streamflow during seven representative melt-induced runoff events sampled at high frequency at two cross sections of the Saldur River. They observed marked reactions of tracers and streamflow both to melt and rainfall inputs, identifying hysteretic loops of contrasting directions. They estimated the maximum contribution of snowmelt during June and July events (up to 33 %) and of glacier melt during the August events (up to 65 %). However, a quantification of the variations of streamflow components not only at the seasonal scale but also at different spatial scales across

the catchment was not performed and a conceptual model of meltwater dynamics was not presented. Therefore, despite the number of studies that have conducted hydrological tracer-based investigations in high-elevation mountain catchments, there is still the need to gain a better comprehension of the factors determining the complex hydrochemical signature of stream water and groundwater in glacierized catchments.

This research builds on the existing database for the Saldur River and on the first results presented in Penna et al. (2014) and Engel et al. (2016) to improve the knowledge of the complex hydrology and the water source dynamics in glacierized catchments. Specifically, we aim to

- assess the controls on the spatial and temporal variability of the isotopic composition and EC in the main stream, tributaries and springs in the Saldur River catchment;

- quantify the proportion of snowmelt and glacier melt in streamflow at different stream locations and at different times of the year, as well as the related uncertainty;

- analyse the relation between the tracer signature and streamflow variability;

- derive a conceptual model of streamflow response to meltwater dynamics.

2 Study area

The research has been conducted in the upper portion of the Saldur River catchment, Vinschgau Valley, eastern Italian Alps (Fig. 1). The catchment size is 61.7 km^2 and altitude ranges between 1632 m a.s.l. at the outlet (46°42′42.37″ N, 10°38′51.41″ E) and 3725 m a.s.l. A glacier lies in the upper part of the catchment, with an extent of 2.28 km^2 in 2013, i.e. approximately 4 % of the total catchment area (Galos and Kaser, 2013). The glacier lost 21 % of its area from 2005 to 2013 (Galos and Kaser, 2013). Several glacier-fed and non-glacier-fed lateral tributaries contribute to the Saldur River streamflow, and various springs, apparently connected or not connected to the main stream, can be found on the valley floor and at the toe of the hillslopes in the mid-upper part of the catchment. Rocks are metamorphic, mainly gneisses, mica-gneisses and schists. Land cover changes with elevation typically varying from Alpine forests (up to about 2200 m a.s.l.) to shrubs to Alpine grassland, bare soil and rocks above 2700 m a.s.l. The area is characterized by a continental climate with an average annual air temperature of 6.6 °C and precipitation as low as 569 mm yr^{-1} (at 1570 m a.s.l.), likely increasing up to 800–1000 mm yr^{-1} in the upper parts of the catchment. At 3000 m a.s.l., the total precipitation can be estimated, using the approach of Mair et al. (2016), to be about 1500 mm, 80 % of which falls as snow. The hydrological regime is typically nivo-glacial with

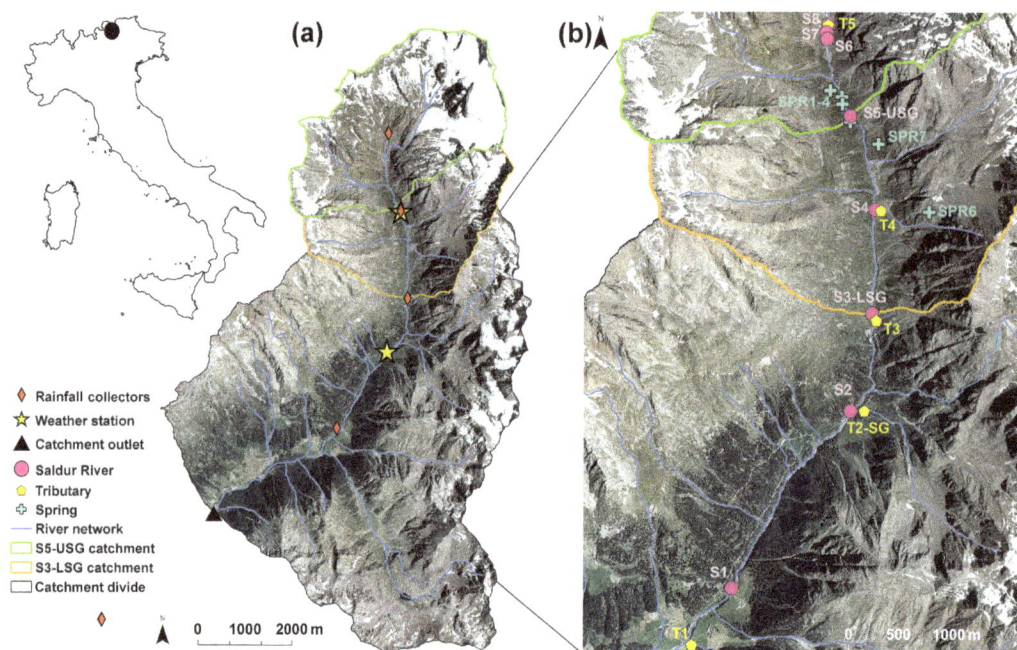

Figure 1. Map of the Saldur catchment, with its localization in the country, and position of field instruments and sampling points. Data from the rainfall collectors were not used in this study but their position is reported for completeness.

minimum streamflow recorded in winter and high flows occurring from late spring to mid-summer, when marked diurnal streamflow cycles occur, related to snowmelt and glacier melt (Mutzner et al., 2015). More detailed information on the study area are reported in Mao et al. (2014) and Penna et al. (2014).

3   Materials and methods

3.1   Hydrological and meteorological measurements

Field measurements were conducted from April 2011 to October 2013. Meteorological data were recorded at 15 min temporal resolution by two stations located at 2332 and 1998 m a.s.l. (Fig. 1a). The stage in the Saldur River was recorded every 10 min by pressure transducers at the catchment outlet and at two river sections labelled lower stream gauge (S3-LSG; 2150 m a.s.l.) and upper stream gauge (S5-USG; 2340 m a.s.l.), which defined two nested subcatchments with an area of 18.6 and 11.2 km^2, respectively (Fig. 1a). Streamflow values were obtained by 82 discharge measurements acquired by the salt dilution method during various hydrometric conditions over the three study years. Water level was also continuously measured on a left tributary (T2-SG; 2027 m a.s.l.; Fig. 1b) draining an area of 1.7 km^2 but a robust rating curve was not available to derive streamflow.

3.2   Tracer sampling and measurement

Samples analysed for the two tracers were collected from snowmelt, glacier melt, stream water and groundwater. Snowmelt was sampled in late spring–early summer from water dripping from the residual snowpack at different elevations and different locations. Snowmelt was sampled on three occasions in summer 2012 (end of June, beginning and end of July), at elevations roughly between 2150 and 2350 m a.s.l., and on nine occasions in summer 2013 (June, July and August) at elevations roughly between 2150 and 2600 m a.s.l. Glacier melt was sampled from small rivulets flowing on the glacier surface, roughly at 2800 m a.s.l. in July and August 2012, and in July, August and September 2013. Grab stream-water samples were taken approximately monthly at eight locations in the Saldur River (labelled from S1 to S8), at elevations spanning from 1809 m a.s.l. (S1) and 2415 m a.s.l. (S8), and from five tributaries (labelled from T1 to T5), at elevations between 1775 m a.s.l. (T1) to 2415 m a.s.l. (T5; Fig. 1b). Samples at T1 were taken only in 2012, and samples at T3 only in 2011. In 2013 samples were collected monthly during clear days only from the river at four sections (S1, S3-LSG, S5-LSG, S8), respectively at the same time of the day on each occasion in order to ensure consistency and comparability between measurements. The representativeness of these samples for the typical melting conditions in the catchment was visually ensured by comparing the hydrographs of the sampled days with the ones of the corresponding months during the three monitored years. No wells are available in the study

Table 1. Sampling years and number of samples collected from the different water sources and used in this study.

Water source	ID of sampling locations	Sampling years	Total no. of samples
Snowmelt	–	2011–2013	24
Glacier melt	–	2012–2013	16
Stream (main river)	S1–S8 S1, S3-LSG, S5-USG, S8	2011–2012 2013	535
Stream (tributaries)	T1 T2, T4, T5 T3	2012 2011–2013 2011	102
Spring	SPR1–SPR4 SPR6, SPR7	2011–2013 2013	84

catchment; thus, spring water was assumed to represent shallow groundwater (Kong and Pang, 2012; Racoviteanu et al., 2013). Four springs (labelled from SPR1 to SPR4) localized near the outlet of USG, between 2334 and 2360 m a.s.l., were sampled monthly during the three study years. On one occasion (17 October 2011) no sample was taken from SPR1 because it was found dry. Additionally, monthly samples were also taken from June to September 2013 from two springs on the left valley hillslope, SPR6 and SPR7 at 2512 and 2336 m a.s.l., respectively (Fig. 1b). A list of all sampling locations with their main characteristics is reported in Penna et al. (2014).

In addition to the monthly sampling, stream water samples were collected at USG and LSG during seven runoff events induced by meltwater in July and August 2011, and June, July and August 2012 and 2013. Samples were collected from 10:00 LT of one day to 10:00 LT (or longer) on the following day at hourly frequency during the day until 22:00 LT, and every 3 h during the night. For those events, two- and three-component mixing models were applied to quantify the fraction of snowmelt and glacier melt in streamflow. Description of the runoff events and hydrograph separation results are reported in Engel et al. (2016). The number of samples collected from the different water sources at the various locations and years used in this study is reported in Table 1.

EC was determined directly in the field by means of a conductivity meter with a precision of $\pm 0.1 \, \mu S \, cm^{-1}$. The EC meter was routinely calibrated to ensure consistency among the measurements. Grab water samples for isotopic determination were taken by 50 mL HDPE (high-density polyethylene) bottles with two caps and completely filled to avoid head space. Isotopic analysis was carried out by an off-axis integrated cavity output spectroscope tested for precision, accuracy and memory effect in previous intercomparison studies (Penna et al., 2010, 2012). The observed instrumental precision, considered as the long-term average standard deviation, is 0.5 ‰ for $\delta^2 H$ and 0.08 ‰ for $\delta^{18} O$. Isotopic values are presented using the δ notation referred to the SMOW2–SLAP2 scale provided by the International Atomic Energy Agency.

3.3 Two- and three-component mixing models and underlying assumptions

A one-tracer, two-component mixing model (Pinder and Jones, 1969; Sklash and Farvolden, 1979) was used to quantify and separate two streamflow components (groundwater and snowmelt), and a two-tracer, three-component mixing model (Ogunkoya and Jenkins, 1993) was used for three streamflow components (groundwater, snowmelt and glacier melt). Mixing models were applied only to 2013 data because in that year water samples were collected at four locations along the main stream (S1, S3-LSG, S5-USG and S8) at the same time of the day on all sampling occasions. This was critical to ensure comparability of the results, given the high diurnal variability of streamflow and associated isotopic composition and EC, especially during the summer. In addition, results from the application of the two- and three-component mixing models to data collected hourly during seven melt-induced runoff events presented in Engel et al. (2016) were also used in this study for comparison purposes (see Sect. 4.3).

The following simplifying assumptions were made for the application of the mixing models:

– Streamflow at each selected sampling location of the Saldur River was a mixture of two components, viz. groundwater and snowmelt, or three components, viz. groundwater, snowmelt and glacier melt. The influence of precipitation was considered negligible because samples were collected during non-rainy periods, and particularly during warm, clear days when the meltwater input to runoff was remarkable and overwhelmed the possible presence of rain water in streamflow.

– The largest contribution of snowmelt to streamflow was assumed to derive from snow melting at an approximate elevation of 2800 m a.s.l. The elevation band between 2800 and 2850 m a.s.l. was the one with the largest area in the catchment (3.4 km^2), where much snow can accumulate, as confirmed by the analysis of snow cover data from Moderate Resolution Imaging Spectroradiometer (MODIS) images (cf. Engel et al., 2016).

The three-component mixing model was based on isotopic and EC data (Maurya et al., 2011; Penna et al., 2015) and first applied to all samples collected in the Saldur River in 2013. When the three-component mixing model yielded inconsistent results, typically in May and June and partially in October, it was inferred that there was no glacier melt component in streamflow; thus, the two-component mixing model was performed to separate the snowmelt from the groundwater component. As a preliminary step, both EC and isotopes were used in the two-component mixing model. The resulting estimates were strongly correlated ($p < 0.01$) but, overall, snowmelt fractions computed for May and June using isotopes were smaller compared to those computed through EC. In agreement with our previous work in the Saldur catchment (Engel et al., 2016), we decided to present EC-based results for the sampling days in May and June because of the large difference between the low EC of the snowmelt end-member and the relatively high EC of the stream that provided lower uncertainties in the estimated fractions compared to isotopes (Genereux, 1998). Conversely, for the sampling day in October, there was a relatively small difference between the EC of the groundwater end-member and the EC of the stream, while the difference in the isotopic signal of the end-members was greater, and thus the uncertainty in the estimated fractions was lower. Therefore, in these cases we used isotopes instead of EC in the two-component mixing model.

Based on the stated assumptions, the following mass balance equations can be written for periods when only snowmelt and groundwater contributed to streamflow:

$$SF = SM + GW, \tag{1}$$

$$1 = sm + gw, \tag{2}$$

$$\delta_{SF} = sm \cdot \delta_{SM} + gw \cdot \delta_{GW}, \tag{3}$$

$$EC_{SF} = sm \cdot EC_{SM} + gw \cdot EC_{GW}, \tag{4}$$

where SM, GW and SF denote snowmelt, groundwater and streamflow, respectively; sm and gw indicate the streamflow fraction due to snowmelt and groundwater, respectively; and the notations δ and EC are used for the isotopic composition and the EC of each component, respectively. Equations (1)–(4) can be solved for the unknown sm as follows:

$$sm(\%) = \frac{\delta_{SF} - \delta_{GW}}{\delta_{SM} - \delta_{GW}} \cdot 100 \tag{5}$$

or, using EC,

$$sm(\%) = \frac{EC_{SF} - EC_{GW}}{EC_{SM} - EC_{GW}} \cdot 100. \tag{6}$$

The gw component can then be calculated by Eq. (2). Analogously, the following mass balance equations can be written for periods when snowmelt, glacier melt and groundwater contributed to streamflow:

$$SF = SM + GM + GW, \tag{7}$$

$$1 = sm + gm + gw, \tag{8}$$

$$\delta_{SF} = sm \cdot \delta_{SM} + gm \cdot \delta_{GM} + gw \cdot \delta_{GW}, \tag{9}$$

$$EC_{SF} = sm \cdot EC_{SM} + gm \cdot EC_{GM} + gw \cdot EC_{GW}, \tag{10}$$

where in additions to the symbols used in Eqs. (1)–(6), GM denotes glacier melt, and gm indicates the streamflow fraction due to glacier melt. Equations (7)–(10) can be solved for the unknown sm and gm as follows:

$$sm(\%) =$$
$$\frac{(\delta_{SF} - \delta_{GW}) \cdot (EC_{GM} - EC_{GW}) - (\delta_{GM} - \delta_{GW}) \cdot (EC_{SF} - EC_{GW})}{(\delta_{SM} - \delta_{GW}) \cdot (EC_{GM} - EC_{GW}) - (\delta_{GM} - \delta_{GW}) \cdot (EC_{SM} - EC_{GW})} \cdot 100, \tag{11}$$

$$gm(\%) =$$
$$\frac{(\delta_{SF} - \delta_{GW}) \cdot (EC_{SM} - EC_{GW}) - (\delta_{SM} - \delta_{GW}) \cdot (EC_{SF} - EC_{GW})}{(\delta_{GM} - \delta_{GW}) \cdot (EC_{SM} - EC_{GW}) - (\delta_{SM} - \delta_{GW}) \cdot (EC_{GM} - EC_{GW})} \cdot 100. \tag{12}$$

The gw component can be then calculated by Eq. (8).

The uncertainty of the end-member fractions calculated through the two-component mixing model was quantified following the method of Genereux (1998) at the 70 % confidence level. The uncertainty of the end-member fractions calculated through the three-component mixing model was determined by varying the isotopic composition and EC of each end-member by ±1 SD (standard deviation) (Carey and Quinton, 2005; Engel et al., 2016). All mixing models were applied using both δ^2H and $\delta^{18}O$ data; however, results based on $\delta^{18}O$ measurements showed a greater uncertainty than those derived from δ^2H data due to the instrumental performance (Penna et al., 2010). Thus, all results related to isotopes reported in this study are based on δ^2H data.

3.4 Scenarios of mixing model application

The spatial and temporal variability of an end-member tracer signal is usually very difficult to characterize at the catchment scale (Hoeg et al., 2000), especially in glacierized catchments (Jeelani et al., 2016), and it can noticeably affect the uncertainty of the results of mixing models. Since field measurements cannot reliably capture such a large spatial and temporal variability, we identified four different scenarios of mixing model application, assuming that they were representative for this variability. The four scenarios differed considering the groundwater end-member based on springs or stream locations during baseflow conditions, and time-invariant or monthly variable isotopic composition and EC

Table 2. Summary of the properties of the end-members used in the four mixing model scenarios for 2013 data.

Scenario	Groundwater end-member	Snowmelt end-member	Glacier melt end-member
A	Average δ^2H and EC of samples taken from selected springs in fall (2011–2013)	Time-invariant isotopic composition and EC (2013)	
B	Average δ^2H and EC of samples taken at each stream location in fall and winter (2011–2013)		Monthly variable isotopic composition and EC (2013)
C	Average δ^2H and EC of samples taken from selected springs in fall (2011–2013)	Monthly variable isotopic composition and EC (2013)	
D	Average δ^2H and EC of samples taken at each stream location in fall and winter (2011–2013)		

Table 3. Isotopic composition (δ^2H) and EC of the groundwater end-member used in the two- and three-component mixing model for the four scenarios for 2013 data. n: number of samples; avg.: average; SD: standard deviation.

| | δ^2H (‰) | | | | | | EC (μS cm^{-1}) | | | | | |
| | Scenarios A and C | | | Scenarios B and D | | | Scenarios A and C | | | Scenarios B and D | | |
Sampling location	n	avg.	SD	n	avg.	SD	n	avg.	SD	n	avg.	SD
S1	7	−101.7	5.7	5	−101.5	2.8	7	317.7	76.6	5	257.0	11.4
S3-LSG				3	−101.7	1.4				3	298.0	6.6
S5-USG	5	−98.5	1.3	4	−101.6	3.0	5	288.2	40.7	4	220.4	19.0
S8				1	−101.8	(−) 0.5*				1	210.0	(−) 0.1*

* For S8 only one sample was collected during baseflow conditions due to the difficult accessibility of the location in fall and winter; therefore, no standard deviation could be computed, and the instrumental precision was used for the computation of the uncertainty of the estimated fractions.

of the snowmelt end-member (Table 2). Particularly, in scenarios A and C, the groundwater end-member was based on the average isotopic composition and EC of samples taken from springs during baseflow conditions in fall of the three study years (springs were not sampled during winter due to limited accessibility of the area), which is consistent with Engel et al. (2016) (Table 3). This assumes a negligible influence of the inter-annual variability of the climatic forcing on the tracer signal of spring water during baseflow. In scenarios B and D, the groundwater end-member was defined as the average of the tracer signal of different stream samples taken during baseflow conditions (late fall and winter of the three study years), at the four Saldur River locations selected in 2013 (Table 3). For the definition of these two groundwater end-members, we selected the samples taken during baseflow conditions when we assumed that there was no or negligible contribution of snowmelt, glacier melt and rainfall to streamflow. It is important to note that we consider as groundwater components both the spring baseflow and the stream baseflow, because the hydrochemistry of streams during baseflow conditions generally integrates and reflects the

hydrochemistry of the (shallow) groundwater at the catchment scale (Sklash, 1990; Klaus and McDonnell, 2013; Fischer et al., 2015).

In scenarios A and B, the tracer signature of the snowmelt end-member was considered time invariant (Maurya et al., 2011) (Table 4). Following Engel et al. (2016), the high-elevation (2800 m a.s.l.) snowmelt isotopic composition was identified through the regression analysis of snowmelt samples collected at different elevations in June 2013, according to Eq. (13) ($R^2 = 0.616$, $n = 7$, $p < 0.05$):

$$\delta^2\text{H}(‰) = -0.0705 \cdot \text{elevation(m a.s.l.)} + 37.261. \quad (13)$$

EC$_{SM}$ was based on the average EC of all snowmelt samples collected in 2013, without applying any regression-based modification.

In scenarios C and D, the isotopic composition of a high-elevation snowmelt end-member was considered seasonally variable, taking into account that water from the melting snowpack typically undergoes progressive fractionation and isotopic enrichment over the season (Taylor et al., 2001; Lee et al., 2010) (cf. Sect. 4.1). A depletion rate of $-7.0‰$ in δ^2H

Table 4. Isotopic composition (δ^2H) and EC of the snowmelt end-member used in the two- and three-component mixing model for the four scenarios for 2013 data. Abbreviations are used as in Table 2.

Sampling day	δ^2H (‰)[a] Scenarios A and B		δ^2H (‰)[a] Scenarios C and D		EC (μS cm^{-1}) Scenarios A and B			EC (μS cm^{-1}) Scenarios C and D		
	n	avg.	n	avg.	n	avg.	SD	n	avg.	SD
23 May			1	−195.4				1	15.3	(−) 0.1[c]
19 Jun			7	−160.1				7	11.9	22.1
16 Jul	7	−160.1	3	−134.3	13	10.9	17.1	3	12.5	14.7
12 Aug										
11 Sept[b]			2	−139.9				2	2.9	0.4
18 Oct[b]										

[a] Because the isotopic composition of the high-elevation snowmelt end-member was derived by a regression (Eq. 11), the standard deviation was not computed. Thus, the computation of uncertainty was based on the standard error of the estimate of the regression (6.0‰) instead of the standard deviation of the samples averaged for each month. [b] Because no snowmelt samples were collected in September and October, the August value was used also for the two sampling days in September and October. [c] In May 2013, only one snowmelt sample was collected; therefore, no standard deviation could be computed, and the instrumental precision was used for the computation of the uncertainty of the estimated fractions.

for 100 m of elevation rise was derived from Eq. (13), and used to estimate the isotopic composition of high-elevation snowmelt from snowmelt samples collected monthly at different elevations from May to August 2013 (Table 4). Analogously, the average EC of snowmelt samples taken monthly was adopted.

In scenarios A and B, Eq. (13) was applied to snowmelt samples collected at different elevations (lower than 2800 m a.s.l.) in order to estimate the average isotopic composition of high-elevation snowmelt, and thus to define a temporally fixed end-member isotopic composition that was used in the calculations of streamflow-component fractions for each sampling date (Table 4, scenarios A and B). In scenarios C and D, Eq. (13) was applied to snowmelt samples collected at different elevations (lower than 2800 m a.s.l.) and at different times of the melting season in order to estimate the seasonally variable isotopic compositions of high-elevation snowmelt, which were used in the calculations of streamflow-component fractions for each sampling (Table 4, scenarios C and D).

For all scenarios, the isotopic signature and EC of the glacier melt end-member was considered monthly variable (Table 5 and Sect. 4.1).

4 Results

4.1 Isotopic composition and EC of the different water sources

Snowmelt sampled from snow patches in summer 2012 and 2013 ranged in δ^2H from −106.1 to −139.5‰ and in EC from 3.2 to 77.0 μS cm^{-1}. Glacier melt displayed a marked enrichment in heavy isotopes over summer, particularly in 2013 (Table 5). The spatial variability in the isotopic composition of glacier melt was generally small, with spatial

Table 5. Isotopic composition (δ^2H) and EC of the glacier melt end-member used in the three-component mixing model for all scenarios for 2013 data. Abbreviations are used as in Table 2.

Sampling day	δ^2H (‰) n	avg.	SD	EC (μS cm^{-1}) n	avg.	SD
16 Jul	3	−110.7	1.5	3	2.0	0.3
12 Aug	2	−104.2	3.8	2	2.2	0.7
11 Sept	2	−92.6	6.5	2	2.5	1.8
18 Oct*	2	−89.6	4.5	2	2.7	1.7

* No samples were collected on 18 October, when the stream was sampled. Therefore, the tracer value of the glacier melt samples collected on 26 September was used in the mixing model calculations.

standard deviations ranging between 1.3 and 6.5‰. The EC of glacier melt was very low and little variable in space and in time (average: 2.1 μS cm^{-1}; standard deviation: 0.7 μS cm^{-1}; $n = 16$) for 2012 and 2013 overall, even though a slight progressive increase in EC was observed in 2013 (Table 5).

The Saldur catchment was characterized by a marked variability of tracer signature within the same water compartment (i.e. main stream water, tributary water, groundwater) both in time and in space (Table 6, Figs. 2 and 3). There was a statistically significant difference in δ^2H and EC between the Saldur River and its sampled tributaries for the entire sampling period (Mann–Whitney test with $p = 0.004$ and $p < 0.001$, respectively). On average, stream water showed more isotopically negative and variable values and had lower EC and higher variability in summer than in fall and winter. Moreover, the main stream had more depleted isotopic composition and lower EC compared to the tributaries (Table 6). Spring water was the most enriched water source during the fall but became more depleted compared to stream water during the summer when it also showed higher EC. The coeffi-

Table 6. Basic statistics of isotopic composition (^{2}H) and EC of stream water in the Saldur catchment for data collected in the three sampling years. CV: coefficient of variation. The other abbreviations are used as in Table 2. Note that for simplicity the negative sign from the coefficient of variation of isotope data was removed.

Period*	Statistic	δ^2H Saldur River (‰)	δ^2H tributaries (‰)	δ^2H springs (‰)	EC Saldur River (μS cm^{-1})	EC tributaries (μS cm^{-1})	EC springs (μS cm^{-1})
Entire period	n	274	102	80	257	102	74
	avg.	−105.3	−103.4	−105.5	166.5	226.8	227.7
	SD	5.2	4.9	6.1	57.1	104.0	77.8
	CV	0.049	0.047	0.058	0.343	0.459	0.342
Summer	n	240	81	68	223	81	62
	avg.	−105.9	−104.5	−107.0	153.7	218.5	229.7
	SD	5.3	4.5	5.1	48.3	100.6	78.3
	CV	0.050	0.043	0.048	0.314	0.460	0.341
Fall–winter	n	34	21	12	34	21	12
	avg.	−101.1	−99.2	−96.9	250.7	258.8	217.2
	SD	2.6	4.0	4.2	32.9	113.0	77.8
	CV	0.026	0.040	0.044	0.131	0.437	0.358

* Summer is considered between mid-June (21) and end of September (23), and fall–winter between end of September and end of March (21).

cient of variations of δ^2H for groundwater were generally slightly higher than those for the stream water in all seasons, but the variability in EC was similar to that of the Saldur River and smaller than that of the tributaries (Table 6).

Overall, the median isotopic composition of stream water in the Saldur River varied slightly with location, but long error bars indicate a great temporal variability (Fig. 2). On the contrary, tributaries showed a wider range in the isotopic composition but a smaller temporal variability compared to the main stream (Fig. 2a). EC showed an increasing trend from upper to lower locations along the Saldur River (although with a slight interruption at S3-LSG) (Fig. 2b). Interestingly, T4 was the stream location with the most negative isotopic composition and highest EC. Groundwater tracer signature was overall intermediate between the main stream and the tributaries with a remarkable difference between SPR1-3 and SPR4.

Despite the strong variability, some spatial and temporal patterns can be observed (Fig. 3). For instance, all locations in June and early July 2012 showed isotopically depleted water and so did, overall, locations T4 and T5. Groundwater in SPR4 was constantly more enriched than in the other springs (Fig. 3a). The increasing trend in EC from the highest Saldur River location (S8) down to the lowest location (S1) in July and August of both years is also clearly visible, as well as the temporally constant and relatively very high EC of tributary water at T4 and very low EC of groundwater in SPR4 (Fig. 3b).

The mixing plot between δ^2H and EC of stream water and groundwater of all sampling locations further highlights the

differences in the tracer signature of the main stream, the tributaries and the springs (Fig. 4). Overall, the main stream showed a wider range in isotopic composition compared to the tributaries, in agreement with the long error bars of locations S1–S8 in Fig. 2. EC of the Saldur River was also more variable than EC in the other waters, except for T5 where plots separately compared to other tributaries and the main stream. The spring data points only partially overlap with the main stream data points: indeed, the tracer signal of the main stream water is upper-bounded by springs SPR1-3 and partially by T2-SG, and laterally, towards the less negative isotopic values, by SPR4. Only the tracer signal of T1, a left tributary flowing into the Saldur River a few hundred meters downstream of S1, lies within the main stream data, but samples were taken only in 2012 and therefore a robust comparison cannot be performed.

4.2 Quantification of snowmelt and glacier melt in streamflow and associated uncertainty

The results of the two- and three-component mixing models applied to 2013 data reveal a seasonally variable influence of snowmelt and glacier melt on streamflow, with estimated fractions generally decreasing from the highest to the lowest sampling location (Fig. 5). Overall, the proportion of snowmelt in stream water was comparable for the four sampling locations in August, September and October. Estimated snowmelt fractions were the highest on 19 June, up to 79 ± 6 % (scenario B) at S8. Field observations and MODIS data (Engel et al., 2016) revealed that the glacier surface was still covered with snow until the end of June. All four mix-

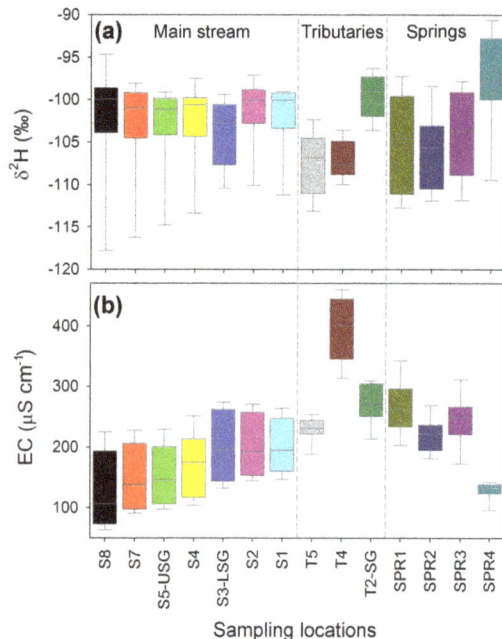

Figure 2. Box plot of δ^2H (**a**) and EC (**b**) for samples taken on the same day at all locations in 2011 and 2012 ($n = 10$ for all locations except for isotope data in T5 and for both tracers at SPR1, for which $n = 9$). Locations T1 and T3 are excluded because sampled only for 1 year. The boxes indicate the 25th and 75th percentile, the whiskers indicate the 10th and 90th percentile, the horizontal line within the box defines the median. In 2013, samples were collected only at some locations (Table 1) and therefore, for consistency, 2013 data are not reported here.

ing model scenarios agree with these observations and estimate no contribution of glacier melt to streamflow on the sampling days in May and June, and only partially on 18 October (Fig. 5). Glacier melt was an important component of streamflow on 16 July, especially according to scenarios A and B, and dominated the streamflow in mid-August according to all scenarios, with peak estimates at S8 ranging from 50–66 % (scenario D) to 68–71 % (scenario A). On 12 August, meltwater was the prevalent streamflow component at the three upper sampling locations and was still relevant at the lowest sampling location.

Overall, the four scenarios provide similar patterns of meltwater dynamics with higher similarities between scenarios A and B, and between scenarios C and D. Indeed, strong correlations exist between the estimates of the same component computed in each scenario, with R^2 for all possible combinations ranging between 0.91 and 0.997 for groundwater, 0.68 and 0.94 for snowmelt, and 0.74 and 0.94 for glacier melt ($n = 22$, $p < 0.01$ for all correlations). Despite the general agreement, differences in the estimated streamflow components among the four scenarios do exist. Particularly, scenarios C and D yield higher overall proportions of snowmelt compared to scenarios A and B, and scenarios A and D pro-

vide the overall highest and smallest fraction of glacier melt, respectively. Furthermore, scenarios C and D provide larger proportions of snowmelt and smaller proportions of glacier melt in July compared to the two other scenarios (Fig. 5). Overall, the uncertainty associated with the computation of the streamflow fractions is larger for scenarios A and C than for scenarios B and D (compare the length of error bars in Fig. 5).

It is worth mentioning that different proportions of meltwater components at the same stream location could be estimated according to the sampling time of the day. For the melt-induced runoff events sampled at high temporal resolution in 2011, 2012 and 2013 (Engel et al., 2016), the maximum contribution of meltwater to streamflow occurred at the streamflow peak or within an hour after the streamflow peak in 79 % of the observations, whereas the maximum contribution of meltwater was observed within 2 h before the streamflow peak in the remaining 21 % of the cases. Therefore, sampling several hours before or after the streamflow peak can lead to an underestimation of the meltwater fractions in streamflow (Fig. 6). However, the differences in meltwater fractions between samples collected at the streamflow peak and samples collected after the streamflow peak are lower and less variable (shorter error bars) than the ones computed before the streamflow peak (Fig. 6).

4.3 Relation between the two tracers, streamflow and meltwater fractions

The relation between δ^2H and EC of stream water samples collected at S5-USG and S3-LSG on the same days in 2011, 2012 and 2013, and grouped by month, shows different behaviours according to the sampling period (Fig. 7). Overall, sampling days in May, June and September were characterized by lower mean daily temperatures and stream discharge, much higher EC and more depleted isotopic composition compared to sampling days in July and August (Table 7). The relation between the two tracers is statistically significant in the colder months, whereas it is more scattered and not statistically significant during the warmest months (Fig. 7). The range of δ^2H values was slightly larger in the mid-summer period compared to May, June and September (16.7 ‰ vs. 15.1 ‰); on the contrary, the range of EC values was much larger in the spring–late summer period compared to July and August (173.9 μS cm^{-1} vs. 77.1 μS cm^{-1}).

Streamflow during the summer-melt runoff events sampled hourly in 2011, 2012 and 2013 at the two monitored cross sections S5-USG and S3-LSG (Engel et al., 2016) is positively correlated with the fraction of meltwater (snowmelt plus glacier melt components) (Fig. 8). Streamflow is presented for comparison purposes both in terms of specific discharge and relative to bankfull discharge, the latter being estimated in the two reaches based on direct observations during high flows. A closer inspection of the figure reveals the occurrence of hysteretic loops between stream-

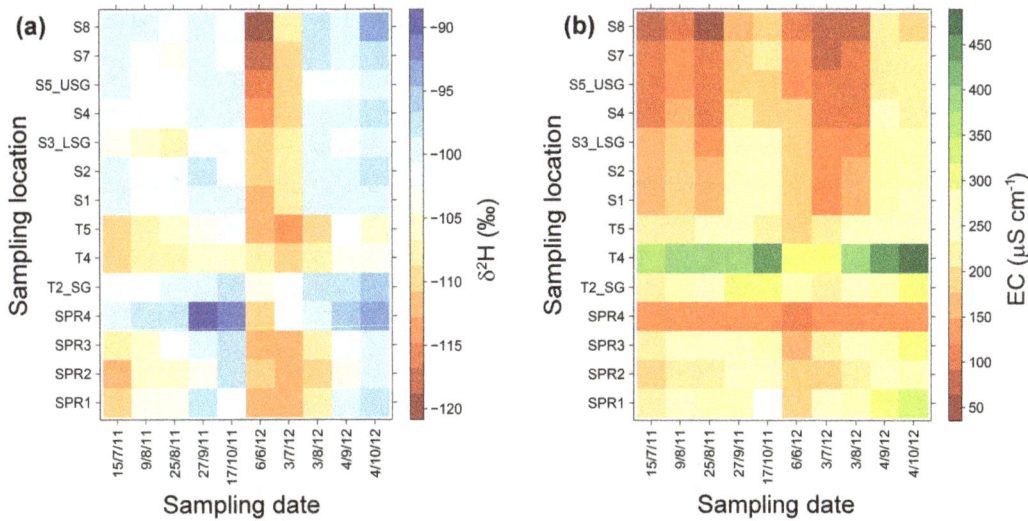

Figure 3. Spatio-temporal patterns of δ^2H (**a**) and EC (**b**) for samples taken on the same day at all locations in 2011 and 2012. Location T1 and T3 are excluded because sampled only for 1 year. White cells indicate no available measurements. In 2013, samples were collected only at some locations (Table 1) and therefore, for consistency, 2013 data are not reported here.

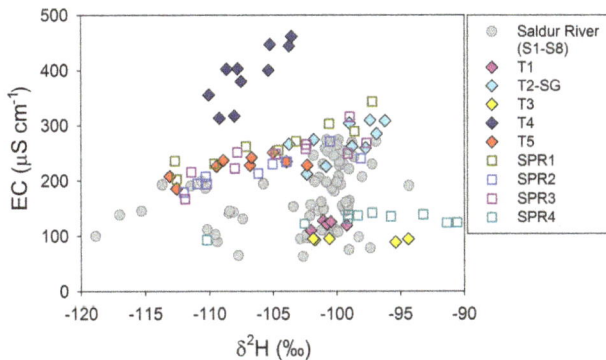

Figure 4. Relation between δ^2H and EC at all locations in the main stream, the tributaries and the springs in 2011 and 2012. Data refer to samples collected at each location on the same days except for T1 and T3, where samples were taken for 1 year only (cf. Table 1). In 2013, samples were collected only at some locations (Table 1) and therefore, for consistency, 2013 data are not reported here.

flow and meltwater at both locations more evident for events on 12–13 July 2011, 10–11 August 2011 and 21–22 August 2013 at S5-USG, due to their magnitude. Nevertheless, a general positive trend between the two variables is observable, with meltwater fractions increasing when streamflow increased ($R^2 = 0.48$, $n = 130$; $p < 0.01$ at S5-USG; $R^2 = 0.26$, $n = 114$; $p < 0.01$ at S3-LSG). The relation between meltwater fractions (computed as average of the results of the four mixing model scenarios) and streamflow is also plotted for the samples taken monthly in 2013, indicated by the stars in Fig. 8. The samples collected during the 2013 campaigns plot consistently with the samples taken during the melt-induced runoff events at both locations, overall

agreeing with the positive trend of the meltwater–streamflow relation (Fig. 8).

5   Discussion

5.1   Controls on the spatio-temporal patterns of the tracer signal

Glacier melt was characterized by similar isotopic composition in 2012 and 2013 and, most of all, by a marked isotopic enrichment and a slight EC increase over the summer season (Table 5). Yde et al. (2016) showed similar trends in the isotopic composition of meltwater draining Mittivakkat Gletscher, Greenland, for two summers, and Zhou et al. (2014) reported an isotopic enrichment in the firn pack during the early melting season on a glacier in the Tibetan Plateau. However, other studies have reported a strong interannual variability in the isotopic signature of glacier melt (Yuanqing et al., 2001) or fairly consistent values over time (Cable et al., 2011; Maurya et al., 2011; Ohlanders et al., 2013; Racoviteanu et al., 2013). In our case, since melting of the surface ice determines no isotopic fractionation (Jouzel and Souchez, 1982), as confirmed by glacier melt samples falling on the local meteorological water line (Penna et al., 2014), the progressive enrichment could be explained by contributions from deeper portions of the glacier surface with increasing ablation over the melting season or sublimation of surface ice (Stichler et al., 2001). More data from this and other glacierized sites should be acquired to better assess this variability that we believe must be taken into account in the application of mixing models for the estimation of glacier melt contribution to streamflow in different seasons.

Table 7. Basic statistics of specific discharge, $\delta^2 H$ and EC for the two groups reported in Fig. 7 for data collected in the three sampling years. Abbreviations are used as in Table 2.

	May, Jun, Sept 2011–2013				Jul, Aug 2011–2013			
	q ($m^3\,s^{-1}\,km^{-2}$)	$\delta^2 H$ (‰)	EC ($\mu S\,cm^{-1}$)	T (°C)	q ($m^3\,s^{-1}\,km^{-2}$)	$\delta^2 H$ (‰)	EC ($\mu S\,cm^{-1}$)	T (°C)
n	12	12	12	12	12	12	12	12
avg.	0.08	−109.3	193.5	5.9	0.15	−107.0	118.3	11.6
SD	0.09	5.2	52.7	5.4	0.04	5.6	25.7	1.0

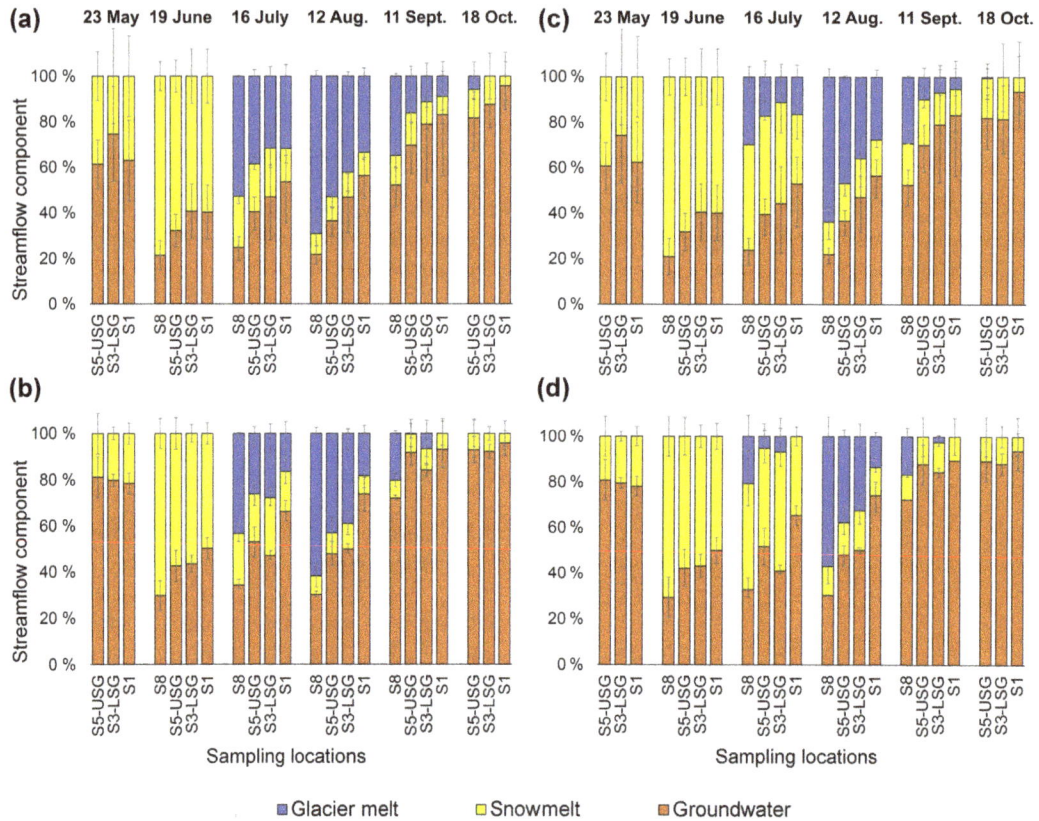

Figure 5. Fractions of groundwater, snowmelt and glacier melt in streamflow for the six sampling days in 2013 at four cross sections along the Saldur River. Left column panels: the isotopic composition and EC of the snowmelt end-member was considered time invariant, and the groundwater end-member was based on spring data (scenario A, **a**) or on stream data (scenario B, **b**). Right column panels: the isotopic composition of the snowmelt end-member was considered monthly variable, and the groundwater end-member was based on spring data (scenario C, **c**) or on stream data (scenario D, **d**) during baseflow conditions. The error bars represent the statistical uncertainty for each component.

More negative $\delta^2 H$ values and lower EC observed in the Saldur River and in its tributaries during the summer than during the winter (Table 6) clearly indicate contributions of meltwater, namely snowmelt, typically isotopically depleted, and glacier melt, typically very diluted in solutes. However, differences exist in the tracer signal among the main stream and the tributaries. The much lower EC of the Saldur River in summer compared to the tributaries (Table 6) suggests important contributions of both snowmelt from high elevations and almost solute-free glacier melt to the main stream, but fewer glacier melt contributions to the tributaries. The larger difference of the coefficients of variation between summer and fall–winter in the Saldur River with respect to the tributaries (Table 6) confirms greater inputs of waters with contrasting isotopic signals (depleted snowmelt and more enriched glacier melt) but relatively similar low EC (Maurya et al., 2011). This observation is corroborated by the larger temporal variability (longer error bars) in the isotopic com-

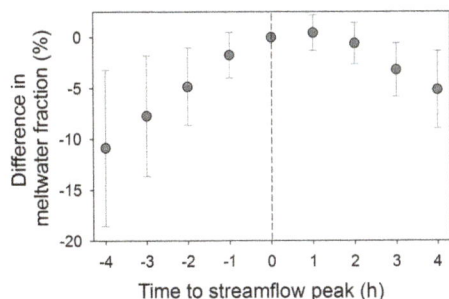

Figure 6. Average difference between the meltwater fraction in streamflow at the time of streamflow peak and the meltwater fraction at different hours from the time of streamflow peak for the melt-induced runoff events at S5-USG and S3-LSG in 2011–2013. Error bars represent the standard deviation. The vertical line indicates the time of streamflow peak.

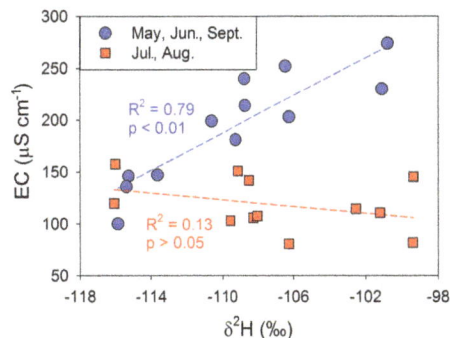

Figure 7. Relation between δ^2H and EC of samples collected at S5-USG and S3-LSG on the same days in 2011, 2012 and 2013, grouped by month.

position of the main stream compared to the tributaries, by the similar temporal variability in EC (expressed by the similar length of error bars in Fig. 2), and by the larger span of δ^2H values in the main stream compared to the tributaries visible in the mixing plot (Fig. 4).

The same isotopic composition of the Saldur River and the springs (Table 6, despite the lack of temporal consistency) and the partial overlap of the spring data points with the stream data points in the mixing plot (Fig. 4) suggest connectivity between the main stream and shallow groundwater, in agreement with observations in other glacierized catchments (Hindshaw et al., 2011; Magnusson et al., 2014). However, a large spatio-temporal variability in the tracer signal of springs was observed (Fig. 2–4) highlighting the complex hydrochemistry of the groundwater system (Brown et al., 2006; Hindshaw et al., 2011; Kong and Pang, 2012). The depleted signal in summer months (Table 6) suggests a role of snowmelt in groundwater recharge (Baraer et al., 2015; Fan et al., 2015; Xing et al., 2015) that was quantified in a previous study (Penna et al., 2014). At the same time, the relatively high EC during summer demonstrates solute concentration and suggests longer residence times and/or flow pathways (and thus long contact with the soil particles) of infiltrating meltwater before recharging the groundwater (Brown et al., 2006; Esposito et al., 2016). The similar coefficients of variations of the two tracers in summer and fall indicate fewer inter-seasonal differences in water inputs to the springs compared to the streams and suggest continuous groundwater recharge even at the end of the melting seasons, pointing out again to relatively long travel times and recharge times.

We mainly attribute the large spatial and temporal variability of tracers in stream water and groundwater to the control exerted by climate (seasonality), topography and geological settings. For instance, the depleted waters at all locations in June and early July 2012 (Fig. 3a) indicate heavy snowmelt contributions, consistent with the results of the mixing models (Fig. 5), clearly reflecting a climatic control (snow ac-

cumulation during the winter–early spring and subsequent melting). The increasing trend in EC from S8 to S1 during summer periods (Fig. 3b), consistent with other works (Kong and Pang, 2012; Fan et al., 2015), reflects the combined effect of lower elevations, smaller snow-covered area, decreasing glacierized area, progressive decrease of meltwater fractions and proportional increase of groundwater contributions (Fig. 5), and inflows by groundwater-dominated lateral tributaries.

The more depleted median isotopic composition and the higher EC of S3-LSG (Fig. 2) reflected the influence of the tributary T4, a few tens of meters upstream of S3-LSG that had a depleted signal and very high EC and that plotted separately in the mixing diagram (Fig. 4). A combination of depleted isotopic composition (typical of snowmelt) and high EC (typical of groundwater) was very rare in the catchment, and we do not have evidence to explain the origin of tributary T4 and the reason of its tracer signature. Analogously, our data did not provide robust explanations about the more enriched isotopic composition and the constantly much lower EC of SPR4 compared to other springs (Figs. 3 and 4). Ongoing and future analyses of major anions and cations will help to shed some light on the origin of T4 and SPR4.

5.2 Seasonal control on the δ^2H–EC relation and on meltwater fractions

As observed elsewhere (e.g. Hindshaw et al., 2011; Maurya et al., 2011; Blaen et al., 2014), streamflow in the main stream increased during melting periods, EC decreased due to the dilution effect and the isotopic composition generally shifted towards depleted values reflecting the meltwater signal. However, the two tracers were strongly correlated only in May, June and September (Fig. 7), when glacier melt was negligible or absent (Fig. 5), because the tracer signal in the stream reflected the low EC and the depleted isotopic composition of snowmelt. Conversely, during mid-summer, when glacier melt significantly contributed to streamflow (Fig. 5), the relation between the two tracers became weak (Fig. 7),

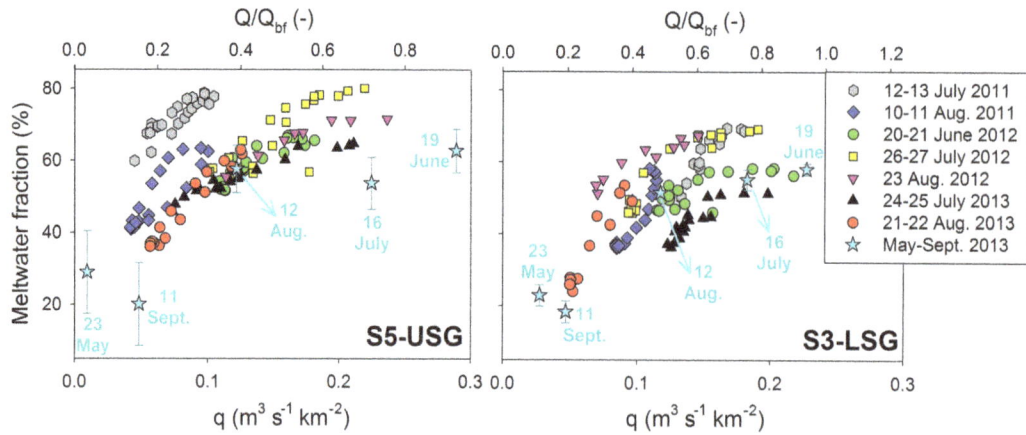

Figure 8. Relation between specific discharge (q) and meltwater fraction (%) in streamflow for the melt-induced runoff events in 2011, 2012 and 2013 sampled at hourly timescale (represented by different coloured symbols), and for the monthly sampling days in 2013 at S5-USG and S3-LSG (represented by stars in cyan). Meltwater fractions for the melt-induced runoff events were taken from Engel et al. (2016), while meltwater fractions for the monthly sampling days in 2013 are given by the average of the four different mixing models scenarios (presented in Fig. 5), and error bars indicate the standard deviation. For the double-peak event on 23–24 August 2012 at S5-USG, where a 9 mm rainstorm superimposed the melt event (cf. Engel et al., 2016), only the melt-induced part of the event was considered. Discharge is reported also as fraction of the bankfull discharge Q_{bf} at the two sections.

because glacier melt had very low EC but was not as isotopically depleted as snowmelt. Having multiple tracers is of certain usefulness when investigating water sources and mixing processes (Barthold et al., 2011), especially in highly heterogeneous environments (Hindshaw et al., 2011), and is essential for the identification of various streamflow components. However, it is important to know the periods when only one tracer could be reliably used, at least for assessing meltwater inputs, especially in glacierized catchments where logistical constraints are always challenging.

The hysteretic behaviour observed between streamflow and meltwater fraction for the melt-induced runoff events (Fig. 8) reflects the hysteresis observed in the relation between streamflow and EC (Engel et al., 2016), suggesting contributions from water sources characterized by different temporal dynamics (Dzikowski and Jobard, 2012). The combination of the highest streamflow and the highest meltwater proportion was obtained at both stream sections in June due to the remarkable contribution of meltwater from the relatively deep snowpack in the upper part of the catchment. It is worth highlighting how the meltwater fraction can frequently represent a substantial (> 50 %) proportion of the bankfull discharge, both during snow and glacier melt flows. This implies that the expected progress of glacier shrinking and future changes in both runoff components will likely have important consequences for the morphological configuration of high-elevation streams like the Saldur River, especially in the wider, braided reaches more responsive to variations in water and sediment fluxes (Wohl, 2010).

5.3 Role of snowmelt and glacier melt on streamflow

The spatial and temporal patterns of meltwater dynamics are consistent with those estimated in other high-elevation catchments worldwide. For instance, the dominant role of snowmelt in late spring–early summer and of glacier melt later in summer was observed across different sites in Asia, North America, South American and Europe (Aizen et al., 1996; Cable et al., 2011; Ohlanders et al., 2013; Blaen et al., 2014, respectively). The decreasing contribution of meltwater from the upper to the lower stream locations from June to October shown almost consistently by all scenarios (Fig. 5) is related to the increasing distance from the glacier and catchment size, and decreasing elevation, in agreement with results from other sites (Cable et al., 2011; Prasch et al., 2013; Racoviteanu et al., 2013; Marshall, 2014). Moreover, lateral contributions from non-glacier-fed tributaries and/or tributaries dominated by groundwater increased the groundwater fraction in streamflow as well and proportionally decreased the meltwater fraction (Marshall, 2014; Fan et al., 2015).

Our estimates of snowmelt contribution to streamflow during the melting season are consistent with those reported in other studies (Carey and Quinton, 2004; Mukhopadhyay and Khan, 2015) and with those found in the same catchment during individual runoff events (Engel et al., 2016). It is more difficult to compare our computed fractions of glacier melt in stream water with estimates in other sites because they can be highly dependent on the yearly climatic variability, on the proportion of glacierized area in the catchment and because they are usually reported at the monthly or yearly scale. However, when considering the total meltwater contribution, the computed fractions for the June–August period agree reasonably well with those recently estimated at the seasonal scale in other high-elevation catchments by Pu et al. (2013) (41–62, 12 % of glacierized area), Fan et al. (2015) (26–69 %), Xing et al. (2015) (almost 60 %) and at the annual scale by Jeelani et al. (2016) (52, 3 % of glacierized area), and are even higher than those computed by Mukhopadhyay and Khan (2015) (25–36 %). These observations stress the importance of water resources stored within the cryosphere even in catchments with limited extent of glacierized area, such as the Saldur catchment.

Overall, our tracer-based results on the influence of snowmelt and glacier melt on streamflow agree with glacier mass balance results, which revealed important losses from the glacier surface (−428 mm in snow water equivalent) for the year 2012–2013 (Galos and Kaser, 2013). Particularly, the first strong heat wave serving as melting input was observed in mid-June, when the glacier was still covered by snow and no glacier melt occurred (Galos and Kaser, 2013), in agreement with our estimates of snowmelt contributions (Fig. 5). Glaciological results also showed that most of the glacier mass loss occurred at the end of July to mid-August 2013, but glacier ablation in the lower part of the glacier (below 3000 m a.s.l.) was observed until the beginning of October (Galos and Kaser, 2013), corroborating our tracer-based estimates (Fig. 5).

5.4 Sources of uncertainties in the estimated streamflow components

Various sources of uncertainty affect the estimate of the streamflow components when using mixing models in complex environments such as mountain catchments (Uhlenbrook and Hoeg, 2003; Ohlanders et al., 2013). In cases of mixing model applications to separate snowmelt from glacier melt and groundwater, thus not considering rainfall, and in the case of no availability of streamflow measurements (in our case at S8 and S1), uncertainty can be mainly ascribed to the precision of the instrument used for the determination of the tracer signal, and the spatio-temporal patterns of the end-member tracer signature. The instrumental precision can be relatively easily taken into account and quantified by adopting statistically based procedures (e.g. Genereux, 1998). However, the spatio-temporal variation in the hydro-chemical signal of the end-members is more challenging to capture and can provide the largest source of uncertainty (Uhlenbrook and Hoeg, 2003; Pu et al., 2013). The isotopic composition and EC of shallow groundwater emerging from springs can be very different within a catchment, especially in cases of heterogeneous geology, as well as the tracer signature of streams at different locations even during baseflow conditions (Jeelani et al., 2010, 2015). Indeed, in our case, the highest uncertainty in the estimated component fractions provided by scenarios A and C can likely be ascribed to the spatial variability of the tracer signature of the sampled springs.

The isotopic composition of snowmelt can mainly change according to (i) macro-topography (e.g. aspect determines different melting rates and so different isotopic compositions); (ii) micro-topography, because small hollows tend to host "older" snow with a more enriched isotopic composition compared to sloping areas; (iii) elevation; and (iv) season, with δ values becoming more negative with increasing elevation and more positive over the melting season (Uhlenbrook and Hoeg, 2003; Holko et al., 2013; Ohlanders et al., 2013). EC of snow, and therefore, snowmelt can change as well due, for instance, to the ionic pulse at the beginning of the melting season (Williams and Melack, 1991) and/or reflecting seasonal inputs of impurities from the atmosphere (Li et al., 2006), although this variability is usually much more limited compared to that of the isotopes.

In our case, the instrumental precision of the isotope analyser and the EC meter is relatively low and was entirely taken into account by the statistical assessment of uncertainty we applied. The spatio-temporal variability of snowmelt was addressed by sampling snowmelt at different elevations, aspects and times of the seasons. Finally, we observed very limited spatial patterns but a marked seasonal change in the tracer signature of glacier melt (Table 5) that was taken into account in the mixing model application (Table 2). Despite these efforts, logistical issues related to the size of the catchment as well as practical and safety issues related to the accessibility of most areas of the catchment, not only in winter, and, not last, economical issues prevent a very detailed characterization and quantification of all sources of uncertainty associated with the estimates of the streamflow components at different times of the year and different stream locations. In addition, an underestimation of meltwater fractions due to sampling time not always corresponding to the streamflow peak should be considered (Fig. 6). Specifically, the samples taken on 19 June at S5-USG and S3-LSG were collected almost 4 h before the streamflow peak. This means that an additional contribution of snowmelt almost up to 20 % could be expected (Fig. 6). As far as we know, these results have not been reported elsewhere and are critical for a proper assessment of the uncertainty in the estimated component fractions. Moreover, these observations suggest that adequate sampling strategies are critical (Uhlenbrook and Hoeg, 2003) and must be considered when planning field campaigns aiming at the quantification of meltwater in glacierized catchments.

5.5 Conceptual model of streamflow components dynamics

The findings from our two previous studies (Penna et al., 2014; Engel et al., 2016) and from the present work allow us to derive a conceptual model of streamflow and tracer response to meltwater dynamics in the Saldur catchment (Fig. 9). To the best of our knowledge, this is the first study to present such a conceptual model of streamflow-component dynamics. Although intuitive, this conceptualization is im-

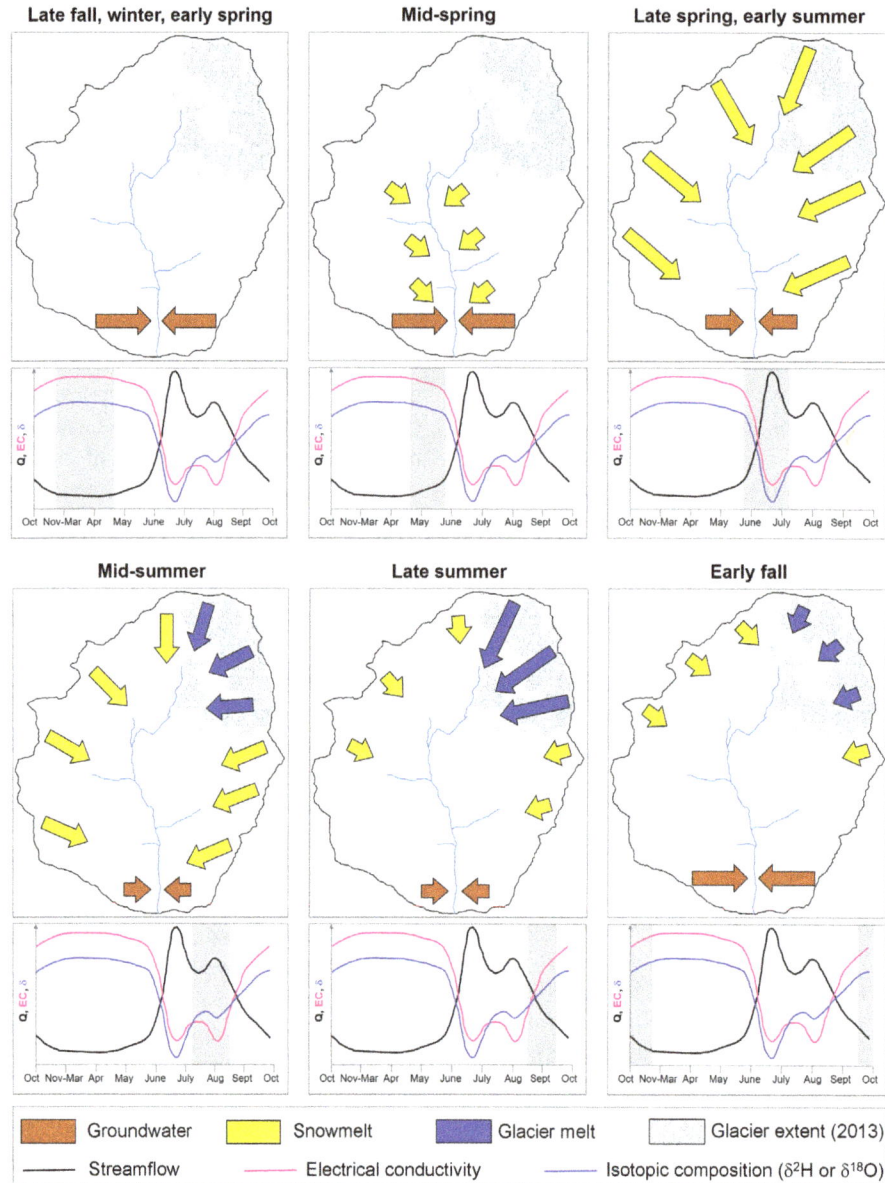

Figure 9. Conceptual model of the seasonal evolution of streamflow contributions in the Saldur River catchment (closed at LSG). The top subplots in each panel represent the water contributions to streamflow, and the size of the arrows is roughly proportional to the intensity of water fluxes. The bottom subplots show a sketch hydrograph along with EC and isotopic composition of stream water, and the shaded areas indicate time periods corresponding to the top subplots. The winter months, approximately between November and March, when the catchment is in a quiescent state and no significant hydrological dynamics is assumed to occur, are compacted in order to give more space to the other seasons.

portant because it represents a paradigm that, given the characteristics of the study site, can be applied to many other glacierized catchments worldwide.

During late fall, winter and early spring, precipitation mainly falls in form of snow, streamflow reaches its minimum and is predominantly formed by baseflow. EC in stream water is highest and the isotopic composition is relatively enriched, reflecting the groundwater signal. In mid-spring the melting season begins. The snowpack starts to melt at the

lower elevations in the catchment and the snow line progressively moves upwards; stream water EC begins to decrease due to the dilution effect and δ values become more negative, reflecting the first contribution of snowmelt (19–39 %). In late spring and early summer the combination of relatively high radiation inputs and still deep snowpack in the middle and upper portion of the catchment provides maximum snowmelt contributions to streamflow (up to 79 ± 6 % in the Saldur River at the highest sampling location) which is characterized by marked diurnal fluctuations and the highest melt-induced peaks. Groundwater fractions in stream water become proportionally smaller. The glacier surface is still totally snow covered; thus, glacier melt does not appreciably contribute to streamflow. EC is very low due to the strong dilution effect and the isotopic composition is most depleted. In mid-summer the snowpack is present only at the highest elevations and the glacier surface is mostly snow free, so that a combined role of snowmelt and glacier melt occurs. Streamflow is characterized by important diurnal fluctuations, but melt-induced peaks tend to be smaller in absolute values than in early summer associated with snowmelt. Although the snowmelt contribution has decreased, EC in the main stream is still very low due to the input of the extremely low EC of glacier melt. On the contrary, the stream water isotopic composition is less depleted compared to late spring and early summer due to the relatively more enriched signal of glacier melt with respect to snowmelt. In late summer snow disappears from most of the catchment and is only limited to residual patches in sheltered locations. The most important inputs to streamflow are provided by glacier melt that reaches its largest contributions (up to 68–71 % in the highest monitored Saldur River location). Diurnal fluctuations are still clearly visible but the decreasing radiation energy combined with lower melting supply limits high flows. EC begins to decrease and the isotopic composition to increase. From late spring to late summer low-intensity rainfall events provide limited contributions to streamflow. However, rainfall events of moderate or relatively high intensity can occur so that rain-induced runoff superimposes the melt-induced runoff and produces the highest observed streamflow peaks. In early fall, meltwater contributions are limited to snowmelt from early snowfalls at high elevations and residual glacier melt and the groundwater proportions become progressively more important. Streamflow decreases significantly and only small diurnal fluctuations are observable during clear days. The two tracers slowly return to their background values.

6   Conclusions and future perspectives

Our tracer-based studies (water isotopes and EC) in the Saldur catchment aimed to investigate the water sources variability and the contribution of snowmelt, glacier melt and groundwater to streamflow in order to contribute to a better comprehension of the hydrology of high-elevation glacierized catchments. We highlighted the highly complex hydrochemical signature of water in the catchment and the main controls on such variability. We applied mixing models to estimate the fractions of meltwater in streamflow over a season, not only at the catchment outlet as usually performed in other studies but also at different locations along the main stream. We found that snowmelt dominated the hydrograph in late spring–early summer, with fractions ranging between 50 ± 5 and 79 ± 6 % at different stream locations and accord-

ing to different model scenarios that took into account the spatial and temporal variability of end-member tracer signature. Glacier melt was a remarkable streamflow component in August, with maximum contributions ranging between 8–15 and 68–71 % at different stream locations and according to different scenarios. These estimates underline the key role of snowpack and glaciers on streamflow and stress their strategic importance as water resources.

From a methodological perspective, our results showed that during mixed snowmelt and glacier melt periods, EC and isotopes were not correlated due to the different tracer signature of the two sources of meltwater, whereas they provided a consistent pattern during snowmelt periods only. Such a behaviour, which we found hardly reported elsewhere, should be better assessed over longer time spans and in other sites, but suggests possible simplified monitoring strategies in snow-dominated catchments or during snowmelt periods in glacierized catchments. We identified the main sources of uncertainty in the computed estimates of streamflow components, mainly related to the spatio-temporal variability of the end-member tracer signature, including a clear seasonal enrichment of glacier melt isotopic composition. This is a pattern that must be considered when applying mixing models on a seasonal basis and that we invite to investigate in other glacierized environments. Furthermore, this is the first study, to our knowledge, which quantified the possible underestimation of meltwater fractions in streamflow occurring when stream water is sampled far from the streamflow peak during melt-induced runoff events. Again, this raises awareness about the need for careful planning of tracer-based field campaigns in high-elevation catchments.

We developed a perceptual model of meltwater dynamics and associated streamflow and tracer response in the Saldur catchment that likely applies to many other glacierized catchments worldwide. However, some limitations intrinsic in our approach should be considered. For instance, the reduced number of rain water samples collected at the rainfall-event scale over the 3 years did not allow us to fully assess the seasonal role of precipitation on streamflow in relation to meltwater. Furthermore, the use of EC, which integrates all water solutes in a single measurement, cannot differentiate well some water sources and their relation with the underlying geology. Finally, the monthly sampling resolution at different locations is useful to obtain a general overview and first estimates of the seasonal variability of streamflow components but high-frequency sampling can certainly help to capture finer hydrological dynamics. In this context, the results of the present work can serve as a very useful basis for modelling applications, particularly to constrain the model parametrization and to reduce the simulation uncertainties, and therefore to obtain more reliable predictions of streamflow dynamics and meltwater contributions to streamflow in high-elevation catchments.

Acknowledgements. This work was supported by the research projects "Effects of climate change on high-altitude ecosystems: monitoring the Upper Match Valley" (Foundation of the Free University of Bozen-Bolzano), "EMERGE: Retreating glaciers and emerging ecosystems in the southern Alps" (Dr. Erich Ritter- und Dr. Herzog-Sellenberg-Stiftung im Stifterverband für die Deutsche Wissenschaft) and partly by the project "HydroAlp", financed by the Autonomous Province of Bozen-Bolzano. We thank the Dept. of Hydraulic Engineering and Hydrographic Office of the Autonomous Province of Bozen-Bolzano for their technical support, G. Niedrist (EURAC) for maintaining the meteorological stations, Giulia Zuecco (University of Padova, Italy) for the isotopic analyses and Stefan Galos (University of Innsbruck, Austria) for sharing glacier mass balance results. The site Matsch/Mazia belongs to the national and international long-term ecological research networks (LTER-Italy, LTER Europe and ILTER).

Edited by: M. Hrachowitz

References

Aizen, V. B, Aizen, E. M., and Melack, J. M.: Precipitation, melt and runoff in the northern Tien Shan, J. Hydrol., 186, 229–251, 1996.

Baraer, M., McKenzie, J., Mark, B. G., Gordon, R., Bury, J., Condom, T., Gomez, J., Knox, S., and Fortner, S. K.: Contribution of groundwater to the outflow from ungauged glacierized catchments: a multi-site study in the tropical Cordillera Blanca, Peru, Hydrol. Process., 29, 2561–2581, doi:10.1002/hyp.10386, 2015.

Barthold, F. K., Tyralla, C., Schneider, K., Vaché, K. B., Frede, H.-G., and Breuer, L.: How many tracers do we need for end member mixing analysis (EMMA)? A sensitivity analysis, Water Resour. Res., 47, W08519, doi:10.1029/2011WR010604, 2011.

Beniston, M.: Impacts of climatic change on water and associated economic activities in the Swiss Alps, J. Hydrol., 412–413, 291–296, doi:10.1016/j.jhydrol.2010.06.046, 2012.

Blaen, P. J., Hannah, D. M., Brown, L. E., and Milner, A. M.: Water source dynamics of high Arctic river basins: water source dynamics of high arctic river basins, Hydrol. Process., 28, 3521–3538, doi:10.1002/hyp.9891, 2014.

Brown, L. E., Hannah, D. M., Milner, A. M., Soulsby, C., Hodson, A. J., and Brewer, M. J.: Water source dynamics in a glacierized alpine river basin (Taillon-Gabiétous, French Pyrénées): alpine basin water source dynamics, Water Resour. Res., 42, W08404, doi:10.1029/2005WR004268, 2006.

Cable, J., Ogle, K., and Williams, D.: Contribution of glacier melt-water to streamflow in the Wind River Range, Wyoming, inferred via a Bayesian mixing model applied to isotopic measurements, Hydrol. Process., 25, 2228–2236, doi:10.1002/hyp.7982, 2011.

Carey, S. K. and Quinton, W. L.: Evaluating snowmelt runoff generation in a discontinuous permafrost catchment using stable isotope, hydrochemical and hydrometric data, Nord. Hydrol., 35, 309–324, 2004.

Carey, S. K. and Quinton, W. L.: Evaluating runoff generation during summer using hydrometric, stable isotope and hydrochemical methods in a discontinuous permafrost alpine catchment, Hydrol. Process., 19, 95–114, doi:10.1002/hyp.5764, 2005.

Carturan, L., Zuecco, G., Seppi, R., Zanoner, T., Borga, M., Carton, A., and Dalla Fontana, G.: Catchment-scale permafrost mapping using spring water characteristics, Permafrost Periglac. Process., 27, 253–270, doi:10.1002/ppp.1875, 2016.

Dzikowski, M. and Jobard, S.: Mixing law versus discharge and electrical conductivity relationships: application to an alpine proglacial stream, Hydrol. Process., 26, 2724–2732, doi:10.1002/hyp.8366, 2012.

Engel, M., Penna, D., Bertoldi, G., Dell'Agnese, A., Soulsby, C., and Comiti, F.: Identifying run-off contributions during melt-induced run-off events in a glacierized alpine catchment, Hydrol. Process., 30, 343–364, doi:10.1002/hyp.10577, 2016.

Engelhardt, M., Schuler, T. V., and Andreassen, L. M.: Contribution of snow and glacier melt to discharge for highly glacierised catchments in Norway, Hydrol. Earth Syst. Sci., 18, 511–523, doi:10.5194/hess-18-511-2014, 2014.

Esposito, A., Engel, M., Ciccazzo, S., Daprà, L., Penna, D., Comiti, F., Zerbe, S., and Brusetti, L.: Spatial and temporal variability of bacterial communities in high alpine water spring sediments, Res. Microbiol., 167, 325–333, doi:10.1016/j.resmic.2015.12.006, 2016.

Fan, Y., Chen, Y., Li, X., Li, W., and Li, Q.: Characteristics of water isotopes and ice-snowmelt quantification in the Tizinafu River, north Kunlun Mountains, Central Asia, Quatern. Int., 380–381, 116–122, doi:10.1016/j.quaint.2014.05.020, 2015.

Finger, D., Hugentobler, A., Huss, M., Voinesco, A., Wernli, H., Fischer, D., Weber, E., Jeannin, P.-Y., Kauzlaric, M., Wirz, A., Vennemann, T., Hüsler, F., Schädler, B., and Weingartner, R.: Identification of glacial meltwater runoff in a karstic environment and its implication for present and future water availability, Hydrol. Earth Syst. Sci., 17, 3261–3277, doi:10.5194/hess-17-3261-2013, 2013

Fischer, B. M. C., Rinderer, M., Schneider, P., Ewen, T., and Seibert, J.: Contributing sources to baseflow in pre-alpine headwaters using spatial snapshot sampling, Hydrol. Process., 29, 5321–5336, doi:10.1002/hyp.10529, 2015.

Galos, S. and Kaser, G.: The mass balance of Matscherferner 2012/13, Report of the research project "A physically based regional mass balance approach for the glaciers of Vinschgau – glacier contribution to water availability", funded by the Autonomous Province of Bozen-Bolzano, Italy, 2013.

Genereux, D.: Quantifying uncertainty in tracer-based hydrograph separations, Water Resour. Res., 34, 915–919, doi:10.1029/98WR00010, 1998.

Hindshaw, R. S., Tipper, E. T., Reynolds, B. C., Lemarchand, E., Wiederhold, J. G., Magnusson, J., Bernasconi, S. M., Kretzschmar, R., and Bourdon, B.: Hydrological control of stream water chemistry in a glacial catchment (Damma Glacier, Switzerland), Chem. Geol., 285, 215–230, doi:10.1016/j.chemgeo.2011.04.012, 2011.

Hoeg, S., Uhlenbrook, S., and Leibundgut, C.: Hydrograph separation in a mountainous catchment – combining hydro-chemical and isotopic tracers, Hydrol. Process., 14, 1199–1216, doi:10.1002/(SICI)1099-1085(200005)14:7<1199::AID-HYP35>3.0.CO;2-K, 2000.

Holko, L., Danko, M., Dóša, M., Kostka, Z., Šanda, M., Pfister, L., and Iffly, J. F.: Spatial and temporal variability of stable water isotopes in snow related hydrological processes, Bodenkultur, 39, 3–4, 2013.

Jeelani, G., Bhat, N. A., and Shivanna, K.: Use of $\delta^{18}O$ tracer to identify stream and spring origins of a mountainous catchment: A case study from Liddar watershed, Western Himalaya, India, J. Hydrol., 393, 257–264, doi:10.1016/j.jhydrol.2010.08.021, 2010.

Jeelani, G., Kumar, U. S., Bhat, N. A., Sharma, S., and Kumar, B.: Variation of $\delta^{18}O$, δD and ^3H in karst springs of south Kashmir, western Himalayas (India), Hydrol. Process., 29, 522–530, doi:10.1002/hyp.10162, 2015.

Jeelani, G., Shah, R. A., Jacob, N., and Deshpande, R. D.: Estimation of snow and glacier melt contribution to Liddar stream in a mountainous catchment, western Himalaya: an isotopic approach, Isotop. Environ. Health Stud., doi:10.1080/10256016.2016.1186671, in press, 2016.

Jost, G., Moore, R. D., Menounos, B., and Wheate, R.: Quantifying the contribution of glacier runoff to streamflow in the upper Columbia River Basin, Canada, Hydrol. Earth Syst. Sci., 16, 849–860, doi:10.5194/hess-16-849-2012, 2012.

Jouzel, J. and Souchez, R. A.: Melting-refreezing at the glacier sole and the isotopic composition of the ice, J. Glaciol., 28, 35–42, 1982.

Kendall, C. and McDonnell, J. J.: Isotope tracers in catchment hydrology, Elsevier Science Limiter, Amsterdam, 1998.

Klaus, J. and McDonnell, J. J.: Hydrograph separation using stable isotopes: Review and evaluation, J. Hydrol., 505, 47–64, doi:10.1016/j.jhydrol.2013.09.006, 2013.

Kong, Y. and Pang, Z.: Evaluating the sensitivity of glacier rivers to climate change based on hydrograph separation of discharge, J. Hydrol., 434–435, 121–129, doi:10.1016/j.jhydrol.2012.02.029, 2012

Lee, J., Feng, X., Faiia, A. M., Posmentier, E. S., Kirchner, J. W., Osterhuber, R., and Taylor, S.: Isotopic evolution of a seasonal snowcover and its melt by isotopic exchange between liquid water and ice, Chem. Geol., 270, 126–134, doi:10.1016/j.chemgeo.2009.11.011, 2010.

Li, Z., Ross, E., Mosley-Thompson, E., Wang, F., Dong, Z., You, X., Li, H., Li, Z., and Chuanjin, Y.: Seasonal variabilities of ionic concentration in surface snow and elution process in snow-firn packs at PGPI Site on Glacier No. 1 in Eastern Tianshan, China, Ann. Glaciol., 43, 2006.

Magnusson, J., Kobierska, F., Huxol, S., Hayashi, M., Jonas, T., and Kirchner, J. W.: Melt water driven stream and groundwater stage fluctuations on a glacier forefield (Dammagletscher, Switzerland): stream-groundwater interactions on a glacier forefield, Hydrol. Process., 28, 823–836, doi:10.1002/hyp.9633, 2014.

Mair, E., Leitinger, G., Della Chiesa, S., Niedrist, G., Tappeiner, U., and Bertoldi, G.: A simple method to combine snow height and meteorological observations to estimate winter precipitation at sub-daily resolution, Hydrolog. Sci. J., doi:10.1080/02626667.2015.1081203, in press, 2016.

Mao, L., Dell'Agnese, A., Huincache, C., Penna, D., Engel, M., Niedrist, G., and Comiti, F.: Bedload hysteresis in a glacier-fed mountain river: bedload hysteresis in a glacier-fed mountain river, Earth Surf. Proc. Land., 39, 964–976, doi:10.1002/esp.3563, 2014.

Marshall, S. J.: Meltwater run-off from Haig Glacier, Canadian Rocky Mountains, 2002–2013, Hydrol. Earth Syst. Sci.,18, 5181–5200, doi:10.5194/hess-18-5181-2014, 2014.

Maurya, A. S., Shah, M., Deshpande, R. D., Bhardwaj, R. M., Prasad, A., and Gupta, S. K.: Hydrograph separation and precipitation source identification using stable water isotopes and conductivity: River Ganga at Himalayan foothills, Hydrol. Process., 25, 1521–1530, doi:10.1002/hyp.7912, 2011.

Mukhopadhyay, B. and Khan, A.: A reevaluation of the snowmelt and glacial melt in river flows within Upper Indus Basin and its significance in a changing climate, J. Hydrol., 527, 119–132, doi:10.1016/j.jhydrol.2015.04.045, 2015.

Mutzner, R., Weijs, S. V., Tarolli, P., Calaf, M., Oldroyd, H. J., and Parlange, M. B.: Controls on the diurnal streamflow cycles in two subbasins of an alpine headwater catchment, Water Resour. Res., 51, 3403–3418, doi:10.1002/2014WR016581, 2015.

Ogunkoya, O. O. and Jenkins, A.: Analysis of storm hydrograph and flow pathways using a three-component hydrograph separation model, J. Hydrol., 142, 71–88, 1993.

Ohlanders, N., Rodriguez, M., and McPhee, J.: Stable water isotope variation in a Central Andean watershed dominated by glacier and snowmelt, Hydrol. Earth Syst. Sci., 17, 1035–1050, doi:10.5194/hess-17-1035-2013, 2013.

Penna, D., Stenni, B., Šanda, M., Wrede, S., Bogaard, T. A., Gobbi, A., Borga, M., Fischer, B. M. C., Bonazza, M., and Chárová, Z.: On the reproducibility and repeatability of laser absorption spectroscopy measurements for δ^2H and $\delta^{18}O$ isotopic analysis, Hydrol. Earth Syst. Sci., 14, 1551–1566, doi:10.5194/hess-14-1551-2010, 2010.

Penna, D., Stenni, B., Šanda, M., Wrede, S., Bogaard, T. A., Michelini, M., Fischer, B. M. C., Gobbi, A., Mantese, N., Zuecco, G., Borga, M., Bonazza, M., Sobotková, M., Čejková, B., and Wassenaar, L. I.: Technical Note: Evaluation of between-sample memory effects in the analysis of δ^2H and $\delta^{18}O$ of water samples measured by laser spectroscopes, Hydrol. Earth Syst. Sci., 16, 3925–3933, doi:10.5194/hess-16-3925-2012, 2012.

Penna, D., Engel, M., Mao, L., Dell'Agnese, A., Bertoldi, G., and Comiti, F.: Tracer-based analysis of spatial and temporal variations of water sources in a glacierized catchment, Hydrol. Earth Syst. Sci., 18, 5271–5288, doi:10.5194/hess-18-5271-2014, 2014.

Penna, D., van Meerveld, H. J., Oliviero, O., Zuecco, G., Assendelft, R. S., Dalla Fontana, G., and Borga, M.: Seasonal changes in runoff generation in a small forested mountain catchment, Hydrol. Process., 29, 2027–2042, doi:10.1002/hyp.10347, 2015.

Pinder, G. F. and Jones, J. F., 1969. Determination of ground-water component of peak discharge from chemistry of total runoff, Water Resour. Res., 5, 438–445, doi:10.1029/WR005i002p00438, 1969.

Prasch, M., Mauser, W., and Weber, M.: Quantifying present and future glacier melt-water contribution to runoff in a central Himalayan river basin, The Cryosphere 7, 889–904, doi:10.5194/tc-7-889-2013, 2013.

Pu, T., He, Y., Zhu, G., Zhang, N., Du, J., and Wang, C.: Characteristics of water stable isotopes and hydrograph separation in Baishui catchment during the wet season in Mt. Yulong region, south western China, Hydrol. Process., 27, 3641–3648, doi:10.1002/hyp.9479, 2013.

Racoviteanu, A. E., Armstrong, R., and Williams, M. W.: Evaluation of an ice ablation model to estimate the contribution of melting glacier ice to annual discharge in the Nepal Himalaya, Water Resour. Res., 49, 5117–5133, doi:10.1002/wrcr.20370, 2013.

Rock, L. and Mayer, B.: Isotope hydrology of the Oldman River basin, southern Alberta, Canada, Hydrol. Process., 21, 3301–3315, doi:10.1002/hyp.6545, 2007.

Schaefli, B., Hingray, B., and Musy, A.: Climate change and hydropower production in the Swiss Alps: quantification of potential impacts and related modelling uncertainties, Hydrol. Earth Syst. Sci., 11, 1191–1205, doi:10.5194/hess-11-1191-2007, 2007.

Sklash, M. G.: Environmental isotope studies of storm and snowmelt runoff generation, in: Process studies in hillslope hydrology, edited by: Anderson, M. G. and Burt, T. P., Wiley, Chichester, 401–35, 1990.

Sklash, M. G. and Farvolden, R. N.: Role of groundwater in storm runoff, J. Hydrol., 43, 45–65, doi:10.1016/0022-1694(79)90164-1, 1979.

Stichler, W., Schotterer, U., Fröhlich, K., Ginot, P., Kull, C., Gäggeler, H., and Pouyaud, B.: Influence of sublimation on stable isotope records recovered from high-altitude glaciers in the tropical Andes, J. Geophys. Res., 106, 22613–22620, 2001.

Taylor, S., Feng, X., Kirchner, J. W., Osterhuber, R., Klaue, B., and Renshaw, C. E.: Isotopic evolution of a seasonal snowpack and its melt, Water Resour. Res., 37, 759–769, 2001.

Uhlenbrook, S. and Hoeg, S.: Quantifying uncertainties in tracer-based hydrograph separations: a case study for two-, three- and five-component hydrograph separations in a mountainous catchment, Hydrol. Process., 17, 431–453, doi:10.1002/hyp.1134, 2003.

Vaughn, B. H. and Fountain, A. G.: Stable isotopes and electrical conductivity as keys to understanding water pathways and storage in South Cascade Glacier, Washington, USA, Ann. Glaciol., 40, 107–112, doi:10.3189/172756405781813834, 2005.

Williams, M. W. and Melack, J. M.: Solute chemistry of snowmelt and runoff in an Alpine Basin, Sierra Nevada, Water Resour. Res., 27, 1575–1588, doi:10.1029/90WR02774, 1991.

Wohl, E.: Mountain rivers revisited, vol. 19, American Geophysical Union, 2010.

Xing, B., Liu, Z., Liu, G., and Zhang, J.: Determination of runoff components using path analysis and isotopic measurements in a glacier-covered alpine catchment (upper Hailuogou Valley) in southwest China: hydrograph separation; path analysis; glacier-covered catchment, Hydrol. Process., 29, 3065–3073, doi:10.1002/hyp.10418, 2015.

Yde, J. C., Knudsen, N. T., Steffensen, J. P., Carrivick, J. L., Hasholt, B., Ingeman-Nielsen, T., Kronborg, C., Larsen, N. K., Mernild, S. H., Oerter, H., Roberts, D. H., and Russell, A. J.: Stable oxygen isotope variability in two contrasting glacier river catchments in Greenland, Hydrol. Earth Syst. Sci., 20, 1197–1210, doi:10.5194/hess-20-1197-2016, 2016.

Yuanqing, H., Theakstone, W. H., Tandong, Y., and Yafeng, S.: The isotopic record at an alpine glacier and its implications for local climatic changes and isotopic homogenization processes, J. Glaciol., 47, 147–151, doi:10.3189/172756501781832601, 2001.

Zhou, S., Wang, Z., and Joswiak, D. R.: From precipitation to runoff: stable isotopic fractionation effect of glacier melting on a catchment scale, Hydrol. Process., 28, 3341–3349, doi:10.1002/hyp.9911, 2014.

11

Reconciling high-altitude precipitation in the upper Indus basin with glacier mass balances and runoff

W. W. Immerzeel[1,3,4]**, N. Wanders**[1,2]**, A. F. Lutz**[3]**, J. M. Shea**[4]**, and M. F. P. Bierkens**[1]

[1]Department of Physical Geography, Utrecht University, Utrecht, the Netherlands
[2]Department of Civil and Environmental Engineering, Princeton University, Princeton, NJ, USA
[3]FutureWater, Wageningen, the Netherlands
[4]International Centre for Integrated Mountain Development, Kathmandu, Nepal

Correspondence to: W. W. Immerzeel (w.w.immerzeel@uu.nl)

Abstract. Mountain ranges in Asia are important water suppliers, especially if downstream climates are arid, water demands are high and glaciers are abundant. In such basins, the hydrological cycle depends heavily on high-altitude precipitation. Yet direct observations of high-altitude precipitation are lacking and satellite derived products are of insufficient resolution and quality to capture spatial variation and magnitude of mountain precipitation. Here we use glacier mass balances to inversely infer the high-altitude precipitation in the upper Indus basin and show that the amount of precipitation required to sustain the observed mass balances of large glacier systems is far beyond what is observed at valley stations or estimated by gridded precipitation products. An independent validation with observed river flow confirms that the water balance can indeed only be closed when the high-altitude precipitation on average is more than twice as high and in extreme cases up to a factor of 10 higher than previously thought. We conclude that these findings alter the present understanding of high-altitude hydrology and will have an important bearing on climate change impact studies, planning and design of hydropower plants and irrigation reservoirs as well as the regional geopolitical situation in general.

1 Introduction

Of all Asian basins that find their headwaters in the greater Himalayas, the Indus basin depends most strongly on high-altitude water resources (Immerzeel et al., 2010; Lutz et al., 2014) The largest glacier systems outside the polar regions are found in this area and the seasonal snow cover is the most extensive of all Asian basins (Immerzeel et al., 2009). In combination with a semi-arid downstream climate, a high demand for water as a result of the largest irrigation scheme in the world and a large and quickly growing population, the importance of the upper Indus basin (UIB) is evident (Immerzeel and Bierkens, 2012).

The hydrology of the UIB ($4.37 \times 10^5 \, \text{km}^2$) is, however, poorly understood. The quantification of the water balance in space and time is a major challenge due the lack of measurements and the inaccessibility of the terrain. The magnitude and distribution of high-altitude precipitation, which is the driver of the hydrological cycle, is one of its largest unknowns (Hewitt, 2005, 2007; Immerzeel et al., 2013; Mishra, 2015; Ragettli and Pellicciotti, 2012; Winiger et al., 2005). Annual precipitation patterns in the UIB result from the intricate interplay between synoptic scale circulation and valley scale topography–atmosphere interaction resulting in orographic precipitation and funnelling of air movement (Barros et al., 2004; Hewitt, 2013). At the synoptic scale, annual precipitation originates from two sources: the south-eastern monsoon during the summer and moisture transported by the westerly jet stream over central Asia (Mölg et al., 2013; Scherler et al., 2011) during winter. The relative contribution of westerly disturbances to the total annual precipitation increases from south-east to north-west, and the anomalous behaviour of Karakoram glaciers is commonly attributed to changes in winter precipitation (Scherler et al., 2011; Yao et al., 2012).

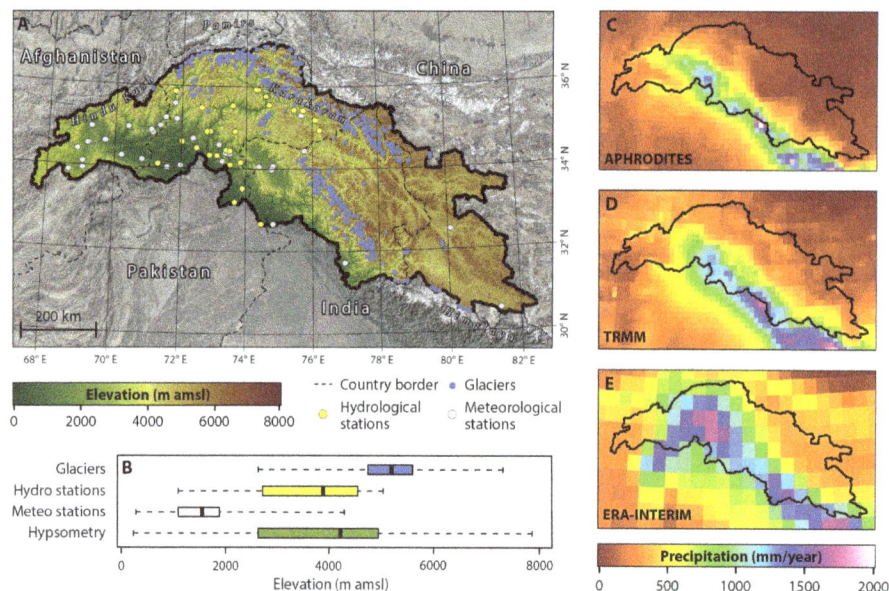

Figure 1. Overview of the UIB (Lehner et al., 2008), basin hypsometry and three gridded precipitation products. (**a**) shows the digital elevation model, the location of the major glacier systems (area > 5 km^2), the available stations in the Global Summary of the Day (GSOD) of the World Meteorological Organization (WMO) and the hydrological stations used for validation. Panel B shows box plots of the elevation distribution of the basin, the large glacier systems, the GSOD meteorological stations and the average elevation of the catchment area of each hydrological station. (**c**) to (**e**) show the average gridded annual precipitation between 1998 and 2007 for the APHRODITE (Yatagai et al., 2012), TRMM (Huffman et al., 2007) and ERA-Interim (Dee et al., 2011) data sets.

At smaller scales the complex interaction between the valley topography and the atmosphere dictates the spatial distribution of precipitation (Bookhagen and Burbank, 2006; Immerzeel et al., 2014b). Valley bottoms, where stations are located, are generally dry and precipitation increases up to a certain maximum altitude (HMAX) above which all moisture has been orographically forced out of the air and precipitation decreases again. In westerly dominated rainfall regimes HMAX is generally higher, which is likely related to the higher tropospheric altitude of the westerly airflow (Harper, 2003; Hewitt, 2005, 2007; Scherler et al., 2011; Winiger et al., 2005).

Gridded precipitation products are the de facto standard in hydrological assessments, and they are either based on observations (e.g. APHRODITES; Yatagai et al., 2012), remote sensing (e.g. the Tropical Rainfall Monitoring Mission; Huffman et al., 2007) or re-analysis (e.g. ERA-Interim; Dee et al., 2011) (Fig. 1c–e). In most cases the station data strongly influence the distribution and magnitude of the precipitation in those data products; however, the vast majority of the UIB is located at elevations far beyond the average station elevation (Fig. 1a–b). The few stations that are at elevations above 2000 m are located in dry valleys and we hypothesise that the high-altitude precipitation is considerably underestimated (Fig. 1c–e). Moreover, remote-sensing-based products, such as the Tropical Rainfall Mea-

suring Mission (TRMM), are insufficiently capable of capturing snowfall (Bookhagen and Burbank, 2006; Huffman et al., 2007) and the spatial resolution (25–75 km^2) of most rainfall products (and the underlying models) is insufficient to capture topography–atmosphere interaction at the valley scale (Fig. 1c–e). Thus, there is a pressing need to improve the quantification of high-altitude precipitation, preferably at large spatial extents and at high resolution.

A possible way to correct mountain precipitation is to inversely close the water balance. Previous studies in Sweden and Switzerland have shown that it is possible to derive vertical precipitation gradients using observed runoff in a physically realistic manner (Valéry et al., 2009, 2010). Earlier work at the small scale in high mountain Asia suggested that the glacier mass balance may be used to reconstruct precipitation in its catchment area (Harper, 2003; Immerzeel et al., 2012a). Figure 1a and b show that UIB glaciers are located at high elevations that are not represented by station data. Therefore, the mass balances of the glaciers may contain important information on high-altitude accumulation in an area that is inaccessible and ungauged, but very important from a hydrological point of view. In this study we further elaborate this approach by inversely modelling average annual precipitation from the mass balance of 550 large (> 5 km^2) glacier systems located throughout the UIB. We perform a rigorous uncertainty analysis and we validate our findings using independent observations of river runoff.

2 Methods

We estimate high-altitude precipitation by using a glacier mass balance model to simulate observed glacier mass balances. We use a gridded data set from valley bottom stations as a basis for our precipitation estimate and we compute a vertical precipitation gradient (PG; $\% \, m^{-1}$) until observed mass balances match the simulated mass balance. We repeat this process for the 550 major glacier systems in the UIB, and the resulting PGs are then spatially interpolated to generate a spatial field that represents the altitude dependence of precipitation. We use this field to update the APHRODITE precipitation and generate a corrected precipitation field that is able to reproduce the observed glacier mass balance. We validate the findings independently with a water balance approach. Estimated (annual) runoff, based on the corrected precipitation, actual evapotranspiration based on four gridded products and the observed glacier mass balance, is compared with an extensive set of UIB runoff observations. We also analyse the physical realism of our simulations by deriving a Turc–Budyko plot using precipitation, measured runoff and potential evapotranspiration. A rigorous uncertainty analysis is also conducted on the six most critical model parameters including potential effects of spatial correlation.

2.1 Data sets

2.1.1 Glacier mass balance and outlines

Glacier mass balance trends based on NASA's Ice, Cloud and land Elevation Satellite (ICESat) (Kääb et al., 2012a) are recomputed for the period 2003 until 2008 for the three major mountain ranges in the UIB: the Karakoram, the Hindu Kush and the Himalaya (Fig. 1). For each zone the mass balance is computed including a regional uncertainty estimate (Kääb et al., 2012a). From the zonal uncertainty (σ_z) we estimate the standard deviation between glaciers within a zone (σ_g) as

$$\sigma_g = \sigma_z \sqrt{n}, \tag{1}$$

where n is the number of glaciers within a zone. The σ_g values used in the uncertainty analysis are shown in Table 1.

The glacier boundaries are based on the glacier inventory of the International Centre for Integrated Mountain Development (Bajracharya and Shrestha, 2011).

2.1.2 Precipitation and temperature

The daily APHRODITE precipitation (Yatagai et al., 2012) and air temperature data sets (Yasutomi et al., 2011) from 2003 until 2007 are used as reference data sets to ensure maximum temporal overlap with the ICESat-based glacier mass balance data set (Kääb et al., 2012a). The precipitation data set is resampled from the nominal resolution of $25 \, km^2$ to a resolution of $1 \, km^2$ using the nearest neighbour algorithm. The air temperature data set is then bias corrected using monthly linear regressions with independent station data

to account for altitudinal and seasonal variations in air temperature lapse rates (Fig. 3).

2.1.3 Runoff and evapotranspiration

We use runoff data, potential evapotranspiration (ETp) and actual evapotranspiration (ETa) data for the validation of our results. For runoff we compiled all available published data, which we complemented with data made available by the Pakistan Meteorological Department and the Pakistan Water and Power Development Authority.

Evapotranspiration is notoriously difficult to monitor and there are few direct measurements of ETa in the upper Indus. In earlier UIB studies, ET was estimated using empirical formulae based on air temperature but was only applied to the Siachen glacier (Bhutiyani, 1999; Reggiani and Rientjes, 2014). We take into account the uncertainty in ET in our streamflow estimates and develop a blended product based on re-analysis data sets, a global hydrological model and an energy balance model. Four gridded ETa and three gridded ETp products were resampled to a $1 \, km^2$ resolution at which we perform all our analyses:

- ERA-Interim re-analysis (Dee et al., 2011): ERA-Interim uses the HTESSEL land-surface scheme (Dee et al., 2011) to compute ETa. For transpiration a distinction is made between high and low vegetation in the HTESSEL scheme and these are parameterised from the Global Land Cover Characteristics database at a nominal resolution of $1 \, km^2$.

- Modern-Era Retrospective Analysis for Research and Applications (MERRA) re-analysis (Rienecker et al., 2011): the MERRA re-analysis product of NASA applies the state-of-the-art GEOS-5 data assimilation system that includes many modern observation systems in a climate framework. MERRA uses the GEOS-5 catchment land surface model (Koster et al., 2000) to compute actual ET. For the MERRA product ETp is not available.

- ET-Look (Bastiaanssen et al., 2012): The ET-Look remote sensing model infers information on ET from combined optical and passive microwave sensors, which can observe the land surface even under persistent overcast conditions. A two-layer Penman–Monteith forms the basis of quantifying soil and canopy evaporation. The data set is available only for the year 2007, but it was scaled to the 2003–2007 average using the ratio between the 2003–2007 average and the 2007 annual ET based on ERA-Interim.

- PCR-GLOBWB (Wada et al., 2014): The global hydrological model PCR-GLOBWB computes actual evapotranspiration using potential evapotranspiration based on Penman–Monteith, which is further reduced based on available soil moisture.

Table 1. Averages (μ) and standard deviations (σ) of predictors for the precipitation gradient. Values and ranges are based on literature as follows: HREF and HMAX: Hewitt (2007, 2011), Immerzeel et al. (2012b, 2014b), Putkonen (2004), Seko (1987) and Winiger et al. (2005); DDFd and DDFdf: Azam et al (2012), Hagg et al. (2008), Immerzeel et al. (2013), Mihalcea et al. (2006) and Nicholson and Benn (2006); MB: Kääb et al. (2012a).

Variable	Acronym	Distribution	μ	σ
Reference elevation (m)	HREF	log-Gaussian	2500	500
Maximum elevation (m)	HMAX	log-Gaussian		
	Himalaya		4500	500
	Karakoram		5500	500
	Hindu Kush		5500	500
Degree day factor debris-covered glaciers (mm $°C^{-1}\,d^{-1}$)	DDFd	log-Gaussian	2	2
Degree day factor debris-free glaciers (mm $°C^{-1}\,d^{-1}$)	DDFdf	log-Gaussian	7	2
Threshold slope (m m^{-1})	TS	log-Gaussian	0.2	0.05
Mass balance (m w.e. yr^{-1})	MB	Gaussian		
	Himalaya		−0.49	0.57
	Hindu Kush		−0.21	0.76
	Karakoram		−0.07	0.61

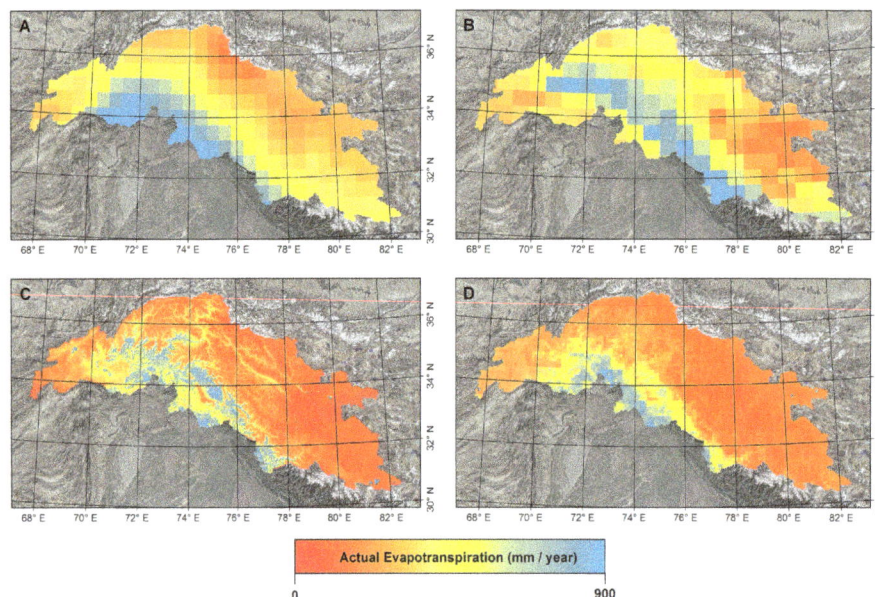

Figure 2. Average annual actual evapotranspiration between 2003 and 2007 for ERA-Interim (**a**), MERRA (**b**), ET-Look (**c**) and PCR-GLOBWB.

The average annual ETa for the period 2003–2007 for each of the four products is shown in Fig. 2. The spatial patterns show good agreement, but the magnitudes differ considerably. The ensemble mean ETa for the entire upper Indus equals 359 ± 107 mm yr^{-1}.

2.2 Model description

We use the PCRaster spatial–temporal modelling environment (Karssenberg et al., 2001) to model the mass balance of the major glaciers in each zone and subsequently estimate precipitation gradients required to sustain the observed mass balance. The model operates at a daily time step from 2003 to 2007 and a spatial resolution of 1 km^2. For each time step the total accumulation and total melt are aggregated over the entire glacier surface. Only glaciers with a surface area above 5 km^2 are included in the analysis (Karakoram is 232 glaciers, Hindu Kush is 119, Himalaya is 204 glaciers), as the ICESat measurements do not reflect smaller glaciers. The model is forced by the spatial precipitation and temperature fields. The precipitation fields are corrected using a PG (% m^{-1}). Precipitation is positively lapsed using a PG between a reference elevation (HREF) to an elevation of maximum precipitation (HMAX). At elevations above HMAX,

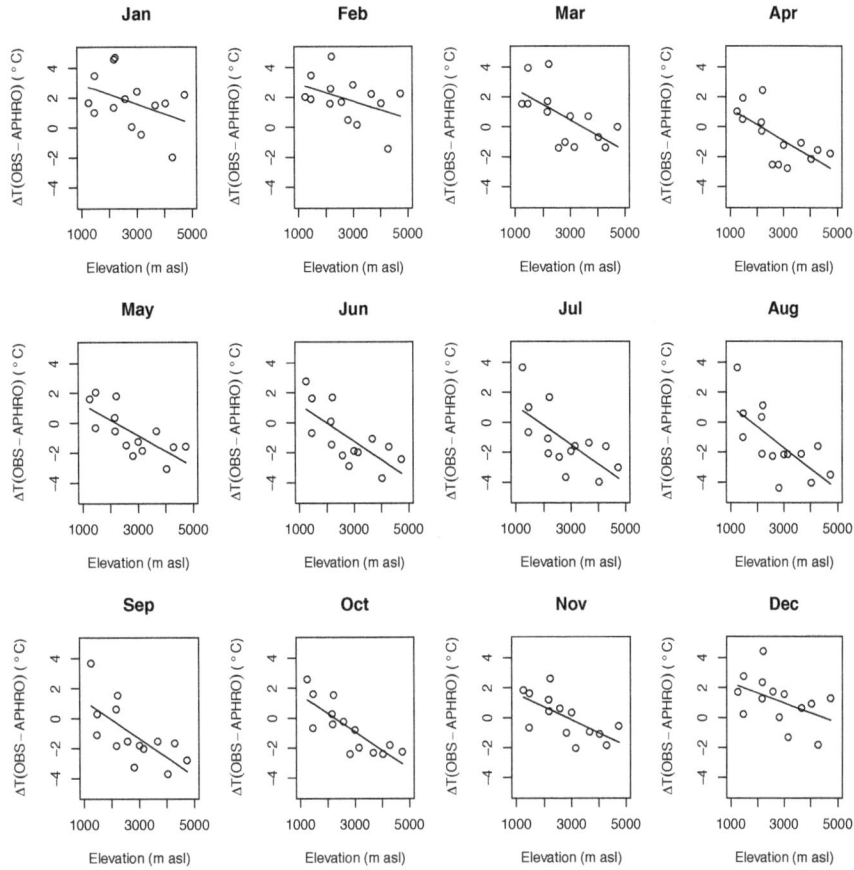

Figure 3. Monthly relation between observed temperatures at meteorological stations (OBS) and the APHRODITE temperature fields (APHRO) (Yasutomi et al., 2011).

the precipitation is negatively lapsed from its maximum at HMAX with the same PG according to

$$P_{cor} = P_{APHRO} \cdot (1 + (H - HREF)) \cdot PG \cdot 0.01) \quad (2)$$

for HREF < H ≤ HMAX, and

$$P_{cor} = P_{APHRO} \cdot (1 + (((HMAX - HREF) + (HMAX - H)) \cdot PG \cdot 0.01) \quad (3)$$

for H > HMAX.

HREF and HMAX values are derived from literature (Table 1) and uncertainty is taken into account in the uncertainty analysis. HMAX varies per zone and lies at a lower elevation in the Himalayas than in the other two zones (Table 1). We spatially interpolate HMAX from the average zonal values to cover the entire UIB.

The melt is modelled over the glacier area using the positive degree day method (Hock, 2005), with different degree day factors (DDFs) for debris-covered (DDFd) and debris-free (DDFdf) glaciers derived from literature (Table 1). To account for uncertainty in DDFs, the DDFd and DDFdf are taken into account separately in the uncertainty analysis.

At temperatures below the critical temperature of 2 °C (Immerzeel et al., 2013; Singh and Bengtsson, 2004), precipitation falls in the form of snow and contributes to the accumulation. Avalanche nourishment of glaciers is a key contributor for UIB glaciers (Hewitt, 2005, 2011) and to take this process into account, we extend the glacier area with steep areas directly adjacent to the glacier with a slope over an average threshold slope (TS) of $0.2 \, \text{m m}^{-1}$. This average threshold slope is derived by analysing the slopes of all glacier pixels in the basin (Fig. 4). To account for uncertainty in TS, this parameter is taken into account in the uncertainty analysis.

For each glacier system, we execute transient model runs from 2003 to 2007 and we compute the average annual mass balance from the total accumulation and melt over this period. We make two different runs for each glacier system with two different PGs (0.3 and $0.6 \, \% \, \text{m}^{-1}$) and we use the simulated mass balances of these two runs and the observed mass balances based on ICESat to optimise the PG per glacier, such that the simulated mass balance matches the observed.

To interpolate the glacier-specific PG values to PG spatial fields over the entire domain we use geostatistical conditional simulation (Goovaerts, 1997). Simulated spatial fields of PGs are thus conditioned on the PGs determined at the glacier's

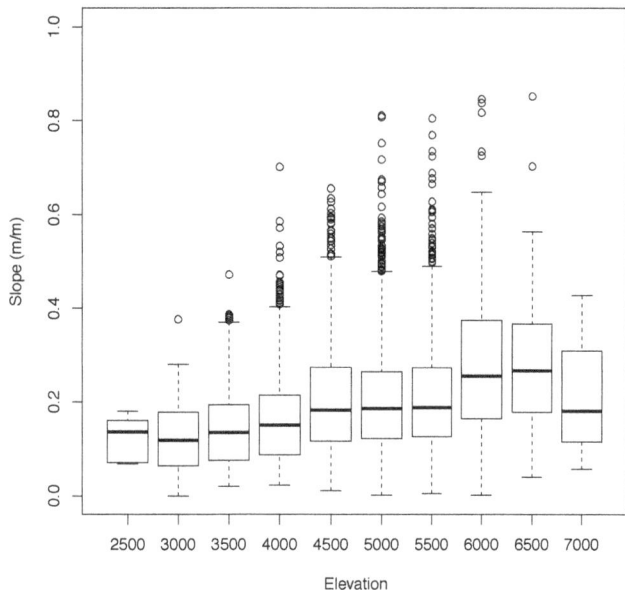

Figure 4. Box plots of slopes of glacierised areas per elevation bin.

centroid. The semi-variogram has the following parameters: nugget = 0, the range = 120 km, sill is the variance of PGs.

2.3 Uncertainty analysis

A rigorous uncertainty analysis is performed to take into account the uncertainty in parameter values and uncertainty in regional patterns. To account for parameter uncertainty, we perform a 10 000 member Monte Carlo simulation on the parameters given in Table 1. For each run we randomly sample the parameter space based on the average (μ) and the standard deviation (σ), which are all based on literature values. For the positively valued parameters, we use a log-Gaussian distribution and a Gaussian distribution in case parameter values can be negative. We take into account uncertainty in the following key parameters (HREF, HMAX, DDFd, DDFdf, TS) for the PG as well as uncertainty in the mass balance against which the PG is optimised (mass balance, MB). We randomly vary the five parameters (HREF, HMAX, DDFd, DDFdf, TS) 10 000 times and calculate the PG for each glacier for each random parameter set drawn, thus resulting in 10 000 PG sets for each glacier considered. For each of the 10 000 PG sets, we then use conditional simulation (see above) to arrive at 10 000 equally probable spatial PG fields, taking account of parameter's uncertainty, mass-balance uncertainty and the interpolation error. Note that for each of the 10 000 sets, the variogram is scaled with the variance of the PGs associated with the specific parameter combination drawn. Finally, based on the results of the 10 000 simulations we derive the average-corrected precipitation field and the associated uncertainty in the estimates

Using the 10 000 combinations of parameters and associated PGs, we ran a multi-variate linear regression analy-

sis to determine relative contribution of each parameter to the spread in the PG to understand which parameter has the largest influence on the PG.

It is possible that certain parameters used in the model are spatially correlated. To account for uncertainty in this spatial correlation and the presence of spatial patterns in the parameters, we perform a sensitivity analysis where we consider three cases:

– Fully correlated: we assume the parameters are spatially fully correlated within a zone, e.g. for each of the 10 000 simulations a parameter has the same value within a zone.

– Uncorrelated: we assume the parameters are spatially uncorrelated and within each zone each glacier system is assigned a random value.

– Intermediate case: we use geostatistical unconditional simulation (Goovaerts, 1997) with a standardised semi-variogram (nugget = 0, sill is the variance of parameter, range = 120 km) to simulate parameter values for each glacier system.

2.4 Validation

We estimate the average annual runoff (Q) for sub-basins in the UIB from

$$Q = P_{\text{cor}} - \text{ET} + \text{MB}, \tag{4}$$

where P_{cor} is the average corrected precipitation, ET the average annual evapotranspiration based on the four products described above and MB the glacier mass balance expressed over the sub-basin area in mm yr^{-1}. We then compare the estimated runoff values to the observed time series (Table 2).

For the three zones (Himalaya, Karakoram and Hindu Kush) we also perform a water balance analysis to verify whether the use of the corrected precipitation product results in a more realistic closure of the water balance. Finally, we test the physical realism of the corrected precipitation product using a non-dimensional Turc–Budyko plot as described in Valéry et al. (2010). This plot is based on two assumptions: (i) the mean annual runoff should not exceed the mean annual precipitation and (ii) the mean annual runoff should be larger than or equal to the difference between precipitation and potential evapotranspiration. By plotting P / ETp versus Q / P on a catchment basis, it is tested whether the use of corrected precipitation results in more physically realistic values.

3 Results and discussion

3.1 Corrected precipitation

The average annual precipitation based on 10 000 conditionally simulated fields reveals a striking pattern of high-altitude

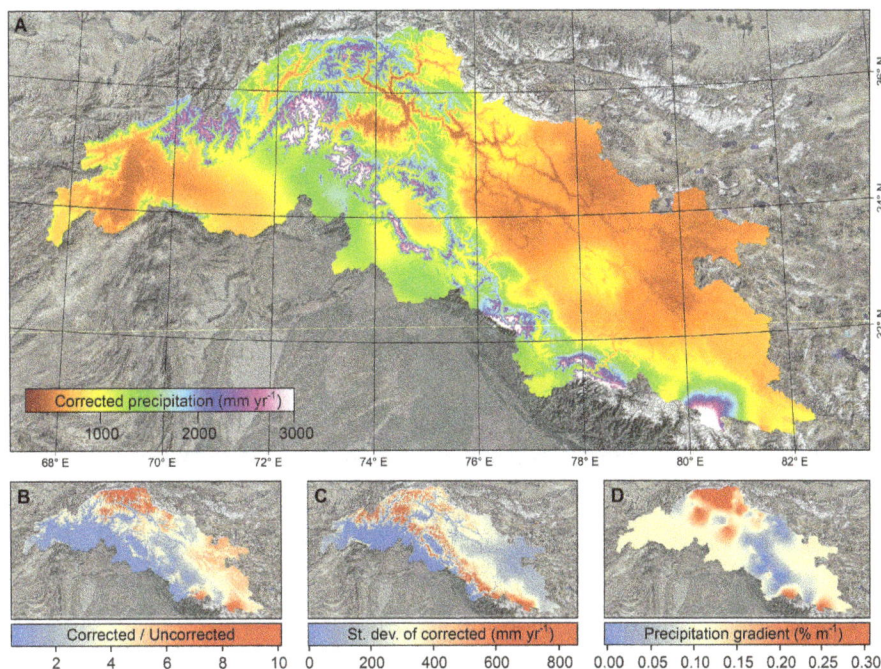

Figure 5. Corrected precipitation and estimated uncertainty for the UIB for the case with intermediate spatial correlation between model parameters. (**a**) shows the average modelled precipitation field based on 10 000 simulations for the period 2003–2007, (**b**) shows the ratio of corrected precipitation to the uncorrected APHRODITE precipitation for the same period, (**c**) shows the standard deviation of the 10 000 simulations and (**d**) shows the average precipitation gradient.

precipitation. The amount of precipitation required to sustain the large glacier systems is much higher than either the station observations or the gridded precipitation products imply. For the entire UIB the uncorrected average annual precipitation (Yatagai et al., 2012) for 2003–2007 is 437 mm yr^{-1} (191 km^3 yr^{-1}), an underestimation of more than 200 % compared with our corrected precipitation estimate of 913 ± 323 mm yr^{-1} (399 ± 141 km^3 yr^{-1}; Fig. 5). The greatest corrected annual precipitation totals in the UIB (1271 mm yr^{-1}) are observed in the elevation belt between 3750 to 4250 m (compared to 403 mm yr^{-1} for the uncorrected case). In absolute terms the main water-producing region is located in the elevation belt between 4250 and 4750 m where approximately 78 km^3 of rain and snow precipitates annually.

In the most extreme case, precipitation is underestimated by a factor 5 to 10 in the region where the Pamir, Karakorum and Hindu Kush ranges intersect (Fig. 5). Our inverse modelling shows that the large glacier systems in the region can only be sustained if snowfall in their accumulation areas totals around 2000 mm yr^{-1} (Hewitt, 2007). This is in sharp contrast to precipitation amounts between 200 and 300 mm yr^{-1} that are reported by the gridded precipitation products (Fig. 1). Our results match well with the few studies on high-altitude precipitation that are available. Annual accumulation values between 1000 and 3000 mm have been reported for accumulation pits above 4000 m in the Karakoram

(The Batura Glacier Investigation Group, 1979; Wake, 1989; Winiger et al., 2005). Our results show that the highest precipitation amounts are found along the monsoon-influenced southern Himalayan arc with values up to 3000 mm yr^{-1}, while north of the Himalayan range the precipitation decreases quickly towards a vast dry area in the north-eastern part of the UIB (Shyok sub-basin). In the north-western part of the UIB, westerly storm systems are expected to generate considerable amounts of precipitation at high altitude.

Our results reveal a strong relation between elevation and precipitation with a median PG for the entire upper Indus of 0.0989 % m^{-1}, but with large regional differences. Median precipitation gradients in the Hindu Kush and Karakoram ranges (0.260 and 0.119 % m^{-1}, respectively) are significantly larger than those observed in the Himalayan range, e.g 0.044 % m^{-1} (Fig. 6). In the Hindu Kush, for example, for every 1000 m elevation rise, precipitation increases by 260 % with respect to APHRODITE, which is based on valley floor precipitation. Higher HMAX in the Hindu Kush and the Karakoram (e.g. 5500 m versus 4500 m in the Himalayas; Hewitt, 2007; Immerzeel et al., 2014a; Putkonen, 2004; Seko, 1987; Winiger et al., 2005) suggests that westerly airflow indeed has a higher tropospheric altitude and that the interplay between elevation and precipitation is stronger for this type of precipitation. Further research should thus focus on the use of high-resolution cloud-resolving weather models (Collier et al., 2014; Mölg et al., 2013) for this region

Table 2. Runoff stations used for validation. Catchment areas are delineated based on SRTM DEM (Shuttle Radar Topographic Mission – digital elevation model). [a] is calculated based on discharge provided by the Pakistan Water and Power Development (WAPDA), [b] is based on Mukhopadhyay and Khan (2014a), [c] is based on Sharif et al. (2013), [d] is based on Archer (2003) and [e] is based on Khattak et al. (2011).

Station	Lat	Long	Area (km^2)	Catchment mean elevation (m)	Observed Q (m^3 s^{-1})	Period
Besham Qila[a]	34.906	72.866	198 741	4598	2372.2	2000–2007
Tarbela inflow[a]	34.329	72.856	203 014	4532	2370.3	1998–2007
Mangla inflow[a]	33.200	73.650	29 966	2494	831.8	1998–2007
Marala inflow[a]	32.670	74.460	29 611	3003	956.5	1998–2007
Dainyor bridge[a]	35.925	74.372	14 147	4468	331.8	1998–2004
Skardu-Kachura[b]	35.435	75.468	146 200	4869	1074.2	1970–1997
Partab Bridge[b]	35.767	74.597	177 622	4747	1787.9	1962–1996
Yogo[b]	35.183	76.100	64 240	5048	359.4	1973–1997
Kharmong[b]	34.933	76.217	70 875	4795	452.3	1982–1997
Gilgit[b]	35.933	74.300	13 174	4039	286.7	1960–1998
Doyian[b]	35.550	74.700	4000	3987	135.7	1974–1997
Chitral[c]	35.867	71.783	12 824	4086	271.9	1964–1996
Kalam[c]	35.467	72.600	2151	3874	89.6	1961–1997
Naran[c]	34.900	73.650	1181	3881	48.1	1960–1998
Alam bridge[c]	35.767	74.600	28 201	4228	644.0	1966–1997
Chakdara[c]	34.650	72.017	5990	2701	178.9	1961–1997
Karora[c]	34.900	72.767	586	2260	21.2	1975–1996
Garhi Habibullah[c]	34.450	73.367	2493	3303	101.8	1960–1998
Muzafferabad[c]	34.430	73.486	7604	3245	357.0	1963–1995
Chinari[c]	34.158	73.831	14 248	2513	330.0	1970–1995
Kohala[c]	34.095	73.499	25 820	2751	828.0	1965–1995
Kotli[c]	33.525	73.890	2907	1901	134.0	1960–1995
Shigar[b]	35.422	75.732	6681	4591	202.6	1985–1997
Phulra[d]	34.317	73.083	1106	1613	19.2	1969–1996
Daggar[d]	34.500	72.467	534	1085	6.9	1969–1996
Warsak[e]	34.100	71.300	74 092	2828	593.0	1967–2005
Shatial Bridge[b]	35.533	73.567	189 263	4667	2083.2	1983–1997

to further resolve seasonal topography–precipitation interaction at both synoptic and valley scales.

The estimated precipitation is considerably higher than what was reported in previous studies. Several studies have used TRMM products to quantify UIB precipitation (Bookhagen and Burbank, 2010; Immerzeel et al., 2009, 2010) and they show average annual precipitation values around 300 mm. It was also noted that the water balance was not closing and average annual river runoff at Tarbela exceeded the TRMM precipitation (Immerzeel et al., 2009). Two possible reasons have been suggested to explain this gap: (i) the high-altitude precipitation is underestimated and (ii) the glaciers are in a significant negative imbalance (Immerzeel et al., 2009). Since the ICESat study and several other geodetic mass balance studies (Gardelle et al., 2013; Kääb et al., 2012b) it has become clear that the glaciers in this region are not experiencing a significant ice loss and that this cannot be the explanation for the missing water in the water balance. This supports our conclusion that it is the high-altitude precipitation that has been underestimated. A study based on long-term observations of Tarbela inflow also confirm our results (Reggiani and Rientjes, 2014). In this

study the total UIB precipitation above Tarbela is estimated to be 675 ± 100 mm yr^{-1} and the difference remaining between our results may stem from the fact that the UIB we consider is twice the size of the area above the Tarbela, the different approach used to estimate actual ET, the different period considered and their assumption that ice storage has not changed between 1961 and 2009.

3.2 Uncertainty

We estimated the uncertainty in the modelled precipitation field with the standard deviation (σ) of the 10 000 realisations (Fig. 5c). The signal-to-noise ratio is satisfactory in the entire domain, e.g. the σ is always considerably smaller than the average precipitation with an average coefficient of variation of 0.35. The largest absolute uncertainty is found along the Himalayan arc and this coincides with the precipitation pattern found here. Strikingly, the region where the underestimation of precipitation is largest, at the intersection of the three mountain ranges in the northern UIB, is also an area where the uncertainty is small even though precipitation gradients are large.

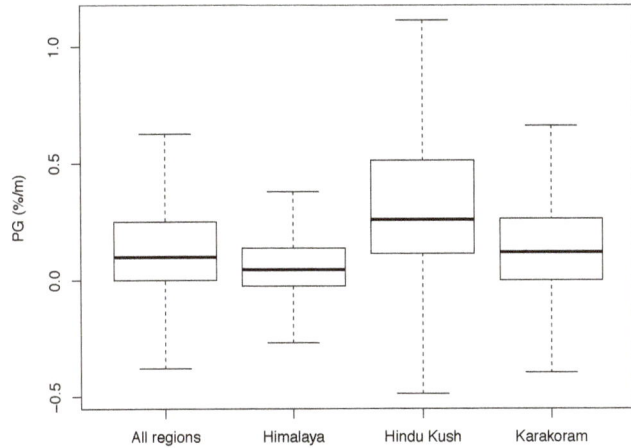

Figure 6. Box plots of precipitation gradients for the entire UIB and for the three regions separately.

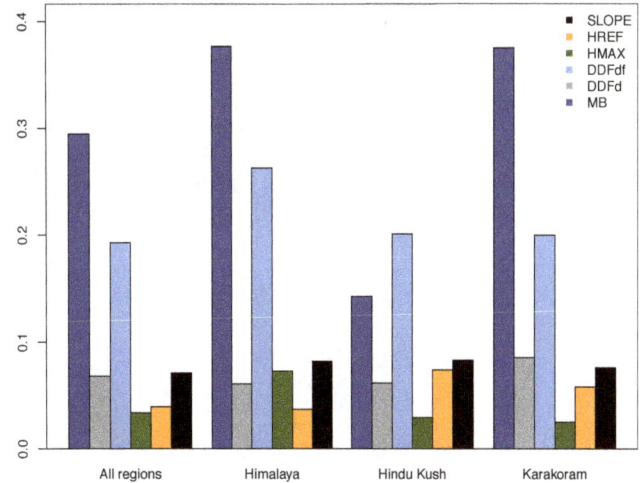

Figure 7. Normalised weights of multiple regression of the precipitation gradients by the predictors slope (slope threshold for avalanching to contribute to accumulation), HREF (base elevation from which lapsing starts), HMAX (elevation with peak precipitation), DDFd (degree day factor for debris-covered glaciers), DDFdf (degree day factor for debris-free glaciers) and the MB (mass balance of the glacier).

By running a multiple regression analysis after optimising the PGs, we quantify the contribution of each parameter to the total uncertainty. The largest source of uncertainty in our estimate of UIB high-altitude precipitation stems from the MB estimates, followed by the DDFdf, DDFd, TS, HREF and HMAX, although regional differences are considerable (Fig. 7). The MB constrains the precipitation gradients and thereby exerts a strong control over the corrected precipitation fields, in particular because the intra-zonal variation in MB is relatively large (Table 1). Thus, improved spatial monitoring techniques of the MB are expected to greatly improve precipitation estimates.

Figure 8 shows the result of uncertainty analysis associated with the spatial correlation of the parameters, which reveals that the impact on the average-corrected precipitation is limited. Locally there are minor differences in the corrected precipitation amounts, but overall the magnitude and spatial patterns are similar. However, there are considerable differences in the uncertainty. The lowest uncertainty is found for the fully uncorrelated case, the fully correlated case has the highest uncertainty whereas the intermediate case is in between both. For the fully correlated case all glacier systems have the same parameter set for the specific realisation and this results in a larger final uncertainty. In the uncorrelated case each glacier system has a different randomly sampled parameter set and this reduces the overall uncertainty as it spatially attenuates the variation in precipitation gradients.

3.3 Validation

The corrected precipitation is validated independently by a comparison to published average annual runoff data of 27 stations (Table 2). Runoff estimates based on the corrected precipitation agree well with the average observed annual runoff (Fig. 9, top panel). It is interesting to note that the higher catchments ($r = 0.98$, red outline) show a better cor-

relation with the observed runoff than the lower catchments ($r = 0.76$, black outline). The runoff estimated for the uncorrected APHRODITE is consistently lower than the observed runoff, and in some occasions even negative. Runoff estimates were also made based on the ERA-Interim and TRMM precipitation products. The TRMM results yield a similar underestimation as the uncorrected APHRODITES product, but the runoff estimates based on the ERA-Interim precipitation agrees reasonably well with the observations. However, the coarse resolution ($\sim 70\,\text{km}^2$) (Fig. 1) is problematic and cannot be used to reproduce the mass balance (Fig. 11). Averaged over large catchments the precipitation may be applied for hydrological modelling, but at smaller scales there are likely very large biases. As a result, the observed glacier mass balances cannot be reproduced when the ERA-Interim data set is used. Although the ERA-Interim data set may not be used to reproduce the glacier mass balances, it can be used to verify the atmospheric convergence as the product is based on a data assimilation scheme and the ECMWF IFS forecast model that includes fully coupled components for atmosphere, land surface and ocean waves, including closure of the atmospheric water balance. The total precipitation sum from 1998 to 2007 of the ERA-Interim data set over the entire UIB is 947 mm, which is very close to our corrected precipitation sum of 913 mm. This indicates that the westerlies and monsoon circulation transport sufficient moisture into the region to account for the precipitation we estimate. The source of precipitation in the upper Indus is a mixture of the Arabian Sea (westerlies), Bay of Bengal (Monsoon) and potentially also intra-basin moisture recycling (Tuinenburg et al., 2012;

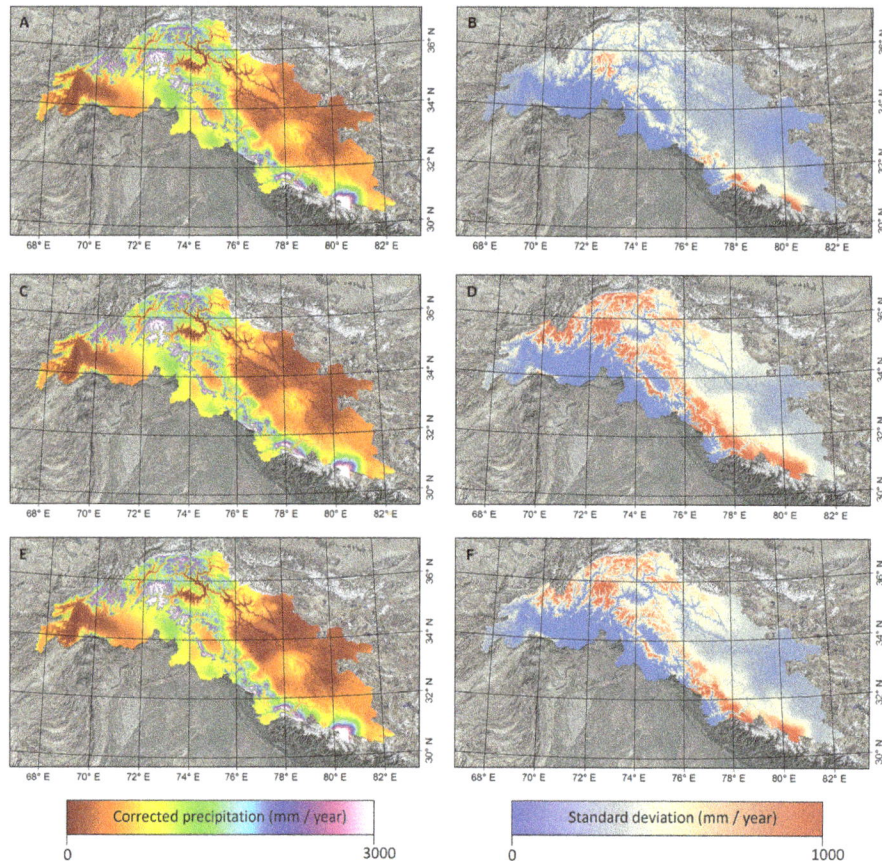

Figure 8. Impact of spatial correlation of parameters on the corrected precipitation field and associated uncertainty. The top panels show the corrected precipitation field (**a**) and uncertainty (**b**) for the fully uncorrelated case. The middle panels (**d, e**) for the fully correlated case and the bottom panels (**e, f**) for the intermediate case.

Wei et al., 2013); however, further research with atmospheric models is required to quantify these contributions further.

The zonal water balance analysis (Fig. 9, bottom panels) reveals that the water balance is much more realistic when the corrected precipitation is used. Although the uncertainties are considerable, our analysis shows that the Himalaya and Hindu Kush zones are about twice as wet as the Karakoram zone. For all three zones the glacier mass imbalance only plays a marginal role in the overall water balance and about 60 % of the total precipitation runs off while 40% is lost through evapotranspiration. Notable the values for Corg, which represents the water balance gap in case the uncorrected precipitation is used, are approximately $500 \, \text{mm yr}^{-1}$ in all three zones. Our validation does not take into account groundwater fluxes and we have assumed that over the observed period from 2003 to 2007 there is no net loss or gain of groundwater in the upper Indus basin. We do acknowledge that groundwater may play an important role in the hydrology. A study in the Nepal Himalaya shows that fractured basement aquifers fill during the monsoon and they purge in the post-monsoon thus causing a natural delay in runoff (Andermann et al., 2012). However, this does not imply sig-

nificant net gains or losses over multiple year periods, which is what we consider. A second component that we have not considered and that may play a role in the high-altitude water cycle is sublimation. There are some indications that wind redistribution and sublimation may play a considerable role in the high-altitude water balance (Wagnon et al., 2013). However, our PGs are constrained on the observed mass balance; hence, our precipitation can be considered as a net precipitation and sublimation losses are thus accounted for.

In Fig. 10 the Turc–Budyko plot is shown to confirm the physical realism of our results. Those dots located in the hatched part of the graph are physically less realistic. For the uncorrected case almost all dots (open dots) are above the $Q / P = 1$ line. For the corrected case the Q / P values are much more plausible; however, there many catchments that are located slightly to the right side of the theoretical Budyko line, meaning that the Q is smaller than the difference between P and ETp. Possible deviation can potentially be explained by uncertainties in observed flows and ETp estimates; the fact that in glacierised catchments the theoretical Budyko curve may be different because of a glacier imbalance can be an additional water balance term that is unac-

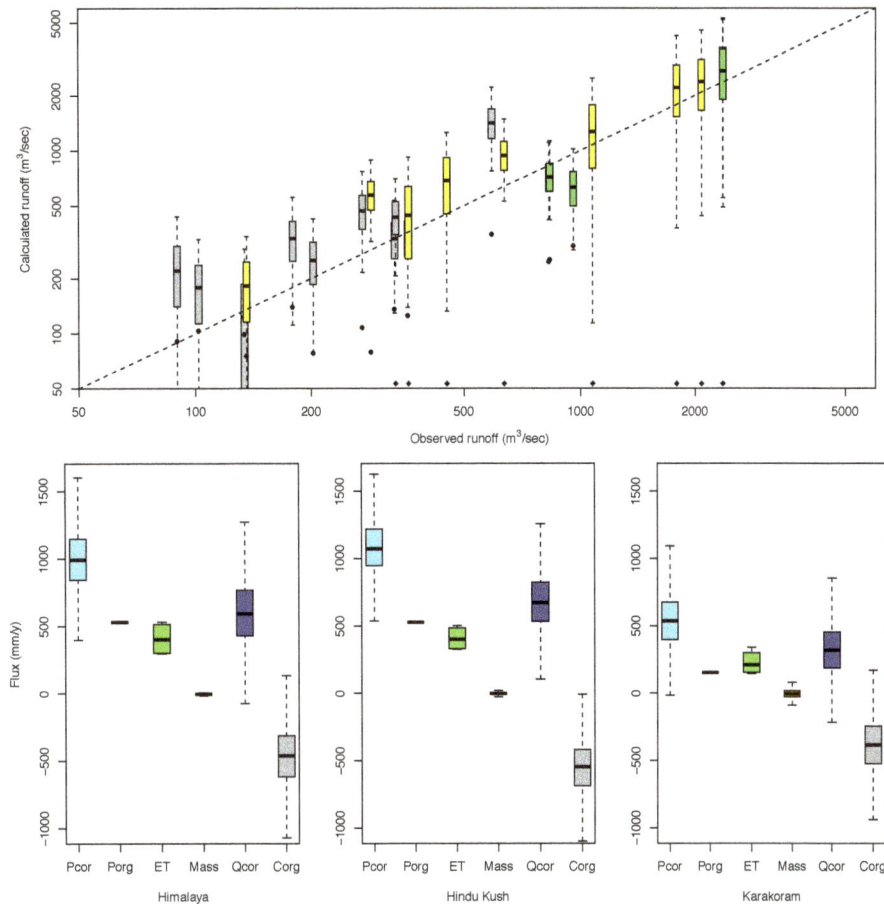

Figure 9. Validation of the precipitation correction using observed discharge (Table 2). Top panel: the box plots are based on the runoff estimate based on 10 000 corrected precipitation fields (grey: stations for which the observed record does not coincide with the 2003–2007 period; yellow: stations for which the 2003–2007 period is part of the observational record; green: stations for which the observations are based precisely on the 2003–2007 period). The black dots and red diamonds (estimated runoff below $50 \, \mathrm{m^3 \, s^{-1}}$) show the estimated runoff based on the uncorrected precipitation. The box plots with a red outline have an average elevation higher than 4000 m. and the box plots with a black outline have an elevation lower than 4000 m. Bottom panels: water balance components of each zone (P_{cor} is corrected precipitation, P_{org} is uncorrected APHRODITES precipitation, ET is actual evapotranspiration, Mass is glacier mass balance, Q_{cor} is estimated runoff and C_{org} is water balance gap in case the P_{org} is used).

counted for, a too short time period that is used to construct the water balance and, finally, that some of the discharge observations do not align in time with the rest of the water balance terms. Overall we conclude though that the use of the corrected precipitation results in physically more realistic results, where the water balance could be closed and no significant amount of precipitation input is missing.

Figure 11 shows how the average simulated zonal glacier mass balance using the corrected, the APHRODITES, the ERA-Interim and the TRMM precipitation data sets. It shows that none of the precipitation products can reproduce the observed mass balance. Mostly the mass balances are underestimated, which is consistent with the underestimation of the precipitation. The ERA-Interim data set overestimates the mass balance in the Himalaya, but underestimates the mass

balances in the other two zones as a result of the coarse resolution.

4 Conclusions

In this study we inversely model high-altitude precipitation in the upper Indus basins from glacier mass balance trends derived by remote sensing. Although there are significant uncertainties, our results unambiguously show that high-altitude precipitation in this region is underestimated and that the large glaciers here can only be sustained if high-altitude accumulation is much higher than most commonly used gridded data products.

Our results have an important bearing on water resources management studies in the region. The observed gap between precipitation and streamflow (Immerzeel et al., 2009) (with

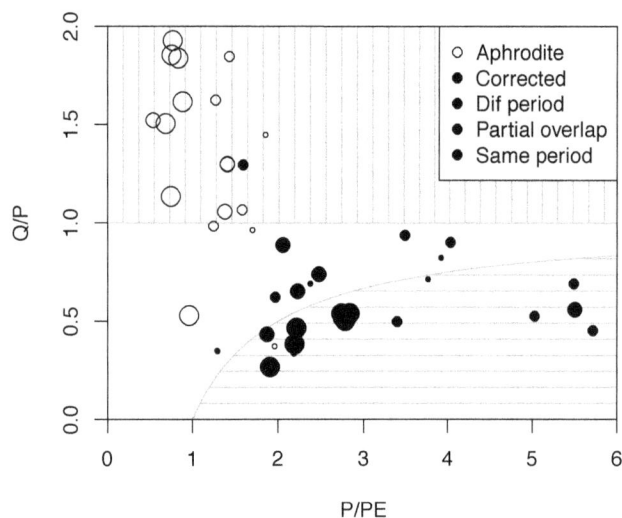

Figure 10. Non-dimensional graphical representation of catchments using their mean runoff, Q, precipitation, P, and potential evapotranspiration, PE. The grey line in the empty centre area represents the theoretical Budyko relationship in the non-dimensional graph. The size of the dots is scaled to the catchment area.

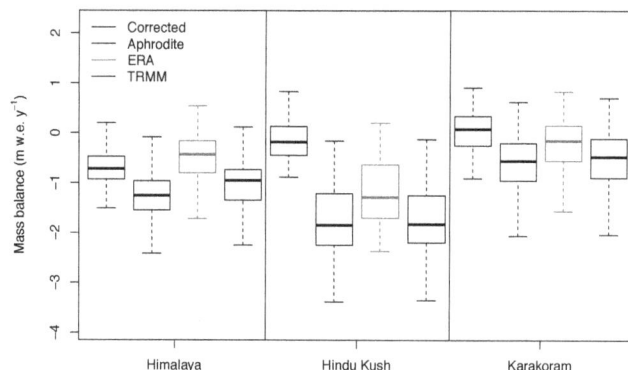

Figure 11. Reconstructed mass balances based on the corrected, APHRODITE, ERA-Interim and TRMM data sets. The black horizontal dotted line shows the observed mass balance for each zone.

streamflow being larger) cannot be attributed to the observed glacier mass balance (Kääb et al., 2012a), but is most likely the result of an underestimation of precipitation, as also follows from this study. With no apparent decreasing trends in precipitation (Archer and Fowler, 2004), the observed negative trends in streamflow in the glacierised parts of the UIB should thus be primarily attributed to decreased glacier and snowmelt (Sharif et al., 2013) and increased glacier storage (Gardelle et al., 2012). In a recent study the notion of negative trends in UIB runoff was contested and based on a recent analysis (1985–2010) it was concluded that runoff of Karakoram rivers is increasing (Mukhopadhyay and Khan, 2014b). The study suggests that increase glacier melt during summer is the underlying reason, which in combination with positive precipitation trends in summer does not contradict the neutral glacier mass balances in the region. From all of these studies it becomes apparent that precipitation is the key to understanding behaviour of glacier and hydrology at large in the UIB. The precipitation we estimate in this study differs considerably, in magnitude and spatial distribution, from data sets that are commonly used in design of reservoirs for hydropower and irrigation and as such it may have a significant impact and improve such planning processes.

The water resources of the Indus River play an important geopolitical role in the region, and our results could contribute to the provision of independent estimates of UIB precipitation. We conclude that the water resources in the UIB are even more important and abundant than previously thought. Most precipitation at high altitude is now stored in the glaciers, but when global warming persists and the runoff

regime becomes more rain dominated, the downstream impacts of our findings will become more evident.

Acknowledgements. This study was funded by the Netherlands Organization for Scientific Research through their VENI program, User Support Program for Space Research and Rubicon program and by the UK Department for International Development (DFID). This study was also partially supported by core funds of ICIMOD contributed by the governments of Afghanistan, Australia, Austria, Bangladesh, Bhutan, China, India, Myanmar, Nepal, Norway, Pakistan, Switzerland, and the United Kingdom. The views and interpretations in this publication are those of the authors and are not necessarily attributable to ICIMOD. The authors acknowledge the Pakistan Meteorological Department and the Pakistan Water and Power Development Authority for providing meteorological and hydrological data. The authors are grateful to Andreas Kääb for calculating the zonal mass balances based on the ICESat data, to Rianne Giesen for organising the glacier mass balance data, to Samjwal Bajracharya for providing the glacier boundaries and to Philip Kraaijenbrink for assisting with the figures.

Edited by: R. Woods

References

Andermann, C., Longuevergne, L., Bonnet, S., Crave, A., Davy, P., and Gloaguen, R.: Impact of transient groundwater storage on the discharge of Himalayan rivers, Nat. Geosci., 5, 127–132, doi:10.1038/ngeo1356, 2012.

Archer, D.: Contrasting hydrological regimes in the upper Indus Basin, J. Hydrol., 274, 198–210, doi:10.1016/S0022-1694(02)00414-6, 2003.

Archer, D. R. and Fowler, H. J.: Spatial and temporal variations in precipitation in the Upper Indus Basin, global teleconnections and hydrological implications, Hydrol. Earth Syst. Sci., 8, 47–61, doi:10.5194/hess-8-47-2004, 2004.

Azam, M. F., Wagnon, P., Ramanathan, A., Vincent, C., Sharma, P., Arnaud, Y., Linda, A., Pottakkal, J. G., Chevallier, P., Singh, V. B., and Berthier, E.: From balance to imbalance: a shift in the dynamic behaviour of Chhota Shigri glacier, western Himalaya, India, J. Glaciol., 58, 315–324, doi:10.3189/2012JoG11J123, 2012.

Bajracharya, S. R. and Shrestha, B.: The Status of Glaciers in the Hindu Kush-Himalayan Region, Kathmandu, International centre for Integrated Mountain Development, 1–127, 2011.

Barros, A. P., Kim, G., Williams, E., and Nesbitt, S. W.: Probing orographic controls in the Himalayas during the monsoon using satellite imagery, Nat. Hazards Earth Syst. Sci., 4, 29–51, doi:10.5194/nhess-4-29-2004, 2004.

Bastiaanssen, W. G. M., Cheema, M. J. M., Immerzeel, W. W., Miltenburg, I. J., and Pelgrum, H.: Surface energy balance and actual evapotranspiration of the transboundary Indus Basin estimated from satellite measurements and the ETLook model, Water Resour. Res., 48, W11512, doi:10.1029/2011WR010482, 2012.

Bhutiyani, M. R.: Mass-balance studies on Siachen Glacier in the Nubra valley, Karakoram Himalaya, India, J. Glaciol., 45, 112–118, 1999.

Bookhagen, B. and Burbank, D. W.: Topography, relief, and TRMM-derived rainfall variations along the Himalaya, Geophys. Res. Lett., 33, 1–5, doi:10.1029/2006GL026037, 2006.

Bookhagen, B. and Burbank, D. W.: Toward a complete Himalayan hydrological budget: Spatiotemporal distribution of snowmelt and rainfall and their impact on river discharge, J. Geophys. Res., 115, 1–25, doi:10.1029/2009JF001426, 2010.

Collier, E., Nicholson, L. I., Brock, B. W., Maussion, F., Essery, R., and Bush, A. B. G.: Representing moisture fluxes and phase changes in glacier debris cover using a reservoir approach, The Cryosphere, 8, 1429–1444, doi:10.5194/tc-8-1429-2014, 2014.

Dee, D. P., Uppala, S. M., Simmons, a. J., Berrisford, P., Poli, P., Kobayashi, S., Andrae, U., Balmaseda, M. a., Balsamo, G., Bauer, P., Bechtold, P., Beljaars, a. C. M., van de Berg, L., Bidlot, J., Bormann, N., Delsol, C., Dragani, R., Fuentes, M., Geer, a. J., Haimberger, L., Healy, S. B., Hersbach, H., Hólm, E. V., Isaksen, L., Kållberg, P., Köhler, M., Matricardi, M., McNally, a. P., Monge-Sanz, B. M., Morcrette, J.-J., Park, B.-K., Peubey, C., de Rosnay, P., Tavolato, C., Thépaut, J.-N., and Vitart, F.: The ERA-Interim reanalysis: configuration and performance of the data assimilation system, Q. J. Roy. Meteor. Soc., 137, 553–597, doi:10.1002/qj.828, 2011.

Gardelle, J., Berthier, E., and Arnaud, Y.: Slight mass gain of Karakoram glaciers in the early twenty-first century, Nat. Geosci., 5, 322–325, doi:10.1038/ngeo1450, 2012.

Gardelle, J., Berthier, E., Arnaud, Y., and Kääb, A.: Region-wide glacier mass balances over the Pamir-Karakoram-Himalaya during 1999–2011, The Cryosphere, 7, 1263–1286, doi:10.5194/tc-7-1263-2013, 2013.

Goovaerts, P.: Geostatistics for Natural Resources Evaluation, Oxford University Press, New York-Oxford, 1–496, 1997.

Hagg, W., Mayer, C., Lambrecht, A., and Helm, A.: Sub-Debris Melt Rates on Southern Inylchek Glacier, Central Tian Shan, Geogr. Ann. Ser. A Phys. Geogr., 90, 55–63, doi:10.1111/j.1468-0459.2008.00333.x, 2008.

Harper, J. T.: High altitude Himalayan climate inferred from glacial ice flux, Geophys. Res. Lett., 30, 3–6, doi:10.1029/2003GL017329, 2003.

Hewitt, K.: The Karakoram anomaly? Glacier expansion and the "elevation effect," Karakoram Himalaya, Mt. Res. Dev., 25, 332–340, doi:10.1659/0276-4741, 2005.

Hewitt, K.: Tributary glacier surges: an exceptional concentration at Panmah Glacier, Karakoram Himalaya, J. Glaciol., 53, 181–188, doi:10.3189/172756507782202829, 2007.

Hewitt, K.: Glacier Change, Concentration, and Elevation Effects in the Karakoram Himalaya, Upper Indus Basin, Mt. Res. Dev., 31, 188–200, doi:10.1659/MRD-JOURNAL-D-11-00020.1, 2011.

Hewitt, K.: The regional context, in Glaciers of the Karakoram Himalaya, 1–33, Springer, Dordrecht, the Netherlands, 2013.

Hock, R.: Glacier melt: a review of processes and their modelling, Prog. Phys. Geogr., 29, 362–391, doi:10.1191/0309133305pp453ra, 2005.

Huffman, G. J., Adler, R. F., Bolvin, D. T., Gu, G., Nelkin, E. J., Bowman, K. P., Hong, Y., Stocker, E. F., and Wolff, D. B.: The TRMM Multisatellite Precipitation Analysis (TMPA): Quasi-Global, Multiyear, Combined-Sensor Precipitation Estimates at Fine Scales, J. Hydrometeorol., 8, 38–55, doi:10.1175/JHM560.1, 2007.

Immerzeel, W. W. and Bierkens, M. F. P.: Asia's water balance, Nat. Geosci., 5, 841–842, doi:10.1038/ngeo1643, 2012.

Immerzeel, W. W., Droogers, P., De Jong, S. M., and Bierkens, M.: Large-scale monitoring of snow cover and runoff simulation in Himalayan river basins using remote sensing, Remote Sens. Environ., 113, 40–49, doi:10.1016/j.rse.2008.08.010, 2009.

Immerzeel, W. W., Van Beek, L. P. H., and Bierkens, M. F. P.: Climate change will affect the Asian water towers, Science, 328, 1382–1385, doi:10.1126/science.1183188, 2010.

Immerzeel, W. W., Pellicciotti, F., and Shrestha, A. B.: Glaciers as a proxy to quantify the spatial distribution of precipitation in the Hunza basin, Mt. Res. Dev., 32, 30–38, doi:10.1659/MRD-JOURNAL-D-11-00097.1, 2012a.

Immerzeel, W. W., Beek, L. P. H., Konz, M., Shrestha, a. B., and Bierkens, M. F. P.: Hydrological response to climate change in a glacierized catchment in the Himalayas, Clim. Change, 110, 721–736, doi:10.1007/s10584-011-0143-4, 2012b.

Immerzeel, W. W., Pellicciotti, F., and Bierkens, M. F. P.: Rising river flows throughout the twenty-first century in two Himalayan glacierized watersheds, Nat. Geosci., 6, 742–745, doi:10.1038/NGEO1896, 2013.

Immerzeel, W. W., Kraaijenbrink, P. D. A., Shea, J. M., Shrestha, A. B., Pellicciotti, F., Bierkens, M. F. P., and de Jong, S. M.: High-resolution monitoring of Himalayan glacier dynamics using unmanned aerial vehicles, Remote Sens. Environ., 150, 93–103, doi:10.1016/j.rse.2014.04.025, 2014a.

Immerzeel, W. W., Petersen, L., Ragettli, S., and Pellicciotti, F.: The importance of observed gradients of air temperature and precipitation for modeling runoff from a glacierised watershed in the Nepalese Himalayas, Water Resour. Res., 50, 2212–2226, doi:10.1002/2013WR014506, 2014b.

Kääb, A., Berthier, E., Nuth, C., Gardelle, J., and Arnaud, Y.: Contrasting patterns of early twenty-first-century glacier mass change in the Himalayas, Nature, 488, 495–498, doi:10.1038/nature11324, 2012a.

Kääb, A., Berthier, E., Nuth, C., Gardelle, J., Arnaud, Y., Kaab, A., Berthier, E., Nuth, C., Gardelle, J., and Arnaud, Y.: Contrasting patterns of early twenty-first-century glacier mass change in the Himalayas, Nature, 488, 495–498, 2012b.

Karssenberg, D., Burrough, P. A., Sluiter, R., and de Jong, K.: The PCRaster Software and Course Materials for Teaching Numerical Modelling in the Environmental Sciences, Trans. GIS, 5, 99–110, doi:10.1111/1467-9671.00070, 2001.

Khattak, M., Babel, M., and Sharif, M.: Hydro-meteorological trends in the upper Indus River basin in Pakistan, Clim. Res., 46, 103–119, doi:10.3354/cr00957, 2011.

Koster, R. D., Suarez, M. J., Ducharne, A., Stieglitz, M., and Kumar, P.: A catchment-based approach to modeling land surface processes in a general circulation model structure, J. Geophys. Res., 105, 24809–24822, 2000.

Lehner, B., Verdin, K., and Jarvis, A.: New global hydrography derived from spaceborne elevation data, Eos Trans. AGU, 89, 93–94, 2008.

Lutz, A. F., Immerzeel, W. W., Shrestha, A. B., and Bierkens, M. F. P.: Consistent increase in High Asia's runoff due to increasing glacier melt and precipitation, Nat. Clim. Chang., 4, 1–6, doi:10.1038/NCLIMATE2237, 2014.

Mihalcea, C., Mayer, C., Diolaiuti, G., Lambrecht, A., Smiraglia, C., and Tartari, G.: Ice ablation and meteorological conditions on the debris-covered area of Baltoro glacier, Karakoram, Pakistan, Ann. Glaciol., 1894, 292–300, 2006.

Mishra, V.: Climatic uncertainty in Himalayan Water Towers, J. Geophys. Res. Atmos., 120, 2689–2705, doi:10.1002/2014JD022650, 2015.

Mölg, T., Maussion, F., and Scherer, D.: Mid-latitude westerlies as a driver of glacier variability in monsoonal High Asia, Nat. Clim. Change, 4, 68–73, doi:10.1038/nclimate2055, 2013.

Mukhopadhyay, B. and Khan, A.: A quantitative assessment of the genetic sources of the hydrologic flow regimes in Upper Indus Basin and its significance in a changing climate, J. Hydrol., 509, 549–572, doi:10.1016/j.jhydrol.2013.11.059, 2014a.

Mukhopadhyay, B. and Khan, A.: Rising river flows and glacial mass balance in central Karakoram, J. Hydrol., 513, 191–203, 2014b.

Nicholson, L. and Benn, D. I.: Calculating ice melt beneath a debris layer using meteorological data, J. Glaciol., 52, 463–470, doi:10.3189/172756506781828584, 2006.

Putkonen, J. K.: Continuous Snow and Rain Data at 500 to 4400 m Altitude near Annapurna, Nepal, 1991–2001, Arctic, Antarct. Alp. Res., 36, 244–248, doi:10.1657/1523-0430(2004)036, 2004.

Ragettli, S. and Pellicciotti, F.: Calibration of a physically based, spatially distributed hydrological model in a glacierized basin: On the use of knowledge from glaciometeorological processes to constrain model parameters, Water Resour. Res., 48, 1–20, doi:10.1029/2011WR010559, 2012.

Reggiani, P. and Rientjes, T. H. M.: A reflection on the long-term water balance of the Upper Indus Basin, Hydrol. Res., 46, 446–462, doi:10.2166/nh.2014.060, 2014.

Rienecker, M. M., Suarez, M. J., Gelaro, R., Todling, R., Bacmeister, J., Liu, E., Bosilovich, M. G., Schubert, S. D., Takacs, L., Kim, G.-K., Bloom, S., Chen, J., Collins, D., Conaty, A., da Silva, A., Gu, W., Joiner, J., Koster, R. D., Lucchesi, R., Molod, A., Owens, T., Pawson, S., Pegion, P., Redder, C. R., Reichle, R., Robertson, F. R., Ruddick, A. G., Sienkiewicz, M., and Woollen, J.: MERRA: NASA's Modern-Era Retrospective Analysis for Research and Applications, J. Clim., 24, 3624–3648, doi:10.1175/JCLI-D-11-00015.1, 2011.

Scherler, D., Bookhagen, B., and Strecker, M. R.: Spatially variable response of Himalayan glaciers to climate change affected by debris cover, Nat. Geosci., 4, 156–159, doi:10.1038/ngeo1068, 2011.

Seko, K.: Seasonal variation of altitudinal dependence of precipitation in Langtang Valley, Nepal Himalayas, Bull. Glacier Res., 5, 41–47, 1987.

Sharif, M., Archer, D. R., Fowler, H. J., and Forsythe, N.: Trends in timing and magnitude of flow in the Upper Indus Basin, Hydrol. Earth Syst. Sci., 17, 1503–1516, doi:10.5194/hess-17-1503-2013, 2013.

Singh, P., and Bengtsson, L.: Hydrological sensitivity of a large Himalayan basin to climate change, Hydrol. Process., 18, 2363–2385, doi:10.1002/hyp.1468, 2004.

The Batura Glacier Investigation Group: The Batura glacier in the Karakoram mountains and its variations, Sci. Sin., 22, 958–974, 1979.

Tuinenburg, O. A., Hutjes, R. W. A., and Kabat, P.: The fate of evaporated water from the Ganges basin, J. Geophys. Res., 117, 1–17, doi:10.1029/2011JD016221, 2012.

Valéry, A., Andréassian, V., and Perrin, C.: Inverting the hydrological cycle?: when streamflow measurements help assess altitudinal precipitation gradients in mountain areas, IAHS Publ., 333, 281–286, 2009.

Valéry, A., Andréassian, V., and Perrin, C.: Regionalization of precipitation and air temperature over high-altitude catchments – learning from outliers, Hydrol. Sci. J., 55, 928–940, doi:10.1080/02626667.2010.504676, 2010.

Wada, Y., Wisser, D., and Bierkens, M. F. P.: Global modeling of withdrawal, allocation and consumptive use of surface water and groundwater resources, Earth Syst. Dynam., 5, 15–40, doi:10.5194/esd-5-15-2014, 2014.

Wagnon, P., Vincent, C., Arnaud, Y., Berthier, E., Vuillermoz, E., Gruber, S., Ménégoz, M., Gilbert, a., Dumont, M., Shea, J. M., Stumm, D., and Pokhrel, B. K.: Seasonal and annual mass balances of Mera and Pokalde glaciers (Nepal Himalaya) since 2007, The Cryosphere, 7, 1769–1786, doi:10.5194/tc-7-1769-2013, 2013.

Wake, C. P.: Glaciochemical investigations as a tool to determine the spatial variation of snow accumulation in the Central Karakoram, Northern Pakistan., Ann. Glaciol., 13, 279–284, 1989.

Wei, J., Dirmeyer, P. A., Wisser, D., Bosilovich, M. G., and Mocko, D. M.: Where Does the Irrigation Water Go? An Estimate of the Contribution of Irrigation to Precipitation Using MERRA, J. Hydrometeorol., 14, 275–289, doi:10.1175/JHM-D-12-079.1, 2013.

Winiger, M., Gumpert, M., and Yamout, H.: Karakorum-Hindukush-western Himalaya: assessing high-altitude water resources, Hydrol. Process., 19, 2329–2338, doi:10.1002/hyp.5887, 2005.

Yao, T., Thompson, L., Yang, W., Yu, W., Gao, Y., Guo, X., Yang, X., Duan, K., Zhao, H., Xu, B., Pu, J., Lu, A., Xiang, Y., Kattel, D. B., and Joswiak, D.: Different glacier status with atmospheric circulations in Tibetan Plateau and surroundings, Nat. Clim. Chang., 2, 663–667, doi:10.1038/nclimate1580, 2012.

Yasutomi, N., Hamada, A., and Yatagai, A.: Development of a Long-term Daily Gridded Temperature Dataset and Its Application to Rain/Snow Discrimination of Daily Precipitation, Global Environ. Res., 3, 165–172, 2011.

Yatagai, A., Yasutomi, N., Hamada, A., Kitoh, A., Kamiguchi, K., and Arakawa, O.: APHRODITE?: constructing a long-term daily gridded precipitation dataset for Asia based on a dense network of rain gauges, Geophys. Res. Abstr., 14, 1401–1415, 2012.

12

Recent evolution and associated hydrological dynamics of a vanishing tropical Andean glacier: Glaciar de Conejeras, Colombia

Enrique Morán-Tejeda[1], Jorge Luis Ceballos[2], Katherine Peña[2], Jorge Lorenzo-Lacruz[1], and Juan Ignacio López-Moreno[3]

[1]Department of Geography. University of the Balearic Islands. Palma, Spain
[2]Instituto de Hidrología, Meteorología y Estudios Ambientales (IDEAM), Bogotá, Colombia
[3]Pyrenean Institute of Ecology. Consejo Superior de Investigaciones Científicas, Zaragoza, Spain

Correspondence: Enrique Morán-Tejeda (e.moran@uib.es)

Abstract. Glaciers in the inner tropics are rapidly retreating due to atmospheric warming. In Colombia, this retreat is accelerated by volcanic activity, and most glaciers are in their last stages of existence. There is general concern about the hydrological implications of receding glaciers, as they constitute important freshwater reservoirs and, after an initial increase in melting flows due to glacier retreat, a decrease in water resources is expected in the long term as glaciers become smaller. In this paper, we perform a comprehensive study of the evolution of a small Colombian glacier, Conejeras (Parque Nacional Natural de los Nevados) that has been monitored since 2006, with a special focus on the hydrological response of the glacierized catchment. The glacier shows great sensitivity to changes in temperature and especially to the evolution of the El Niño–Southern Oscillation (ENSO) phenomenon, with great loss of mass and area during El Niño warm events. Since 2006, it has suffered a 37 % reduction, from 22.45 ha in 2006 to 12 ha in 2017, with an especially abrupt reduction since 2014. During the period of hydrological monitoring (June 2013 to December 2017), streamflow at the outlet of the catchment experienced a noticeable cycle of increasing flows up to mid-2016 and decreasing flows afterwards. The same cycle was observed for other hydrological indicators, including the slope of the rising flow limb and the monthly variability of flows. We observed an evident change in the daily hydrograph, from a predominance of days with a purely melt-driven hydrograph up to mid-2016, to an increase in the frequency of days with flows less influenced by melt after 2016. Such a hydrological cycle is not directly related to fluctuations of temperature or precipitation; therefore, it is reasonable to consider that it is the response of the glacierized catchment to retreat of the glacier. Results confirm the necessity for small-scale studies at a high temporal resolution, in order to understand the hydrological response of glacier-covered catchments to glacier retreat and imminent glacier extinction.

1 Introduction

1.1 Andean glaciers and water resources

Glacier retreat is one the most prominent signals of global warming; glaciers from most mountain regions in the world are disappearing or have already disappeared due to atmospheric warming (Vaughan et al., 2013). Of the retreating mountain glaciers worldwide, those located within the tropics are particularly sensitive to atmospheric warming (Chevallier et al., 2011; Kaser and Omaston, 2002). Their locations in the tropical region involve a larger energy forcing, in terms of received solar radiation, compared to other latitudes. Unlike glaciers in middle and high latitudes, which are subject to freezing temperatures during a sustained season, tropical glaciers may experience above-zero temperatures all year round, especially at the lowest elevations, involving constant ablation and rapid response of the glacier snout to climate variability and climate change (Francou et al., 2004; Rabatel et al., 2013). As a result of atmospheric warming since the mid-20th century, glaciers in the tropics are seriously threatened, and many of them have already disappeared (Vuille et al., 2008). Of the tropical glaciers, 99 % are located in the Central Andes and constitute a laboratory for glaciology (see review in Vuille et al., 2017), including studies of glacier response to climate forcing (e.g., Favier et al., 2004; Francou et al., 2003, 2004; López-Moreno et al., 2014), hydrological and geomorphological consequences of glacier retreat (Bradley et al., 2006; Chevallier et al., 2011; Kaser et al., 2010; López-Moreno et al., 2017; Ribstein et al., 1995; Sicart et al., 2011), and the vulnerability of populations to risks associated with glacier retreat (Mark et al., 2017). The glaciers in the most critical situation in the Andean mountains are perhaps those located in the inner tropics, including the countries of Ecuador, Venezuela and Colombia (Klein et al., 2006; Rekowsky, 2016). In the latter country, a constant glacier recession since the 1970s has been reported, with an acceleration since the 2000s (Ceballos et al., 2006; Rabatel et al., 2013), and most glaciers are in danger of disappearing in the coming years (Poveda and Pineda, 2009; Rabatel et al., 2017). In the outer tropics, the variability of glacier mass balance is highly dependent on seasonal precipitation; thus, during the wet season (December–February), freezing temperatures ensure seasonal snow cover that increases the glaciers' surface albedo and compensates mass balance losses of the dry season. In contrast, for glaciers of the inner tropics, ablation rates remain more or less constant throughout the year due to the absence of seasonal fluctuations of temperature and to a freezing level that is constantly oscillating within the glaciers' elevation ranges. Therefore, the mass balance of these glaciers is more sensitive to interannual variations of temperature, and they are much more sensitive to climate warming (Ceballos et al., 2006; Favier et al., 2004; Francou et al., 2004; Rabatel et al., 2013, 2017). In Colombia, this situation is further aggravated by the location of glaciers near or on the top of active volcanos. The hot pyroclastic material emitted during volcanic eruptions and the reduced albedo of glacier surfaces by the deposition of volcanic ash have notably contributed to rapid glacier recession in these areas (Granados et al., 2015; Huggel et al., 2007; Rabatel et al., 2013; Vuille et al., 2017).

Current glacier recession in the Andes involves the loss of natural scientific laboratories (Francou et al., 2003) and of landscape and cultural emblems of mountainous areas (IDEAM, 2012; Rabatel et al., 2017). But in more practical terms, the vanishing of glaciers has a major impact on the livelihoods of communities living downstream, including potential reduction of freshwater storage and changes in the seasonal patterns of water supply by downstream rivers (Kaser et al., 2010). Glaciers constitute natural water reservoirs in the form of ice accumulated during cold and wet seasons, and they provide water when ice melts during above-freezing temperature seasons. The hydrological importance of glaciers for downstream areas depends on the availability of other sources of runoff, including snowmelt and rainfall. Therefore, water supply by glaciers becomes critical for arid or semiarid regions downstream of the glacierized areas, buffering the lack of sustained precipitation or water provided by seasonal melt of snow cover (Rabatel et al., 2013; Vuille et al., 2008). Such is the case for the western slopes of the tropical Andes: in countries like Peru or Bolivia, with a high variability in precipitation and a sustained dry season, the contribution of glacier melt is crucial for socioeconomic activities and for water supply, especially since it is one of the main sources of water for the highly populated capital cities such as La Paz (Kaser et al., 2010; López-Moreno et al., 2014; Soruco et al., 2015; Vuille et al., 2017). In more humid or temperate regions (i.e., the Alps or western North America) the melt of seasonal snow cover provides the majority of water during the melt season (Beniston, 2012; Stewart et al., 2004) and glacier melt is a secondary contributor. However, even in this region, water availability can be subject to climate variability, and the occurrence of dry and warm periods that comprise thin and brief snow cover may involve glacier melt as the main source of water during such events (Kaser et al., 2010). In the inner tropics, glaciers may not constitute the main source of water for downstream populations, as the seasonal shift of the Intertropical Convergence Zone (Poveda et al., 2006) assures two humid seasons every year; however, the loss of water from glacier melt can affect the eco-hydrological functioning of the wetland ecosystems called "*páramos*", which are positioned in the altitudinal tier located below that of the periglacial ecosystem (Rabatel et al., 2017). Agriculture and livestock in Colombian mountain communities are partly dependent on water from these important water reservoirs that provide water flow to downstream rivers, even during periods of less precipitation.

1.2 Hypothesis and objectives

The present work is focused on the hydrological dynamics of a Colombian glacier near extinction due to prolonged deglaciation. Hock et al. (2005) presented a summary of the effects of glaciers on streamflow compared to nonglacierized areas. The main characteristics of streamflow can be summarized as follows (Hock et al., 2005):

- Specific runoff dependence on variability of glacier mass balance (in years of mass balance loss, total streamflow will increase as water is released from glacier storage; the opposite will happen in years of positive mass balance).

- Seasonal runoff variation dependent on ablation and accumulation periods at latitudes with markedly variable temperature and/or precipitation seasonal patterns (in the case of temperature, this does not apply to glaciers in the inner tropics).

– Large diurnal fluctuation in the absence of precipitation, as a result of the daily cycle of temperature and derived glacier melt.

– Moderation of year-to-year variability (moderate percentages, of 10 % to 40 %, of ice cover fraction within the basin reduces variability to a minimum, but it becomes greater at both higher and lower glacierization levels).

– Large glacierization involves a high correlation between runoff and temperature, whereas low levels of glacier cover increase runoff correlation with precipitation.

However, under warming conditions that lead to glacier retreat, the hydrological contribution of the glacier may notably change from the aforementioned characteristics. The retreat of a glacier is a consequence of prolonged periods of negative mass balance, the result of a disequilibrium in the accumulation–ablation ratio that involves an upward shift of the equilibrium line (the elevation at which accumulation and ablation volumes are equal) and an increase of the ablation area with respect to the accumulation area (Chevallier et al., 2011). As a result, the glacierized area is increasingly smaller compared to the nonglacierized area within the catchment in which the glacier is settled. Under such conditions of sustained negative mass balance, the hydrological response of the glacier will be a matter of timescales (Chevallier et al., 2011; Hock et al., 2005).The total runoff production of the retreating glacier comprises a tradeoff between two processes: on one side, an acceleration of glacier melt that will increase the volume of glacier outflows, independent of the volume precipitated as snowfall or rainfall; on the other side, water discharges from the catchment decrease because the water reservoir that represents the glacier is progressively emptying (Huss and Hock, 2018). Thus, the contribution of glacier melt to total water discharge will initially increase, as the first process will dominate over the other; however, after reaching a discharge peak, the second process dominates, leading to a decrease in water discharge until the glacier vanishes. In terms of runoff variability, there is also a different signal between initial and final stages of glacier retreat: on a daily basis, the typical diurnal cycle of glacier melt will exacerbate at the initial stages (larger difference between peak and base runoff) and will moderate at the final stages. However, in terms of year-to-year variability, there can be a reduction or increase at the initial stages, depending on the original glacierized area. And for the long term, increasing variability should be expected, as the water discharge will correlate with precipitation instead of temperature because the percentage of runoff from glacier melt decreases with decreasing glacierization (Hock et al., 2005).

It is expected that changes will be observed in the hydrological dynamics of vanishing glaciers, independently of climate drivers. Such hydrological changes may serve as indicators of glacier shrinkage, complementing others such as

mass balance or areal observations. The objective of this work is to provide a comprehensive analysis of the hydrological dynamics of a glacierized basin, with the glacier in its last stages prior to extinction. Considering the abovementioned characteristics of the hydrology of retreating glaciers, the specific aim is to explore changes over time in streamflow dynamics, focusing on the daily cycle, and to discern whether such changes are driven by climate or are a result of the diminishing glacierized area within the basin.

The case study is a small glacier (see description in Sect. 2) in the central Colombian Andes and the catchment that drains the water at the snout of the glacier. It is one of the very few monitored glaciers in the tropical Andes (Mölg et al., 2017; Rabatel et al., 2017) and represents an ideal case, where the hydrological signal of the glacier can be studied in isolation from any environmental factors that may occur in the downstream areas. For this reason, the approach used (see Sect. 3.3) can be applied to similar environments, and the obtained results can be representative of expected hydrological dynamics in other glacierized areas in the Andes, with glaciers close to extinction.

2   Study site

Our study focuses on the Conejeras glacier, a very small ice mass (14 ha in 2017) that forms part of a larger glacier system called Nevado de Santa Isabel (1.8 km^2), one of the six glaciers that still persist in Colombia. It is located in the Cordillera Central (the central range of the three branches of the Andean chain in Colombia) and, together with the glaciers of Nevado del Ruiz and Tolima, comprises the protected area called Parque Nacional Natural de los Nevados (Fig. 1). The summit of the Santa Isabel glacier reaches 5100 m, being the lowest glacier in Colombia. As a result, it is also the most sensitive to atmospheric warming and so it has been monitored since 2006 as part of the world network of glacier monitoring (IDEAM, 2012). The Santa Isabel glacier has been retreating since the 19th century, with an intensification of deglaciation since the middle of the 20th century. As a result, the glacier is now a set of separated ice fragments instead of a continuous ice mass, as it was a decade ago (IDEAM, 2012). One of the fragments, located at the northeast sector of the glacier, is the Conejeras glacier, which is the object of this study, whose elevation ranges between 4700 and 4895 m. In 2006, at the glacier terminus, hydro-meteorological stations were installed in order to measure glacier contribution to runoff, as well as air temperature and precipitation.

The Conejeras water stream is a tributary of one of the *quebradas* (Spanish for small mountain rivers in South American countries) flowing into the river Rio Claro. Thus, the

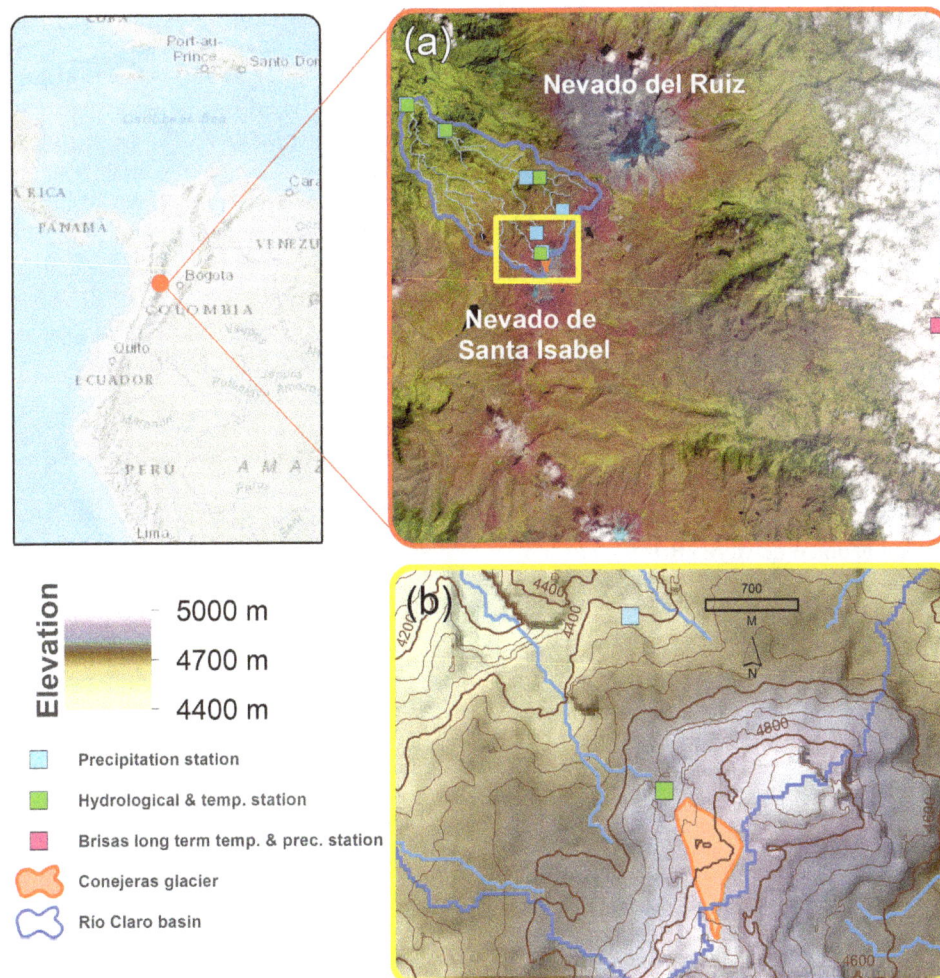

Figure 1. Study area, showing the glaciers of the Parque Nacional Natural de los Nevados, and the Río Claro basin (**a**) and the Conejeras glacier with hydro-meteorological stations (**b**).

Conejeras glacier corresponds to the uppermost headwaters of the Rio Claro basin (Fig. 1). The Rio Claro basin comprises an elevation range of 2700 to 4895 m and, from highest to lowest, presents a succession of typical Andean ecosystems: glacial (4700 to 4894), periglacial (4300–4700 m), *páramo* wetland ecosystem (3600 to 4300 m) and high-elevation tropical forest, *bosque altoandino* (2700 to 3600 m). Mean annual temperature at the glacier base is $1.3 \pm 0.7\,°C$, with very little seasonal variation, and precipitation sums reach $1025 \pm 50\,mm$ annually, with two contrasted seasons (see Fig. 2) resulting from the seasonal migration of the Intertropical Convergence Zone (ITCZ, Poveda et al., 2006). During the dry seasons (December to January and June to August), mean precipitation barely reaches 75 mm per month, whereas during the wet seasons (March to May and September to October), values exceed 150 mm per month.

3   Data and methods

3.1   Hydrological and meteorological data

Meteorological and hydrological data used in the present work have been collected by the Institute for Hydrological, Meteorological and Environmental Studies of Colombia (IDEAM, Instituto de Hidrología, Meteorología y Estudios Ambientales), thanks to the automatic meteorological and gauge stations network in the Río Claro basin (Fig. 1).

The experimental site of the Río Claro basin has been monitored since 2009, with a network of meteorological and hydrological stations located at different tributaries of the Río Claro, covering an altitudinal gradient of 2700–4900 m a.s.l. As this research is focused on the upper catchment in which the glacier is located for the present study, we only used data from the stations located at the Conejeras glacier snout (Fig. 1, bottom map). This includes one stream gauge (with associated rating curve) measuring 15 min resolution water discharge ($m^3 s^{-1}$), one temperature station measuring hourly temperature (°C) (both stations located at 4662 m a.s.l.), and one rain gauge measuring 10 min precipitation (mm; the station is located at 4413 m a.s.l.). Even though these data have been available since 2009, the sensors and loggers experienced technical problems; thus, numerous inhomogeneities, out-of-range values and empty records were present in the data series. From 2013, the technical problems were solved and the data are suitable for analysis. The period covered for analysis ranges from June 2013 to December 2017, a total of 56 months, and data were aggregated hourly, daily and monthly to perform statistical analyses. However, in order to obtain a wider perspective and to take advantage of the effort made by the IDEAM glaciologist, who conscientiously took mass balance measurements every month since 2006, also shown are trends and variability in climate – from a nearby meteorological station of the Colombian national network (Brisas) that contains data since 1982 – and glacier mass evolution for the longest time period available. The multivariate ENSO index (Wolter and Timlin, 1993, 1998), used for characterizing influence of the ENSO phenomenon on glacier evolution, has been downloaded from NOAA: https://www.esrl.noaa.gov/psd/enso/mei/table.html (last access: 15 December 2017).

3.2 Glacier evolution data

The evolution of the Conejeras glacier (Fig. 3) has been monitored by the Department of Ecosystems of IDEAM. Since March 2006, a network of 14 stakes was installed on the Conejeras glacier to measure ablation and accumulation area. The 6–12 m long stakes are PVC pipes of 2 m in length. These 14 stakes are vertically inserted into the glacier at a depth not less than 5 m and they are roughly organized into six cross profiles at approximately 4670, 4700, 4750, 4780, 4830 and 4885 m a.s.l. Accumulation and ablation measurements are performed monthly. Typical measurements of the field surveys include stake readings (monthly), density measurement in snow and firn pits (once per year) and re-drilling of stakes (if required) to the former position. The entire methodology can be found in Mölg et al. (2017) and Rabatel et al. (2017). The mass balance data are calculated using the classical glaciological method that represents the water equivalent that a glacier gains or loses in a given time. These data are used to generate yearly mappings of mass balance and calculate the equilibrium line altitude (ELA), which is the altitude point where mass balance is equal to zero equivalent meters of water and separates the ablation and accumulation area in the glacier (Francou and Pouyaud, 2004).

Changes in glacier surface during the study period were computed by means of satellite imagery (Landsat and Sentinel constellations) for the years 2006, 2010, 2013 and 2017. Cloud-free cover Landsat TM images were selected for 2006 and 2010, and Landsat OLI and Sentinel images for 2013 and 2017 respectively. TOA (top of the atmosphere) reflectance

was obtained using specific radiometric calibration coefficients for each image and sensor (Chander et al., 2009; Padró et al., 2017). BOA (bottom of the atmosphere) Reflectance was based on the dark object subtraction (DOS) approach (Chavez, 1988). The Normalized Difference Snow Index (NDSI) was used to discriminate snow and ice-covered areas from snow-free areas. The NDSI is expressed as the relationship between reflectance in the visible region and reflectance in the medium-infrared region (the specific bands vary among different sensors; e.g., TM bands 2 and 5). Pixels in the different images were classified as snow- or ice-covered areas when the NDSI was greater than 0.4 (Dozier, 1989).

3.3 Statistical analyses

A number of indices were extracted from the streamflow, temperature and precipitation hourly series in order to assess changes in time in the hydrological output of the glacier and their relation to climate (Table 1). These daily indices were subject to statistical analyses, including correlation tests, monthly aggregation and assessment of changes over time.

Since one of the main objectives of the paper is to characterize daily dynamics of streamflow and changes in time, a principal component analysis (PCA) was conducted in order to extract the main patterns of daily streamflow cycles. The data matrix for the PCA was then composed by streamflow hourly values in 1614 columns as variables (number of days) and 24 rows as cases (hours in a day). As the PCA does not allow the number of variables to exceed the number of cases, PCAs were performed on 25 bootstrapped random samples of days ($n = 23$, with replacement); results showed that three principal components were stable throughout the samples (see Table 3 in Results sections). After the main PCs were extracted, calculation of correlation between each day of the time series and the selected PCs was determined. The PC that best correlated with the correspondent day was assigned to every day, obtaining a time series of the three PCs. This allowed assessment of changes in time of the main patterns of daily streamflow cycles observed.

4 Results

4.1 Climatology and glacier evolution

The long-term climatic evolution of the study area is depicted in Fig. 2. The temperature and precipitation series (Fig. 2a, c, d) correspond to the Brisas meteorological station, which is located 25 km from the glacier, at 2721 m elevation. It therefore does not accurately represent the climate conditions at the glacier. It is, however, the closest meteorological station with available meteorological data to study long-term climate. The temperature record measured at the glacier snout (blue line) is included. It can be observed that despite the different range of values (temperatures at the glacier are 3.2 °C

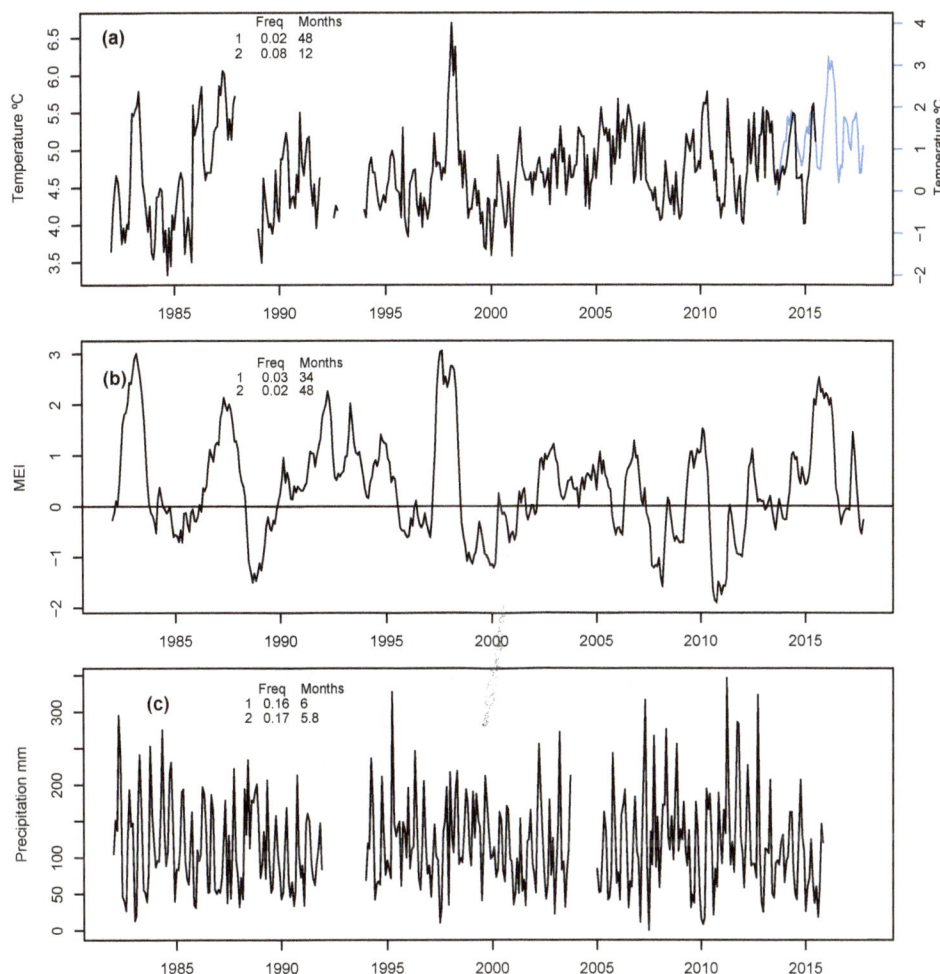

Figure 2. Long-term evolution of climate variables in the study area: **(a)** monthly air temperature at the Brisas meteorological station (2721 m a.s.l.), 1982–2015 (black line), and the temperature at the glacier snout (note the difference in the range of values), 2013–2017 (blue line); **(b)** multivariate ENSO index; **(c)** monthly precipitation at the Brisas station, 1982–2015; The frequency and its equivalent in months (1/frequency) of the two top spectral densities from spectral analysis is shown for temperature, MEI and precipitation monthly series.

lower than at Brisas), there is a match in variability for the common period.

Long-term evolution of temperature does not show any significant trend or pattern from 1982 to 2015; however, a spectral analysis shows that the frequency with higher spectral density corresponds with a seasonality of 48 months, indicating a recurrent cycle every 4 years. By comparing Fig. 2a with Fig. 2b, there is a close match between temperature and evolution of the Multivariate ENSO Index ($R = 0.49$), which also shows a high value of power spectra in the 48-month frequency cycle. Notwithstanding other factors whose analysis is far beyond the scope of this paper, it is evident that the evolution of temperature in the study area is highly driven by the ENSO phenomenon. Regarding precipitation (Fig. 2c), no long-term trend is observed, and the most evident pattern is the bi-modal seasonal regime, which

is confirmed by the frequency analysis showing the highest power spectra in the 6-month cycle.

The evolution of the glacier since 2006 is shown in Fig. 3. Almost every month since measurements began in 2006, the glacier has lost mass (113 months), and very few months (20) recorded a positive mass balance. The global balance in this period is a loss of 34.4 m of water equivalent. For the sake of visual comparison, we have included the time series of MEI, and a close correspondence between the variables is observed (Fig. 3a). During the warm phases of ENSO (Niño events, values of MEI above 0.5), the glacier loses up to 600 mm w.e. per month, as in the Niño event of 2009–2010, when the glacier lost a total of 7000 mm w.e. One could surmise that during La Niña (cold phases of ENSO, MEI values < -0.5) the glacier could recuperate mass. In fact, when MEI values are negative, the glacier experiences much less decrease; however, even during the strongest La

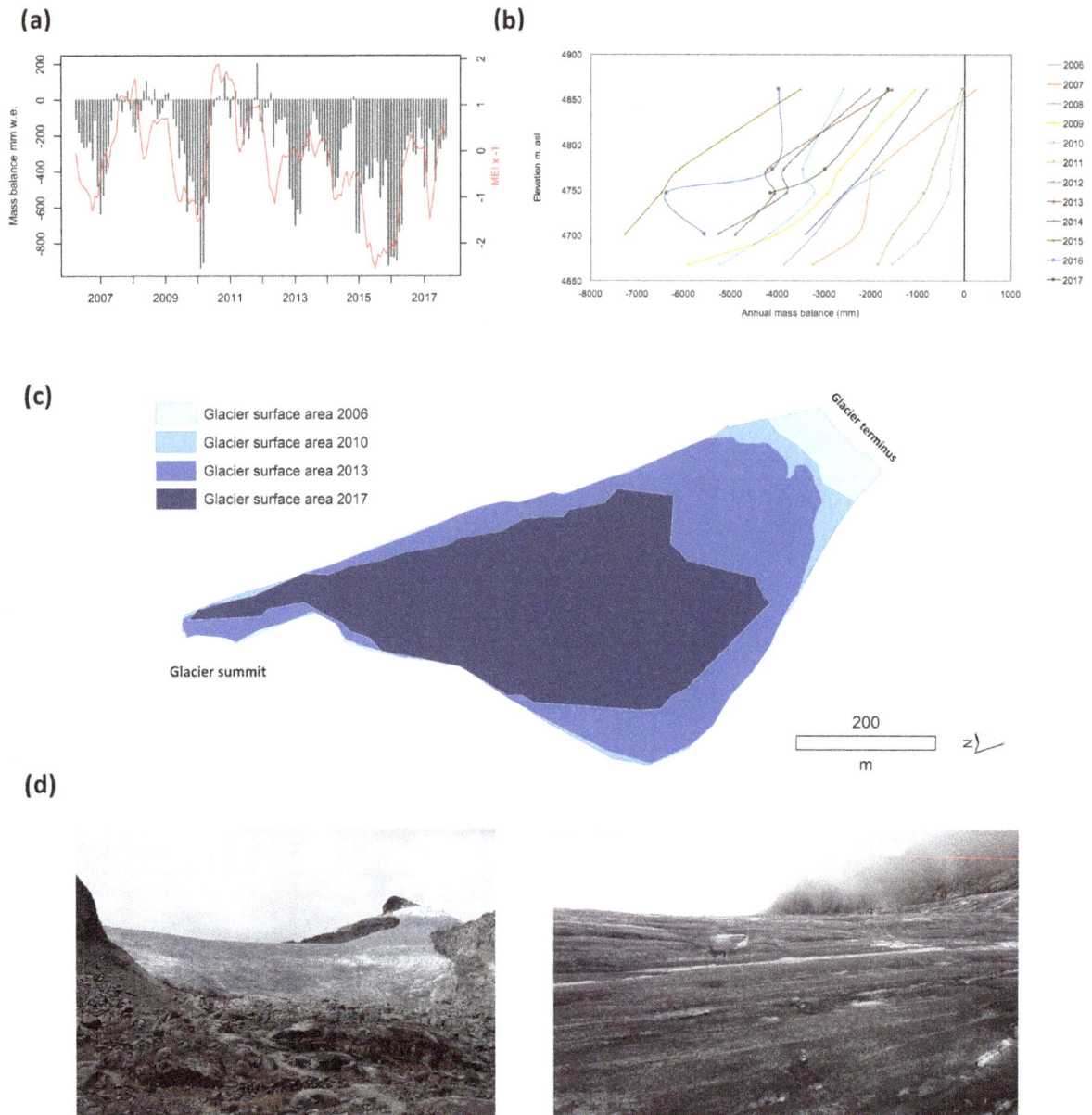

Figure 3. Evolution of the Conejeras glacier: **(a)** monthly mass balance (mm w.e.) and Multivariate ENSO Index (not the inverted axis); **(b)** annual mass balance per altitudinal range; **(c)** extension of the glacier in hectares in 2006, 2010, 2013 and 2017. **(d)** Photographs of the glacier surface covered by volcanic ashes, taken in 2015 and 2016.

Niña events, the balance is negative, with just a few months having a positive balance (e.g., in the 2010–2011 La Niña, the glacier lost 1000 mm w.e.). This occurs because even during La Niña mean temperatures at the glacier are above zero (0.8 ± 0.3 °C). The aforementioned agreement between ENSO and mass balance appears to break from 2012 onwards. There were two events of large mass balance loss around 2013–2014 that do not match with El Niño events. A local factor that can affect the glacier's mass balance independent of climatology is reduced albedo of the surface caused by the quantity of deposited ash that comes from the

nearby Santa Isabel volcano. This variable has not been considered in the present study but there are two pictures of the glacier for visual evidence (Fig. 3d). This fact, together with prevalence of above-zero degrees at the elevation in which the glacier is located (see Fig. 2, top plot) has induced the large glacier recession observed between 2006 and 2017 (Fig. 3c). During this period, there has been a 37 % reduction, from 22.45 ha in 2006 to 12 ha in 2017. However, this reduction has been far from linear. As shown in Fig. 3b, mass balance losses during the first years of the monitoring period were, in general, less pronounced than in the latest years. In

Table 1. Hydrologic and climatic indices computed from the hourly streamflow, temperature and precipitation series. * h_{pulse} is computed as the hourly equivalent of the melting–runoff spring pulse proposed by Cayan et al. (2001) for daily data, i.e., the time of the day when the minimum cumulative streamflow anomaly occurs, which is equivalent to finding the hour after which most flows are greater than the daily average.

Index	Explanation	Unit
$total_Q$	total daily water discharge	$m^3 \, day^{-1}$
Q_{max}	value of maximum hourly water discharge per day	$m^3 \, h^{-1}$
h_{pulse}*	hour of the day when the melting streamflow pulse starts	hour of the day
Q_{base}	mean water discharge value between the start of the day (00:00 h) and the hour when h_{pulse} occurs	$m^3 \, h^{-1}$
$h_{Q_{max}}$	hour of the day when	hour of the day
Q_{range}	difference between Q_{base} and Q_{max}	$m^3 \, h^{-1}$
Q_{slope}	slope of the streamflow rising limb between h_{pulse} and $h_{Q_{max}}$	slope in %
decayslope	slope of the streamflow decaying limb between $h_{Q_{max}}$ and 23:00 h	slope in %
T_{max}	value of maximum hourly temperature per day	$°C \, h^{-1}$
T_{min}	value of minimum hourly temperature per day	$°C \, h^{-1}$
T_{mean}	mean daily temperature	$°C \, day^{-1}$
T_{range}	difference between T_{min} and T_{max}	$°C \, h^{-1}$
$h_{T_{max}}$	hour of the day when the T_{max} occurs	hour of the day
Diffh	time difference between $h_{T_{max}}$ and $h_{Q_{max}}$	Hours
P_{max}	value of maximum hourly precipitation per day	$mm \, h^{-1}$
$h_{P_{max}}$	hour of the day when the P_{max} occurs	hour of the day
pp	daily precipitation sum	$mm \, day^{-1}$

2012, the ice mass retreated up the 4700 m elevation curve, and from then on the years with larger mass loss were 2015, 2016 and 2014.

4.2 Hydrological dynamics

The water discharge of the Conejeras glacier is measured at a gauging station located 300 m from the glacier snout (when the station was installed in 2009, it was only 10 m away from the glacier snout). The water volume measured at this station is a combination of water from glacier melt and water from precipitation into the watershed area, although the former exerts a larger control in water discharge variability. Table 2 shows the correlation between hydrological and temperature indices for samples of days with precipitation, independent of the amount of fallen precipitation (left), and for samples of days without precipitation (right). On days without precipitation, most hydrological indices show significant correlation with temperature, except for the baseflow and $h_{Q_{max}}$. The highest correlation values are found between Q_{max}, Q_{range}, Q_{slope} and $total_Q$, with T_{max} and T_{mean} (correlation values in the range of 0.5–0.65) indicating that the higher the temperatures, the more prominent the melting pulse of runoff. T_{min} shows smaller and less significant correlation values. The h_{pulse} also shows high correlation with temperature, but in this case in a negative fashion, indicating a later occurrence of the daily melting pulse when minimum temperatures and maximum temperatures are lower. On days with precipitation, correlation values are generally smaller, but in some

cases they are still significant, such as those for Q_{max}, Q_{range} and Q_{slope}.

A PCA performed on hourly streamflow data (in a recursive fashion; see Sect. 3.3 for explanation of the method) allowed procurement of the main patterns of daily flow, as well as changes in time during the study period. Three principal components were obtained, whose values of explained variance were stable throughout the 25 bootstrapped samples (Table 3). The first PC explained an average of $48 \pm 6\%$ of the variance throughout the 25 samples, and the second PC an average of $35 \pm 5.7\%$. Together they account for 83 % of variance and they both showed a neat pattern of daily streamflows (Fig. 4a). The main difference between PC1 and PC2 is the time of the day when peak flows are reached and, hence, the time range when most daily flows occur. Thus, PC1 corresponds to days with an earlier melt pulse (towards 10:00 LT) and earlier peak flows (towards 14:00 LT), compared to PC2, with days of melt pulse at 13:00 LT and peak flows at 18:00 LT. The remaining PC explains a residual percentage of the variance and, unlike PC1 and PC2, does not show a stable streamflow pattern across the samples. However, it was decided to keep it, as it can help explain some peculiarities in the results. In Fig. 4b the evolution of the frequency (days per month) of days corresponding to each PC is shown. Although there is some degree of variability, the frequency of days with PC1 streamflow pattern increases over time and dominates over the frequency of PC2 and PC3 days. This is especially significant between 2015 and 2016, coinciding with an El Niño event. However, by mid-2016 the fre-

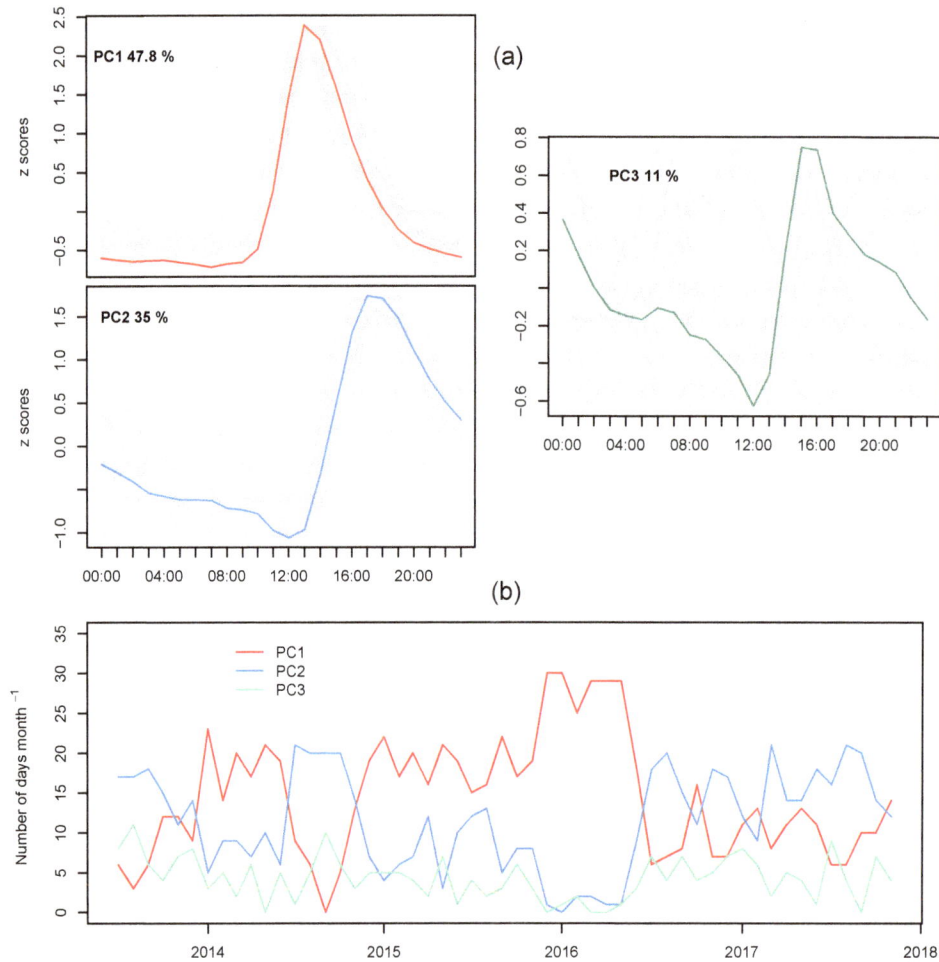

Figure 4. Principal component analysis on hourly streamflow: **(a)** scores of the three main principal components (patterns of daily stream-flow), with gray lines indicating the scores for each one of the 25 bootstrapped samples in the recursive PCA, and colored lines indicating the average. **(b)** Evolution of the number of days per month that show maximum correlation with each PC. Red corresponds to PC1, blue corresponds to PC2 and green corresponds to PC3.

Table 2. Pearson correlation coefficient between daily hydrological indices and temperature for days with and without precipitation (left) and for days without precipitation (right) between July 2013 and June 2017. The correlation values correspond to the average obtained by 100 resampling iterations ($n = 99$) of the correlation test. The symbols * and ** indicate that correlations are significant at 95 % and 99 % confidence respectively (two-tailed test).

Index	Days with and without precipitation ($n = 99$)				Days without precipitation ($n = 99$)			
	T_{min}	T_{max}	T_{mean}	T_{range}	T_{min}	T_{max}	T_{mean}	T_{range}
total$_Q$	0.25**	0.12	0.19	0.02	0.31**	0.54**	0.53**	−0.39**
Q_{max}	0.25**	0.30**	0.33**	−0.18	0.24*	0.64**	0.57**	−0.54**
Q_{base}	0.13	−0.13	−0.05	0.22*	0.18	0.05	0.11	0.06
Q_{range}	0.25**	0.36**	0.37**	−0.25**	0.22*	0.65**	0.58**	−0.57**
Q_{slope}	0.18	0.40**	0.38**	−0.34**	0.12	0.58**	0.48**	−0.55**
hQ_{max}	0.06	−0.03	0.00	0.06	0.04	0.00	0.02	0.02
Hpulse	−0.18	−0.17	−0.21*	0.08	−0.36**	−0.50**	−0.52**	0.31**

Table 3. Mean and standard deviation of variance explained (%) by each PC throughout the 25 bootstrapped samples.

PC	Mean	Standard deviation
PC1	47.78	5.91
PC2	34.99	5.66
PC3	11.82	6.77

quency of PC1 days drops considerably and the frequency of PC2 days increases at the same ratio. Thus, from mid-2016 to the end of the study period, they both maintain similar levels of frequency.

In order to understand the underlying factors of each PC, the frequency distribution of the climatic and hydrological indices for the days corresponding to each PC was computed, in the form of box plots (Fig. 5). From a hydrological point of view, PC1 better corresponds to days with higher total runoff and maximum runoff, and with a more pronounced slope in both the rising and decreasing limbs of the peak flow volume than PC2 and PC3. The variability (expressed by the amplitude of boxes in the box plots) of such hydrological indicators is, as well, higher amongst days of PC1, compared to days of PC2 and PC3. Base runoff is higher in PC1 but not significantly. The contrasted weight of climate may explain such hydrological differences between PCs: days of PC1 present significantly higher mean temperature (median $= 1.7\,°C$) and maximum temperature (median $= 3.8\,°C$) than days of PC2 (0.9 and 2.4 °C respectively) and PC3 (0.5 and 1.6 °C respectively). In contrast, precipitation is notably higher (and shows greater variability) in days grouped within PC3 (median $= 1.9\,\mathrm{mm\,day^{-1}}$) and PC2 (2.2 mm day^{-1}) compared to days of PC1 (0.3 mm day^{-1}). To summarize, PC1 corresponds to a daily regimen of pure glacier melting, whereas PC2 and PC3 correspond to days with a lower glacier melting pulse with more (PC3) or less (PC2) influence of precipitation.

In Fig. 4, a notable change occurs in the frequency of the two main patterns of hourly streamflow, PC1 and PC2, by mid-2016. Further details regarding changes in the hydrological yield of the glacier are shown in Fig. 6, which presents the evolution of the main hydrological indices computed, along with temperature, precipitation and glacier mass balance during the study period and averaged monthly. Total and maximum daily streamflow (total$_Q$ and Q_{max}) depict an increase up to mid-2016, where they begin to decrease. During the last 18 months, they remain at low levels compared to previous months. This turning point seems to coincide in time with the 2015–2016 El Niño event, with higher-than-average temperatures and low levels of precipitation that led to an increasing mass balance loss and, therefore, increased flows. It is remarkable that streamflow increases and decreases in direct proportion to mass balance change,

indicating the strong dependence of runoff to glacier melt. Similar evolution is observed in the difference between base flows and maximum flows (Q_{range}), as well as the slope of the rising limb of diurnal flows (Q_{slope}), which are indicators of diurnal variability: they increase up to 2016 and decrease afterwards, which coincides with the change in the frequency of daily streamflow patterns in Fig. 6. The mean hour of the day at which maximum flows are reached ($h_{Q_{max}}$) shows a steady evolution until mid-2016, when it begins to rise. This seems surprising when comparing it to the evolution of $h_{T_{max}}$ (i.e., the hour of the day when maximum temperature is reached), which does not show any particular temporal pattern. Regarding the monthly variability of flows (third panel on the right, Fig. 6) the same turning point is observed, with a clear decrease in the coefficient of variation until 2016 and an increase afterwards. It is clear that a hydrological change has occurred at the outlet of the glacier, but the two most plausible drivers of change (temperature and precipitation, bottom plots of Fig. 6) do not seem to be responsible for it. They both are affected by the El Niño event, when temperatures increased and precipitation decreased; however, they do not show an increasing–decreasing temporal pattern before and after such an event. This leads to the hypothesis that the hydrological change observed at these last stages of the glacier's life is independent of climate.

4.3   Changes in the runoff–climate relationship

In this section, the runoff is isolated from temperature and precipitation in order to determine if observed hydrological dynamics are driven by climate or are related to shrinkage of the glacier. Figure 7 shows the mean monthly runoff for days with temperatures lower and higher than 2 °C, i.e., water discharge series independent of temperature. Precipitation has also been added to the plot. It was noted that water discharge for days warmer than 2 °C is significantly higher than water discharge on days cooler than 2 °C. The characteristic evolution of runoff, with increasing amounts during most of the study period up to mid-2016 and decreasing runoff from that point onwards, was also observed. The same evolution occurs for both days below and days above 2 °C, and it occurs for very similar amounts of precipitation. This indicates that flows from the melting glacier are becoming less dependent on temperature, or climate in general, and more dependent on the size of the glacier. The box plots of Fig. 8 (bottom) confirm this observation by showing water volumes significantly higher before than after the breaking point, but also because the differences between water discharge at < 2 °C and water discharge at > 2 °C are also smaller (and not significant) after the breaking point, indicating the decreasing importance of temperature in the process of runoff production in the shrinking glacier.

Finally, Fig. 8 shows correlations between temperature–precipitation and monthly flows for different time periods. In Fig. 8a, two years are compared that can be considered

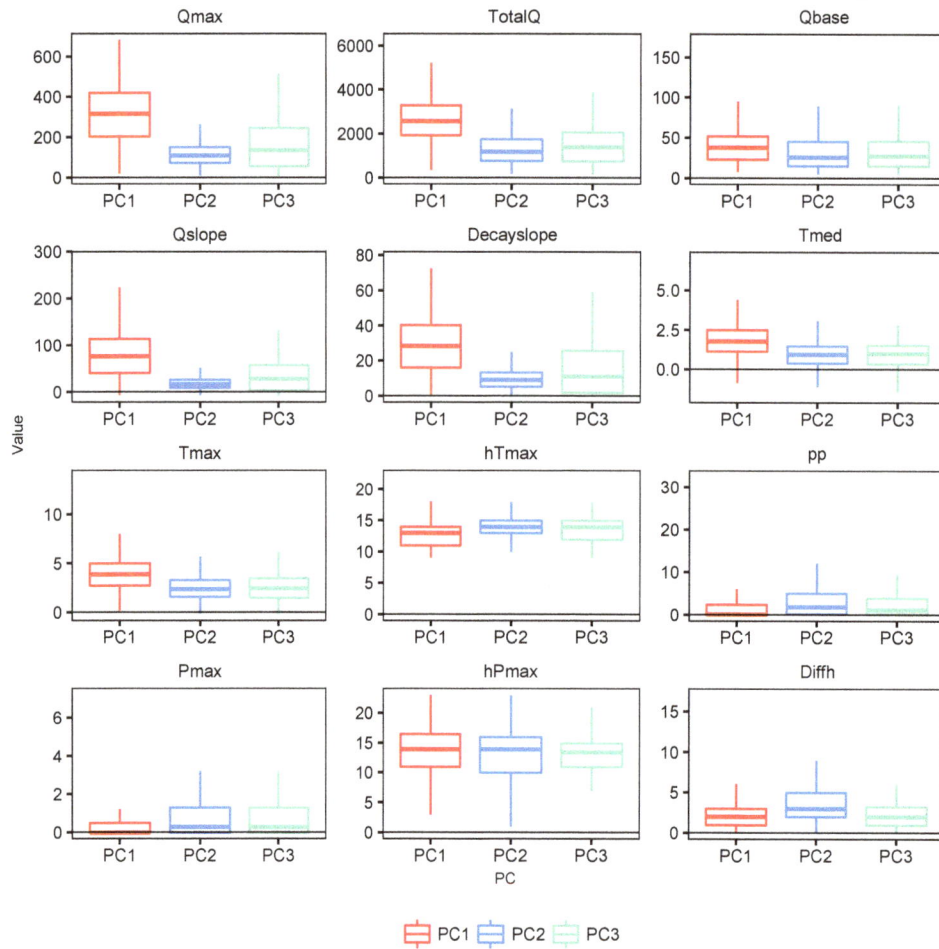

Figure 5. Summary of the frequency distributions (box plots) of the hydrological and meteorological indicators for days grouped within PC1, PC2 and PC3.

analogues in terms of total flow (similar amounts of monthly flow; see Fig. 6), but one year (2013–2014) belongs to the period of increasing flows, before the 2016 breakpoint, and the other year (2017) belongs to the period of decreasing flows after the breakpoint. Correlation between temperature and flow is much higher ($R = 0.65$) for 2013–2014 than for 2017 ($R = 0.35$), which would corroborate the previous observation. However, precipitation also shows higher correlation with flow for 2013–2014 ($R = 0.67$) than for 2017 ($R = 0.42$), which would contradict the hypothesis. One year, however, may not be representative of general trends, and so the same analysis is repeated, not for individual years but for the whole periods pre- and post-2016 breakpoint (Fig. 8b). The pattern seems more clear and corroborates the aforementioned hypothesis: correlation between temperature and flow is significant for the pre-2016 period ($R = 0.55$) but is nonexistent for the post-2016 period ($R = -0.1$). Correlation between precipitation and flow is insignificant ($R = -0.23$) for the pre-2016 period, and it is positive and significant for

the post-2016 period ($R = 0.32$). These previous observations lead to reasoning that during the years of hydrological monitoring (2013–2017), the observed hydrological dynamic, with a marked breakpoint in 2016, is a result of the vanishing glacier process and not a response to climate variability.

5 Discussion and conclusions

The present paper shows a comprehensive analysis of the dynamics of an Andean glacier that is close to extinction, with special focus on its hydrological yield. This research has benefited from a hydro-climatic monitoring network located in the surroundings of the glacier terminus that has been fully operative since 2013 and from monthly and annual estimations of mass balance and glacier extent respectively, derived from ice depth measurements and topographical surveys since 2006. Everything has been managed by the Institute of Hydrology Meteorology and Environmental

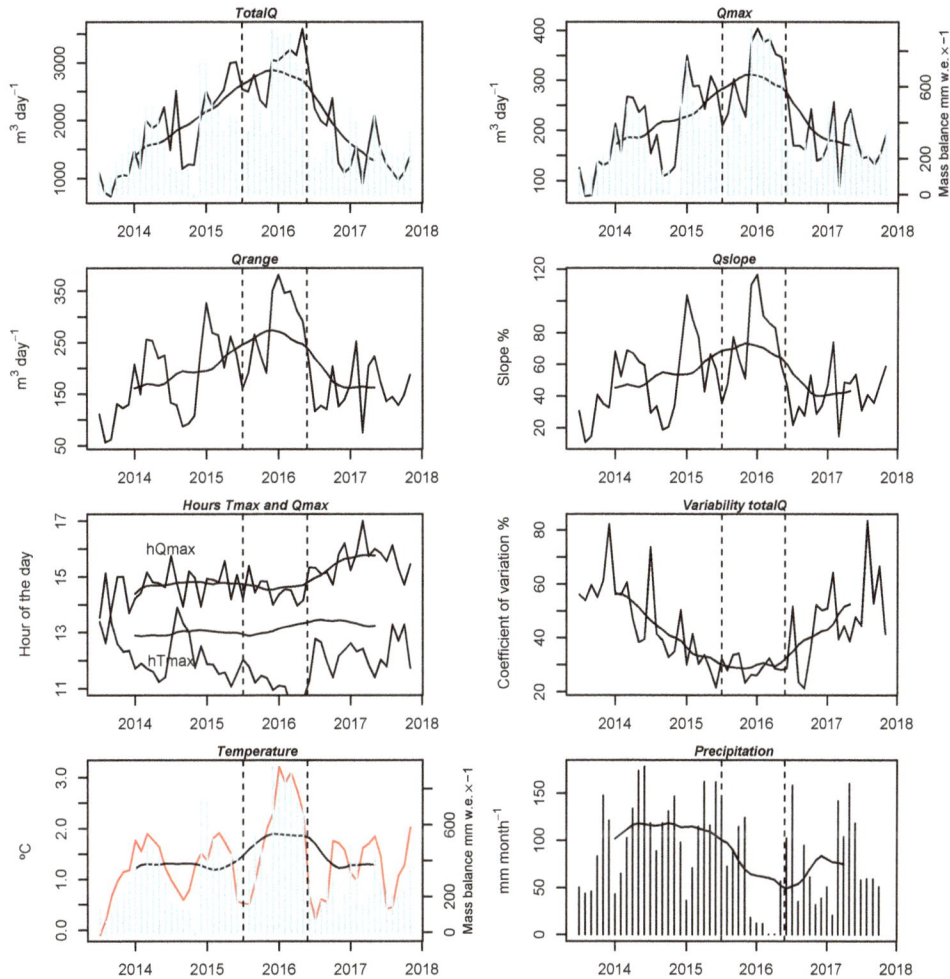

Figure 6. Evolution of monthly averaged hydrological indices, temperature, precipitation and glacier mass balance (in blue bars), for the study period. Dashed lines indicate the 2015–2016 strong El Niño event. The 12-month window moving averages (black smooth lines) are shown to represent trends.

Studies (IDEAM) of Colombia. The Conejeras glacier is currently an isolated small glacier that used to be part of a larger ice body called Nevado de Santa Isabel. Since measurements have been available, the glacier has constantly lost mass and, consequently, a reduction in its area is evident. The extinction of Colombian glaciers has been documented since 1850, with an average loss of 90 % of their area, considering current values (IDEAM, 2012). This reduction, of about 3 % per year, has been much larger during the last 3 decades (57 %) compared to previous decades (23 %), which is directly related to the general increase in temperatures in the region and to re-activation of volcanic activity (IDEAM, 2012; Rabatel et al., 2017). Since direct measurements began in 2006, the glacier studied has constantly lost area; however, until 2014, the area loss was gradual and restricted to the glacier front; from 2014, the sharp retreat also involved higher parts of the glacier. The main reason for this strong shrinkage is the existence of above-zero temperatures during most of the

year and less precipitation fallen as snow. This involves a constant migration of the equilibrium line to higher positions, and decreasing albedo of the ice surface that involves greater energy absorption, the latter accelerated by intense activity of Nevado de el Ruiz in the last years. In terms of mass balance, very few months exhibit a gain of ice during the period studied, and these tend to coincide with La Niña events (negative MEI episodes). These episodes cannot compensate for the great losses that occurred during the majority of months, which are especially large during El Niño events (positive MEI episodes), when above-normal temperatures are recorded. The ENSO phenomenon exerts great influence on the evolution of the glacier, similar to that reported for most inner tropical glaciers (Francou et al., 2004; Rabatel et al., 2013; Vuille et al., 2008); however, some episodes of great mass balance loss, such as that of 2014, cannot be explained by the ENSO. Observations of glacier surface during field surveys showed that, during some periods of mass

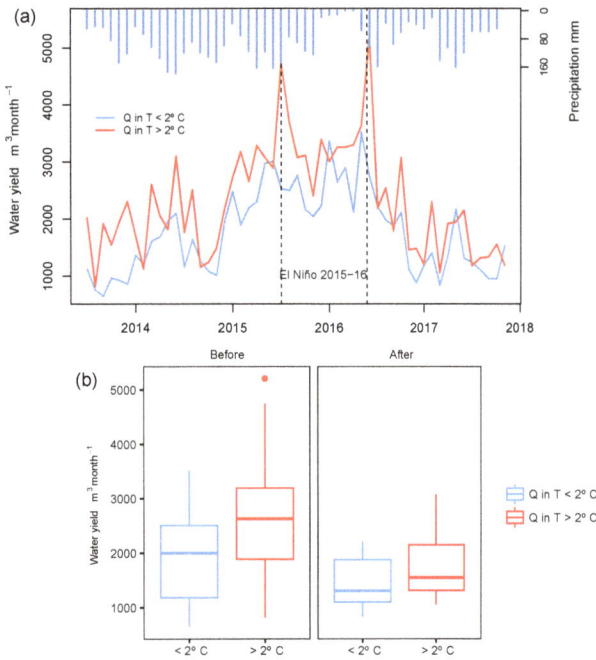

Figure 7. Mean monthly water discharge (Q), for days with temperature lower than 2 °C (blue) and days with temperature higher than 2 °C (red): **(a)** interannual evolution with indication of El Niño 2015–2016 event (grey shading), breakpoint in water discharge evolution (dashed line) and monthly precipitation (blue bars); **(b)** comparative box plots for water discharge before and after breakpoint in May 2016.

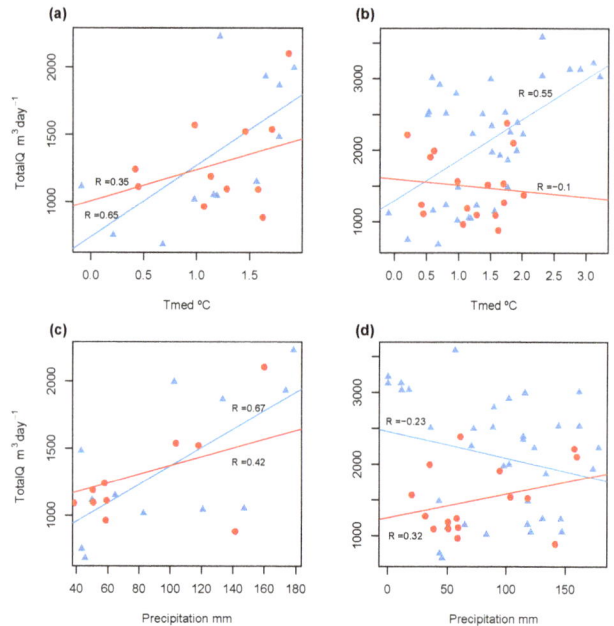

Figure 8. Correlations between monthly flow and monthly temperature **(a, b)** and precipitation **(c, d)** for **(a, c)** 2013–2014 (blue triangles) and 2017 (red circles), which are considered as analogues in terms of amounts of flow, and **(b, d)** months before the May 2016 breakpoint (blue triangles) and months after May 2016 breakpoint (red circles).

loss, surface ice retreat left ancient layers of volcanic ash exposed. The reduced energy reflectance caused by such ash layers might have triggered positive feedback that led to increasing melting and large ice retreat.

Glacier retreat is a worldwide phenomenon, currently linked to global warming (IPCC, 2013). Amongst the environmental issues related to glacier retreat, the issue concerning water resources has produced a vast amount of research. This is because glaciers constitute water reservoirs in the form of accumulated ice over thousands of years, and they provide water supply to downstream areas for the benefit of life, ecosystems and human societies. The rapid decrease in glacier extent during the last decades involves a change in water availability in glacier-dominated regions, and, thus, changes in water policies and water management are advisable (Huss, 2011; Kundzewicz et al., 2008). In the short term, glacier retreat involves increasing runoff in downstream areas but, after reaching a peak, runoff will eventually decrease until the contribution of the glacier melt is zero, when the glacier completely disappears. From a global perspective, such a tipping point is referred to as *peak water* and has given rise to concern from the scientific community (Gleick and Palaniappan, 2010; Huss and Hock, 2018; Kundzewicz et al., 2008; Mark et al., 2017; Sorg et al., 2014).

Research regarding the occurrence of such a runoff peak related to glacier retreat demonstrates that it will not occur concurrently worldwide. In some mountain areas, it has already occurred, i.e., the Peruvian Andes (Baraer et al., 2012), the western US mountains (Frans et al., 2016) or Central Asia (Sorg et al., 2012). At the majority of studied glacier basins, it is expected to occur in the course of the present century (Immerzeel et al., 2013; Ragettli et al., 2016; Sorg et al., 2014; Soruco et al., 2015). In recent global-scale research, Huss and Hock (2018) state that in nearly half of the 56 large-scale glacierized drainage basins studied, peak water has already occurred. In the other half, it will occur in the next decades, depending on extension of the ice cover fraction.

It was not the aim of this study to allocate such a tipping point in our studied glacier; however, observations on the characteristics of streamflow along the period studied suggest that it may have occurred during our study period. Our observations corroborate glacier melt being the main contributor to runoff in the catchment. However, even when correlations between runoff and temperature are mostly significant, the values are not as high as could be expected for a glacierized catchment. This is due to decreasing dependence of runoff on temperature, and therefore to glacier melt, as at a specific point during the study period. We observed a changing dynamic in most hydrological indicators, with a turning point in mid-2016, whereas climate variables, i.e., temperature and precipitation, do not show such evident variation (besides the exceptional conditions during an El Niño event). Both the PCA and the monthly aggregation of hydrological indices point to a less glacier-induced hydrological yield once the runoff peak of 2016 was reached. According to the literature (see Sect. 1.2.) this change from increasing to decreasing runoff, and from lesser importance of glacier contributions to total water discharge, must be expected in glacierized catchments with glaciers close to extinction. The short length of our hydrological series (five years) does not allow long-term analysis to determine water discharge in years of less glacier loss (i.e., from 2006 to 2012; see Fig. 3), which could verify or refute such a hypothesis. However, when we isolated total runoff from climate variables before and after the 2016 breakpoint (Figs. 8 and 9), we observed that the increase and later decrease of flows was mostly independent of temperature and precipitation, which would involve a glacier-driven hydrological change. Summarizing, streamflow measured at the glacier's snout showed the following characteristics: increasing trend in flow volume until mid-2016 and decreasing trend thereafter; increasing diurnal variability (given by the range between high flows and low flows and by the slope of the rising flow limb) up to mid-2016 and decreasing thereafter; decreasing and increasing monthly variability (given by the coefficient of variation of flows within a given month) before and after mid-2016; and high dependence of flow on temperatures before 2016 and low or null dependence after 2016, with increasing dependence on precipitation. In addition, this is supported by an evident change in the type of hydrograph, from a prevalence of days with melt-driven hydrographs (low baseflows, a sharp melting pulse, and great difference between high flows and low flows) before 2016, to an increase in the occurrence of days with less influence of melt and more influence by precipitation. All these characteristics support the idea of a hydrological change driven by the glacier recession in the catchment, as summarized by Hock et al. (2005; see Sect. 1.2). This observation cannot be taken conclusively, because the time period of hydrological observation is not long enough to deduce long-term trends and to explore hydrological dynamics before the great decline in glacier extent in 2014. However, given the current reduced size of the glacier (14 ha, which represents 35 % of the catchment that drains into the gauge station), it is likely that water discharge will continue to decrease in the upcoming years, until glacier contribution ends and runoff depends only on the precipitation that falls within the catchment. Like this glacier, other small glaciers in Colombia are expected to disappear in the coming decades (Rabatel et al., 2017); thus, a similar hydrological response can be expected.

Unlike glaciers in the western semiarid slopes of the Andes (i.e., Peru, Bolivia), Colombian glaciers do not constitute the main source of freshwater for downstream populations (IDEAM, 2012). The succession of humid periods provides enough water in mountain areas, most of which is stored in the deep soils of Páramos. These wetland ecosystems are mainly fed by rainfall (the contribution of glacier melt is mostly unknown, IDEAM, 2012) and act as water buffers, ensuring water availability during not-so-humid periods. Therefore, the role of glaciers in Colombia regarding water resources, including the ice body studied, is more marginal, and the occurrence of the peak water from glacier melt is not a current concern, as it is in Peru or Bolivia (Francou et al., 2014). Yet this does not diminish the relevance of the results of this work because they may be taken as an example of what can happen to the hydrology of glacierized basins in the tropics whose glaciers are in the process of disappearing. The glacier studied has a very small size compared to other ice bodies in the region. This makes it respond rapidly to variations in climate, as well as involving a rapid hydrological response of the catchment to the loss of ice, as was observed in this work. The increasing–decreasing flow dynamic observed as the glacier retreated occurred in roughly 5 years, and this is most likely related to the reduced size of the glacier studied. Most studies on the hydrological response to glacier retreat consider large river basins with large glacier coverage, usually by modeling approaches (i.e., Huss and Hock, 2018; Immerzeel et al., 2013; Ragettli et al., 2016; Sorg et al., 2012, 2014; Stahl et al., 2008), and the response times reported on either increasing flow at the initial stages or decreasing flow at the final stages are always on the scale of decades.

The added value of studying the hydrology related to this small-sized and near-extinct glacier is that the changes observed in the hydrology of the catchment could be directly attributed to the dynamics of the glacier and the climate that occurs at the same timescale, contrary to catchments containing large glaciers that respond with a larger temporal inertia to environmental changes. Hydrological analyses were restricted to the upper catchment because the streamflows measured at the snout of the glacier are not influenced by the signals of other environmental processes that may occur downstream (e.g., forest clearing or increased grazing). The methodological approach, including the PCA and the hydrological indices computed over subdaily resolution data, demonstrated itself as viable for detecting changes on the diurnal cycle of the glacier and can be applied to other small glaciers of the tropical Andes that respond rapidly (at subannual scales) to environmental forcing. The necessity for in situ observations on a fine scale in order to improve the accuracy of future estimations of water availability related to glacier retreat is emphasized.

Data availability. Hydro-meteorological data for the Rio Claro Basin as well as Conejeras glacier mass balance data were collected by the Institute for Hydrological, Meteorological and Environmental Studies of Colombia (IDEAM).

Hydrological and meterological data are managed by the departments of hydrology and meteorology respectively and can be accessed upon formal request at http://www.ideam.gov.co/solicitud-de-informacion (last access: 12 May 2018).

Glacier mass balance data are managed by the department of ecosystems and the environmental information system of Colombia, SIAC. To access these data, within SIAC site http://www.siac.gov.co (last access: 1 July 2018), click on "Biodiversidad", then on "Ecosistemas de importancia ambiental", and finally on "Glaciares".

The Multivariate Enso Index can be downloaded from National Oceanic and Atmospheric Administration of the United States NOAA https://www.esrl.noaa.gov/psd/enso/mei/ (Wolter, 2018).

Author contributions. EMT was responsible for the conceptualization, data processing and validation, statistical analyses, writing of the original draft. JLC took part in the conceptualization, experimental design, field surveys and data collection. KP carried out the field surveys, data collection and data processing. JLL carried out the cartography and remote sensing analysis. JILM was responsible for the conceptualization, statistical analyses and editing and review of the paper.

Competing interests. The authors declare that they have no conflict of interest.

Special issue statement. This article is part of the special issue "Assessing impacts and adaptation to global change in water resource systems depending on natural storage from groundwater and/or snowpacks". It is not associated with a conference.

Acknowledgements. This work has been possible thanks to the monitoring network installed by the Department of Ecosystems of the Colombian Institute for Hydrology, Meteorology and Environmental Studies (*Instituto de Hidrología, Meteorología y Estudios Ambientales*, IDEAM) and to the monthly field surveys on the Conejeras glacier and Río Claro basin, carried out by employed staff. We give our sincere gratitude to them, with special thanks to Yina Paola Nocua. The following projects gave financial support to this paper: "*Estudio hidrológico de la montaña altoandina (Colombia) y su respuesta a procesos de cambio global*" financed by Banco Santander, through the program of exchange scholarships for young researchers in Ibero-America "*Becas Iberoamérica Jóvenes Profesores e Investigadores*" (2015); and CGL2017-82216-R (HIDROIBERNIEVE), funded by the Spanish Ministry of Economy and Competitiveness. We are thankful to the anonymous referees for their valuable comments and suggestions that helped improve the final version of this paper.

Edited by: David Pulido-Velazquez

References

Baraer, M., Mark, B. G., McKenzie, J. M., Condom, T., Bury, J., Huh, K.-I., Portocarrero, C., Gómez, J., and Rathay, S.: Glacier recession and water resources in Peru's Cordillera Blanca, J. Glaciol., 58, 134–150, https://doi.org/10.3189/2012JoG11J186, 2012.

Beniston, M.: Impacts of climatic change on water and associated economic activities in the Swiss Alps, J. Hydrol., 412–413, 291–296, https://doi.org/10.1016/J.JHYDROL.2010.06.046, 2012.

Bradley, R. S., Vuille, M., Diaz, H. F., and Vergara, W.: Climate change. Threats to water supplies in the tropical Andes, Science, 312, 1755–1756, https://doi.org/10.1126/science.1128087, 2006.

Cayan, D. R., Dettinger, M. D., Kammerdiener, S. A., Caprio, J. M., Peterson, D. H., Cayan, D. R., Dettinger, M. D., Kammerdiener, S. A., Caprio, J. M., and Peterson, D. H.: Changes in the Onset of Spring in the Western United States, B. Am. Meteorol. Soc., 82, 399–415, https://doi.org/10.1175/1520-0477(2001)082<0399:CITOOS>2.3.CO;2, 2001.

Ceballos, J. L., Euscátegui, C., Ramírez, J., Cañon, M., Huggel, C., Haeberli, W., and Machguth, H.: Fast shrinkage of tropical glaciers in Colombia, Ann. Glaciol., 43, 194–201, https://doi.org/10.3189/172756406781812429, 2006.

Chander, G., Markham, B. L., and Helder, D. L.: Summary of current radiometric calibration coefficients for Landsat MSS, TM, ETM+, and EO-1 ALI sensors, Remote Sens. Environ., 113, 893–903, https://doi.org/10.1016/J.RSE.2009.01.007, 2009.

Chavez, P. S.: An improved dark-object subtraction technique for atmospheric scattering correction of multispectral data, Remote Sens. Environ., 24, 459–479, https://doi.org/10.1016/0034-4257(88)90019-3, 1988.

Chevallier, P., Pouyaud, B., Suarez, W., and Condom, T.: Climate change threats to environment in the tropical Andes: glaciers and water resources, Reg. Environ. Change, 11, 179–187, https://doi.org/10.1007/s10113-010-0177-6, 2011.

Dozier, J.: Spectral signature of alpine snow cover from the landsat thematic mapper, Remote Sens. Environ., 28, 9–22, https://doi.org/10.1016/0034-4257(89)90101-6, 1989.

Favier, V., Wagnon, P., and Ribstein, P.: Glaciers of the outer and inner tropics: A different behaviour but a common response to climatic forcing, Geophys. Res. Lett., 31, L16403, https://doi.org/10.1029/2004GL020654, 2004.

Francou, B. and Pouyaud, B.: Metodos de observacion de glaciares en los Andes tropicales?: mediciones de terreno y procesamiento de datos?: version-1?: 2004, available at: https://www.researchgate.net/profile/Bernard_Pouyaud/publication/282171220_Metodos_de_observacion_de_glaciares_en_los_Andes_tropicales_mediciones_de_terreno_y_procesamiento_de_datos_version-1_2004/links/561ba9b808ae78721fa0f8ad.pdf (last access: 12 March 2018), 2004.

Francou, B., Vuille, M., Wagnon, P., Mendoza, J., and Sicart, J.: Tropical climate change recorded by a glacier in the central Andes during the last decades of the twentieth century: Chacaltaya, Bolivia, 16° S, J. Geophys. Res., 108, 4154, https://doi.org/10.1029/2002JD002959, 2003.

Francou, B., Vuille, M., Favier, V., and Cáceres, B.: New evidence for an ENSO impact on low-latitude glaciers: Antizana 15, Andes of Ecuador, 0°28′ S, J. Geophys. Res., 109, D18106, https://doi.org/10.1029/2003JD004484, 2004.

Francou, B., Rabatel, A., Soruco, A., Sicart, J. E., Silvestre, E. E., Ginot, P., Cáceres, B., Condom, T., Villacís, M., Ceballos, J. L., Lehmann, B., Anthelme, F., Dangles, O., Gomez, J., Favier, V., Maisincho, L., Jomelli, V., Vuille, M., Wagnon, P., Lejeune, Y., Ramallo, C., and Mendoza, J.: Glaciares de los Andes tropicales:

víctimas del cambio climático, Comunidad Andina, PRAA, IRD, available at: http://bibliotecavirtual.minam.gob.pe/biam/handle/minam/1686 (last access: 22 February 2018), 2014.

Frans, C., Istanbulluoglu, E., Lettenmaier, D. P., Clarke, G., Bohn, T. J., and Stumbaugh, M.: Implications of decadal to century scale glacio-hydrological change for water resources of the Hood River basin, OR, USA, Hydrol. Process., 30, 4314–4329, https://doi.org/10.1002/hyp.10872, 2016.

Gleick, P. H. and Palaniappan, M.: Peak water limits to freshwater withdrawal and use, P. Natl. Acad. Sci. USA, 107, 11155–11162, https://doi.org/10.1073/pnas.1004812107, 2010.

Granados, H. D., Miranda, P. J., Núñez, G. C., Alzate, B. P., Mothes, P., Roa, H. M., Cáceres Correa, B. E., and Ramos, J. C.: Hazards at Ice-Clad Volcanoes: Phenomena, Processes, and Examples From Mexico, Colombia, Ecuador, and Chile, in: Snow Ice-Related Hazards, edited by: Shroder, J. F., Haeberli, V., and Whiteman, C., Risks Disasters, 607–646, https://doi.org/10.1016/B978-0-12-394849-6.00017-2, 2015.

Hock, R., Jansson, P., and Braun, L. N.: Modelling the Response of Mountain Glacier Discharge to Climate Warming, in Global Change and Mountain Regions (A State of Knowledge Overview), Springer, Dordrecht, 243–252, 2005.

Huggel, C., Ceballos, J. L., Pulgarín, B., Ramírez, J., and Thouret, J.-C.: Review and reassessment of hazards owing to volcano–glacier interactions in Colombia, Ann. Glaciol., 45, 128–136, https://doi.org/10.3189/172756407782282408, 2007.

Huss, M.: Present and future contribution of glacier storage change to runoff from macroscale drainage basins in Europe, Water Resour. Res., 47, W07511, https://doi.org/10.1029/2010WR010299, 2011.

Huss, M. and Hock, R.: Global-scale hydrological response to future glacier mass loss, Nat. Clim. Chang., 8, 135–140, https://doi.org/10.1038/s41558-017-0049-x, 2018.

IDEAM: Glaciares de Colombia, más que montañas con hielo, edited by: Comité de Comuniaciones y Publicaciones del IDEAM, Bogotá, 2012.

Immerzeel, W. W., Pellicciotti, F., and Bierkens, M. F. P.: Rising river flows throughout the twenty-first century in two Himalayan glacierized watersheds, Nat. Geosci., 6, 742–745, https://doi.org/10.1038/ngeo1896, 2013.

IPCC: Climate Change 2013: The Physical Science Basis. Contribution of Working Group I to the Fifth Assessment Report of the Intergovernmental Panel on Climate Change, edited by: Stocker, T. F., Qin, D., Plattner, G.-K., Tignor, M., Allen, S. K., Boschung, J., Nauels, A., Xia, Y., Bex, V., and Midgley, P. M., Cambridge University Press, Cambridge, United Kingdom and New York, NY, USA, 2013.

Kaser, G. and Omaston, H.: Tropical glaciers, Cambridge University Press, available at: https://books.google.es/books?hl=es&lr=&id=ZEB-I3twN_gC&oi=fnd&pg=PR11&dq=tropical+glaciers&ots=WLwn1fdjig&sig=897EG6q4Pyc113vo9Qb2bnyUo7g#v=onepage&q=tropicalglaciers&f=false (last access: 21 November 2017), 2002.

Kaser, G., Grosshauser, M., and Marzeion, B.: Contribution potential of glaciers to water availability in different climate regimes., P. Natl. Acad. Sci. USA, 107, 20223–20227, https://doi.org/10.1073/pnas.1008162107, 2010.

Klein, A. G., Morris, J. N., and Poole, A. J.: Retreat of Tropical Glaciers in Colombia and Venezuela from 1984 to 2004 as Measured from ASTER and Landsat Images, in 63 rd EASTERN SNOW CONFERENCE, Newark, Delaware USA, available at: https://www.researchgate.net/publication/228492383 (last access: 5 July 2018), 2006.

Kundzewicz, Z. W., Mata, L. J., W., A. N., Döll, P., Jimenez, B., Miller, K., Oki, T., Şed, Z., and Shiklomanov, I.: The implications of projected climate change for freshwater resources and their management, Hydrolog. Sci. J., 53, 3–10, https://doi.org/10.1623/hysj.53.1.3, 2008.

López-Moreno, J. I., Fontaneda, S., Bazo, J., Revuelto, J., Azorin-Molina, C., Valero-Garcés, B., Morán-Tejeda, E., Vicente-Serrano, S. M., Zubieta, R., and Alejo-Cochachín, J.: Recent glacier retreat and climate trends in Cordillera Huaytapallana, Peru, Global Planet. Change, 112, 1–11, https://doi.org/10.1016/j.gloplacha.2013.10.010, 2014.

López-Moreno, J. I., Valero-Garcés, B., Mark, B., Condom, T., Revuelto, J., Azorín-Molina, C., Bazo, J., Frugone, M., Vicente-Serrano, S. M., and Alejo-Cochachin, J.: Hydrological and depositional processes associated with recent glacier recession in Yanamarey catchment, Cordillera Blanca (Peru), Sci. Total Environ., 579, 272–282, https://doi.org/10.1016/J.SCITOTENV.2016.11.107, 2017.

Mark, B. G., French, A., Baraer, M., Carey, M., Bury, J., Young, K. R., Polk, M. H., Wigmore, O., Lagos, P., Crumley, R., McKenzie, J. M., and Lautz, L.: Glacier loss and hydro-social risks in the Peruvian Andes, Global Planet. Change, 159, 61–76, https://doi.org/10.1016/J.GLOPLACHA.2017.10.003, 2017.

Mölg, N., Ceballos, J. L., Huggel, C., Micheletti, N., Rabatel, A., and Zemp, M.: Ten years of monthly mass balance of Conejeras glacier, Colombia, and their evaluation using different interpolation methods, Geogr. Ann. A, 99, 155–176, https://doi.org/10.1080/04353676.2017.1297678, 2017.

Padró, J.-C., Pons, X., Aragonés, D., Díaz-Delgado, R., García, D., Bustamante, J., Pesquer, L., Domingo-Marimon, C., González-Guerrero, Ò., Cristóbal, J., Doktor, D., and Lange, M.: Radiometric Correction of Simultaneously Acquired Landsat-7/Landsat-8 and Sentinel-2A Imagery Using Pseudoinvariant Areas (PIA): Contributing to the Landsat Time Series Legacy, Remote Sens., 9, 1319, https://doi.org/10.3390/rs9121319, 2017.

Poveda, G. and Pineda, K.: Reassessment of Colombia's tropical glaciers retreat rates: are they bound to disappear during the 2010–2020 decade?, Adv. Geosci., 22, 107–116, https://doi.org/10.5194/adgeo-22-107-2009, 2009.

Poveda, G., Waylen, P. R., and Pulwarty, R. S.: Annual and inter-annual variability of the present climate in northern South America and southern Mesoamerica, Palaeogeogr. Palaeocl., 234, 3–27, https://doi.org/10.1016/j.palaeo.2005.10.031, 2006.

Rabatel, A., Francou, B., Soruco, A., Gomez, J., Cáceres, B., Ceballos, J. L., Basantes, R., Vuille, M., Sicart, J.-E., Huggel, C., Scheel, M., Lejeune, Y., Arnaud, Y., Collet, M., Condom, T., Consoli, G., Favier, V., Jomelli, V., Galarraga, R., Ginot, P., Maisincho, L., Mendoza, J., Ménégoz, M., Ramirez, E., Ribstein, P., Suarez, W., Villacis, M., and Wagnon, P.: Current state of glaciers in the tropical Andes: a multi-century perspective on glacier evolution and climate change, The Cryosphere, 7, 81–102, https://doi.org/10.5194/tc-7-81-2013, 2013.

Rabatel, A., Ceballos, J. L., Micheletti, N., Jordan, E., Braitmeier, M., González, J., Mölg, N., Ménégoz, M.,

Huggel, C., and Zemp, M.: Toward an imminent extinction of Colombian glaciers?, Geogr. Ann. A, 100, 75–95, https://doi.org/10.1080/04353676.2017.1383015, 2017.

Ragettli, S., Immerzeel, W. W., and Pellicciotti, F.: Contrasting climate change impact on river flows from high-altitude catchments in the Himalayan and Andes Mountains., P. Natl. Acad. Sci. USA, 113, 9222–9227, https://doi.org/10.1073/pnas.1606526113, 2016.

Rekowsky, I. C.: Variações de área das geleiras da Colômbia e da Venezuela entre 1985 e 2015, com dados de sensoriamento remoto, available at: https://www.lume.ufrgs.br/handle/10183/149546 (last access: 5 July 2018), 2016.

Ribstein, P., Tiriau, E., Francou, B., and Saravia, R.: Tropical climate and glacier hydrology: a case study in Bolivia, J. Hydrol., 165, 221–234, https://doi.org/10.1016/0022-1694(94)02572-S, 1995.

Sicart, J. E., Hock, R., Ribstein, P., Litt, M., and Ramirez, E.: Analysis of seasonal variations in mass balance and meltwater discharge of the tropical Zongo Glacier by application of a distributed energy balance model, J. Geophys. Res., 116, D13105, https://doi.org/10.1029/2010JD015105, 2011.

Sorg, A., Bolch, T., Stoffel, M., Solomina, O., and Beniston, M.: Climate change impacts on glaciers and runoff in Tien Shan (Central Asia), Nat. Clim. Change, 2, 725–731, https://doi.org/10.1038/nclimate1592, 2012.

Sorg, A., Huss, M., Rohrer, M., and Stoffel, M.: The days of plenty might soon be over in glacierized Central Asian catchments, Environ. Res. Lett., 9, 104018, https://doi.org/10.1088/1748-9326/9/10/104018, 2014.

Soruco, A., Vincent, C., Rabatel, A., Francou, B., Thibert, E., Sicart, J. E., and Condom, T.: Contribution of glacier runoff to water resources of La Paz city, Bolivia (16° S), Ann. Glaciol., 56, 147–154, https://doi.org/10.3189/2015AoG70A001, 2015.

Stahl, K., Moore, R. D., Shea, J. M., Hutchinson, D., and Cannon, A. J.: Coupled modelling of glacier and streamflow response to future climate scenarios, Water Resour. Res., 44, W02422, https://doi.org/10.1029/2007WR005956, 2008.

Stewart, I. T., Cayan, D. R., and Dettinger, M. D.: Changes in Snowmelt Runoff Timing in Western North America under a "Business as Usual" Climate Change Scenario, Clim. Change, 62, 217–232, https://doi.org/10.1023/B:CLIM.0000013702.22656.e8, 2004.

Vaughan, D. G., Comiso, J. C., Allison, I., Carrasco, J., Kaser, G., Kwok, R., Mote, P., Murray, T., Paul, F., Ren, J., Rignot, E., Solomina, O., Steffen, K., and Zhang, T.: Observations: Cryosphere, in Climate Change 2013: The Physical Science Basis. Contribution of Working Group I to the Fifth Assessment Report of the Intergovernmental Panel on Climate Change, edited by: Stocker, T. F., Qin, D., Plattner, G.-K., Tignor, M., Allen, S. K., Boschung, J., Nauels, A., Xia, Y., Bex, V., and Midgley, P. M., 317–382, Cambridge University Press, Cambridge, United Kingdom and New York, NY, USA, 2013.

Vuille, M., Francou, B., Wagnon, P., Juen, I., Kaser, G., Mark, B. G., and Bradley, R. S.: Climate change and tropical Andean glaciers: Past, present and future, Earth-Sci. Rev., 89, 79–96, https://doi.org/10.1016/J.EARSCIREV.2008.04.002, 2008.

Vuille, M., Carey, M., Huggel, C., Buytaert, W., Rabatel, A., Jacobsen, D., Soruco, A., Villacis, M., Yarleque, C., Timm, O. E., Condom, T., Salzmann, N., and Sicart, J.-E.: Rapid decline of snow and ice in the tropical Andes – Impacts, uncertainties and challenges ahead, Earth-Sci. Rev., 176, 195–213, https://doi.org/10.1016/j.earscirev.2017.09.019, 2017.

Wolter, K.: Multivariate ENSO Index webpage, available at: https://www.esrl.noaa.gov/psd/enso/mei/ (last access: 15 December 2017), 2018.

Wolter, K. and Timlin, M. S.: Monitoring ENSO in COADS with a seasonally adjusted principal component index, Proc. of the 17th Climate Diagnostics Workshop, Norman, OK, NOAA/NMC/CAC, NSSL, Oklahoma Clim. Survey, CIMMS and the School of Meteor., Univ. of Oklahoma, 52–57, 1993.

Wolter, K. and Timlin, M. S.: Measuring the strength of ENSO events – how does 1997/98 rank?, Weather, 53, 315–324, 1998.

13

Modelling liquid water transport in snow under rain-on-snow conditions – considering preferential flow

Sebastian Würzer[1,2], **Nander Wever**[2,1], **Roman Juras**[1,3], **Michael Lehning**[1,2], **and Tobias Jonas**[1]

[1]WSL Institute for Snow and Avalanche Research SLF, Flüelastrasse 11, 7260 Davos Dorf, Switzerland
[2]École Polytechnique Fédérale de Lausanne (EPFL), School of Architecture, Civil and Environmental Engineering, Lausanne, Switzerland
[3]Faculty of Environmental Sciences, Czech University of Life Sciences Prague, Kamýcká 129, 165 21, Prague, Czech Republic

Correspondence to: Sebastian Würzer (sebastian.wuerzer@slf.ch)

Abstract. Rain on snow (ROS) has the potential to generate severe floods. Thus, precisely predicting the effect of an approaching ROS event on runoff formation is very important. Data analyses from past ROS events have shown that a snowpack experiencing ROS can either release runoff immediately or delay it considerably. This delay is a result of refreeze of liquid water and water transport, which in turn is dependent on snow grain properties but also on the presence of structures such as ice layers or capillary barriers. During sprinkling experiments, preferential flow was found to be a process that critically impacted the timing of snowpack runoff. However, current one-dimensional operational snowpack models are not capable of addressing this phenomenon. For this study, the detailed physics-based snowpack model SNOWPACK is extended with a water transport scheme accounting for preferential flow. The implemented Richards equation solver is modified using a dual-domain approach to simulate water transport under preferential flow conditions. To validate the presented approach, we used an extensive dataset of over 100 ROS events from several locations in the European Alps, comprising meteorological and snowpack measurements as well as snow lysimeter runoff data. The model was tested under a variety of initial snowpack conditions, including cold, ripe, stratified and homogeneous snow. Results show that the model accounting for preferential flow demonstrated an improved overall performance, where in particular the onset of snowpack runoff was captured better. While the improvements were ambiguous for experiments on isothermal wet snow, they were pronounced for experiments on cold snowpacks, where field experiments found preferential flow to be especially prevalent.

1 Introduction

The flooding potential of rain-on-snow (ROS) events has been reported for many severe floods in the US (Kattelmann, 1997; Kroczynski, 2004; Leathers et al., 1998; Marks et al., 2001; McCabe et al., 2007), but also in Europe (Badoux et al., 2013; Freudiger et al., 2014; Rössler et al., 2014; Sui and Koehler, 2001; Wever et al., 2014b) where for example up to 55 % of peak flow events could be attributed to ROS events for some parts of Austria (Merz and Blöschl, 2003). With rising air temperature due to climate change, the frequency of ROS is likely to increase in high-elevation areas (Surfleet and Tullos, 2013) as well as in high latitudes (Ye et al., 2008). Besides spatial heterogeneity of the snowpack and uncertainties in meteorological forcing, deficits in process understanding make the consequences of extreme ROS events very difficult to forecast (Badoux et al., 2013; Rössler et al., 2014). For hydro-meteorological forecasters, it is particularly important to know a priori how much and when snowpack runoff is to be expected. Particularly, a correct temporal representation of snowpack processes is crucial to identify whether the presence of a snowpack will attenuate or amplify the generation of catchment-wide snowpack runoff. Most studies investigating ROS only consider the generation of snowpack runoff on a daily or multi-day timescale, where an exact description of water transport processes is less important than for sub-daily timescales (Wever et al., 2014a). Water transport processes are further usually described for snowmelt conditions, but not for ROS conditions, where high rain intensities may fall onto a cold snowpack below the freezing point. In this study however, we particularly focus on snowpack runoff generation at sub-daily scales with special attention to the timing of snowpack runoff which is influenced by preferential flow (PF).

Many studies have shown that flow fingering or PF is an important water transport mechanism in both laboratory experiments (Hirashima et al., 2014; Katsushima et al., 2013; Waldner et al., 2004) and under natural conditions, using dye tracer (Gerdel, 1954; Marsh and Woo, 1984; Schneebeli, 1995), temperature investigations (Conway and Benedict, 1994) or by measuring the spatial variability of snowpack runoff (Kattelmann, 1989; Marsh and Pomeroy, 1993, 1999; Marsh and Woo, 1985). The variability of snowpack runoff is defined by the distribution and size of preferential flow paths (PFPs), which are dependent on the structure of the snowpack and weather conditions (Schneebeli, 1995). Beyond its importance for hydrological implications, PF may also be crucial for wet snow avalanche formation processes, where snow stability can be depending on the exact location of liquid water ponding (Wever et al., 2016a).

Most snow models describe the water flow in snow as a uniform wetting front, thereby implicitly only considering the matrix flow component. The history of quantitative modelling of water transport in snow starts with Colbeck (1972), who first described a gravity drainage water transport model for isothermal, homogeneous snow. This was done by applying the general theory of Darcian flow of two-fluid phases flowing through porous media, neglecting capillary forces. Because water transport is not just occurring in isothermal conditions and snow can therefore not be treated as a classical porous medium, Illangasekare et al. (1990) were the first to introduce a 2-D model being able to describe water transport in subfreezing and layered snow. A detailed multi-layer physics-based snow model, where water transport was governed by the gravitational part of the Richards equation (RE) described in Colbeck (1972), was introduced by Jordan (1991). With the implementation of the full RE described by Wever et al. (2014a), the influence of capillary forces on the water flow was firstly represented in an operationally used snowpack model.

A model accounting for liquid water transport through multiple flow paths was developed by Marsh and Woo (1985), but was not able to explicitly account for structures like ice layers and capillary barriers. Recently, multi-dimensional water transport models have been developed, which allow for the explicit simulation of PFPs (Hirashima et al., 2014). These models are valuable for describing spatial heterogeneities and persistence of PFPs, but have not yet been shown to be suitable for hydrological or operational purposes. In general, multi-dimensional

models are limited by the fact that they are computationally intensive, thus not thoroughly validated for seasonal snowpacks, and still lack the description of crucial processes such as snow metamorphism and snow settling.

In snowpack models which are used operationally, PFPs are not yet considered. The recently introduced RE solver for SNOWPACK led to a significant improvement of modelled sub-daily snowpack runoff rates. For this paper, we further modified the transport scheme for liquid water by implementing a dual-domain approach to represent PFPs. This new approach is validated against snow lysimeter measurements which were recorded during both natural and artificial ROS events.

This study aims to better describe snowpack runoff processes during ROS events within snowpack models that can be used for operational purposes such as avalanche warning and hydrological forecasting. This requires that the model results remain reliable, i.e. that improvements are not realized at the expense of a decreased model performance during periods without ROS, and that the model must not be too computationally expensive. This is the first study to test a water transport scheme accounting for PF which has been implemented in a snowpack model that meets the above requirements.

Our analysis of simulations of over 100 ROS events targets the following research questions:

- Is snowpack runoff during ROS in a 1-D model better reproduced with a dual-domain approach to account for PF than with traditional methods considering matrix flow only?

- Are there certain snowpack or meteorological conditions, for which the performance specifically benefits if PF is represented in the model?

This paper is structured as follows: Sect. 2 describes the snowpack model setup, the water transport models, input data and the event definition. Results of the simulations are shown in Sect. 3. This includes data of sprinkling experiments of ROS (3.1), natural ROS events (3.2) and the validation of the model on a long-term dataset from two alpine snow measurement sites (3.3). The results are discussed in Sect. 4, followed by the general conclusions found in Sect. 5.

2 Methods

All results in this study are derived from simulations with the one-dimensional physics-based snowpack model SNOWPACK (Bartelt and Lehning, 2002; Lehning et al., 2002a, b; Wever et al., 2014a) using three different water transport schemes, described in Sect. 2.2. The model was applied to four experimental sites that were set up for this study in the vicinity of Davos (Sect. 2.3). These sites were maintained over two winter seasons between 2014 and 2016 where data were recorded for several natural ROS events. At the same sites, we conducted a set of six sprinkling experiments to simulate ROS events for given rain intensities (Sect. 2.4). Furthermore, we conducted simulations for two extensive datasets from the European Alps: Weissfluhjoch (Switzerland, 46.83° N, 9.81° E, 2536 m a.s.l., WSL Institute for Snow and Avalanche Research SLF (2015), abbreviated as WFJ in the following) and Col de Porte (France, 45.30° N, 5.77° E, 1325 m a.s.l., Morin et al. (2012), abbreviated as CDP in the following). These datasets provide meteorological input data for running SNOWPACK as well as validation data, including snowpack runoff. Both datasets have already been used for simulations with SNOWPACK (Wever et al., 2014a) and provide data over more than 10 years each.

Below, the SNOWPACK model and the different water transport models are described first, followed by the description of the field sites for ROS observation in the vicinity of Davos. Then, we detail the setup of the artificial sprinkling experiments. After summarizing the WFJ and CDP dataset, we finally present the definition of ROS events that is used in this study. Most analyses were performed in R 3.3.0 (R Development Core Team, 2016) and figures were created with base graphics or ggplot2 (Wickham, 2009).

2.1 Snowpack model setup

The setup of the SNOWPACK model is similar to the setup used for simulations in Würzer et al. (2016). For all simulations, snow depth was constrained to observed values, which means that the model interprets an increase in observed snow depth at the stations as snowfall (Lehning et al., 1999; Wever et al., 2015). Because the study focuses on the event-scale and snowpack runoff is essentially dependent on the properties of the available snow, this approach was chosen such that we have the most accurate initial snow depth at the onset of the events to achieve the best comparability between the three water transport models. The temperature used to determine whether precipitation should be considered rain (measurements from rain gauges) or snow (from the snow depth sensors) was set to achieve best results for reproducing measured snow height for precipitation driven simulations for the Davos field sites (between 0 and 1.0 °C). For WFJ and CDP, this threshold temperature was set to 1.2 °C, where mixed precipitation occurred proportionally between 0.7 and 1.7 °C. Turbulent surface heat fluxes are simulated using a Monin–Obukhov bulk formulation with stability correction functions of Stearns and Weidner (1993), as described in Michlmayr et al. (2008). At the Davos field sites (Sect. 2.3) incoming longwave radiative flux is simulated using the parameterization from Unsworth and Monteith (1975), coupled with a clear-sky emissivity following Dilley and O'Brien (1998), as described in Schmucki et al. (2014). For the roughness length z_0, a value of 0.002 m was used for all simulations at the Davos field sites and WFJ, whereas a value of 0.015 was used for CDP. The model was

initialized with a soil depth of 1.4, 2.2 and 2.14 m (for WFJ, CDP and Davos field sites, respectively) divided into layers of varying thickness. For soil, typical values for coarse material were chosen to avoid ponding inside the snowpack due to soil saturation. The soil heat flux at the lower boundary is set to a constant value of 0.06 W m^{-2}, which is an approximation of the geothermal heat flux.

2.2 Water transport models

The two previously existing methods for simulating vertical liquid water movement within SNOWPACK are either a simple so-called bucket approach (BA) (Bartelt and Lehning, 2002) or solving the RE, a recently introduced method for SNOWPACK (Wever et al., 2014a, b).

The BA represents liquid water dynamics by an empirically determined irreducible water content θ_r which defines whether water stays in the corresponding layer or will be transferred to the layer below. This irreducible water content varies for each layer according to Coléou and Lesaffre (1998). The RE represents the movement of water in unsaturated porous media. Its implementation in SNOWPACK and a detailed description can be found in Wever et al. (2014a).

The PF model presented in this study is based on the RE model, but follows a dual-domain approach, dividing the pore space of the snowpack into a part representing matrix flow and a part representing PF. For both domains the RE is solved subsequently. The PF model is described by (i) a function for determining the size of the matrix and preferential flow domain, (ii) the initiation of PF (i.e., water movement from matrix flow to PF) and (iii) a return flow condition from PF to matrix flow.

The area of the preferential domain (F) is as a function of grain size (Eq. 1), which has been determined by results of laboratory experiments presented by Katsushima et al. (2013):

$$F = 0.0584 r_{\mathrm{g}}^{-1.109}, \tag{1}$$

where r_{g} is grain radius (mm). F is limited between 1 and 90 % for reasons of numerical stability. The matrix domain is then accordingly defined as $(1 - F)$. Water is transferred from the matrix domain to the preferential domain if the water pressure head for a layer in the matrix domain is higher than the water entry pressure of the layer below, which can, according to Katsushima et al. (2013), also be expressed as a function of grain size. This condition is expected to be met if water is ponding on a microstructural transition (i.e. capillary barriers, ice lenses) inside the snowpack. Additionally, saturation was equalized between the matrix and the preferential domain, in case the saturation of the matrix domain exceeded the one in the preferential domain. To move water back into the matrix part, we apply a threshold in saturation of the PF domain and water will flow back to the matrix domain once this threshold is exceeded. This threshold is used as a tuning parameter in the model.

Refreezing of liquid water in the snowpack is crucial for modelling water transport in subfreezing snow and may also be important for modelling PF. The presented PF model has also been used to simulate ice layer formation under the presence of PF by Wever et al. (2016b). Thereby, a sensitivity study on the role of refreeze in the PF domain and the return flow condition from PF to matrix flow was conducted. It was found that neglecting refreeze led to the best results for reproducing ice layer formation, but did not significantly affect the performance in reproducing measured hourly snowpack runoff. Therefore, refreeze in the preferential domain is neglected in the presented study. The threshold in saturation for PF (return flow condition) was also determined by the sensitivity study described in Wever et al. (2016b). While they determined a threshold in saturation of 0.1 to reproduce ice-layer observations at WFJ best, a value of 0.06 was determined to reproduce observed seasonal runoff best. We therefore used the value of 0.06. In contrast to Wever et al. (2016b), we did not set the hydraulic conductivity in soil to 0, because this can lead to an inaccurate representation of observed lysimeter runoff due to modelled ponding on soil, which is not expected to happen on a snow lysimeter. Further details on the implementation of the PF model and its performance can be found in Wever et al. (2016b).

In summary, the PF model accelerates liquid water transport in the preferential domain by concentrating water mass in a smaller area, representing the area fraction of flow fingers in the snowpack. The saturation in the preferential domain is hence higher and unsaturated conductivity is larger. Further acceleration is achieved by disabling refreeze in the preferential domain.

2.3   Davos field sites

Four field sites have been installed within an elevational range of 950 to 1850 m a.s.l. in the vicinity of Davos, Switzerland, with one meteorological station and 3–4 snow lysimeters each (15 in total, 0.45 m diameter). The meteorological stations provided most data necessary for running the SNOWPACK model and missing parameters were estimated as described in Sect. 2.1. Lysimeters were installed at ground level with an approximate spacing of 10 m horizontal distance. The lysimeters consisted of a funnel attached to a precipitation gauge buried in the ground, which monitored snowpack runoff with a tipping bucket. To block lateral inflow at the snow-soil interface, each lysimeter was equipped with a rim of 5 cm height around the inlet. The multiple snow lysimeter setups allowed analysing the spatial heterogeneity of snowpack runoff. Snowpack properties (SWE, LWC, HS, TS) were manually measured directly before each natural ROS event so that the initial conditions of the snowpack are known in detail. LWC was measured with the "Denoth meter", a device introduced by Denoth (1994). The onset of

runoff was defined as the time when cumulative snowpack runoff (measured and simulated, respectively) has reached 1 mm.

2.4   Sprinkling experiment description

During winter 2014/15, a total of six artificial sprinkling experiments were performed on all four Davos field sites described above to be able to investigate snowpack runoff generation for different snowpack properties (Table 1). For each experiment, a sprinkling device was placed above a snow lysimeter, covered by an undisturbed natural snowpack, i.e. each lysimeter was only used for one experiment. The device used for sprinkling was a refined version of the portable sprinkling device described in Juras et al. (2013, 2016a). The water used for sprinkling was mixed with the dye tracer Brilliant Blue FCF (concentration $0.4\,g\,L^{-1}$) to be able to observe PFPs within the snowpack. Sprinkling was performed in four bursts of 30 min each, interrupted by 30 min breaks. Sprinkling was conducted over a 2×2 m plot centred above the lysimeters, and with an intensity of $24.7\,mm\,h^{-1}$, leading to a total of 49.4 mm artificial rain in each of the experiments. The intensities were determined by calibration experiments on lysimeters not covered by snow and are valid for a certain distance between the nozzle and the sprinkled surface and water pressure at the nozzle. Despite the fact that this value still represents a very intense ROS event, it is within range of natural ROS events and similar or much lower compared to previous studies ($19\,mm\,h^{-1}$; Eiriksson et al., 2013; 48–$100\,mm\,h^{-1}$; Singh et al., 1997). For the sprinkling experiments, the exact timing of rain and intensities are known and the snowpack runoff measured at 1 min intervals allowed precise analysis of the performance of model simulations. Figure 1 shows a vertical cut of a snowpack after the sprinkling experiment and a top view of the lysimeter after the snowpack was removed for cold and wet conditions, respectively. The blue colour indicates where water transport took place and where sprinkled water was held by capillary forces or refrozen.

2.5   Extensive dataset for in situ validation

Two long-term datasets from two study sites in the European Alps providing snow lysimeter data and high-quality meteorological forcing data for running the energy balance model SNOWPACK were chosen to validate the different water transport models systematically. Datasets of both study sites used for the extensive in situ validation are publicly available. The CDP site, located in the Chartreuse range in southeastern France, has been described in Morin et al. (2012) and the Weissfluhjoch site (WFJ) in the Swiss Alps has been described in Wever et al. (2015). WFJ (46.83° N, 9.81° E) is located at an elevation of 2536 m a.s.l. and CDP (45.30° N, 5.77° E) is located at 1325 m a.s.l. CDP experiences a warmer climate than WFJ and as a consequence the

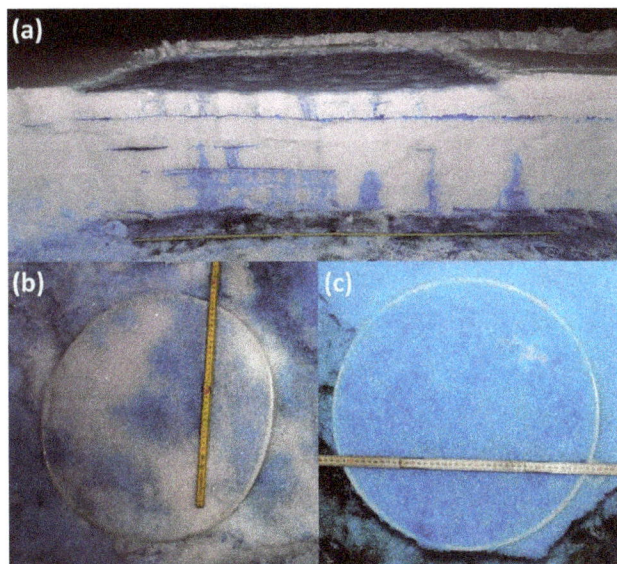

Figure 1. (a) Vertical cut of a snowpack after the sprinkling experiment Sertig Ex3 (28 February 2015). Lateral flow and the presence of PFP were observed. PFP were generated at regions with rain water ponding at ice layers and layer boundaries with a change in grain size (creating capillary barriers). **(b)** Lysimeter area after sprinkling during winter conditions (Serneus Ex1, 26 February 2015): coloured areas indicate the area where water percolated due to PF. **(c)** Lysimeter area after sprinkling during spring conditions (Klosters Ex4, 26 March 2015): coloured area shows that water percolated uniformly, indicating dominating matrix flow.

snowpack produces snowpack runoff more often throughout the entire snow season and ROS events are more frequent than at WFJ. A multi-week snowpack builds up every winter season at CDP, but is, in contrast to WFJ, interrupted by complete melt in some years. The WFJ site is equipped with a $5\,m^2$ snow lysimeter, which measures the liquid water runoff from the snowpack. It has a 60 cm high rim to reduce lateral flow effects near the soil–snow interface (Wever et al., 2014a). CDP is equipped with both a 5 and a $1\,m^2$ lysimeter. Here we use data from the $5\,m^2$ lysimeter, but include data from the $1\,m^2$ lysimeter to discuss the uncertainty associated with measurements of the snowpack runoff. The studied period for WFJ is from 1 October 1999 to 30 September 2013 (14 hydrological years). Because of possible errors in the lysimeter data in the winter seasons of 1999/00 and 2004/05 as described in Wever et al. (2014a), these data were excluded from the study. For CDP the studied period is from 1 October 1994 to 31 July 2011 (17 winter seasons) according to the data availability from the $5\,m^2$ lysimeter. The temporal resolution of lysimeter data is 1 h for CDP and 10 min for WFJ. Simulation results for CDP and WFJ as well as lysimeter data for WFJ were aggregated to an hourly timescale.

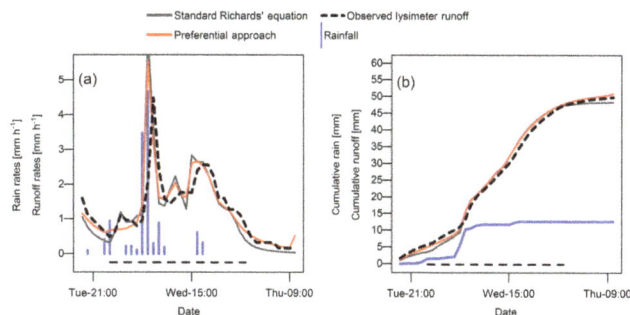

Figure 2. (a) Example of a ROS event occurring at WFJ. The entire extent of the x axis refers to the evaluation period; the bar above the x axis refers to the event length. **(b)** Cumulative version of the plot.

2.6 CDP+WFJ event definition

As the number and characteristics of ROS events are strongly dependent on the event definition, special care needs to be taken to determine beginning and end of a ROS event. Being interested in the temporal characteristics of snowpack runoff during ROS, we need to include the entire period from the onset of rain to the end of ROS-induced snowpack runoff. Here we use an event definition according to Würzer et al. (2016) with slightly decreased thresholds to identify ROS events. According to this definition, a ROS event requires a minimum amount of 10 mm rainfall to fall within 24 h on a snowpack with a height of at least 25 cm at the onset of rainfall. While the event is defined to begin once the first 1 mm of rain has fallen, the event ends once there is less than 3 mm of cumulative snowpack runoff recorded within the following 5 h. This definition resulted in a selection of 61 events at CDP and 40 events at WFJ. The model simulations were subsequently evaluated over a time window that extends the event length by 5 and 10 h at the beginning and end, respectively (Fig. 2). These extended evaluation periods allowed us to also investigate a possible temporal mismatch between modelled and observed snowpack runoff.

3 Results

3.1 Experimental sprinkling experiments

During the winter period 2014/15, six sprinkling experiments (Ex1–Ex6) were conducted on four different sites to be able to investigate snowpack runoff generation for different snowpack properties. With distinct differences in snowpack properties but controlled rain intensities, these experiments were expected to reveal the influence of snow cover properties and differences between the water transport models best. For all experiments, initial snow height (HS), snowpack temperature (TS) and LWC profiles were measured (Table 1 and Fig. 3). According to these measurements, the snowpack conditions on which the sprinkling experiments were conducted can be

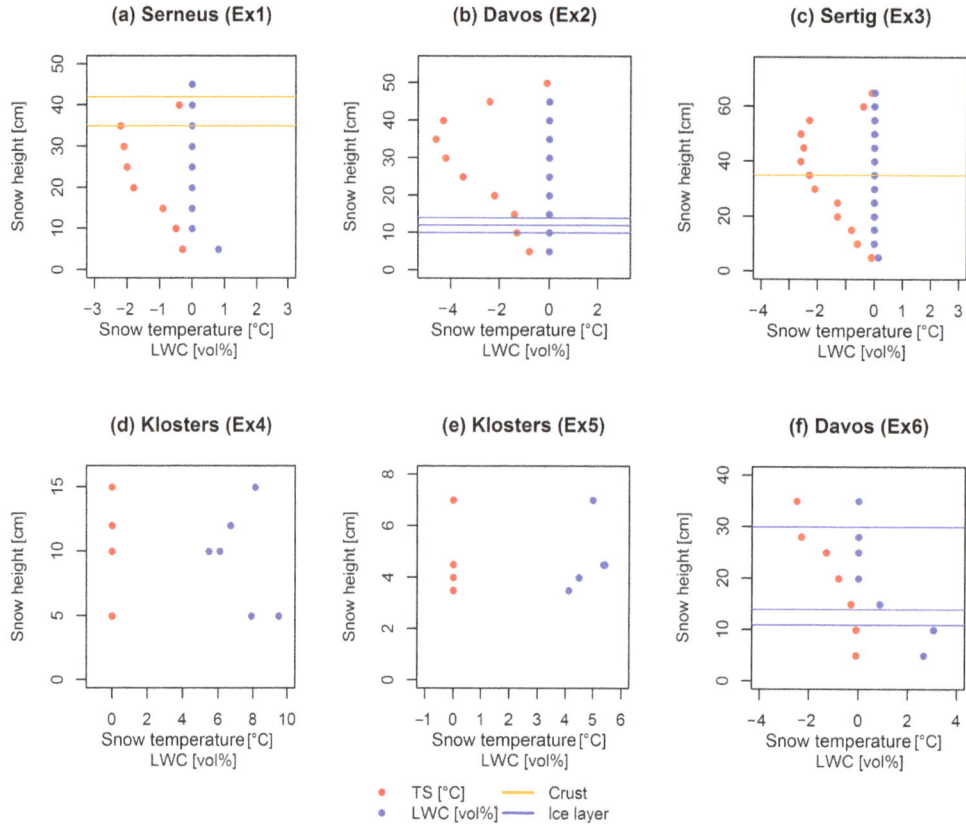

Figure 3. Snow temperature and LWC profiles measured directly before the sprinkling experiment started. The lines represent observed ice layers (blue) and crusts (orange).

Table 1. Snowpack pre-conditions and execution dates for the sprinkling experiments as well as R^2 values for the different model simulations. Measured values are snow height (HS), bulk liquid water content (LWC), bulk snow temperature (TS). No snowpack runoff measurements were available for Sertig (Ex3).

Experiment	Initial snowpack conditions				R^2 of hourly runoff of the simulations		
	HS [cm]	LWC [% vol]	TS [°C]	Date	RE	PF	BA
Serneus (Ex1)	48.5	0.1	−1.3	26-Feb-15	0.14	0.59	0.09
Davos (Ex2)	54.5	0.4	−2.5	27-Feb-15	0.24	0.62	0.08
Sertig (Ex3)	71.5	0	−1.6	28-Feb-15	−	−	−
Klosters (Ex4)	15.7	6.9	0	26-Mar-15	0.75	0.96	0.86
Klosters (Ex5)	7	4.9	0	8-Apr-15	0.70	0.84	0.88
Davos (Ex6)	39.3	0.9	−0.6	10-Apr-15	0.58	0.83	0.36

separated into two cases: the first three experiments were conducted on dry and cold (i.e. below the freezing point) snow and will be called winter experiments. The snowpack of Ex4 and Ex5 was isothermal and in a wet state. At the onset of Ex6 however, part of the snowpack was below freezing and had just little LWC. Nevertheless, the snowpack already passed peak SWE and was in its ablation phase. Therefore

the later three experiments (Ex4–Ex6) will be referred to as spring experiments in the following.

For all winter experiments (Figs. 4 and 5a, b, c), both modelled and observed total event runoff remained below the amount of sprinkling water. Energy input estimated by the SNOWPACK simulations suggests that snowmelt was insignificant for the winter experiments, but refreeze led to significant retention of liquid water. Additionally some sprinkled rain was retained as LWC at the end of the experiments. During Ex3 no snowpack runoff was observed, visual inspection afterwards revealed an impermeable ice layer covering both the lysimeter and the adjacent ground. During spring conditions, on the other hand, snowmelt (5.1, 8.4 and 27.4 mm respectively) led to snowpack runoff exceeding total sprinkling input, except for measured snowpack runoff in Ex6 (Figs. 4 and 5d, e, f).

Additionally, Fig. 5 shows that only the PF model was able to reproduce all four peaks of observed snowpack runoff for winter conditions (Ex1 + 2), and even the magnitude of the first peak of Ex1 was captured well. For spring conditions however, all three models managed to represent four peaks corresponding to the four sprinkling bursts, but the PF model showed best correspondence with observed snowpack runoff (Figs. 4 and 5d, e, f; Table 1). Regarding the onset of snowpack runoff, the PF model especially led to faster snowpack runoff for the first two winter experiments, where the RE and BA models showed delayed snowpack runoff onset. For spring conditions the faster snowpack runoff response of the PF model led to a slightly early snowpack runoff. Maximal snowpack runoff rates for dry and cold conditions were generally overestimated by all models, whereas wetter conditions led to a minor underestimation (except for Ex3, where no snowpack was measured).

Regarding the overall correlation between measured and simulated snowpack runoff, PF outperformed the other models (Table 1), in particular during winter conditions. Summarizing, this initial assessment suggests that the PF approach has potential advantages in particular (a) as to the timing of snowpack runoff and (b) for cold snowpacks which are not yet entirely ripened.

3.2 Natural occurring ROS events

In January 2015, two ROS events occurred in the vicinity of Davos. They were observed over an elevational range of 950 to 1560 m a.s.l. on the same sites on which also the sprinkling experiments were conducted. Figure 6 shows the course of cumulative rainfall and snowpack runoff for both dates and all sites. Pre-event conditions (HS, LWC, TS) were measured shortly before the onset of rain for both events and are shown together with coefficients of determination (R^2) for hourly snowpack runoff of the different models Table 2.

For the event of 3 January 2015 (Fig. 6, upper row) the lower sites Serneus and Klosters (950 and 1200 m a.s.l.) showed a similar snowpack runoff dynamics regarding the delayed onset and the total amount (cumulative sum averaged over the three corresponding lysimeters: 20.3 and 21.1 mm, respectively). Also, the heterogeneity between data from the individual lysimeters was relatively low (Range of 3.1 and 3.9 mm, respectively). For the highest located site (Davos), however, the snowpack runoff measured by all four lysimeters showed a greater variability (Fig. 6c) in the delayed onset of snowpack runoff (0 to 7h) and the total amount of

snowpack runoff (mean 24.7 mm; range of 57.9 mm). The snow cover mostly built up within 1 week before the event. Cold temperatures led to a light melt refreeze crust at the top, but no distinct ice layers were observed. For the lower sites (Serneus and Klosters), the PF and RE models generated snowpack runoff too early (PF: approx. 3 h; RE: 0.2 to 1.4 h). The BA model generated snowpack runoff rather too late (1.3 to 2 h), but still within range of the variability of observed snowpack runoff for Serneus. However, the cumulative lysimeter snowpack runoff showed good accordance with modelled PF and RE snowpack runoff at Serneus, whereas PF led to an overestimation at Klosters and BA to an underestimation of cumulative snowpack runoff at all sites. At the higher-elevation site Davos, the RE model led to a better representation of mean observed snowpack runoff amount, when compared with BA and PF. The mean observed snowpack runoff onset however was represented best by the PF model (0.3 h early) when compared to the BA (3.4 h delay) and RE (1.1 h delay).

For the event of 9 January 2015 (Fig. 6, bottom row) the lower sites showed again little temporal and spatial heterogeneity in lysimeter runoff (range of 1 and 2.2 mm, respectively), whereas this was more the case for Davos again (range of 13.3 mm) probably owing to ice layers that were formed after the event on 3 January. Observed mean event snowpack runoff was more diverse for all elevations, where Klosters had the highest cumulative snowpack runoff (Serneus 13.3 mm; Klosters 17.7 mm; Davos 7.8 mm). If compared to observed total snowpack runoff, the PF model overestimated snowpack runoff for Serneus and Klosters, whereas the RE and especially the BA model underestimate event snowpack runoff for both sites. For Davos, all models were overestimating event snowpack runoff and led to early snowpack runoff. Apart from the RE model, which represented onset of snowpack runoff correctly for Serneus, none of the models were able to model snowpack runoff onset correctly for any of the sites.

3.3 Validation on an long-term dataset

3.3.1 Modelled and observed snowpack runoff for the whole dataset

Given the partly contradictory findings on the performance of the three model variants based on the above assessment for artificial ROS simulations under controlled conditions (Sect. 3.1), as well as natural ROS events (Sect. 3.2), further more systematic model tests were needed. Therefore we validate the different models based on extensive datasets from the two sites WFJ and CDP, as described in Sect. 2.4.

Before we focus on the specific performance of the PF model for a large number of individual ROS events, we first analysed the overall model performance throughout the whole study period, i.e. over entire winter seasons. For this, we analysed observed and modelled hourly snowpack runoff

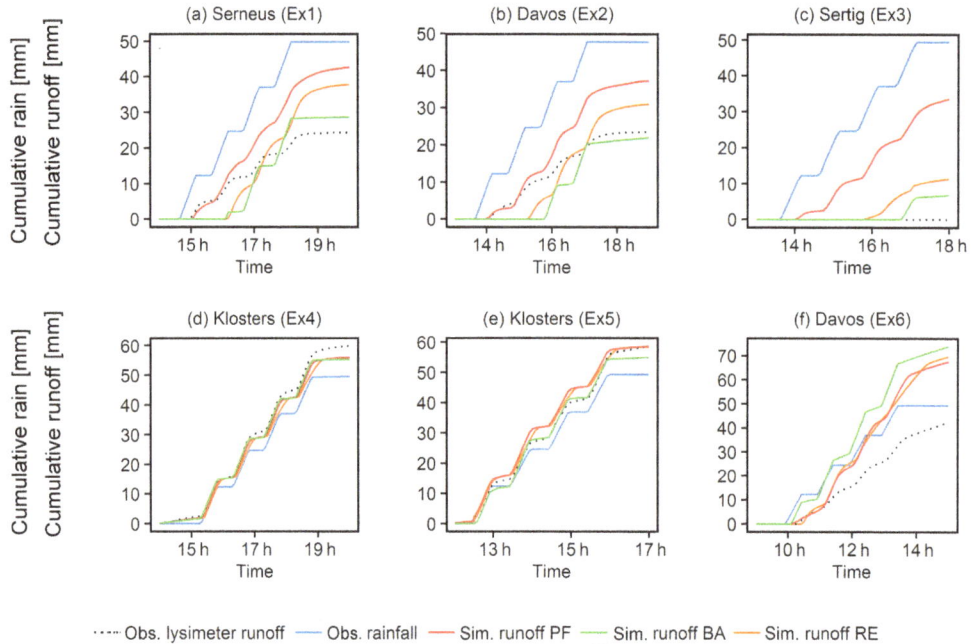

Figure 4. Cumulative rain and snowpack runoff displayed for the six sprinkling events. Ex1 **(a)**–Ex3 **(c)** were conducted during winter conditions, Ex4 **(d)**–Ex6 **(f)** were conducted during spring conditions.

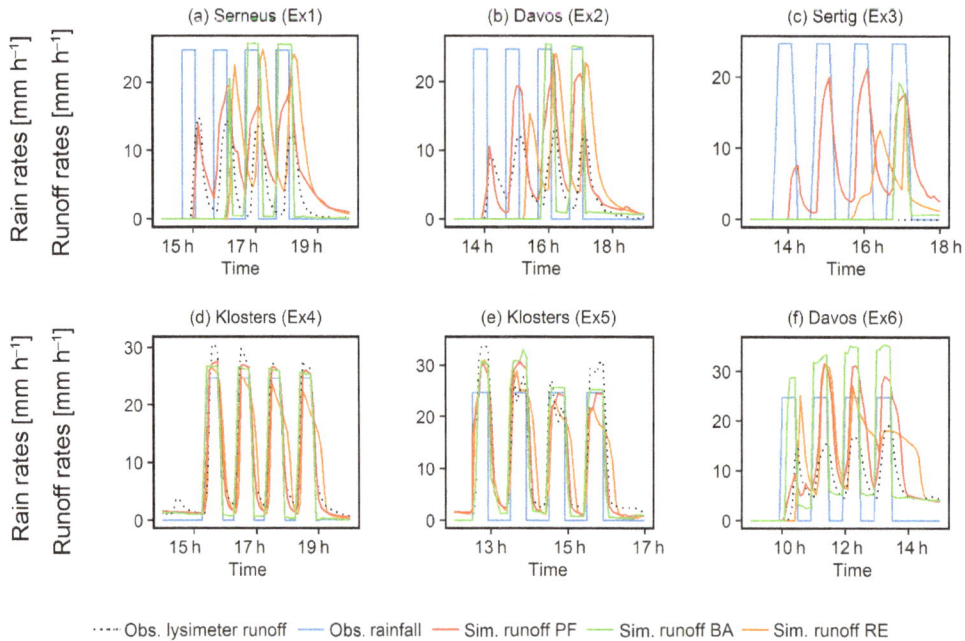

Figure 5. Rain and snowpack runoff displayed as hydrographs for the six sprinkling events. Ex1 **(a)**–Ex3 **(c)** were conducted during winter conditions, Ex4 **(d)**–Ex6 **(f)** were conducted during spring conditions.

provided snow heights exceeded 10 cm to ensure that lysimeter runoff was caused by snowpack runoff and not rainfall. For both sites, R^2 values for PF were slightly higher than for RE (Table 3), which both clearly outperformed the BA.

The root mean squared errors (RMSEs) of the PF model were also lower compared to RE and BA. We can therefore conclude that the implementation of the PF approach slightly improves water transport over entire winter seasons.

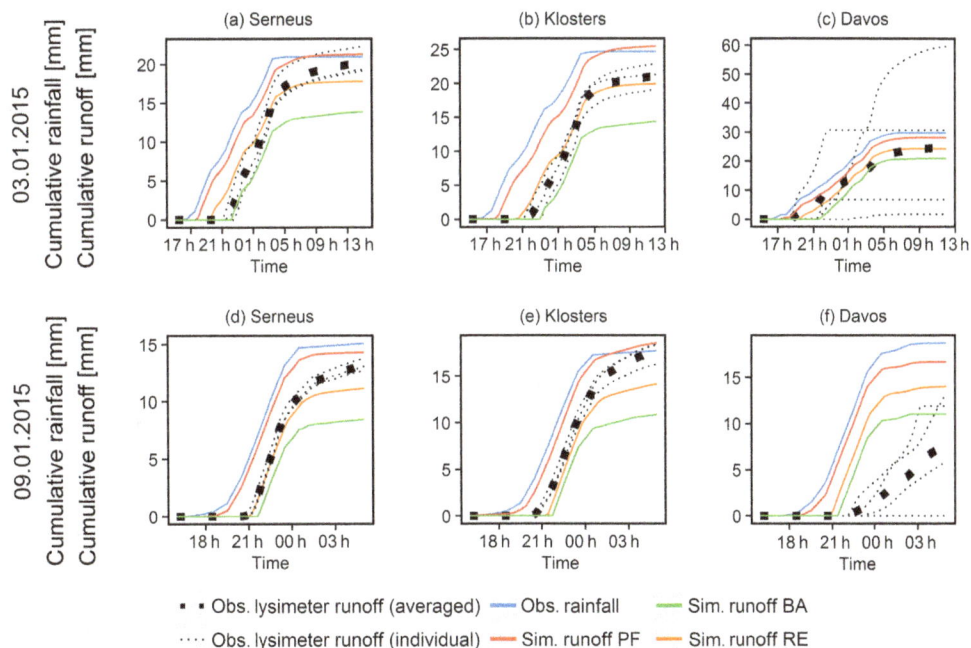

Figure 6. Natural ROS events on 3 and 9 January 2015 in **(a, d)** Serneus, **(b, e)** Klosters and **(c, f)** Davos.

Table 2. Snowpack pre-conditions and R^2 for hourly snowpack runoff for natural events on 3 and 9 January.

	Site	HS (cm)	LWC (% vol)	TS (°C)	RE	PF	BA
03-Jan-2015	Serneus	19	0	0	0.63	0.35	0.83
	Klosters	24	0	−0.1	0.72	0.39	0.78
	Davos	20	0	−0.4	0.27	0.33	0.17
09-Jan-2015	Serneus	14.5	0.1	−0.2	0.94	0.57	0.79
	Klosters	18	0.1	−0.2	0.84	0.73	0.73
	Davos	19.5	0.1	−0.6	0.00	0.04	0.00

The header spans: Pre-event snowpack conditions (HS, LWC, TS); R^2 for hourly snowpack runoff (RE, PF, BA).

3.3.2 ROS event characteristics of the extensive dataset

Median characteristics of the individual ROS events at CDP and WFJ are summarized in Fig. 7. The temporal course of median rain and snowpack runoff rates of all events at WFJ (40 individual events) and CDP (61 individual events) are shown in Fig. 7a, b. ROS events at WFJ showed generally higher maximum rain intensities than at CDP, leading to higher median snowpack runoff intensities at the beginning of the events. Whereas at WFJ, ROS events tended to be short and intense, at CDP the event rainfall extended over a longer period of time. Interestingly, we observed relatively high initial snowpack runoff rates before the actual beginning of the ROS event, especially for WFJ, which suggests that

many ROS events at this site occurred during the snowmelt period. Median snowpack runoff reached a peak after 1 and 3 h after the onset of rain for WFJ and CDP, respectively. At WFJ snowpack runoff and rain rates at the beginning of the events were generally higher than at CDP. The course of the median air temperature during ROS events at both sites is shown in Fig. 7c. Especially for WFJ, median air temperature (TA) dropped with the onset of rain and median TA was higher than at CDP. The mean initial ROS event snow height (HS) for WFJ was 95 cm, which is approximately the average snow height during mid-June (for 70 years of measurements). The mean initial HS for CDP is 67 cm. With a SD of 42 cm, the variability of initial HS for WFJ was higher than for CDP (29 cm).

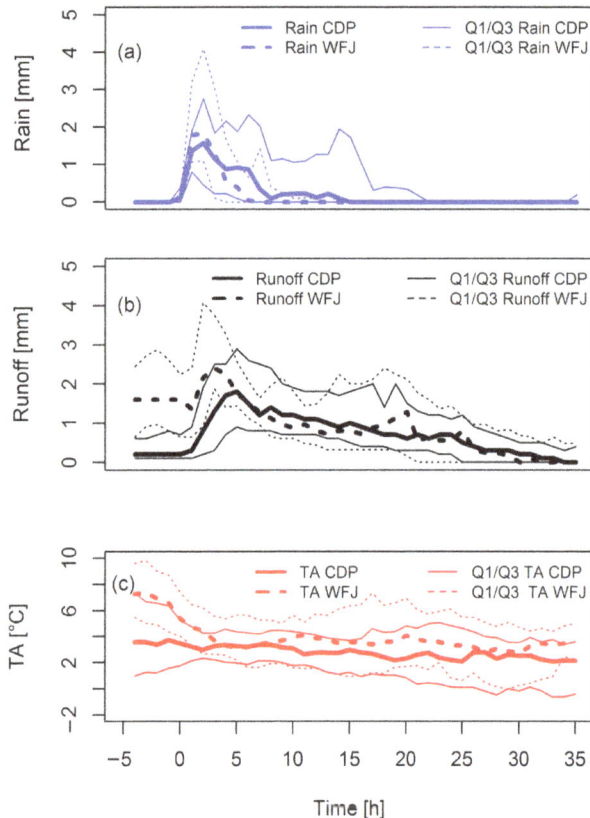

Figure 7. Temporal course of median rain (**a**), measured snowpack runoff (**b**) and air temperature (**c**) for WFJ (dotted) and CDP (solid) aggregated over all 40 and 61 events respectively. The thinner lines represent the lower and upper quartiles, respectively. The displayed period is extended by 5 h prior to event commencement according to the event definition (0 h).

3.3.3 Modelled and observed snowpack runoff at the event scale

Below we investigate the performance of the three water transport schemes at the event scale. Modelled snowpack runoff was assessed against observations by the coefficient of determination (R^2) and the RMSEs. To further analyse the representation of snowpack runoff timing, we defined an absolute time lag error (TLE) as the difference between the onsets of modelled and observed snowpack runoff in hours. The onset of snowpack runoff is defined as the time when cumulative snowpack runoff has reached 10 % of total event-snowpack runoff.

Figure 8 shows box plots of R^2 (a, d), RMSE (b, e) and absolute TLE (c, f) for all 40 ROS events at WFJ (a, b, c) and 61 events at CDP (d, e, f), respectively. For both sites, R^2 values show that the BA model performance was inferior to the RE model which was in turn slightly outperformed by the PF model. The interquartile range of R^2 values for CDP was generally higher than for WFJ and increased from BA to RE, whereas it was decreasing for PF. The PF also led to a re-

duction in RMSE by approximately 50 % if compared to the BA, but less (9 % for WFJ and 25 % for CDP) if compared to the RE model. Whereas the median of TLEs for all models at WFJ was 0 and therefore all models reproduced the onset of snowpack runoff very well, the interquartile range decreased from BA to the RE and PF models. The same behaviour in interquartile range decrease could be observed for CDP, where the magnitude of TLE was higher than for WFJ and mostly negative. The median TLE was again 0 for the PF and -1 h in the case of BA and RE, indicating that for these models, snowpack runoff was on average a bit delayed compared to the observations. For WFJ, TLE for BA was more often positive (early modelled snowpack runoff), which led to a very good median for BA, but also a larger interquartile range. Hence, the PF model showed the most consistent results, especially if regarding the interquartile range. For CDP we added the comparison between the 1 and 5 m^2 lysimeters installed at CDP (Sect. 2.5) as a reference to Fig. 8, referred to as RL. This comparison can be seen as a benchmark performance, as it represents the measurement uncertainty of the validation dataset. As expected, RL shows the highest overall performance measures, but while the results for both PF and RE were reasonably close to those of RL, the BA model performed considerably worse.

The results shown in Fig. 8 may be influenced by both a time lag as well as the degree of reproduction of temporal dynamics. To separate both effects, we conducted a cross-correlation analysis, allowing a shift of up to 3 h to find the best R^2 value. Figure 9 shows both the time lag, as well as the best R^2 value achieved. Interestingly, the BA model showed best correlations if the modelled snowpack runoff was shifted by 1 or 2 h (consistently too early compared to observations). The RE model, on the other hand, showed best correlations for a shift in the other direction (consistently too late compared to observations). Neither was the case for PF with lags centred around 0.

The R^2 of the cross-correlation analysis gives some indication of how well the temporal dynamics of the observed snowpack runoff can be reproduced, neglecting a possible time lag. The results in Fig. 9 show an improvement in R^2 values for both sites and all models if a time lag is applied. Greatest improvements were observed for the BA model for both sites. The good timing with the PF model is confirmed by almost no lag for WFJ and only a small lag for CDP needed to maximize R^2. For CDP, both RE and PF had maximized R^2 values in range of the lysimeter comparison (RL).

4 Discussion

Even though PF of liquid water through snow is a phenomenon that has been known and investigated for a long time, it has not yet been accounted for in 1-D snow models that are in use for operational applications. The results of this study show that including this process into the water trans-

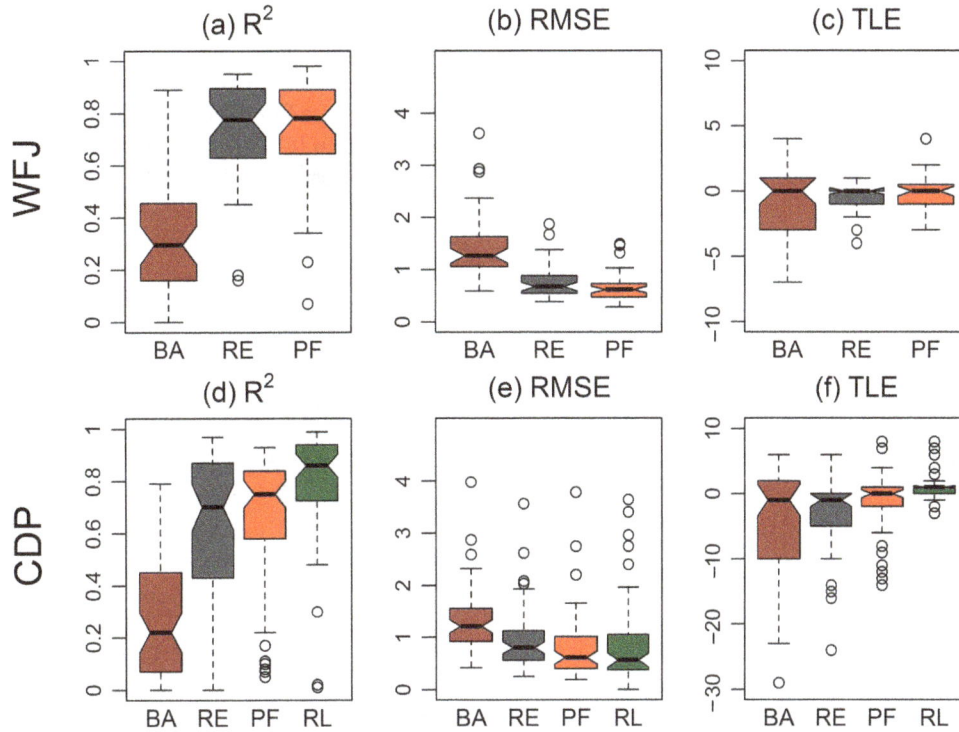

Figure 8. RMSE, R^2 and TLE for simulations of 61 ROS events at the CDP site and of 40 ROS events at the WFJ site for all models (BA, RE, PF) and the reference lysimeter (RL) available only for CDP.

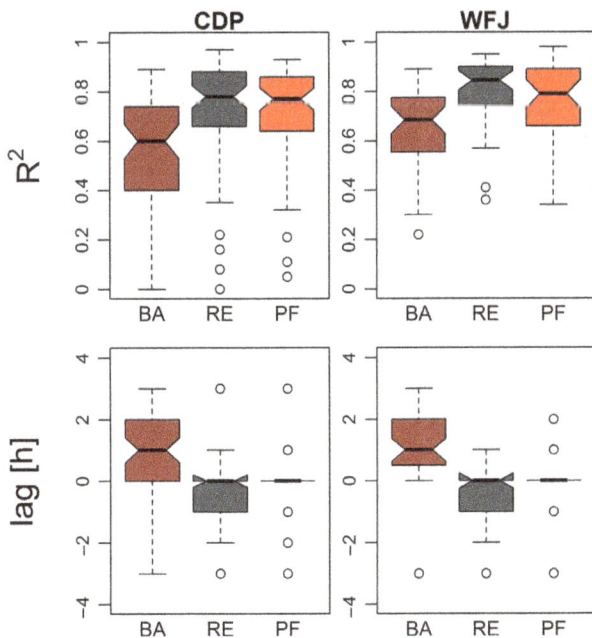

Figure 9. Best R^2 values and corresponding lags using a cross-correlation function allowing a time shift (lag) of max ± 3 h.

port scheme can improve the prediction of snowpack runoff dynamics for individual ROS events as well as for the snowpack runoff of entire snow seasons. Moreover, the representation of the onset of snowpack runoff is improved. This is particularly important at the catchment scale, where a delay of snowpack runoff relative to the start of rain may affect the catchment runoff generation, especially if the time lag varies across a given catchment.

During the sprinkling experiments, sprinkling intensities were higher than average rain intensities during ROS but still within range of peak rain intensities during naturally occurring ROS events in the Swiss Alps (Rössler et al., 2014; Würzer et al., 2016) and the Sierra Nevada, California (Osterhuber, 1999). The use of the PF model clearly led to a better representation of the runoff dynamics for all experiments, including shallow and ripe snowpacks during spring conditions as well as cold and dry snowpacks representing winter conditions. The improvements were strongest for winter conditions, suggesting that under these conditions accounting for PF is most relevant. This is supported by observations of PFPs during winter conditions (Fig. 1a), which were not visible after the spring experiments. During winter conditions just a fraction of the lysimeter area was coloured with tracer, indicating PF of the sprinkled water (Fig. 1b), whereas spring conditions left the whole cross-section of the lysimeter coloured (Fig. 1c). While a fast runoff response can be expected for wet and shallow snowpack and may be easier to handle for all models tested, it is the cold snowpacks that both RE and BA models did not manage to represent well: runoff from these models was more than 1 h delayed (Ex1

Table 3. R^2 and mean absolute errors for hourly snowpack runoff for 17 and 14 years, for CDP and WFJ, respectively.

	R^2 hourly snowpack runoff			RMSE of snowpack runoff ($mm\,h^{-1}$)		
	BA	RE	PF	BA	RE	PF
CDP	0.33	0.50	0.52	0.56	0.44	0.40
WFJ	0.48	0.77	0.78	0.51	0.30	0.28

and Ex2), and missed approx. 10 mm of snowpack runoff within the first hour of observed runoff. This can partly be explained by the fact that BA and RE need to heat up the subfreezing snowpack before they can generate snowpack runoff, whereas refreezing is neglected in the preferential domain of the PF model and runoff can occur even in a not yet isothermal snowpack. Adjusting parameters like the irreducible water content θ_r for the BA model could probably lead to earlier runoff under these conditions, but thereby lead to earlier runoff, for example for WFJ events, where TLE already is positive for several events.

Despite the improved representation of the temporal runoff dynamics of the PF model (Table 1), the total event runoff of both RE and PF models is very similar for most conditions. Notably, the total event runoff for dry snowpacks is mostly overestimated by all models, suggesting an underestimation of water held in the capillarities. In cold snowpacks, dendricity of snow grains may still be high, such that water retention curves developed for rounded grains underestimate the suction. Additionally, high lateral flow was observed during the experiment for those conditions (Fig. 1a). This leads to an effective loss of sprinkling water per surface area of the lysimeter, which of course cannot be reproduced by the models. Therefore, observed snowpack runoff likely underestimates the snowpack runoff that would have resulted from an equivalent natural ROS event and we assume that the performance of the PF and RE models to capture the event runoff is probably better than reported in Table 1. Note that neglecting refreeze in the PF model should not be accountable for differences in the total event runoff between the RE and PF model, if we assume that the cold content is depleted by the end of the event.

Interestingly, despite having the coldest snowpack, time lag for the first natural ROS event at Davos was shorter than for the other two sites. This relationship where a cold and non-ripe snowpack with low bulk density led to smaller lag times was also found during sprinkling experiments conducted by Juras et al. (2016b). We assume that this is an indication for the presence of pronounced PFPs under those conditions, which is also supported by the high spatial variability of snowpack runoff. Glass et al. (1989) state that the fraction of PF per area is decreasing with increasing permeability, which itself was found to be increasing with porosity

(Calonne et al., 2012). Therefore, with a decreasing PF area due to lower densities, the cold content of a snowpack loses importance, but saturated hydraulic conductivity is reached faster within the PFPs. The combination of those effects then is suspected to lead to earlier runoff. This behaviour should be ideally reproduced by the PF model and indeed the onset of runoff is caught well for this event. Here, our multi-lysimeter setup raises the awareness that the observed processes can show considerably spatial heterogeneity as documented, for example, in Fig. 6. The formation of ice layers also underlies spatial heterogeneity. Moreover, the creation of PFPs is strongly dependent on structural features like grain size transitions leading to capillary barriers. Unfortunately, no detailed information about grain size is available in the observations to verify this.

The PF model led to improvements in reproducing hourly runoff rates at CDP and WFJ for a dataset comprising several years of runoff measurements. This is an important finding, demonstrating that the new water transport scheme aimed at a better representation of PF during ROS events, did not negatively impact on the overall robustness of the model. To the contrary, the overall performance over entire seasons could even be improved. All three models represent the overall seasonal runoff better for WFJ than for CDP (Table 3), which was also found on the event scale (Fig. 8). Moreover, the CDP simulations exhibit a larger interquartile range in R^2 values and are therefore generally less reliable. The observed differences in model performance between both sites may either be caused by differences in snowpack or meteorological conditions or by issues with the observational data. Moreover, SNOWPACK developments have in the past often been tested with WFJ data, which could lead to an unintended calibration favouring model applications at this site. Despite an obvious contrast in the elevation of both sites, the average conditions during ROS events seem to vary. Figure 7 suggests that at WFJ short and rather intense rain events dominate. The higher maximum rain intensities at WFJ, compared to CDP, are probably due to the later occurrence of ROS at this site (May–June), where air temperatures and therefore rain intensities are usually higher than earlier in the season (Molnar et al., 2015). Regarding mean intensities over the event scale, data shown in Fig. 7 further imply that short and intense ROS events typically attenuate the rain input (ratio runoff to rain < 1), whereas long ROS events rather lead to additional runoff from snowmelt, which is in line with results presented in Würzer et al. (2016).

Snow height is generally higher at WFJ where the average initial snow height for the ROS events analysed was approximately 30 cm higher than at CDP. Ideally, the performance of the water transport scheme in the snowpack should not be affected by the snow depth. At both sites, the snowpack undergoing a ROS event is mostly isothermal with a mean initial LWC of 1.8 % vol (CDP) and 3.0 % vol (WFJ). The initial snowpack densities at both sites were quite different. At WFJ, densities for all ROS events are around

Table 4. Pearson correlation coefficients between event-R^2 and stratigraphic features at WFJ and CDP. Stratigraphic features are marked grain size changes (bigger than 0.5 mm) and density changes (bigger than 100 kg m^{-3}) in two adjacent simulated layers as well as the wet layer ratio (percentage of layers exceeding 1 % vol over layers below 1 % vol) and the percentage of melt forms.

| | | Pearson correlation coefficient between event R^2 and the following: | | | |
		no. of grain size changes	no. of density changes	ratio of melt forms	wetting ratio
WFJ	PF	−0.44	−0.45	−0.16	−0.20
	RE	−0.54	−0.47	0.17	0.13
	BA	−0.56	0.16	−0.11	−0.09
CDP	PF	−0.14	0.07	0.37	0.39
	RE	−0.19	0.12	0.57	0.66
	BA	−0.11	−0.26	0.15	0.14

450–500 kg m^{-3}, whereas for CDP densities are spread from around 200 kg m^{-3} up to 500 kg m^{-3}. This suggests that the variable performance of all models at CDP (Fig. 8d) may be associated with early season ROS events. At CDP, a linear regression fit suggests a positive, albeit weak correlation between snowpack bulk densities and event-R^2 for the RE (R^2 of 0.2), but no correlation for both the PF and the BA model. It seems that the RE model had some difficulties with low-density snow, which was not the case for the PF model (Fig. 10). This may explain why PF outperformed RE at CDP, but not for WFJ.

Remaining inaccuracies in the representation of runoff for low densities for both models applying the RE may be explained by the fact that the water retention curve have been derived by laboratory measurements with high-density snow samples (Yamaguchi et al., 2012). The parameters defining the PF area (F) have also been developed from snow samples with a density mostly above 380 kg m^{-3} (Katsushima et al., 2013).

We further analysed snowpack stratigraphy derived from the SNOWPACK simulations, such as marked grain size changes (bigger than 0.5 mm) and density changes (bigger than 100 kg m^{-3}) in two adjacent simulated layers as well as the wet layer ratio (percentage of layers exceeding 1 % vol over layers below 1 % vol) and the percentage of melt forms (Table 4). These stratigraphy measures represent possible capillary barriers having implications on the single event-R^2 and might help understanding the advantages and disadvantages of the different models. Any considerable correlation between the abundance of stratigraphy features and event-R^2 would be indicative of potential errors in the respective model. Negative albeit small correlations could be found between the number of grain size changes and the event-R^2 for WFJ. Similar correlations were noted with regards to the number of changes in density between layers for the RE and PF model. In both cases correlations were less negative for the PF model indicating a more balanced and ultimately less degraded performance with increasing number of potential capillary barriers. While at WFJ most events occurred with

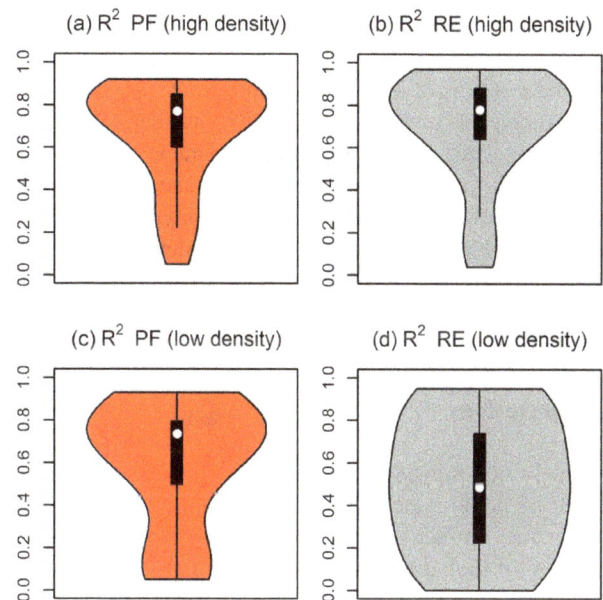

Figure 10. Distribution of event-R^2 for CDP events for the PF (**a, c**) and RE (**b, d**) model. The sample is split into initial bulk snow densities above 350 kg m^{-3} (**a, b**) and below 350 kg m^{-3} (**c, d**).

ripe snow this was not the case for CDP. There, positive correlations were found between the ratio of melt forms and the wet layer ratio with event-R^2 for the RE model (Pearson's R of 0.57 and 0.66) and for the PF model (Pearson's R of 0.37 and 0.39). In this case the PF model also showed more balanced results that were less influenced by the initial LWC, which is in line with our findings of the sprinkling experiments.

System input rates (sum of melt rates and rain rates) are known to significantly affect water transport processes. For example, the area of PF (Eq. 1) is likely to depend on the water supply rate. Data using sandy soils from Glass et al. (1989), shown in DiCarlo (2013), suggest that with increasing system input rates the finger width of PF is increasing. Even though we have used the lowest influx rates from Katsushima et al. (2013), these rates still exceeded what seems representative of natural ROS events. We therefore analysed the effect of system input rates on the performance of our water transport models. Positive, albeit weak correlations (R^2 of 0.07 to 0.21) could be observed between event-R^2 and system input rates for all models, suggesting that they generally performed (slightly) better for higher influx rates. For the PF model this could probably be explained by the PF parameters depending on laboratory measurements with high influx rates.

In combination with the hydraulic properties for lower-density snow samples, additional laboratory experiments might be able to determine the number and size of PFPs for lower input intensities and snow densities. Especially the calibrated parameters threshold for saturation (Θ_{th}) and the number of PFPs for refreeze (N) could benefit from such experimental studies. Even though CDP and WFJ provide long-term measurements on an adequate temporal resolution, these data give little information about spatial variability of snowpack runoff limiting further validation opportunities. Large area multi-compartment lysimeter setups might help to improve estimating size, amount and spatial heterogeneity of flow fingers. Sprinkling experiments with preferably low sprinkling intensities on such a device could fill a knowledge gap about water transport in snow under naturally occurring conditions.

5 Conclusions

A new water transport model is presented that accounts for PF of liquid water within a snowpack. The model deploys a dual-domain approach based on solving the Richards equation for each domain separately (matrix and preferential flow). It has been implemented as part of the physics-based snowpack model SNOWPACK which enables us for the first time to account for PFPs within a model framework that is used operationally for avalanche warning purposes and snow melt forecasting.

The new model was tested for sprinkling experiments over a natural snowpack, dedicated measurements during natural ROS events, and an extensive evaluation over 101 historic ROS events recorded at two different alpine long-term research sites. This assessment led to the following main conclusions.

Compared to alternative approaches, the model accounting for preferential flow (PF) demonstrated an improved overall performance, particularly for lower densities and initially dry snow conditions. This led to smallest interquartile ranges for R^2 values and considerably decreased RMSEs for a set of more than 100 ROS events. When evaluated over entire winter seasons, the performance statistics were superior to those of a single domain approach (RE), even if the differences were small. Both PF and RE models, however, outperformed the model using a bucket approach (BA) by a large margin

(increasing median R^2 by 0.49 and 0.48 for WFJ and 0.53 and 0.48 for CDP). In sprinkling experiments with 30 min bursts of rain at high intensity, the PF model showed a substantially improved temporal correspondence to the observed snowpack runoff, in direct comparison to the RE and BA models. While the improvements were small for experiments on isothermal wet snow, they were pronounced for experiments on cold snowpacks.

Model assessments for over 100 ROS events recorded at two long-term research sites in the European Alps revealed rather variable performance measures on an event-by-event basis between the three models tested. The BA model tended to predict too early onset of snowpack runoff for wet snowpacks and a delayed onset of runoff for cold snowpacks, whereas RE was generally too late, especially for CDP. Combined with results from a separate cross-correlation analysis, results suggested the PF model to provide the best performance concerning the timing of the predicted runoff.

While there is certainly room for improvements of our approach to account for PF of liquid water through a snowpack, this study provides a first implementation within a model framework that is used for operational applications. Adding complexity to the water transport module did not negatively impact on the overall performance and could be done without compromising the robustness of the model results.

Improving the capabilities of a snowmelt model to accurately predict the onset of snowpack runoff during a ROS event is particularly relevant in the context of flood forecasting. In mountainous watersheds with variable snowpack conditions, it may be decisive if snowpack runoff occurs synchronously across the entire catchment, or if the delay between onset of rain and snowpack runoff is spatially variable, e.g. with elevation. In this regard, accounting for PF is a necessary step to improve snowmelt models, as shown in this study.

Competing interests. The authors declare that they have no conflict of interest.

Acknowledgements. We thank the Swiss Federal Office for the Environment FOEN and the scientific exchange program Sciex-NMSch (project code 14.105) for the funding of the project. Special thanks go to Jiri Pavlasek for making it possible to conduct the sprinkling experiments, the extensive work and valuable exchange of ideas during the experiments. We would also like to thank Timea Mareková and Pascal Egli for their help during the experiments. Finally, we acknowledge the two anonymous referees for their the helpful comments, which helped to improve the paper.

Edited by: C. De Michele

References

Badoux, A., Hofer, M., and Jonas, T.: Hydrometeorologis-che Analyse des Hochwasserereignisses vom 10. Oktober 2011, Tech. Rep., WSL/SLF/MeteoSwiss, available at: http://www.wsl.ch/fe/gebirgshydrologie/wildbaeche/projekte/unwetter2011/Ereignisanalyse_Hochwasser_Oktober_2011.pdf (last access: 6 February 2017), 92 pp., 2013 (in German).

Bartelt, P. and Lehning, M.: A physical SNOWPACK model for the Swiss avalanche warning Part I: numerical model, Cold Reg. Sci. Technol., 35, 123–145, doi:10.1016/S0165-232x(02)00074-5, 2002.

Calonne, N., Geindreau, C., Flin, F., Morin, S., Lesaffre, B., Rolland du Roscoat, S., and Charrier, P.: 3-D image-based numerical computations of snow permeability: links to specific surface area, density, and microstructural anisotropy, The Cryosphere, 6, 939–951, doi:10.5194/tc-6-939-2012, 2012.

Colbeck, S. C.: A theory of water percolation in snow, J. Glaciol., 11, 369–385, 1972.

Coléou, C. and Lesaffre, B.: Irreducible water saturation in snow: experimental results in a cold laboratory, Ann. Glaciol., 26, 64–68, 1998.

Conway, H. and Benedict, R.: Infiltration of water into snow, Water Resour. Res., 30, 641–649, doi:10.1029/93WR03247, 1994.

Denoth, A.: An electronic device for long-term snow wetness recording, Ann. Glaciol., 19, 104–106, 1994.

DiCarlo, D. A.: Stability of gravity-driven multiphase flow in porous media: 40 Years of advancements, Water Resour. Res., 49, 4531–4544, doi:10.1002/wrcr.20359, 2013.

Dilley, A. and O'Brien, D.: Estimating downward clear sky long-wave irradiance at the surface from screen temperature and precipitable water, Q. J. Roy. Meteor. Soc., 124, 1391–1401, doi:10.1002/qj.49712454903, 1998.

Eiriksson, D., Whitson, M., Luce, C. H., Marshall, H. P., Bradford, J., Benner, S. G., Black, T., Hetrick, H., and McNamara, J. P.: An evaluation of the hydrologic relevance of lateral flow in snow at hillslope and catchment scales, Hydrol. Process., 27, 640–654, doi:10.1002/hyp.9666, 2013.

Freudiger, D., Kohn, I., Stahl, K., and Weiler, M.: Large-scale analysis of changing frequencies of rain-on-snow events with flood-generation potential, Hydrol. Earth Syst. Sci., 18, 2695–2709, doi:10.5194/hess-18-2695-2014, 2014.

Gerdel, R. W.: The transmission of water through snow, EOS T. AGU, 35, 475–485, 1954.

Glass, R., Steenhuis, T., and Parlange, J.: Wetting Front Instability, 2, Experimental Determination of Relationships Between System Parameters and Two-Dimensional Unstable Flow Field Behavior in Initially Dry Porous Media, Water Resour. Res., 25, 1195–1207, doi:10.1029/WR025i006p01195, 1989.

Hirashima, H., Yamaguchi, S., and Katsushima, T.: A multi-dimensional water transport model to reproduce preferential flow in the snowpack, Cold. Reg. Sci. Technol., 108, 80–90, doi:10.1016/j.coldregions.2014.09.004, 2014.

Illangasekare, T. H., Walter, R. J., Meier, M. F., and Pfeffer, W. T.: Modeling of meltwater infiltration in subfreezing snow, Water Resour. Res., 26, 1001–1012, 1990.

Jordan, R.: A one-dimensional temperature model for a snow cover, Technical documentation for SNTHERM, No. CRREL-SR-91-16, Cold Regions Research and Engineering Lab Hanover NH, 89, DTIC Document, 1991.

Juras, R., Pavlásek, J., Děd, P., Tomášek, V., and Máca, P.: A portable simulator for investigating rain-on-snow events, Z. Geomorphol. Supp., 57, 73–89, 2013.

Juras, R., Pavlásek, J., Vitvar, T., Šanda, M., Holub, J., Jankovec, J., and Linda, M.: Isotopic tracing of the outflow during artificial rain-on-snow event, J. Hydrol., 541, 1145–1154, doi:10.1016/j.jhydrol.2016.08.018, 2016a.

Juras, R., Würzer, S., Pavlásek, J., Vitvar, T., and Jonas, T.: Rainwater propagation through snow pack during rain-on-snow events under different snow condition, Hydrol. Earth Syst. Sci. Discuss., doi:10.5194/hess-2016-612, in review, 2016b.

Katsushima, T., Yamaguchi, S., Kumakura, T., and Sato, A.: Experimental analysis of preferential flow in dry snowpack, Cold Reg. Sci. Technol., 85, 206–216, doi:10.1016/j.coldregions.2012.09.012, 2013.

Kattelmann, R.: Spatial Variability of Snow-Pack Outflow at a Site in Sierra Nevada, U.S.A., Ann. Glaciol., 13, 124–128, 1989.

Kattelmann, R.: Flooding from rain-on-snow events in the Sierra Nevada, IAHS-AISH P., 239, 59–66, 1997.

Kroczynski, S.: A comparison of two rain-on-snow events and the subsequent hydrologic responses in three small river basins in Central Pennsylvania, Eastern Region Technical Attachment, 4, 1–21, 2004.

Leathers, D. J., Kluck, D. R., and Kroczynski, S.: The severe flooding event of January 1996 across north-central Pennsylvania, B. Am. Meteorol. Soc., 79, 785–797, doi:10.1175/1520-0477(1998)079<0785:TSFEOJ>2.0.CO;2, 1998.

Lehning, M., Bartelt, P., Brown, B., Russi, T., Stockli, U., and Zimmerli, M.: SNOWPACK model calculations for avalanche warning based upon a new network of weather and snow stations, Cold Reg. Sci. Technol., 30, 145–157, doi:10.1016/S0165-232X(99)00022-1, 1999.

Lehning, M., Bartelt, P., Brown, B., and Fierz, C.: A physical SNOWPACK model for the Swiss avalanche warning Part III: Meteorological forcing, thin layer formation and evaluation, Cold ,Reg. Sci. Tech., 35, 169–184, doi:10.1016/S0165-232x(02)00072-1, 2002a.

Lehning, M., Bartelt, P., Brown, B., Fierz, C., and Satyawali, P.: A physical SNOWPACK model for the Swiss avalanche warning Part II: Snow microstructure, Cold Reg. Sci. Technol., 35, 147–167, doi:10.1016/S0165-232x(02)00073-3, 2002b.

Marks, D., Link, T., Winstral, A., and Garen, D.: Simulating snowmelt processes during rain-on-snow over a semi-arid mountain basin, Ann. Glaciol., 32, 195–202, doi:10.3189/172756401781819751, 2001.

Marsh, P. and Woo, M.-K.: Wetting Front Advance and Freezing of Meltwater Within a Snow Cover 1. Observations in the Canadian Arctic, Water Resour. Res., 20, 1853–1864, 1984.

Marsh, P. and Woo, M. K.: Meltwater movement in natural heterogeneous snow covers, Water Resour. Res., 21, 1710–1716, 1985.

Marsh, P. and Pomeroy, J.: The impact of heterogeneous flow paths on snowmelt runoff chemistry, Proc. East. Snow. Conf, 1993, 231–238, 1993.

Marsh, P., and Pomeroy, J.: Spatial and temporal variations in snowmelt runoff chemistry, Northwest Territories, Canada, Water Resour. Res., 35, 1559–1567, 1999.

McCabe, G. J., Clark, M. P., and Hay, L. E.: Rain-on-Snow Events in the Western United States, B. Am. Meteorol. Soc., 88, 319–328, doi:10.1175/bams-88-3-319, 2007.

Merz, R. and Blöschl, G.: A process typology of regional floods, Water Resour. Res., 39, 1340, doi:10.1029/2002wr001952, 2003.

Michlmayr, G., Lehning, M., Koboltschnig, G., Holzmann, H., Zappa, M., Mott, R., and Schöner, W.: Application of the Alpine 3D model for glacier mass balance and glacier runoff studies at Goldbergkees, Austria, Hydrol. Process., 22, 3941–3949, doi:10.1002/hyp.7102, 2008.

Molnar, P., Fatichi, S., Gaál, L., Szolgay, J., and Burlando, P.: Storm type effects on super Clausius-Clapeyron scaling of intense rainstorm properties with air temperature, Hydrol. Earth Syst. Sci., 19, 1753–1766, doi:10.5194/hess-19-1753-2015, 2015.

Morin, S., Lejeune, Y., Lesaffre, B., Panel, J.-M., Poncet, D., David, P., and Sudul, M.: An 18-yr long (1993–2011) snow and meteorological dataset from a mid-altitude mountain site (Col de Porte, France, 1325 m alt.) for driving and evaluating snowpack models, Earth Syst. Sci. Data, 4, 13–21, doi:10.5194/essd-4-13-2012, 2012.

Osterhuber, R.: Precipitation intensity during rain-on-snow, in: Proceedings of the 67th Annual Western Snow Conference, 67th Annual Western Snow Conference, South Lake Tahoe California, 1999, 153–155, 1999.

R Development Core Team: R: A Language and Environment for Statistical Computing, Vienna, Austria, R Foundation for Statistical Computing, available at: http://www.R-project.org/ (last access: 6 February 2017), 2016.

Rössler, O., Froidevaux, P., Börst, U., Rickli, R., Martius, O., and Weingartner, R.: Retrospective analysis of a nonforecasted rain-on-snow flood in the Alps – a matter of model limitations or unpredictable nature?, Hydrol. Earth Syst. Sci., 18, 2265–2285, doi:10.5194/hess-18-2265-2014, 2014.

Schmucki, E., Marty, C., Fierz, C., and Lehning, M.: Evaluation of modelled snow depth and snow water equivalent at three contrasting sites in Switzerland using SNOWPACK simulations driven by different meteorological data input, Cold Reg. Sci. Technol., 99, 27–37, doi:10.1016/j.coldregions.2013.12.004, 2014.

Schneebeli, M.: Development and stability of preferential flow paths in a layered snowpack, in: Biogeochemistry of Seasonally Snow-Covered Catchments, Proceedings of a Boulder Symposium July 1995, edited by: Tonnessen, K., Williams, M., and Tranter, M., 89–96, 1995.

Singh, P., Spitzbart, G., Hübl, H., and Weinmeister, H.: Hydrological response of snowpack under rain-on-snow events: a field study, J. Hydrol., 202, 1–20, doi:10.1016/S0022-1694(97)00004-8, 1997.

Stearns, C. R. and Weidner, G. A.: Sensible and Latent Heat Flux Estimates in Antarctica, in: Antarctic Meteorology and Climatology: Studies Based on Automatic Weather Stations, edited by: Bromwich, D. H. and Stearns, C. R., Antarctic Research Series, Vol. 61, American Geophysical Union, 109–138, 1993.

Sui, J. and Koehler, G.: Rain-on-snow induced flood events in Southern Germany, J. Hydrol., 252, 205–220, doi:10.1016/S0022-1694(01)00460-7, 2001.

Surfleet, C. G. and Tullos, D.: Variability in effect of climate change on rain-on-snow peak flow events in a temperate climate, J. Hydrol., 479, 24–34, doi:10.1016/j.jhydrol.2012.11.021, 2013.

Unsworth, M. H. and Monteith, J.: Long-wave radiation at the ground I. Angular distribution of incoming radiation, Q. J. Roy. Meteor. Soc., 101, 13–24, doi:10.1002/qj.49710142703, 1975.

Waldner, P. A., Schneebeli, M., Schultze-Zimmermann, U., and Flühler, H.: Effect of snow structure on water flow and solute transport, Hydrol. Process., 18, 1271–1290, doi:10.1002/hyp.1401, 2004.

Wever, N., Fierz, C., Mitterer, C., Hirashima, H., and Lehning, M.: Solving Richards Equation for snow improves snowpack meltwater runoff estimations in detailed multi-layer snowpack model, The Cryosphere, 8, 257–274, doi:10.5194/tc-8-257-2014, 2014a.

Wever, N., Jonas, T., Fierz, C., and Lehning, M.: Model simulations of the modulating effect of the snow cover in a rain-on-snow event, Hydrol. Earth Syst. Sci., 18, 4657–4669, doi:10.5194/hess-18-4657-2014, 2014b.

Wever, N., Schmid, L., Heilig, A., Eisen, O., Fierz, C., and Lehning, M.: Verification of the multi-layer SNOWPACK model with different water transport schemes, The Cryosphere, 9, 2271–2293, doi:10.5194/tc-9-2271-2015, 2015.

Wever, N., Vera Valero, C., and Fierz, C.: Assessing wet snow avalanche activity using detailed physics based snowpack simulations, Geophys. Res. Lett., 43, 5732–5740, doi:10.1002/2016GL068428, 2016a.

Wever, N., Würzer, S., Fierz, C., and Lehning, M.: Simulating ice layer formation under the presence of preferential flow in layered snowpacks, The Cryosphere, 10, 2731–2744, doi:10.5194/tc-10-2731-2016, 2016b.

Wickham, H.: ggplot2: Elegant Graphics for Data Analysis, Use R!, Springer-Verlag New York, VIII, 213 pp., 2009.

WSL Institute for Snow and Avalanche Research SLF: Meteorological and snowpack measurements from Weissfluhjoch, Davos, Switzerland, doi:10.16904/1, 2015.

Würzer, S., Jonas, T., Wever, N., and Lehning, M.: Influence of initial snowpack properties on runoff formation during rain-on-snow events, J. Hydrometeorol., 17, 1801–1815, doi:10.1175/JHM-D-15-0181.1, 2016.

Yamaguchi, S., Watanabe, K., Katsushima, T., Sato, A., and Kumakura, T.: Dependence of the water retention curve of snow on snow characteristics, Ann. Glaciol., 53, 6–12, doi:10.3189/2012AoG61A001, 2012.

Ye, H., Yang, D., and Robinson, D.: Winter rain on snow and its association with air temperature in northern Eurasia, Hydrol. Process., 22, 2728–2736, doi:10.1002/hyp.7094, 2008.

PERMISSIONS

LIST OF CONTRIBUTORS

T. Jonas and C. Fierz
WSL Institute for Snow and Avalanche Research SLF, Flüelastrasse 11, 7260 Davos Dorf, Switzerland

N. Wever and M. Lehning
WSL Institute for Snow and Avalanche Research SLF, Flüelastrasse 11, 7260 Davos Dorf, Switzerland
CRYOS, School of Architecture, Civil and Environmental Engineering, EPFL, Lausanne, Switzerland

Marc J. P. Vis
Department of Geography, University of Zurich, Zurich, 8057, Switzerland

Jan Seibert
Department of Geography, University of Zurich, Zurich, 8057, Switzerland
Department of Aquatic Sciences and Assessment, Swedish University of Agricultural Sciences, Uppsala, Sweden

Irene Kohn, Markus Weiler and Kerstin Stahl
Faculty of Environment and Natural Resources, University of Freiburg, 79098 Freiburg, Germany

Tobias Jonas
WSL Institute for Snow and Avalanche Research SLF, Davos, Switzerland

Nena Griessinger
WSL Institute for Snow and Avalanche Research SLF, Davos, Switzerland
Department of Geography, University of Zurich, Zurich, Switzerland

Jan Seibert
Department of Geography, University of Zurich, Zurich, Switzerland

Jan Magnusson
Norwegian Water Resources and Energy Directorate (NVE), Oslo, Norway

Siraj Ul Islam and Stephen J. Déry
Environmental Science and Engineering Program, University of Northern British Columbia, 3333 University Way, Prince George, BC, V2N 4Z9, Canada

Anne F. Van Loon
School of Geography, Earth and Environmental Sciences, University of Birmingham, Birmingham, UK

Marit Van Tiel
School of Geography, Earth and Environmental Sciences, University of Birmingham, Birmingham, UK
Hydrology and Quantitative Water Management Group, Wageningen University and Research, Wageningen, the Netherlands

Adriaan J. Teuling
Hydrology and Quantitative Water Management Group, Wageningen University and Research, Wageningen, the Netherlands

Niko Wanders
Department of Civil and Environmental Engineering, Princeton University, Princeton, NJ, USA
Department of Physical Geography, Utrecht University, Utrecht, the Netherlands

Kerstin Stahl
Faculty of Environment and Natural Resources, University of Freiburg, Freiburg, Germany

Xun Liu and Cheng Kou
Jiangsu Provincial Key Laboratory of Geographic Information Science and Technology, Nanjing University, Nanjing 210023, China
Key Laboratory for Satellite Mapping Technology and Applications of State Administration of Surveying, Mapping and Geo information of China, Nanjing University, Nanjing 210023, China

Chang-Qing Ke
Jiangsu Provincial Key Laboratory of Geographic Information Science and Technology, Nanjing University, Nanjing 210023, China
Key Laboratory for Satellite Mapping Technology and Applications of State Administration of Surveying, Mapping and Geo information of China, Nanjing University, Nanjing 210023, China
Collaborative Innovation Center of South China Sea Studies, Nanjing 210023, China

Xiu-Cang Li
National Climate Center, China Meteorological Administration, Beijing 100081, China
Collaborative Innovation Center on Forecast and Evaluation of Meteorological Disasters Faculty of Geography and Remote Sensing, Nanjing University of Information Science & Technology, Nanjing 210044, China

Dong-Hui Ma
Jiangsu Provincial Key Laboratory of Geographic Information Science and Technology, Nanjing University, Nanjing 210023, China Key Laboratory for Satellite Mapping Collaborative Innovation Center of South China Sea Studies, Nanjing 210023, China

Hongjie Xie
Collaborative Innovation Center of South China Sea Studies, Nanjing 210023, China

David R. Casson and Micha Werner
IHE Delft Institute of Water Education, Hydro informatics Chair Group, 2601 DA, Delft, the Netherlands
Deltares, Operational Water Management, 2600 MH, Delft, the Netherlands

Dimitri Solomatine
IHE Delft Institute of Water Education, Hydro informatics Chair Group, 2601 DA, Delft, the Netherlands
Delft University of Technology, Water Resources Section, 2600 GA, Delft, the Netherlands

Albrecht Weerts
Deltares, Operational Water Management, 2600 MH, Delft, the Netherlands
Wageningen University and Research, Hydrology and Quantitative Water Management group, 6700 AA, Wageningen, the Netherlands

Travis R. Roth and Anne W. Nolin
Water Resource Sciences, Oregon State University, Corvallis, OR 97331, USA

Rose Petersky
Graduate Program of Hydrologic Sciences, University of Nevada, 1664 N Virginia St., Reno, NV 89557, USA

Adrian Harpold
Graduate Program of Hydrologic Sciences, University of Nevada, 1664 N Virginia St., Reno, NV 89557, USA
Natural Resources Environmental Science Department, University of Nevada, 1664 N Virginia St., Reno, NV 89557, USA

Global Water Center, University of Nevada, 1664 N Virginia St., Reno, NV 89557, USA

Daniele Penna
Department of Agricultural, Food and Forestry Systems, University of Florence, via San Bonaventura 13, 50145 Florence, Italy

Michael Engel and Francesco Comiti
Faculty of Science and Technology, Free University of Bozen-Bolzano, Piazza dell' Università 5, 39100 Bolzano, Italy

Giacomo Bertoldi
Institute for Alpine Environment, EURAC – European Academy of Bolzano/Bozen, viale Druso 1, 39100 Bolzano, Italy

M. F. P. Bierkens
Department of Physical Geography, Utrecht University, Utrecht, the Netherlands

N. Wanders
Department of Physical Geography, Utrecht University, Utrecht, the Netherlands
Department of Civil and Environmental Engineering, Princeton University, Princeton, NJ, USA

W. W. Immerzeel
Department of Physical Geography, Utrecht University, Utrecht, the Netherlands
Future Water, Wageningen, the Netherlands
International Centre for Integrated Mountain Development, Kathmandu, Nepal

A. F. Lutz
Future Water, Wageningen, the Netherlands

J. M. Shea
International Centre for Integrated Mountain Development, Kathmandu, Nepal

Enrique Morán-Tejeda and Jorge Lorenzo-Lacruz
Department of Geography. University of the Balearic Islands. Palma, Spain

Jorge Luis Ceballos and Katherine Peña
Instituto de Hidrología, Meteorología y Estudios Ambientales (IDEAM), Bogotá, Colombia

Juan Ignacio López-Moreno
Pyrenean Institute of Ecology. Consejo Superior de Investigaciones Científicas, Zaragoza, Spain

Tobias Jonas
WSL Institute for Snow and Avalanche Research SLF, Flüelastrasse 11, 7260 Davos Dorf, Switzerland

Sebastian Würzer and Michael Lehning
WSL Institute for Snow and Avalanche Research SLF, Flüelastrasse 11, 7260 Davos Dorf, Switzerland
École Polytechnique Fédérale de Lausanne (EPFL), School of Architecture, Civil and Environmental Engineering, Lausanne, Switzerland

Roman Juras
WSL Institute for Snow and Avalanche Research SLF, Flüelastrasse 11, 7260 Davos Dorf, Switzerland
Faculty of Environmental Sciences, Czech University of Life Sciences Prague, Kamýcká 129, 165 21, Prague, Czech Republic

Index